Lecture Notes in Computer Science 6449

Commenced Publication in 1973
Founding and Former Series Editors:
Gerhard Goos, Juris Hartmanis, and Jan van Leeuwen

José M. Laginha M. Palma
Michel Daydé   Osni Marques
João Correia Lopes (Eds.)

High Performance Computing for Computational Science – VECPAR 2010

9th International Conference
Berkeley, CA, USA, June 22-25, 2010
Revised Selected Papers

 Springer

Volume Editors

José M. Laginha M. Palma
Faculdade de Engenharia da Universidade do Porto
Rua Dr. Roberto Frias s/n, 4200-465 Porto Portugal
E-mail: jpalma@fe.up.pt

Michel Daydé
University of Toulouse, INP (ENSEEIHT); IRIT
2 rue Charles-Camichel, 31071 Toulouse CEDEX 7, France
E-mail: Michel.Dayde@enseeiht.fr

Osni Marques
Lawrence Berkeley National Laboratory, Berkeley, USA
E-mail: oamarques@lbl.gov

João Correia Lopes
University of Porto, Faculty of Engineering
Rua Dr. Roberto Frias, s/n, 4200-465, Porto, Portugal
E-mail: jlopes@fe.up.pt

ISSN 0302-9743   e-ISSN 1611-3349
ISBN 978-3-642-19327-9   ISBN 978-3-642-19328-6 (eBook)
DOI 10.1007/978-3-642-19328-6
Springer Heidelberg Dordrecht London New York

Library of Congress Control Number: 2011921123

CR Subject Classification (1998): D, F, C.2, G, J.2-3

LNCS Sublibrary: SL 1 – Theoretical Computer Science and General Issues

Typesetting: Camera-ready by author, data conversion by Scientific Publishing Services, Chennai, India

Printed on acid-free paper

Springer is part of Springer Science+Business Media (www.springer.com)

Preface

VECPAR is an international conference series dedicated to the promotion and advancement of all aspects of high performance computing for computational science, as an industrial technique and academic discipline, extending the frontier of both the state of the art and the state of practice. The audience and participants of VECPAR are seen as researchers in academic departments, government laboratories, and industrial organizations. There is now a permanent website for the conference series at http://vecpar.fe.up.pt, where the history of the conference is described.

The 9th edition of VECPAR was organized in Berkeley (USA), June 22–25, 2010. It was the 4th time the conference was celebrated outside Porto after Valencia (Spain) in 2004, Rio de Janeiro (Brazil) in 2006, and Toulouse (France) in 2008.

The whole conference program consisted of 6 invited talks, 45 papers, and 5 posters.

The major themes were:

- Large Scale Simulations in CS&E
- Linear Algebra on GPUs and FPGAs
- Linear Algebra on Emerging Architectures
- Numerical Algorithms
- Solvers on Emerging Architectures
- Load Balancing
- Parallel and Distributed Computing
- Parallel Linear Algebra
- Numerical Algorithms on GPUs

Three workshops were organized before the conference:

iWAPT — Fifth international Workshop on Automatic Performance Tuning
PEEPS — Workshop on Programming Environments for Emerging Parallel Systems
HPC Tools — Tutorial on High Performance Tools for the Development of Scalable and Sustainable Applications

The most significant contributions have been made available in the present book, edited after the conference, and after a second review of all orally presented papers at the conference.

Henricus Bouwmeester, from the University of Colorado Denver received the Best Student Presentation award for his talk on "Towards an Efficient Tile Matrix Inversion of Symmetric Positive Definite Matrices on Multicore Architectures".

VECPAR 2010 took place at the Sutardja Dai Hall of the Center for Information Technology Research in the Interest of Society (CITRIS), University of California, Berkeley, USA. The logistics and organizational details were dealt with by Yeen Mankin, with the kind support of Dany DeCecchis and Jean Piero Suarez (students at San Diego State University).

Paper submission and selection were managed via the conference management system, hosted and operated by the Faculty of Engineering of the University of Porto (FEUP)[1]. Websites were maintained by both FEUP and the Lawrence Berkeley National Laboratory; registrations were managed by the Lawrence Berkeley National Laboratory.

The success of the VECPAR conferences and the long life of the series result from the collaboration of many people. As before, given the widespread organization of the meeting, a large number of collaborators were involved. Here we mention only a few. Through them we thank many others who offered their time and commitment to the success of the conference workshops and tutorial: Takahiro Katagiri, Richard Vuduc, Reiji Suda, Jonathan Carter, John Cavazos, Kengo Nakajima, Lenny Oliker, Nick Wright, Tony Drummond, Sameer Shende, and Jose Roman.

For their contributions to the present book, we must thank all the authors for meeting the deadlines and all members of the Scientific Committee who helped us so much in selecting the papers. We also thank the members of the committees involved in the organization of the workshops held before the conference.

November 2010

José M.L.M. Palma
Michel Daydé
Osni Marques
J. Correia Lopes

---

[1] The VECPAR series of conferences has been organized by the Faculty of Engineering of Porto (FEUP) since 1993.

Organization

Organizing Committee

Osni Marques	LBNL, USA (Chair)
Jonathan Carter	LBNL, USA
Tony Drummond	LBNL, USA
Masoud Nikravesh	LBNL, USA
Erich Strohmaier	LBNL, USA
J. Correia Lopes	FEUP/INESC Porto, Portugal (Web Chair)

Steering Committee

José Palma	University of Porto, Portugal (Chair)
Álvaro Coutinho	COPPE/UFRJ, Brazil
Michel Daydé	University of Toulouse/IRIT, France
Jack Dongarra	University of Tennessee,USA
Inês Dutra	University of Porto, Portugal
José Fortes	University of Florida, USA
Vicente Hernandez	Technical University of Valencia, Spain
Ken Miura	National Institute of Informatics, Japan

Scientific Committee

Michel J. Daydé	France (Chair)
P. Amestoy	France
Ben Allen	USA
Reza Akbarinia	France
Jacques Bahi	France
Carlos Balsa	Portugal
Valmir Barbosa	Brazil
Xiao-Chuan Cai	USA
Jonathan Carter	USA
Olivier Coulaud	France
José Cardoso e Cunha	Portugal
Rudnei Cunha	Brazil
Frédéric Desprez	France
Jack Dongarra	USA
Tony Drummond	USA

Inês de Castro Dutra	Portugal
Nelson F.F. Ebecken	Brazil
Jean-Yves L'Excellent	France
Omar Ghattas	USA
Luc Giraud	France
Serge Gratton	France
Ronan Guivarch	France
Daniel Hagimont	France
Abdelkader Hameurlain	France
Bruce Hendrickson	USA
Vicente Hernandez	Spain
Vincent Heuveline	Germany
Jean-Pierre Jessel	France
Takahiro Katagiri	Japan
Jacko Koster	Norway
Dieter Kranzlmueller	Germany
Stéphane Lanteri	France
Kuan-Ching Li	USA
Sherry Li	USA
Thomas Ludwig	Germany
Osni Marques	USA
Marta Mattoso	Brazil
Kengo Nakajima	Japan
José Laginha Palma	Portugal
Christian Perez	France
Serge G. Petiton	France
Thierry Priol	France
Heather Ruskin	Ireland
Mitsuhisa Sato	Japan
Satoshi Sekiguchi	Japan
Sameer Shende	USA
Claudio T. Silva	USA
António Augusto Sousa	Portugal
Mark A. Stadtherr	USA
Domenico Talia	Italy
Adrian Tate	USA
Francisco Tirado	Spain
Miroslav Tuma	Czech Rep.
Paulo Vasconcelos	Portugal
Xavier Vasseur	France
Richard (Rich) Vuduc	USA
Roland Wismuller	Germany

Invited Speakers

Charbel Farhat	Stanford University, USA
David Mapples	Allinea Software Inc., USA
David Patterson	UC Berkeley, USA
John Shalf	Lawrence Berkeley National Laboratory, USA
Thomas Sterling	Louisiana State University and CALTECH, USA
Takumi Washio	University of Tokyo, Japan

Additional Reviewers

Ignacio Blanquer
Jonathan Bronson
Vitalian Danciu
Murat Efe Guney
Linh K. Ha
Wenceslao Palma
Francisco Isidro Massetto
Manuel Prieto Matias
Silvia Knittl
Andres Tomas
Erik Torres
Johannes Watzl

Sponsoring Organizations

The Organizing Committee is very grateful to the following organizations for
their support:

Allinea	Allinea Software, USA
Meyer Sound	Meyer Sound Laboratories Inc., USA
ParaTools	ParaTools Inc., USA
Berkeley Lab	Lawrence National Berkeley Laboratory, USA
U. Porto	Universidade do Porto, Portugal

Table of Contents

Parallel and Distributed Computing

Numerical Algorithms

Exascale Computing Technology Challenges

John Shalf[1], Sudip Dosanjh[2], and John Morrison[3]

[1] NERSC Division, Lawrence Berkeley National Laboratory,
1 Cyclotron Road, Berkeley, California 94611
[2] Sandia National Laboratories,
New Mexico 87185
[3] Los Alamos National Laboratory,
Los Alamos, New Mexico 87544
jshalf@lbl.gov, sudip@sandia.gov, jfm@lanl.gov

Abstract. High Performance Computing architectures are expected to change dramatically in the next decade as power and cooling constraints limit increases in microprocessor clock speeds. Consequently computer companies are dramatically increasing on-chip parallelism to improve performance. The traditional doubling of clock speeds every 18-24 months is being replaced by a doubling of cores or other parallelism mechanisms. During the next decade the amount of parallelism on a single microprocessor will rival the number of nodes in early massively parallel supercomputers that were built in the 1980s. Applications and algorithms will need to change and adapt as node architectures evolve. In particular, they will need to manage locality to achieve performance. A key element of the strategy as we move forward is the co-design of applications, architectures and programming environments. There is an unprecedented opportunity for application and algorithm developers to influence the direction of future architectures so that they meet DOE mission needs. This article will describe the technology challenges on the road to exascale, their underlying causes, and their effect on the future of HPC system design.

Keywords: Exascale, HPC, codesign.

1 Introduction

Node architectures are expected to change dramatically in the next decade as power and cooling constraints limit increases in microprocessor clock speeds (which are expected to remain near 1 GHz). Consequently computer companies are dramatically increasing on-chip parallelism to improve performance. The traditional doubling of clock speeds every 18-24 months is being replaced by a doubling of cores, threads or other parallelism mechanisms. During the next decade the amount of parallelism on a single microprocessor will rival the number of nodes early massively parallel supercomputers that were built in the 1980s.

Applications and algorithms will need to change and adapt as node architectures evolve. They will need to manage locality and perhaps resilience to achieve high performance. In addition, hardware breakthroughs will be needed to achieve useful Exascale computing later this decade, at least within any reasonable power budget. A

J.M.L.M. Palma et al. (Eds.): VECPAR 2010, LNCS 6449, pp. 1–25, 2011.

key element of the strategy as we move forward is the co-design of applications, architectures and programming environments as shown in Figure 1. Much greater collaboration between these communities will be needed to overcome the key Exascale challenges. There is an unprecedented opportunity for application and algorithm developers to influence the direction of future architectures so that they meet DOE mission needs.

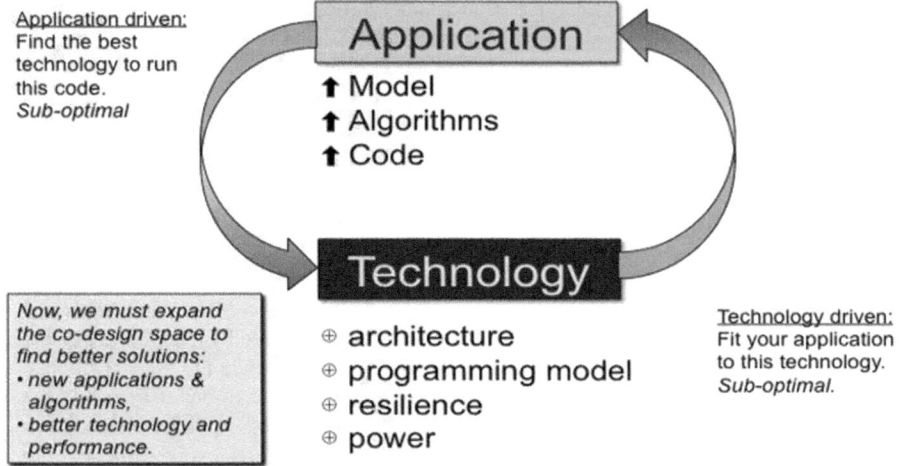

Fig. 1. Schematic of application-driven hardware/software co-design process

These trends are illustrated in Figure 2, which shows the energy cost of moving data to different levels of system memory relative to the cost of a floating point operation. The cost of data movement will not improve substantially whereas the cost of performing a floating -point operation will likely improve between 5x to 10x. Past attempts to exploit intra-node parallelism did not show significant benefits primarily because the cost of data movement within a node was not substantially lower than the cost of moving data across the interconnect because the cost of moving data off-chip dominated the energy costs. However, modern chip multiprocessors have CPU's co-located on the same chip. Consequently, there is a huge opportunity to capture energy-efficiency and performance benefits by directly taking advantage of intra-chip communication pathways.

2 Metrics, Cost Functions, and Constraints

For Exascale systems, the primary constraints are (for the purposes of discussion) platform capital costs under $200M, less than 20MW power consumption, and delivery in 2018. All other system architectural choices are free parameters, and are optimized to deliver maximum application performance subject to these very challenging constraints.

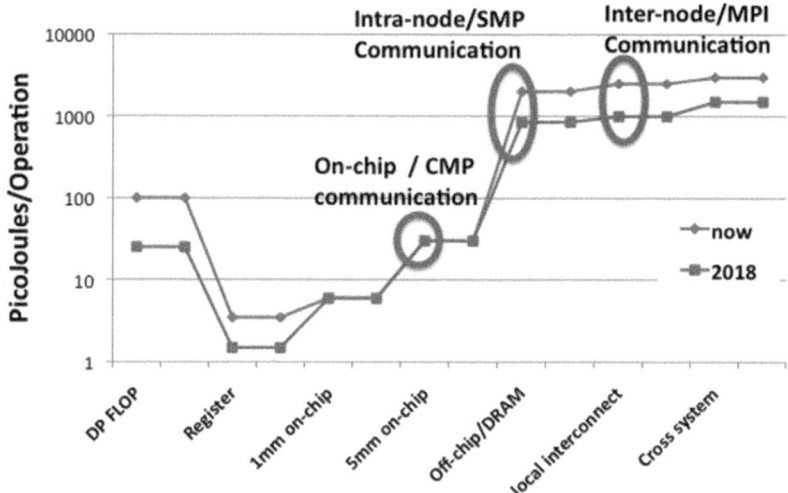

Fig. 2. Energy cost of data movement relative to the cost of a flop for current and 2018 systems (the 2018 estimate is conservative and doesn't account for the development of an advanced memory part). The biggest change in energy cost is moving data off-chip. Therefore, future programming environments must support the ability of algorithms and applications to exploit locality which will, in turn, be necessary to achieve performance and energy efficiency.

In an ideal world, we would design systems that would never subject applications to any performance constraints. However, component costs and power usage force system architects to consider difficult trade-offs that balance the actual cost of system components against their effect on application performance. For example, if doubling floating point execution rate nets a 10% gain in overall application performance, but only increases system costs by 5%, then it is a net benefit despite degrading system balance. It is important to have an open dialog to fully understand the cost impacts of key design choices so that they can be evaluated against their benefit to the application space.

Cost Functions

The Cost of Power: Even with the least expensive power available in the US, the cost of electricity to power supercomputing systems is a substantial part of the Total Cost of Ownership (TCO). When burdened with cooling and power distribution over-heads, even the least expensive power in the U.S. (< 5cents/KWH) ultimately costs $1M per Megawatt per year to operate a system. To keep the TCO manageable DOE's Exascale Initiative Steering Committee adopted 20MW as the upper limit for a reasonable system design [1,2]. This limit is movable, but at great cost and design risk.

The Cost of a FLOP: Floating point used to be the most costly component of a system both in terms of design cost and power. However, today, FPUs consume a very small fraction of the area of a modern chip design and a much smaller fraction of the power consumption. On modern systems, a double-precision FMA (fused multiply add) consumes 100 picoJoules. By contrast, reading the double precision operands from DRAM costs about 2000 pJ. By 2018 floating point operations will consume about ~10.6pJ/op on 11nm lithography technology [3], and the cost of reading from DRAM will only improve modestly to 1000pJ unless more energy-efficient memory technology is developed.

With these figures of merit, it would only consume 100W to put 10 Teraflops on a chip, which is easily achievable. However, it would require 2000W of power required to supply memory bandwidth to those floating point units at a modest memory bandwidth to floating point ratio of 0.2. The consequence is that we can engineer far more floating point capability onto a chip than can reasonably be used by an application. Engineering FLOPs is not a design constraint – data movement presents the most daunting engineering and computer architecture challenge.

The Cost of Moving Data: Memory interfaces and communication links on modern computing systems are currently dominated by electrical/copper technology. However, wires are rapidly being subsumed by optical technology because of the limits of bit rate scaling as we shrink wires length scales as observed by David A. B. Miller of Stanford [4-5]. Miller observes that for a conventional electrical line (without repeaters or equalization) can be modeled as a simple RC circuit by virtue of the simplified Telegrapher's equation for lossy transmission line. The wire must be charged and discharged at a rate governed by the RC time constant, which is given by equation 1 where R_l is the resistance of the wire, C_l is the capacitance and l is the length of the wire. As the wire length increases, the risetime (given by the RC time constant) increases by the square of the length – thereby reducing the bit-rate.

$$\text{risetime} = R_l C_l \, l^2 \tag{1}$$

Miller observes that if you shrink the wire proportionally in all dimensions by a factor of s, the resistance (R_l) increases proportionally to the reduced wire aspect ratio, which reduces by a factor of s^2, but capacitance (C_l) remains the same. The consequence is that for constant voltage, the bit-rate carrying capacity of an RC line scales proportional to $B \approx A/\, l^2$, where B is the bandwidth of the wire and A is the cross-sectional area of the wire and l^2 is the length of the wire. The consequence of this observation is that natural bit rate capacity of the wire depends on the aspect ratio of the line, which is the ratio of the length to the cross-sectional area for a constant input voltage and does not improve as we shrink the wires down with smaller lithographic processes. We can push to a higher bitrate by increasing the drive voltage to the wire, but this also increases power consumption. These effects are summarized in equation 2, which assumes a simple RC model of the wire and no re-amplification (*long-haul wires on-chip are normally re-amplified at regular intervals to maintain a linear power profile as a function of length, but at a cost of more power consumption*).

$$\text{Power} \approx B \times l^2 /\, A \tag{2}$$

This has the following consequences to system design [6, 16]:

- Power consumed increases proportionally to the bit-rate, so as we move to ultra-high-bandwidth links, the power requirements will become an increasing concern.

- Power consumption is highly distance-dependent (quadratically with wire length without re-amplification), so bandwidth is likely to become increasingly localized as power becomes a more difficult problem.
- Improvements in chip lithography (making smaller wires) will not improve the energy efficiency or data carrying capacity of electrical wires.

In contrast, optical technology does not have significant distance-dependent energy consumption. It costs nearly the same amount of energy to transmit an optical signal 1 inch as it does to transmit it to the other end of a room. Also, signaling rate does not strongly affect the energy required for optical data transmission. Rather, the fixed cost of the laser package for optical systems and the absorption of light to receive a signal are the dominant power costs for optical solutions.

As the cost and complexity of moving data over copper will become more difficult over time, the cross-over point where optical technology becomes more cost-effective than electrical signaling has been edging closer to the board and chip package at a steady pace for the past 2 decades. Contemporary short-distance copper links consume about 10-20 pJ/bit, but could be improved to 2pJ/bit for short-haul 1 cm length links by 2018. However, the efficiency and/or data carrying capacity of the copper links will fall off rapidly with distance (as per equation 2) that may force a movement to optical links. Contemporary optical links consume about 30-60pJ/bit, but solutions that consume as little as 2.5pJ/bit have been demonstrated in the lab. In the 2018 timeframe optical links are likely to operate at 10pJ/bit efficiency [7]. Moreover, silicon photonics offers the promise of breaking through the limited bandwidth and packaging constraints of organic carriers using electrical pins.

Another serious barrier to future performance growth is cost of signals that go off-chip as we rapidly approach pin-limited bandwidth. Due to the skin effect [19], and overheads of more complex signal equalization, it is estimated that 10-15GHz is likely the maximum feasible signaling rate for off-chip differential links that are 1-2cm in length. A chip with 4000 pins would be a very aggressive, but feasible design point for 2018. If you consider that half of those pins (2000) are power and ground, while the remaining 2000 pins are differential pairs, then the maximum feasible off-chip bandwidth would be ~1000 × 10GHz, which comes to approximately 1 Terabyte/second (*10 Terabits/sec with 10/8 encoding*). Breaking through this 1 TB/s barrier would require either more expensive, exotic packaging technology (ceramics rather than organic packages), or migration to on-chip optics, such as silicon-photonic ring-resonator technology [20, 21].

Without major breakthroughs in packaging technology or photonics, it will not be feasible to support globally flat bandwidth across a system. Algorithms, system software, and applications will need to aware of data locality. The programming environment must enable algorithm designers to express and control data locality more carefully. The system must have sufficient information and control to make decisions that maximally exploit information about communication topology and locality. Flat models of parallelism (e.g. flat MPI or shared memory/PRAM models) will not map well to future node architectures.

3 Memory Subsystem

Ultimately, memory performance is primarily constrained by the dynamics of the commodity market. One key finding of DOE's Architecture and Technology workshop [8] was that memory bandwidth is primarily constrained by power & efficiency of the memory interface protocols, whereas memory capacity is primarily constrained by cost. Early investments in improving the efficiency of DRAM interfaces and packaging technology may result in substantially improved balance between memory bandwidth and floating point rate. Investments in packaging (mainly chip-stacking technology) can also provide some benefit in the memory capacity of nodes, but it is unclear how much the price of the components can be affected by these investments given commodity market forces.

3.1 Memory Bandwidth

The power consumed by data movement will dominate the power consumption profile of future systems. Chief among these concerns is the power consumed by memory technology, which would easily dominate the overall power consumption of future systems if we attempt to maintain historical bandwidth/flop ratios of 1 byte/flop. A 20 MW power constraint on an Exascale system will limit the breadth of applications that can execute effectively on such systems unless there are fundamental breakthroughs in memory and communications technologies.

For example, today's DDR-3 memory interface technology consumes about 70picoJoules/bit, resulting in approximately 5000 pJ of energy to load a double-precision operand (accounting for ECC overhead). If we extrapolate the energy-efficiency of memory interfaces to DDR-5 in 2018, the efficiency could be improved to 30pJ/bit. A system with merely 0.2 bytes/flop of memory bandwidth would consume > 70Megawatts of power, which is not considered a feasible design point. Keeping under the 20MW limit would force the memory system to have less than 0.02 bytes/flop, which would severely constrain the number of applications that could run efficiently on the system as illustrated in Figures 4 and Figure 5.

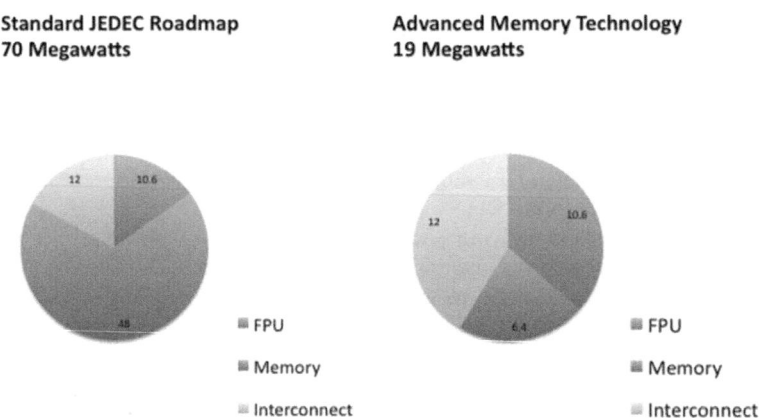

Fig. 3. If we follow standard JEDEC memory technology roadmap, the power consumption of a feasible Exascale system design (using 0.2 bytes/flop memory bandwidth balance) will be >70Megawatts due to memory power consumption, which is an impractical design point. Keeping memory power under control will either require substantial investments in more efficient memory interface protocols, or substantial compromises in memory bandwidth and floating point performance (< 0.02 bytes/flop).

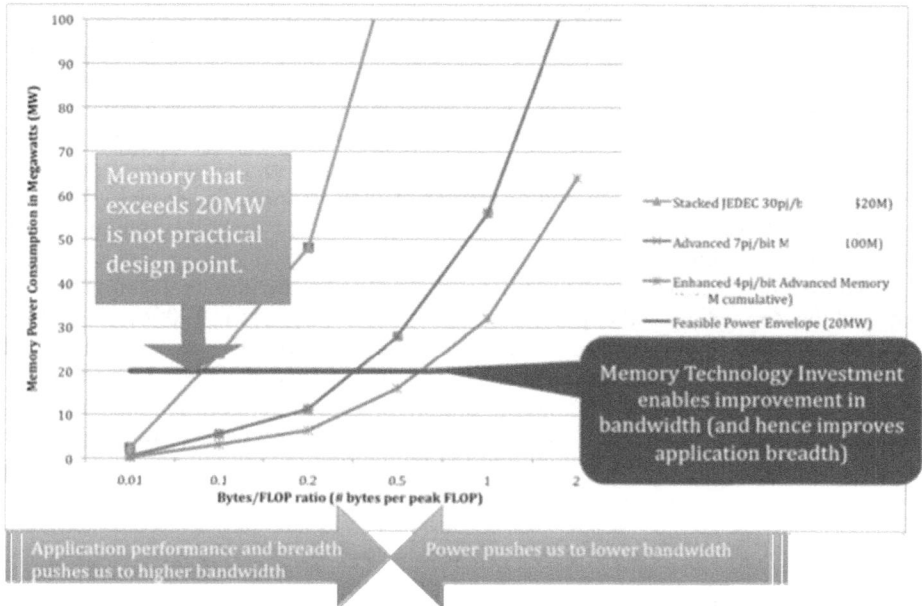

Fig. 4. This figure illustrates the trade-offs between memory power consumption and the desire for a more broadly applicable Exascale system design under different assumptions about investments in advanced memory technology.

We cannot reach reasonable memory energy efficiency by following the JEDEC roadmap. Getting to a reasonable energy efficiency requires development of new, more efficient interface designs and memory protocols. Advanced memory technology can get to about 7 pJ/bit with investments to bring the technology to market. The limit of this new technology is estimated to be 4pJ/bit (excluding memory queues and controller logic). Therefore, in order to maintain 0.2 byte/flop system balance and stay under a 20MW design limit for power requires either substantial investments in advanced memory technology, or a substantial degradation in system memory balance, as illustrated in Figure 5. As always, these ratios are movable. For example, the power limit could be relaxed, but would put the feasibility of fielding siting such a system in jeopardy and increase the total cost of ownership.

3.2 Memory Capacity

One figure of merit for improvements to HPC systems is the total memory capacity. More aggregate memory enables systems to solve problems that have either proportionally higher resolution, or more physics fidelity/complexity – or both. However, cost considerations may limit an exascale system to a memory capacity that improves only by a factor of 100x in comparison to the system peak floating point rate which will improve by 1000x. This is a movable parameter in the design space of the machine, but the consequence of moving this parameter is increased cost for the memory subsystem and the total cost of the system.

The DRAM capacity of a system is primarily limited by cost, which is defined by the dynamics of a broad-based high-volume commodity market. The commodity market for memory makes pricing of the components highly volatile, but the centroid of the market is approximately $1.80/chip. Figure 4 illustrates that the rate of memory density improvement has gone from a 4x improvement every 3 years to a 2x improvement every 3 years (a 30% annual rate of improvement). Consequently the cost of memory technology is not improving as rapidly as the cost of Floating Point capability. Given the new rate of technology improvement, 8 gigabit memory parts will be widely available in the commodity market in the 2018 timeframe and 16 gigabit parts will also have been introduced. It is unclear which density will be the most cost-effective in that timeframe.

If we assume that memory should not exceed 50% of the cost of a computer system, and that the anticipated capital cost of an Exascale system is $200M, then Table 5 shows that the memory capacity we could afford lies somewhere between 50 and 100Petabytes. Again, these are not hard limits on capacity, but they do have a substantial effect on the cost of the system, and the trade-off between memory capacity and other system components must be considered carefully given a limited procurement budget.

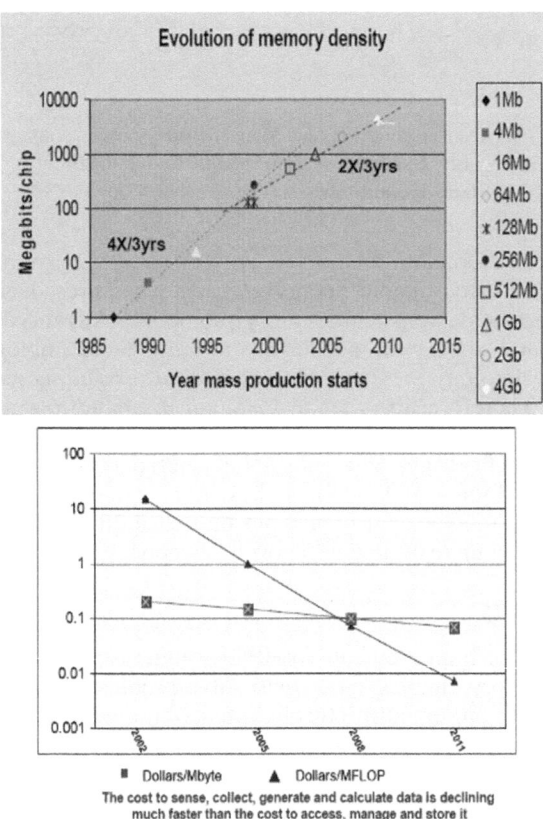

Fig. 5. The rate of improvement in memory technology improving at slower rates -- now approaching 30% per year. (Figure courtesy of David Turek, IBM).

3.3 Latency

Off-chip latencies are unlikely to improve substantially over existing systems. With a fixed clock rate of 2 GHz, the distance to off-chip memory on modern systems is approximately 100ns (200 clock cycles away), and will potentially improve to 40-50ns (100 clock cycles away from memory) in the 2018 timeframe. A modern interconnect has a messaging latency of 1 microsecond. Most of that latency is on the end-points for the message (message overhead of assembling a message and interrupt handling to receive it). By 2018, this could improve to as little as 200-500ns for message latency, which is at that point limited by the speed of light (0.75c in optical fiber) and comes to about 5ns latency per meter of cable.

Lastly, the message injection rates of modern systems (an indirect measure of the overhead of sending messages) is approximately tens of thousands of messages/second on leading-edge designs. If the interconnect NIC is moved on-chip, it may be feasible to support message injection rates of hundreds of millions of messages per second for lightweight messaging (such as one-sided messages for PGAS languages).

With no substantial improvements in off-chip and cross-system latency, the bandwidth-latency product for future systems (which determines the number of bytes that must be in flight to fully saturate bandwidth) will be large. This means there must be considerable attention to latency hiding support in both algorithms and in hardware designs. The approach to latency hiding has not yet been determined.

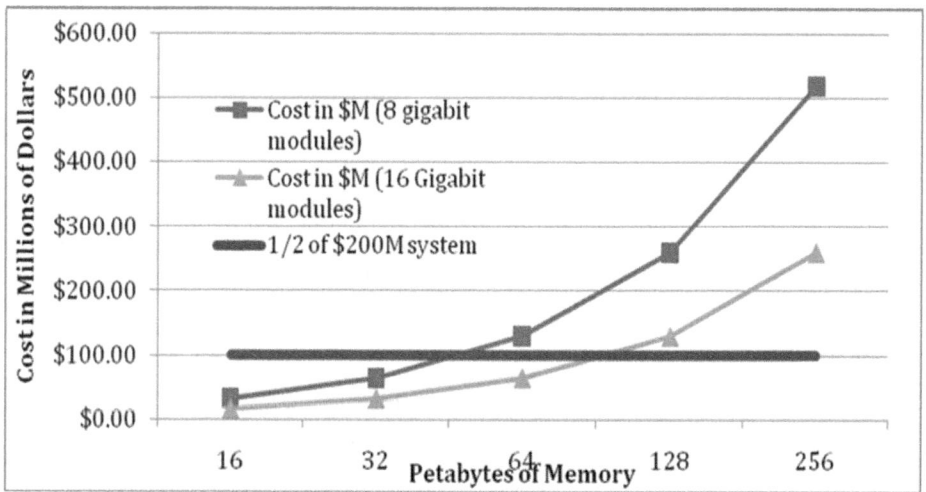

Fig. 6. There are two possible memory chip densities in the 2018 timeframe. It is less certain which option will be most cost-effective.

4 Node Architecture Projections for 2018

There are many opportunities for major reorganization of our computation model to take better advantage of future hardware designs. However, much of the discussion to-date of inter-processor communication semantics and node organization has focused on evolutionary rather than revolutionary features.

4.1 Clock Rate

Technology projections[1,2,3,9] indicate that clock-speeds will not change appreciably by 2018 and will remain near 1-2 GHz. This sets clear design constraints for the number of floating point functional units that will be on a future chip design. In order to keep component counts for future systems within practical limits (< 1M nodes), a node must perform between 1-10 Teraflops. At 1 GHz, that means there will be between 1000 and 10,000 discrete Floating Point Units on a chip.

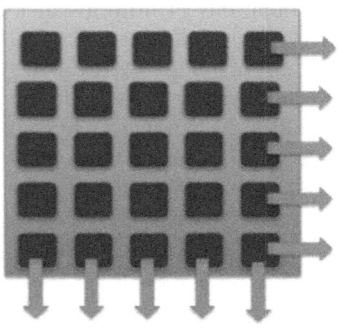

Scale-out for Planar geometry

- ~1000-10k simple cores
- 4-8 wide SIMD or VLIW bundles
- Either 4 or 50+ HW threads
- On-chip communication Fabric
  - Low-degree topology for on-chip communication (torus or mesh)
  - *Scale cache coherence?*
  - cache coherent clusters
  - HW msg. passing
  - Global (nonCC memory)
  - Shared register file (clusters)
- Off-chip communication fabric
  - Integrated directly on an SoC
  - Reduced component counts
  - Coherent with TLB (no pinning)

Fig. 7. Schematic of a future node architecture. The number of functional units on the chip will need to scale out in a 2-D planar geometry and communication locality between the functional units will be increasingly important for efficient computation.

4.2 Instruction Level Parallelism

Up until recently, microprocessors depended on Instruction Level Parallelism and out-of-order execution to make implicit parallelism available to a programmer and to hide latency. Power and complexity costs make it clear that we cannot depend on out-of-order instruction streams to hide latency and improve performance. Instead, we must move to more explicit forms of exposing parallelism such as SIMD units and chips with many independent CPUs.

4.3 Instruction Bundling (SIMD and VLIW)

One way to organize floating point functional units to get implicit parallelism is to depend on grouping multiple operations together into SIMD or VLIW bundles. The benefit of such bundling is that they enable finer-grained data sharing among the instructions, which lowers energy costs and controls complexity. Although SIMD is the most popular approach to organizing FPUs today, there may be movement towards a VLIW organization because it is more flexible in instruction mixing.

The number of SIMD units on x86 chips has doubled in recent years, but the ability to fully exploit evern greater widths is questionable. GPUs also depend on very wide SIMD units, but the semantics of the GPU programming model (CUDA for example) make it easier to automatically use SIMD or VLIW lanes. Currently, Nvidia uses 32-wide SIMD lanes, but there is a pressure to shrink down to 4-8. Current CPU designs have a SIMD width of 4 slots, but will likely move to 8 slots. Overall, this indicates a convergence in the design space towards 4-8 wide instruction bundles (whether it be SIMD or VLIW).

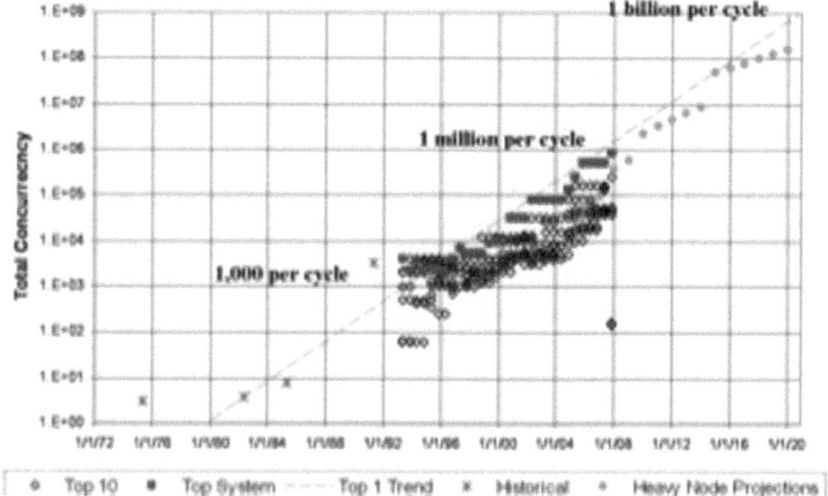

Fig. 8. Due to the stall in clock speeds, future performance improvements will be from increased explicit parallelism. 2018 systems may have as much as 1 billion way parallelism (from DARPA Exascale Report)[2].

4.4 Multithreading to Hide Latency

Little's Law (equation 3) is derived from general information theory, but has important application to understanding the performance of memory hierarchies.

$$\#outstanding\_memory\_requests = bandwidth * latency \qquad (3)$$

In order to fully utilize the available bandwidth of a memory interface, this equation must be balanced. If you have a high bandwidth memory interface, bandwidth will be underutilized if there are not enough outstanding memory requests to hide the latency term of this equation (latency limited). Since we will no longer be depending on complex out-of-order instruction processors to hide latency in the memory hierarchy, there will be increased dependence on hardware multithreading to achieve latency hiding. The latency to local memory is 100ns, but typically you don't have to hide all of the time due to cache reuse.

In swim-lane #1, manycore chip architectures currently support 2-4-way multi-threading, and this may increase to 4-8 way multithreading in future architectures depending on energy cost. GPUs currently depend on 48-64-way hardware multi-threading and will likely support these many threads in the future.

The consequence for programming models is that the baseline expression of parallelism will require 1 billion-way parallelism to achieve an Exaflop if a 1 GHz clock-rate is used. Additional hardware threading required to hide latency will increase the amount of parallelism by a factor of 10-100x.

4.5 FPU Organization

Floating point used to be the most costly component of a system both in terms of design cost and power. However, today, FPUs use a very small fraction of the area of a modern chip design and consume an even smaller fraction of power. On modern systems, a double-precision FMA (fused multiply add) consumes 100 pJ per FMA in 65nm lithography. For 11nm technology anticipated for a 2018 system, a double precision FMA will consume approximately 10.6pJ/op and take 0.02 mm^2 of chip surface area. The FPUs of modern CPUs consume a very small fraction of chip surface area (5-10% of a 400 mm^2 die), whereas GPUs see a larger fraction of their surface area developed to FPUs and general ALUs. A CPU design that consists of many lightweight cores (a manycore chip akin to Larrabee, or Tilera) would likely see the fraction of die area devoted to FPUs close to that observed on modern GPUs.

In order to reduce failure rates and component counts, it is desirable to build a system that reduces the total number of nodes by maximizing the performance of each node. Placing 10,000 FPUs on a chip would only consume 100Watts in this time-frame, and is entirely reasonable in terms of area and power consumption. However supplying memory bandwidth and capacity to a 10Teraflop chip is the primary barrier to this design point. Without advanced packaging technology and substantial improvements in DRAM interface energy efficiency, the upper limit for per-chip performance will likely be 1-2 Teraflops/chip.

We consider two design points to represent this range.

  – Swim Lane 1: 1,000 FPUs per chip
  – Swim Lane 2: 10,000 FPUs per chip

To support full floating point performance, the on-chip register file bandwidth would need to supply 64 bytes per op. Therefore, a 10 Teraflops chip requires 320TB/s of register file bandwidth and 64TB/s register file bandwidth is needed for a 1TF chip. The upper limit of feasible off-chip memory bandwidth will be 4TB/s. Therefore, the design point for Swim Lane 2 would require O(100) data reuse on chip and the design point for Swim Lane 1 would require O(10) data reuse on chip if a 4TB/s memory interface is used. In both cases, the assumed quantity of on-chip memory is on the order of 0.5-1GB/chip, so all temporal recurrences necessary to achieve on-chip data reuse would need to be captured within this memory footprint.

For node organizations that use more than one chip for a node, the bandwidth would likely be more on the order of 0.5 to 1TB/s to remote DRAM (1/4 to 1/8 of local DRAM BW). Therefore, NUMA effects on a multi-chip node will have a substantial performance impact.

4.6 System on Chip (SoC) Integration

To reduce power, and improve reliability it is useful to minimize off-chip I/O by integrating peripheral functions, such as network interfaces and memory controllers, directly onto the chip that contains the CPUs. There are fringe benefits, such as having the communication adaptor be TLB-coherent with the processing elements, which eliminates the need for expensive memory pinning or replicated page tables that is required for current high-performance messaging layers. It also reduces exposure to hard-errors caused by mechanical failure of solder joints. From a packaging standpoint, the node design can be reduced to a single chip surrounded by stacked memory packages, which increases system density. SoC integration will play an increasingly important role in future HPC node designs.

4.7 Alternative Exotic Functional Unit Organizations

Accelerators and Heterogenous Multicore Processors: Accelerators and heterogeneous processing offers some opportunity to greatly increase computational performance within a fixed power budget, while still retaining conventional processors to manage more general purpose components of the computation such as OS services. Currently, such accelerators have disjoint memory spaces that are at the other end of a PCIe interface, which makes programming them very difficult.

There is a desire to have these accelerators fully integrated with the host processor's memory space. At low end, accelerators already are integrated in a unified memory space, but such integration is difficult at the high-end because of differences in the specialized memory technology used for the accelerator and the host processor. By 2015 it will be feasible from a market standpoint to integrate scalar cores with accelerators to obviate the need to copy data between disjoint memory spaces. This was true for NVidia GPU solutions and possibly for heterogeneous manycore architectures like Intel's Larrabee/Knight's Corner[10].

FPGAs and Application-Specific Accelerators: Application specific functional unit organizations may need to be considered to tailor computation and power utilization profiles to more closely match application requirements. However, the scope of such systems may be limited and therefore impact the cost-effectiveness of the resulting system design. FPGAs enable application-tailored logic to be created on-the-fly, but are currently too expensive. Otherwise, FPGA's could be used to implement application-specific primitives.

There is some evidence that power considerations will force system architects to rely on application-tailored processor designs in the 2020 timeframe. Economics will likely constrain the number of application tailored processor designs to a small number and the high performance computing marketplace may not be of sufficient size to warrant its own application-tailored processor.

5 Cache Hierarchy

5.1 Levels of Cache Hierarchy

There has been general agreement among computer companies that there will be 2-4-levels of on-chip hierarchy that can be managed explicitly or flipped to implicit state. The reason for a multi-level hierarchy is mostly governed by the cost of data movement across the chip. Moving data 1 mm across a chip costs far less than a floating point operation, but movement of 20mm (to the other end of the chip) costs substantially more than a floating point operation. Consequently, computation and memory hierarchy on a chip will likely be grouped into clusters or hierarchies of some form to exploit spatial locality of data accesses. There will need to be more effort to create Hardware Block Transfer support to copy data between levels of the memory hierarchy with gather/scatter (multi-level DMA).

5.2 Private vs. Shared Caches

Most codes make no use of cache coherence. So it is likely the cache hierarchy will be organized to put most of the on-chip memory into private cache. Performance analysis indicate less sharing is best (ie. Code written in threads to look like MPI generally performs better).

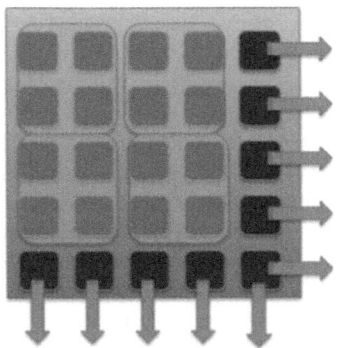

- Cost of moving long-distances on chip motivates clustering on-chip
    - 1mm costs ~6pj (today & 2018)
    - 20mm costs ~120 pj (today & 2018)
    - FLOP costs ~100pj today
    - FLOP costs ~25pj in 2018

- Different Architectural Directions
    - GPU: WARPs of hardware threads clustered around shared register file
    - CMP: limited area cache-coherence
    - CMT: hardware multithreading clusters

Fig. 9. Processor cores or functional units will likely be organized into groups or a hierarchy in order to exploit spatial locality of data accesses

5.3 Software Managed Caches vs. Conventional Caches

Automatically managed caches virtualize the notion of on-chip and off-chip memory, and are therefore invisible to current programming models. However, the cost of moving data off-chip is so substantial, that virtualizing data location in this manner wastes energy and substantially reduces performance. Therefore, there has been increasing interest in explicit software management of memory, such as the Local-stores used by the STI Cell processor and by GPUs. Over the next decade, explicitly managed on-chip memory will become mainstream in conventional CPU designs as well.

However, we have not found the right abstraction for exposing software-controlled memories in our existing programming models. To support an incremental path for existing applications, these explicitly managed memory hierarchies will need to co-exist with conventional automatically managed caches. These software-managed caches may depend on the ability to switch dynamically from automatically managed caches to software-managed caches. Switchable, and dynamically partitionable caches are already demonstrated in the Fermi GPUs, but will likely be seen in conventional multicore architectures as well.

When data is placed into an explicitly controlled cache, it can be globally visible to other processors on the chip, but is not visible to the cache-coherence protocol. Therefore, if the path to higher performance involves keeping more data in these explicitly managed caches, then it means cache-coherence (and the notion of an SMP with it) cannot be part of the high-performance path. Programming language designers must consider how to enable expression of on-chip parallelism without an SMP/cache-coherent model.

6   Intra-node Communication (Networks-on-Chip)

The primary area of growth in parallelism is explicit parallelism on-chip. Whereas the number of nodes in an Exascale system is expected to grow by a factor of 10x over the next decade, on-chip parallelism is expected to grow by a factor of 100x. This requires reconsideration of on-chip organization of CPU cores, and the semantics of inter-processor communication.

6.1   Cache Coherence (or Lack Thereof)

It is likely that cache-coherence strategies can scale to dozens of processing elements, but the cost and latency of data movement on chip would make cache-coherence an inefficient method for interprocessor communication for future chip designs. In all likelihood cache-coherence could be used effectively in clusters or sub-domains of the chip (as illustrated in figure 7), but is unlikely be effective if extended across a chip containing thousands of cores. It is more likely that global memory addressing without cache-coherence will be supported with synchronization primitives to explicitly manage memory consistency.

It is unlikely that cache-coherence will be eliminated completely, but there will need to be careful consideration of the trade-offs of the size of the coherency domain with the magnitude of NUMA (Non-Uniform Memory Access) effects. For a fixed power budget, you can offer users a cluster of cache-coherent domains that have minimal NUMA effects, or very large numbers of cores in the cache-coherent domain that expose the programmer to large NUMA effects. A chip with minimal NUMA effects and small coherence domain could be programmed without substantial attention to data locality, but would derive less benefit from surface-to volume ratios if the coherence-domain is small. There is some opportunity in language support for better implicit locality management in both cases. Creating a chip that has a large coherence domain and minimal NUMA effects would require a substantial increase in power budget to over-design the on-chip interconnection network.

6.2   Global Address Space

Partitioned Global Address Space (PGAS) programming models, including the HPCS programming languages benefit from Global Address Space (GAS) to ensure a compact way to reference remote memory across the machine. PGAS models are willing to accept global addressing without SMP cache-coherence on the node. Therefore, there will likely be support for incoherent global addressing for small-scale systems, but will require hardware investment to scale to larger systems. It is not clear how many address bits will be supported in mainstream implementation. From a technology standpoint, it is entirely feasible to support global addressing within context of Exascale. However, larger scale global addressing schemes will not naturally occur without investment. Global addressing only makes sense with hardware support for sync, which is also investment dependent.

6.3 Fine Grained Synchronization Support

Future programming models will need much finer-grained synchronization features that could directly map to programming language primitives. These features could greatly improve the efficiency of fine-grained on-chip parallelism.

One option is moving atomics memory operations (AMOs) to memory controllers and full empty bits on-chip. Moving atomics as close to memory as possible makes sense from a power and performance standpoint, but would force us to give up some temporal recurrences since the data operated on by the atomics would not pass through the cache hierarchy.

An alternative approach to supporting these atomics is to use an intermediate level of the memory hierarchy where synchronization constructs get enforced/resolved. For example, you could imagine an L2 cache on-chip that is specifically dedicated to fine-grained inter-processor synchronization and atomic memory operations. This approach would potentially encode synchronization state information or other coordinating state using the ECC words of the memory system, because it cannot be held on-chip. All of these options are feasible, but would require close interaction with application developers and programming model designers to determine which approach will be most effective.

7 Power Management

Thermally limited designs force compromises that lead to highly imbalanced computing systems (such as reduced global system bandwidth). The design compromises required for power-limited logic will reduce system bandwidth and consequently reduce delivered application performance and greatly limit the scope and effectiveness of such systems.

From an applications perspective, active power management techniques improve application performance on systems with a limited power budget by dynamically directing power usage only to the portions of the system that require it. For example, a system without power management would melt if it operated memory interfaces at full performance while also operating the floating point unit at full performance — forcing design compromises that limit the memory bandwidth to 0.01 bytes/flop according to the DARPA projections. However, in this thermally limited case you can deliver higher memory bandwidth to the application for the short periods of time by shifting power away from other components. Whereas the projected bandwidth ratio for a machine would be limited to 0.01 bytes/flop without power management, the delivered bandwidth could be increased to 1 byte/flop for the period of time when the application is bandwidth limited by shifting the power away from floating point (or other components that are under-utilized in the bandwidth-limited phase of an algorithm). Therefore, power management is an important part of enabling better delivered application performance through dynamic adjustment of system balance to fit within a fixed power budget.

Currently, changes between power modes take many clock-cycles to take effect. In a practical application code that contains many solvers, the power modes cannot switch fast enough to be of use. Technology that would enable power management systems to switch to low-power modes within a single clock cycle may emerge in the 2015 timeframe. However, there is still a lot of work required to coordinate switching across a large-scale HPC system. Without system scale coordination of power modes, this approach will not be effective.

Current power management features are primarily derived from consumer technology, where the power savings decisions are all made locally. For a large parallel system, locally optimal solutions can be tremendously non-optimal at the system scale. When nodes go into low-power modes opportunistically based on local decisions, it creates jitter that can substantially reduce system-scale performance. For this reason, localized automatic power management features are often turned *off* on production HPC systems. Moreover, the decision to change system balance dynamically to conserve power requires advance notice because there is latency for changing between different power modes. The control loop for such a capability requires a predictive capability to make optimal control decisions. Therefore, new mechanisms that can coordinate these power savings technologies at system scale will be required to realize an energy-efficiency benefit without a corresponding loss in delivered performance.

A complete adaptive control system requires a method for sensing current resource requirements, making a control decision based on an accurate model for how the system will respond to the control decision, and then distributing that control decision in a coordinated fashion. Currently the control loop for accomplishing this kind of optimal control for power management is fundamentally broken. Predictive models for response to control decisions are generally hand-crafted (a time-consuming process) for the few examples that currently[11]. There is no common expression of policy or objective. There is no comprehensive monitoring or data aggregation. More importantly, there is almost NO tool support for integration of power management into libraries and application codes.

Without substantial investments to create system-wide control systems for power management, standards to enable vertical and horizontal integration of these capabilities, and the tools to facilitate easier integration of power management features into application codes, there is little chance that effective power management technologies will emerge. The consequence will be systems that must compromise system balance (and hence delivered application performance) to fit within fixed power constraints, or systems that have impractical power requirements.

7.1 Node-Scale Power Management

Operating systems must support Quality-of-Service management for node-level access to very limited/shared resources. For example, the OS must enable coordinated/fair sharing of the memory interface and network adaptor by hundreds or even thousands of processors on the same node. Support for local and global control decisions require standardized monitoring interfaces for energy and resource utilization (PAPI for energy counters). Standard control and monitoring interfaces enable adaptable software to handle diversity of hardware features/designs. Future OS's must also manage heterogeneous computing resources, and manage data movement and locality in memory hierarchy [13].

7.2   System-Scale Power Management

We need to develop power Performance monitoring and aggregation that scales to 1B+ core system. System management services require standard interfaces to enable coordination across subsystems and international collaboration on component development. Many power management decisions must be executed too rapidly for a software implementation, so must be expressed as a declarative policy rather than a procedural description of actions. Therefore, policy descriptions must be standardized to do fine-grained management on chip. This requires standards for specifying reduced models of hardware power impact and algorithm performance to make logistical decisions about when and where to move computation as well as the response to adaptations. This includes analytical power models of system response and empirical models based on advanced learning theory. We must also develop scalable control algorithms to bridge gap between global and local models. Systems to aggregate sensor data from across the system (scalable data assimilation and reduction), make control decisions and distribute those control decisions in a coordinated fashion across large scale machines are needed. Both online and offline tuning options based on advanced search pruning heuristics should be considered.

7.3   Energy Aware Algorithms

New algorithms must base order of complexity on energy cost of operations rather than FLOPs. A good example of this approach is communication-avoiding algorithms, which trade-off FLOPS for communication to save energy. However, the optimal trade-off is very context specific. There would need to be some methodology to annotate code with a parameterized model of energy consumption for different architectures so that the trade-offs could be computed analytically for different systems. Alternatively, a persistent database could collect runtime information to build up an empirical model of energy consumption for each basic-block of code. Standardizing the approach to specifying or building lightweight analytical models to predict response to resource adjustment will be important to this effort.

7.4   Library Integration with Power Management Systems

Library designers need to use their domain-specific knowledge of the algorithm to provide power management and policy hints to the power management infrastructure. This research agenda requires performance/energy efficiency models and power management interfaces in software libraries to be standardized. This ensures compatibility of the management interfaces and policy coordination across different libraries (horizontal integration) as well as supporting portability across different machines (vertical integration).

7.5   Compiler Assisted Power Management

Compilers and code generators must be able to automatically instrument code for power management sensors and control interfaces to improve the programmability of such systems. Compiler technology can be augmented to automatically expose "knobs for control" and "sensors" for monitoring of non-library code. A more advanced research topic would be to find ways to automatically generate reduced performance and energy consumption models to predict response to resource adaptation.

7.6 Application-Directed Power Management

Applications require more effective declarative annotations for policy objectives and interfaces to coordinate with advanced power-aware libraries and power management subsystems.

7.7 System "Aging"

Today's systems operate with clock rates and voltages in guard margins to account for chip "wear-out". By employing slight clock speed reduction over the lifetime of the system, a 5% power savings can be achieved instead of using guard bands to account for silicon aging effects.

7.8 Voltage Conversion and Cooling Efficiency

Another key area for power reduction is to design hardware to minimize the losses in voltage regulation and power conversion components. For example, the D.E. Shaw system had 30% efficiency loss just from the power conversion stages going from 480V to lowest voltage level delivered to chips.

There are opportunities to use smart-grid strategies to reduce energy consumption. Improve data center efficiencies (5-10% savings in total power consumption) have been demonstrated using this approach [13]. Smart grid technology can rapidly shift power distribution to balance power utilization across the system.

Exascale systems should be water cooled (some may be warm water cooled) because it is substantially more efficient that air cooling.

8 Fault Detection and Recovery

There is a vibrant debate regarding how much responsibility for fault resilience will need to be handled by applications. As a baseline, nearly all applications running on extreme-scale platforms already incorporate some form of application-based defensive I/O (checkpointing). The discussion is primarily concerns shifting balance of responsibility between hardware and software, and its effect on how much additional burden beyond conventional application-driven checkpointing will be required. System architects are keenly aware that applications writers prefer not to have have additional burdens placed upon them.

The circuit hardening techniques required to handle resiliency entirely in hardware are well understood by industry circuit designers for milspec/radiation-hardened parts. Shifting the responsibility more toward the hardware will have a cost in performance or in power consumption (for example, if you add redundancy to harden critical data paths). However, the biggest concern is how far such parts will depart from high-volume mainstream components that will benefit from sharing NRE costs across a larger set of applications. The current failure rates of nodes are primarily defined by market considerations rather than technology. So perhaps it is better to project reliability based on market pressure rather than technology scaling.

From the standpoint of technology scaling, the sources of transient errors will increase by a factor of 100 to 1000x. However, offering a laptop or even cell phone that fails at a 1000x higher rate than today is wholly and entirely impractical from the standpoint of a mainstream technology vendor. Therefore industry will be highly motivated to keep per-node soft error rates from degrading.

1. Moore's law die shrinks will deliver a 100x increase in processors per node in the 11 years between the debut of Petascale systems and the debut of Exascale in 2018.
2. We will need to increase the number of nodes by 10x to get to an Exaflop by 2018.
3. Therefore, market pressure will likely result in a system that is 10x worse than today's extreme-scale systems because of the increased node (and hence component) count.

With 10x, localized checkpointing techniques (such as LLNL's SCR[15]) may be sufficient. As long as users move to a standardized API for user-level checkpointing, these techniques would be comparatively non-invasive since most user codes already understand the importance of defensive I/O (and message logging/replay techniques are transparent to the application).

HPC traditionally acquires network switches, disks, etc from a marketplace that isn't as focused on reliability. We still need a better understanding of the reliability cost trade-offs of these choices. An MTTI of 1 day is achievable for an Exascale system in the 2018 timeframe if the FIT rate *per node* (Failures in time per billion hours of operation for transient uncorrectable errors) stays constant.

8.1 Hard (Permanent) Errors

Hard errors, which are also known as permanent errors, depend on a different mitigation strategy than soft errors. Hard errors might be partly accommodated by incorporating redundant or spare components. For example, building extra cores into a processor chip that can be pressed into service to replace any failed processors on chip. System on Chip designs, described in the Node Architecture section above, can greatly reduce the hard-error rate by reducing the number of discrete chips in the system. Both sockets and solder-joints are a large source of hard-failures – both of which are minimized if all peripheral components are integrated onto a single chip. This approach has been employed successfully on BlueGene systems to achieve a 10-15x lower hard-error rate than conventional clusters.

8.2 Soft (Transient) Errors

The soft (transient) error rate refers to transient errors that affect the Mean time between application interruption (MTTI). The MTTI is any failure that requires application remedial action as opposed to errors that are hidden from the application by resilience mechanism in the hardware or the system software. The MTTI can be much better, using mechanisms a supplier can provide.

It can be useful if the application does some self-checking with a common API to facilitate error detection. Defining a common API for error detection and resilience would help provide uniformity of semantics and innovation of mechanism across multiple vendor platforms. Software approaches for managing error detection and resilience can reduce dependence on hardware checking mechanisms, which can save on power and cost of the system. For example a code could run duplex calculations to self-check could be alternative approach to error detection.

8.3 Node Localized Checkpointing

Localized checkpointing to node-integrated non-volatile storage can accommodate O(10 day) uncorrectable soft errors, but failure characteristics of nonvolatile node-localized storage must be far lower that current commodity parts would support. Using increased redundancy and extensions to Reed-Solomon error correction encodings could make high-volume commodity NVRAM components suitable for node-localized checkpointing.

9 Interconnection Networks

The path towards realizing next-generation petascale and exascale computing is increasingly dependent on building supercomputers with unprecedented numbers of processors. To prevent the interconnect from dominating the overall cost of these ultra-scale systems, there is a critical need for scalable interconnects that capture the communication requirements of ultrascale applications. Future computing systems must rely on development of interconnect topologies that efficiently support the underlying applications' communication characteristics. It is therefore essential to understand high-end application communication characteristics across a broad spectrum of computational methods, and utilize that insight to tailor interconnect designs to the specific requirements of the underlying codes.

9.1 Topology

Throughout the 1990's and early 2000's, high performance computing (HPC) systems implementing *fully-connected networks* (FCNs) such as fat-trees and crossbars have proven popular due to their excellent bisection bandwidth and ease of application mapping for arbitrary communication topologies. However, as supercomputing systems move towards tens or even hundreds of thousands of nodes, FCNs quickly become unfeasibly expensive in terms of wiring complexity, power consumption, and cost[15]. The two leading approaches discussed at the meeting were multi-dimensional Torii and Dragonfly[17] as feasible scalable interconnect topologies. Both approaches present feasible wiring and cost-scaling characteristics for an exascale system. However, it is unclear what portion of scientific computations have communication patterns that can be efficiently embedded onto these types of networks.

The Dragonfly depends on availability of high-radix (radix 64 or greater) router technology to implement a tapered CLOS interconnect topology. The Dragonfly organizes the wiring pattern for the CLOS to localize the high-density wiring within individual cabinets and taper bandwidth for the longer-haul connections. The high-density wiring within a cabinet is amenable to lower-cost copper backplanes to minimize use of discrete wires. Long-haul connections between cabinets would rely on optical transceivers. The tapering of bandwidth for the long-haul connections keeps wiring complexity & cost within practical limits, and results in power and bisection bandwidth characteristics that are similar to the Torus and hypercube.

Another viable technology option is low-radix torus and hypercube interconnects, which rely on low-degree (6-12 port) routers and exploit spatial locality in application communication patterns. The growth in system parallelism has renewed interest in networks with a lower topological degree, such as mesh and torus interconnects (like those used in the IBM BlueGene and Cray XT series), whose costs rise linearly with system scale. Indeed, the number of systems using lower degree interconnects such as the BG/L and Cray Torus interconnects has increased from 6 systems in the November 2004 list to 58 systems in the more recent Top500 list of June 2009[18]. Although there has been a move towards higher-dimensional torus and hypercube networks, in the 2018 timeframe computing system designs may be forced back towards lower-dimensional (4D or 3D) designs in order to keep control of wiring complexity & wire lengths (maximizing the use of wire paths that can be embedded into board designs.

9.2   Effect of Interconnect Topology on Interconnect Design

Practical wiring, cost and power constraints force us away from fully-connected networks. Both networks (Dragonfly[17] and Torus), will require algorithms and other support software that are more aware of the underlying network topology to make the most efficient use of the available network bandwidth at different levels of the network hierarchy. Both networks have similar bisection bandwidth characteristics when compared with similar link performance and message injection bandwidth.

10   Conclusions

Addressing the technology challenges discussed in this report and accelerating the pace of technology development will require focused investments to achieve Exascale computing by 2018. Achieving an Exascale level of performance by the end of the decade will require applications to exploit on the order of a billion-way parallelism provided by an envisioned exascale system. This is in sharp contrast to the approximately quarter million-way parallelism in today's petascale systems. Node architectures are expected to change dramatically in the next decade as power and cooling constraints limit increases in microprocessor clock speeds. Consequently computer companies are dramatically increasing on-chip parallelism to improve performance. The traditional doubling of clock speeds every 18-24 months is being replaced by a doubling of cores, threads or other parallelism mechanisms. Exascale systems will be designed to achieve the best performance within both power and cost constraints. In addition, hardware breakthroughs will be needed to achieve useful exascale computing later this decade, at least within any reasonable power budget. Applications and algorithms will need to change and adapt as node architectures evolve. They will need to manage locality and perhaps resilience to achieve high performance. A key element of the strategy as we move forward is the co-design of applications, architectures and programming environments as shown in Figure 1. Much greater collaboration between these communities will be needed to overcome the key Exascale challenges. There is an unprecedented opportunity for application and algorithm developers to influence the direction of future architectures to reinvent computing for the next decade.

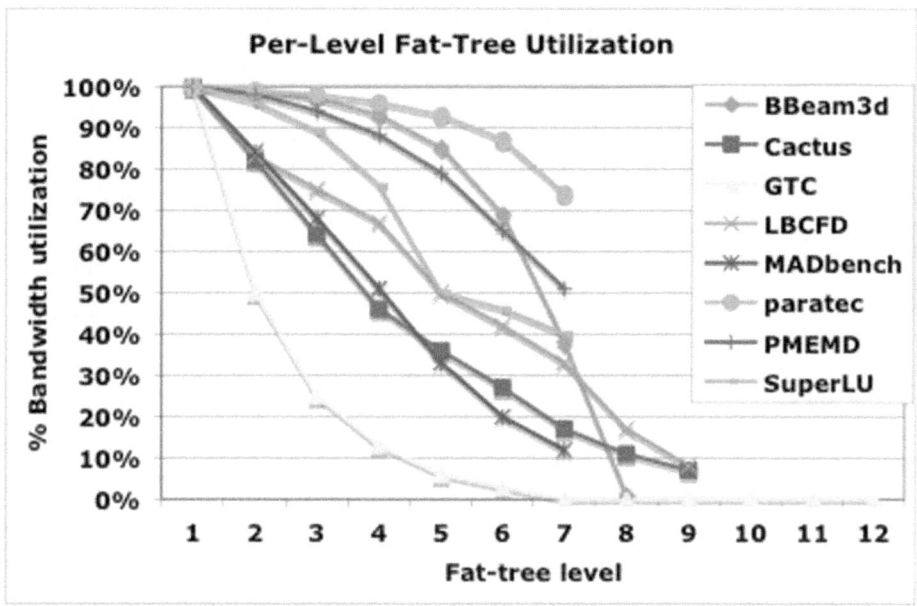

Fig. 10. This figure illustrates the bandwidth tapering characteristics of the communication patterns of 8 key DOE applications when mapped optimally to a multi-layer hierarchical network. Many applications do not fully utilize the upper-layers of the interconnect, meaning that full bisection is not required. [15]

Table 1. Overview of technology scaling for exascale systems. Swimlane 1 represents an extrapolation of manycore system design point whereas swimlane 2 represents scaling of a GPU design point.

Systems	2009	2018 Swimlane 1	2018 Swim-Lane 2
System peak	2 Peta	1 Exa	Same as Swim-lane 1
Power	6 MW	~20 MW	Same as SL1
System memory	0.3 PB	32 - 64 PB	Same as SL1
Node performance	125 GF	1,2TF	10TF

Table 1. (*continued*)

Systems	2009	2018 Swimlane 1	2018 Swim-Lane 2
Interconnect Latency (for longest path)	1-5usec (limited by overhead at endpoints)	0.5-1usec (speed of light)	Same
Memory Latency	150-250 clock cycles (~70-100ns)	100-200 clock cycles (~50ns)	same
Node memory BW	25 GB/s	0.4TB/s	4-5TB/s
Node concurrency	12	O(1k)	O(10k)
Total Node Interconnect BW	3.5 GB/s	100-400GB/s (1:4 or 1:8 from memory BW)	2TB/s
System size (nodes)	18,700	O(1M)	O(100,000)
Total concurrency	225,000	O(100M)*10 for latency hiding	O(100M)*100 for latency hiding
Storage	15 PB	500-1000 PB (>10x system memory is min)	Same as SL1
IO	0.2 TB	60 TB/s	Same as SL1

References

[1] DOE E3 Report,
http://www.er.doe.gov/ascr/ProgramDocuments/ProgDocs.html
[2] A Platform Strategy for the Advanced Simulation and Computing Program (NA-ASC-113R-07-Vol. 1-Rev. 0)
[3] DARPA Exascale Computing Study (TR-2008-13),
http://www.cse.nd.edu/Reports/2008/TR-2008-13.pdf
[4] Miller, D.A., Ozaktas, H.M.: Limit to the bit-rate capacity of electrical interconnects from the aspect ratio of the system architecture. J. Parallel Distrib. Comput. 41(1), 42–52 (1997), DOI http://dx.doi.org/10.1006/jpdc.1996.1285
[5] Miller, D.A.B.: Rationale and challenges for optical interconnects to electronic chips. Proc. IEEE, 728–749 (2000)
[6] Horowitz, M., Yang, C.K.K., Sidiropoulos, S.: High-speed electrical signaling: Overview and limitations. IEEE Micro. 18(1), 12–24 (1998)
[7] IAA Interconnection Network Workshop, San Jose, California, July 21-22 (2008),
http://www.csm.ornl.gov/workshops/IAA-IC-Workshop-08/

[8] Architectures and Technology for Extrame Scale Computing Workshop, San Diego, Cali-
      fornia, December 8-10 (2009),
      http://extremecomputing.labworks.org/hardware/index.stm
[9] Asanovic, K., et al.: The Landscape of Parallel Computing Research: A View from
      Berkeley, Electrical Engineering and Computer Sciences. University of California at
      Berkeley, Technical Report No. UCB/EECS-2006-183, December 18 (2006)
[10] Seiler, L., Carmean, D., Sprangle, E., Forsyth, T., Abrash, M., Dubey, P., Junkins, S.,
      Lake, A., Sugerman, J., Cavin, R., Espasa, R., Grochowski, E., Juan, T., Hanrahan, P.:
      Larrabee: a many-core x86 architecture for visual computing. ACM Trans. Graph. 27(3),
      1–15 (2008)
[11] Liu, Y., Zhu, H.: A survey of the research on power management techniques for high-
      performance systems. Softw. Pract. Exper. 40(11), 943–964 (2010)
[12] Colmenares, J.A., Bird, S., Cook, H., Pearce, P., Zhu, D., Shalf, J., Hofmeyr, S., Asano-
      vić, K., Kubiatowicz, J.: Resource Management in the Tesselation Manycore OS. In:
      HotPar 2010, Berkeley (2010), http://www.usenix.org/event/hotpar10/
      final_posters/Colmenares.pdf
[13] U.S. Department of Energy, DOE Data Center Energy Efficiency Program (April 2009)
[14] Moody, A., Bronevetsky, G., Mohror, K., de Supinski, B.R.: Design, Modeling, and
      Evaluation of a Scalable Multi-level Checkpointing System. In: IEEE/ACM Supercom-
      puting Conference (SC) (November 2010)
[15] Kamil, S., Oliker, L., Pinar, A., Shalf, J.: Communication Requirements and Interconnect
      Optimization for High-End Scientific Applications. IEEE Transactions on Parallel and
      Distributed Systems (2009)
[16] Balfour, J., Dally, W.J.: Design tradeoffs for tiled CMP on-chip networks. In: Proceed-
      ings of the 20th Annual International Conference on Supercomputing, ICS 2006, Cairns,
      Queensland, Australia, June 28-July 01, pp. 187–198. ACM, New York (2006)
[17] Kim, J., Dally, W., Scott, S., Abts, D.: Cost-Efficient Dragonfly Topology for Large-
      Scale Systems. IEEE Micro. 29(1), 33–40 (2009)
[18] Top500 List Home, http://www.top500.org/
[19] Hayt, W.H.: Engineering Electromagnetics, 7th edn. McGraw Hill, New York (2006)
[20] Guha, B., Kyotoku, B.B.C., Lipson, M.: CMOS-compatible athermal silicon microring
      resonators. Optics Express 18(4) (2010)
[21] Hendry, G., Chan, J., Kamil, S., Oliker, L., Shalf, J., Carloni, L.P., Bergman, K.: Silicon
      Nanophotonic Network-On-Chip Using TDM Arbitration. In: IEEE Symposium on High
      Performance Interconnects (HOTI) 5.1 (August 2010)

The Parallel Revolution Has Started: Are You Part of the Solution or Part of the Problem?

An Overview of Research at the Berkeley Parallel Computing Laboratory

David Patterson

University of California at Berkeley
465 Soda Hall
Computer Science Division
Berkeley, CA 94720-1776, USA
pattrsn@cs.berkeley.edu

Abstract. The Par Lab started in 2008, based on an earlier technical report "The Berkeley View" on the parallel computing challenge. (K. Asanovic, R. Bodik, B. C. Catanzaro, J. J. Gebis, P. Husbands, K. Keutzer, D. A. Patterson, W. L. Plishker, J. Shalf, S. W. Williams, and K. A. Yelick. The landscape of parallel computing research: A view from Berkeley. Technical Report UCB/EECS-2006-183, EECS Department, University of California, Berkeley, December 18 2006.) This talk gives an update on where we are two years in the Par Lab. We picked five applications to drive our research, and believe they collectively capture many of the important features of future client applications even if they themselves do not become the actual future "killer app". The Personalized Medicine application focuses on detailed modeling of individual's responses to treatments, representing the important health market. The Music application emphasizes real-time responsiveness to rich human input, with high-performance many-channel audio synthesis. The Speech application focuses on making speech input work well in the real-world noisy environments where mobile devices will be operated. The Content-Based Image Recognition (CBIR) application represents the growing practical use of machine vision. Finally, the Parallel Web Browser is currently perhaps the most important single application on client devices, as well as representative of many other interactive rich-document processing tasks.

Our first step in attacking the parallel programming challenge was to analyze a wide range of applications, including workloads from embedded computing, desktop computing, games, databases, machine learning, and scientific computing, as well as our five driving applications. We discovered a surprisingly compact set of recurring computational patterns, which we termed "motifs". We have greatly expanded on this work, and now believe that any successful software architecture, parallel or serial, can be described as a hierarchy of patterns. We divide patterns into either computational patterns, which describe a computation to be performed, or structural patterns, which describe how computations are composed. The patterns have proven central to ourresearch effort, serving as both a common human vocabulary for multidisciplinary discussions spanning application developers to hardware architects, as well as an organizing structure

J.M.L.M. Palma et al. (Eds.): VECPAR 2010, LNCS 6449, pp. 26–27, 2011.

for software development. Another organizing principle in our original proposal was to divide the software development stack into two layers: efficiency and productivity. Programmers working in the efficiency layer are generally experts in achieving high performance from the underlying hardware, but are not necessarily knowledgeable of any given application domain. Programmers working in the productivity layer are generally knowledgeable about an application domain, but are less concerned with hardware details. The patterns bridge these two layers. Efficiency programmers develop libraries and frameworks that efficiently implement the standard patterns, and productivity programmers can decompose an application into patterns and use high-level languages to compose corresponding libraries and frameworks to form applications.

To improve the quality and portability of efficiency-level libraries, we proposed to leverage our earlier work on autotuning. Autotuning is an automatic search-based optimization process whereby multiple variants of a routine are generated and empirically evaluated on the hardware platform. We have also included a major effort on parallel program correctness to help programmers test, verify, and debug their code. Different correctness techniques apply at the efficiency layer, where low-level data races and deadlocks are of concern, and at the productivity layer, where we wish to ensure semantic determinism and atomicity. Our whole pattern-based component approach to the software stack hinges on the ability to efficiently and flexibly compose software modules. We developed a low-level user-level scheduling substrate called "Lithe" to support efficient sharing of processing resources between arbitrary modules, even those written in different languages and to different programming models.

Our operating system and architecture research is devoted to supporting the software stack. The OS is based on space-time partitioning, which exports stable partitions of the machine resources with quality-of-service guarantees to an application, and two-level scheduling, which allows a user-level scheduler, such as Lithe, to perform detailed application-specific scheduling within a partition. Our architecture research focuses on techniques to support OS resource partitioning, performance counters to support application adaptivity, software-managed memory hierarchies to increase memory efficiency, and scalable coherence and synchronization mechanisms to lower parallel system overheads. To experiment with the behavior of our new software stack on our new OS and hardware mechanisms, we have developed an FPGA-based simulation environment, "RAMP Gold". By running our full application and OS software environment on our fast architectural simulator, we can quickly iterate across levels in our system stack.

HPC Techniques for a Heart Simulator

Takumi Washio[1], Jun-ichi Okada[1], Seiryo Sugiura[1], and Toshiaki Hisada[2]

[1] Graduate School of Frontier Sciences
The University of Tokyo
5-1-5, Kashiwanoha, Kashiwa, Chiba 277-0882, Japan
[2] Graduate School of Frontier Sciences
The University of Tokyo
7-3-1 Hongo, Bunkyo-ku, Tokyo 113-8656, Japan
washio@sml.k.u-tokyo.ac.jp, hisada@mech.t.u-tokyo.ac.jp

Abstract. In the post-genome era, the integration of molecular and cellular findings in studies into the functions of organs and individuals is recognized as an important field of medical science and physiology. Computational modeling plays a central role in this field, which is referred to as Physiome. However, despite advancements in computational science, this task remains difficult. In addition to coupling multiple disciplines, including electricity, physical chemistry, solid mechanics and fluid dynamics, the integration of events over a wide range of scales must also be accomplished. Our group, including clinical practitioners, has been tackling this problem over several years, with a focus on the human heart.

The morphology of our heart model is reconstructed from human multi-detector computed tomography data and discretized using the finite element method. For the electrophysiology simulation, a composite voxel mesh with fine voxels in and around the heart and coarse voxels covering the torso is adopted to solve the bidomain equation. Following the excitation of a sinus node, the simulator reproduces the excitation propagation and depolarization of the membrane potentials of virtual cells   sequentially in the atria, conduction system, and ventricles. The mechanical simulation for the interaction between the heart wall and intracavitary blood flow is performed on a tetrahedral mesh. The Ca^{2+} concentration data obtained from the electrophysiology model are applied to the molecular model of sarcomere dynamics to compute the contraction force of every element. This results in the synchronous contraction of the heart and blood flow.

Thus far, we have been able to retrieve and present the data in the same way as clinical diagnostic tools, such as ECG, UCG, and magneto-cardiogram in our simulation studies. These data are in good agreement with the clinical data for both normal and diseased heart models, thus suggesting their potentials for diagnostic support.

However, a more important aspect of the simulation involves modeling the underlying mechanism driving the myocardium, i.e., the origin of the pulsation of the heart, which includes electrophysiological regulation and cross-bridge kinetics in the cardiac cells. To integrate such microscopic phenomena with the macroscopic function of the organ in a seamless manner, the cardiac cells are also modeled using the finite element method, based on the cell physiology for

J.M.L.M. Palma et al. (Eds.): VECPAR 2010, LNCS 6449, pp. 28–29, 2011.

every finite element in the heart model. The mathematical linkage is realized using the so-called homogenization method. All the cell models and the heart model are then solved simultaneously, because the instantaneous states of the macroscopic model, such as the strains and strain rates over the heart wall, also regulate each cell response. It is apparent that the total number of degrees of freedom of all the cell models becomes prohibitively large.

We will introduce basic algorithms and parallel computational techniques applied to the above mentioned multi-physics and multi-scale simulations.

Game Changing Computational Engineering Technology

Charbel Farhat

Stanford University
William F. Durand Building, Rm. 257
496 Lomita Mall, Stanford, CA 94305-4035, USA
cfarhat@stanford.edu

Abstract. During the last two decades, giant strides have been achieved in many aspects of Computational Engineering. Higher-fidelity mathematical models, better approximation methods, and faster algorithms have been developed for many time-dependent applications. SIMD, SPMD, MIMD, coarse-grain, and fine-grain parallel processors have come and gone. Linux clusters are now ubiquitous, cores have replaced CEs, and GPUs have shattered computing speed barriers. Most importantly, the potential of high-fidelity physics-based simulations for providing deeper understanding of complex engineering systems and enhancing system performance has been recognized in almost every field of engineering. Yet, in many engineering applications, high-fidelity time-dependent numerical simulations are not performed as often as needed, or are more often performed in special circumstances than routinely. The reason is very simple: these simulations remain too computationally intensive for time-critical operations such as design, design optimization, and active control. Consequently, the impact of computational sciences on such operations has yet to materialize. Petascale or exascale computing alone is unlikely to make this happen. Achieving this objective demands instead a game-changing computational technology that bridges both ends of the computing spectrum. This talk will attempt to make the case for this pressing need and outline a candidate computational technology for filling it that is based on model reduction, machine learning concepts, trained data bases, and rigorous interpolation methods. It will also illustrate it with preliminary results obtained from its application to the support of the flutter flight testing of a fighter aircraft and the aerodynamic optimization of Formula 1 car.

J.M.L.M. Palma et al. (Eds.): VECPAR 2010, LNCS 6449, p. 30, 2011.
© Springer-Verlag Berlin Heidelberg 2011

HPC in Phase Change: Towards a New Execution Model

Thomas Sterling

Louisiana State University
320 Johnston Hall
Baton Rouge, LA 70803, USA
tron@cct.lsu.edu

Abstract. HPC is entering a new phase in system structure and operation driven by a combination of technology and architecture trends. Perhaps foremost are the constraints of power and complexity that as a result of the at-lining of clock rates relies on multicore as the primary means by which performance gain is being achieved with Moore's Law. Indeed, for all intense and purposes, "multicore" is the new "Moore's Law" with steady increases in the number of cores per socket. Added to this are the highly multithreaded GPU components moving HPC into the heterogeneous modality for additional performance gain. These dramatic changes in system architecture are forcing new methods of use including programming and system management. Historically HPC has experienced five previous phase changes involving technology, architecture, and programming models. The current phase of two decades is exemplified by the communicating sequential model of computation replacing previous vector and SIMD models. HPC is now faced with the need for new effective means of sustaining performance growth with technology through rapid expansion of multicore with anticipated structures of hundreds of millions of cores by the end of this decade delivering Exaflops performance. This presentation will discuss the driving trends and issues of the new phase change in HPC and will discuss the ParalleX execution model that is serving as a pathfinding framework for exploring an innovative synthesis of semantic constructs and mechanisms that may serve as a foundation for computational systems and techniques in the Exascale era. This talk is being given just as DARPA is initiating its UHPC program and DOE is launching additional programs such as their X-stack all aimed at catalyzing research in to the challenging area.

J.M.L.M. Palma et al. (Eds.): VECPAR 2010, LNCS 6449, p. 31, 2011.
© Springer-Verlag Berlin Heidelberg 2011

Factors Impacting Performance of Multithreaded Sparse Triangular Solve

Michael M. Wolf, Michael A. Heroux, and Erik G. Boman

Scalable Algorithms Dept., Sandia National Laboratories, Albuquerque, NM, USA
{mmwolf,maherou,egboman}@sandia.gov

Abstract. As computational science applications grow more parallel with multi-core supercomputers having hundreds of thousands of computational cores, it will become increasingly difficult for solvers to scale. Our approach is to use hybrid MPI/threaded numerical algorithms to solve these systems in order to reduce the number of MPI tasks and increase the parallel efficiency of the algorithm. However, we need efficient threaded numerical kernels to run on the multi-core nodes in order to achieve good parallel efficiency. In this paper, we focus on improving the performance of a multithreaded triangular solver, an important kernel for preconditioning. We analyze three factors that affect the parallel performance of this threaded kernel and obtain good scalability on the multi-core nodes for a range of matrix sizes.

1 Introduction

1.1 Motivation

With the emergence of multi-core processors, most supercomputers are now hybrid systems in that they have shared memory multi-core nodes that are connected together into a larger distributed memory system. Although for many numerical algorithms the traditional message programming model is sufficient to obtain good scalability, some numerical methods can benefit from an hybrid programming model that uses message passing between the nodes with a shared memory approach (e.g., threads) within the node. Scalable threaded algorithms that run efficiently on the node are essential to such a hybrid programming model and are the emphasis of our work in this paper.

Solvers are a good example of numerical algorithms that we believe can benefit from a hybrid approach. Solver implementations based on a flat MPI programming model (where subcommunicators are not utilized) often suffer from poor scalability for large numbers of tasks. One difficulty with these approaches is that with domain decomposition based preconditioners, the number of iterations per linear solve step increase significantly as the number of MPI tasks (and thus the number of subdomains) becomes particularly large. We also see this behavior with scalable preconditioners such as an algebraic multilevel preconditioner. Figure 1 shows an example of this difficulty for Charon, a semiconductor device simulation code [1,2,3], with a three level multigrid preconditioner. As the

J.M.L.M. Palma et al. (Eds.): VECPAR 2010, LNCS 6449, pp. 32–44, 2011.
© Springer-Verlag Berlin Heidelberg 2011

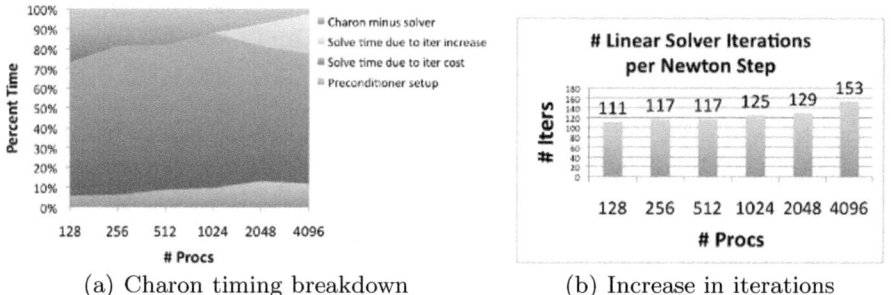

(a) Charon timing breakdown   (b) Increase in iterations

Fig. 1. Strong scaling analysis of Charon on Sandia Tri-Lab Linux Capacity Cluster for 28 million unknowns

number of MPI tasks increases, the number of linear solver iterations increases (Figure 1(b)). Figure 1(a) shows that these extra iterations require an increasingly higher percentage of the total runtime as the number of MPI tasks increase, resulting in a degradation in the parallel performance.

By having fewer (but larger) subdomains, better convergence can be obtained for the linear solver. With fewer subdomains, the solvers for these larger subdomains must be parallel in order to maintain the overall scalability of the algorithm. This leads to a two-level model of parallelism, where MPI is used to communicate between subdomains and a second level of parallelism is used within each subdomain. One approach is to also use MPI to obtain parallelism at the subdomain level (e.g., [4]). Another approach, which we explore in this paper, utilizes multithreading to obtain parallelism at the subdomain level. This approach is limited in that each subdomain does not extend beyond the processor boundaries. However, we feel that as the number of cores per processor continues to increase, this will become less important and threads may be a better approach for exploiting the shared memory architecture on the node.

Keeping the iteration count low is not sufficient, however, to obtain performance gains over MPI-only implementations. The shared memory numerical kernels that run on each multi-core node also need to be scalable. It is particularly important to have a scalable shared memory implementation of a triangular solver to run on each node since this kernel will be executed for each iteration of the linear solver. The focus of this paper is to study the various factors that affect the performance of this shared memory triangular solver kernel in the pursuit of a sufficiently scalable algorithm.

1.2   Level-Set Triangular Solver

We focus our attention on improving the performance of a level-set triangular solver for sparse matrices as described in [5]. Below we describe the process for lower triangular matrices, but the upper triangular case is analogous. First, we express the data dependencies of the triangular solve for the lower triangular matrix $\mathbf{L}$ as a directed acyclic graph (DAG). A vertex v_i of this DAG correspond to the vector entry x_i that will be calculated in the triangular solve. The directed edges in the DAG represent data dependencies between the $\mathbf{x}$ vector entries, with a directed edge connecting v_i to v_j if and only if x_i is needed to compute x_j. The level-sets of this DAG represent sets of row operations in the triangular solve operation that can be performed independently. Specifically, the ith level-set is a set of vertices that have incoming edges only from vertices of the previous $i-1$ levels (the corresponding x_i entries are only dependent on x vector entries in previous levels). We calculate these level-sets from the DAG using a variant of breadth-first search. Permuting the system symmetrically so that the rows/columns are in order of the level-sets, we obtain the following permuted matrix,

$$\tilde{\mathbf{L}} = \mathbf{P}\mathbf{L}\mathbf{P}^T = \begin{bmatrix} \mathbf{D}_1 & & & \\ \mathbf{A}_{2,1} & \mathbf{D}_2 & & \\ \mathbf{A}_{3,1} & \mathbf{A}_{3,2} & \mathbf{D}_3 & \\ \vdots & \vdots & \vdots & \ddots \\ \mathbf{A}_{l,1} & \mathbf{A}_{l,2} & \mathbf{A}_{l,3} & \dots & \mathbf{D}_l \end{bmatrix},$$

where l is the number of level-sets. This symmetrically permuted matrix is still triangular since $\tilde{x}_i$ can only depend on those $\tilde{x}_j$ calculated in a previous level with this dependency corresponding to a lower triangular nonzero in the permuted matrix. Since there are no data dependencies within a level in the permuted matrix (i.e., no edges connecting vertices within a level-set), the $\mathbf{D}_i$ must be diagonal matrices.

With this basic level-set permuted matrix structure, we can use either a forward-looking or backward-looking algorithm. After a diagonal solve determines a set of vector entries $\tilde{\mathbf{x}}_i$, the forward-looking algorithm uses $\tilde{\mathbf{x}}_i$ to immediately update $\tilde{\mathbf{x}}_j, j > i$ with the matrix-vector product operations in the ith column block. A backward-looking algorithm uses all previously computed $\tilde{\mathbf{x}}_i, i = 1, \dots, l-1$, in a series of matrix-vector products updates immediately before computing $\tilde{\mathbf{x}}_l$. Both algorithms have the same operation counts but have different memory access patterns for the matrices and vectors. In particular, the forward-looking algorithm exploits the temporal locality of the previously calculated $\tilde{\mathbf{x}}_i$ that are used in the matrix-vector products while the backward-looking algorithm exploits the temporal locality of $\tilde{\mathbf{x}}_i$ that are being determined/stored. While both algorithms have different advantages, we chose to implement the backward-looking algorithm, since we were able to use our compressed row storage matrices in a more natural manner. The operations needed to solve the permuted system for $\tilde{\mathbf{x}}$ in this backward-looking algorithm are as follows

$$\tilde{\mathbf{x}}_1 = \mathbf{D}_1^{-1}\tilde{\mathbf{y}}_1$$
$$\tilde{\mathbf{x}}_2 = \mathbf{D}_2^{-1}\left(\tilde{\mathbf{y}}_2 - \mathbf{A}_{2,1}\tilde{\mathbf{x}}_1\right)$$
$$\vdots \quad \vdots \qquad \vdots$$
$$\tilde{\mathbf{x}}_l = \mathbf{D}_l^{-1}\left(\tilde{\mathbf{y}}_l - \mathbf{A}_{l,1}\tilde{\mathbf{x}}_1 - \dots - \mathbf{A}_{l,l-1}\tilde{\mathbf{x}}_{l-1}\right).$$

(Note that the above operations were written to elucidate this algorithm but in practice the sparse matrix-vector product (SpMV) operations for each level can be combined into one SpMV.) $\mathbf{x}$ can be subsequently recovered by the operation $\mathbf{P}^T \tilde{\mathbf{x}}$.

The vector entries at each level (those in $\tilde{\mathbf{x}}_i$ above) can be calculated independently. Thus, in our threaded solver kernel, the computation in each level can be distributed across threads (e.g., with an OpenMP/TBB like **parallel for** operation) without the need for synchronization. However, synchronization is needed between the levels of the algorithm to ensure that the vector entries $\tilde{\mathbf{x}}_j$ needed for matrix-vector product portion of each level of computation (i.e., $\mathbf{A}_{i,j}\tilde{\mathbf{x}}_j$) have been previously determined. In this context, we examine how specific factors affect the performance of this multithreaded triangular solve kernel.

This approach is most beneficial for solving triangular systems resulting from incomplete factorizations, where the resulting matrix factors are sufficiently sparse to yield sufficiently large levels. For matrices that do not result in sufficiently large levels, this approach to parallelism will not be particularly effective (as we will see in the subsequent section). However, for matrices where the resulting levels are sufficiently large, the synchronization costs in our multithreaded algorithm should be small enough to allow for good parallel performance.

1.3 Related Work

Saltz described the usage of directed acyclic graphs and wavefront (level-set) methods for obtaining parallelism for the sparse triangular solve operation in [5]. In this paper, his work focused on sparse triangular systems generated by incomplete factorizations arising from discretization of partial differential equations. This approach was applicable to both shared memory and message passing systems. Rothberg and Gupta addressed the sparse triangular solve bottleneck in the context of the incomplete Cholesky conjugate gradient algorithm ([6]). They argued that the two most promising algorithms at that time (one of which was the level-set method) performed poorly on the more modern shared memory machines that utilized deep memory hierarchies. One of the problems they found was that the level-set algorithm's performance was negatively impacted by the poor spatial locality of the data. This is not a major difficulty for our implementation since we explicitly permute the triangular matrix unlike the original implementation that simply accessed the rows in a permuted order. In more recent work ([7]), Mayer has developed two new algorithms for solving triangular systems on shared memory architectures. The first algorithm uses a block partitioning to split the triangular matrix across both rows and columns in a 2D Cartesian manner. Given this partitioning, the operations using blocks on the same anti-diagonal can be computed in parallel on different threads. The difficulty with this method is finding a good partitioning of blocks that balances the work. The second method is a hybrid method combining the first method with a method of solving by column blocks (after a diagonal block has been solved, the updates in the column block below may be done in parallel). The hybrid method is easier to partition than the first method. Mayer's results were somewhat modest in the speedups that were obtained. However, the methods are more general than the level-set method and may be effective over a larger range of matrices.

2   Factors Affecting Performance

One factor we examine is data locality for the matrices. First, we experiment on two special types of matrices (Figure 2), where the number of rows is the same for each level and the matrices are already ordered by level-sets. One of these special matrices results in good data locality (Figure 2(a)) in that threads will not use vector entries that another thread has computed in its computation. This can be seen in the precisely placed off-diagonal bands in these matrices that ensure that if a vector entry x_i is calculated by a thread, any subsequent computation involving x_i (i.e., computation corresponding to nonzeros in column i) will assigned to that same thread. We can enforce this good data locality since these matrices have the same number of rows per level-set and we assign row operations to threads in a contiguous block manner. The other matrix results in bad data locality (Figure 2(b)) in that threads will often use vector entries calculated by another thread. Again, this can be seen in the precisely placed off-diagonal nonzero blocks in these matrices that ensure that if a vector entry x_i is calculated by a thread, subsequent computation involving x_i in the next level will not be assigned to that same thread.

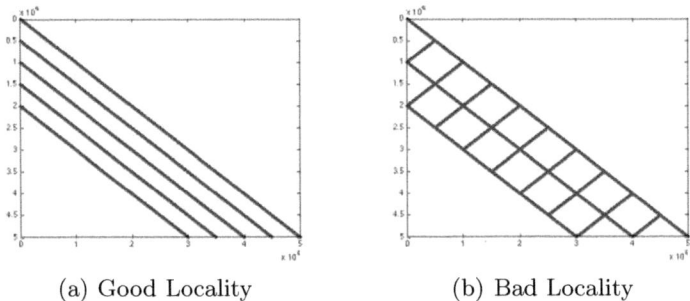

(a) Good Locality   (b) Bad Locality

Fig. 2. Nonzero patterns for matrix types

We also look at variants of the triangular solve algorithm with different barriers and thread affinity settings. The barrier is an important part of this level set method, providing synchronization between the levels of our triangular solve. The first type of barrier is somewhat passive and uses mutexes and conditional wait statements. All the threads wait (**pthread_cond_wait**) until every thread has entered the barrier. Then a signal is broadcast (**pthread_cond_broadcast**) that allows the threads to exit the barrier. The disadvantage with this more passive barrier is a thread calling this barrier might be switched to a different computational core while waiting for a signal in the conditional wait statement. The second type of barrier is more active and uses spin locks and active polling.

A thread entering this barrier will actively poll until all threads have reached the barrier. This makes it less likely for the threads to be switched, which is a good thing assuming there is nothing else being computed simultaneously with the triangular solve. Thread affinity describes how likely a thread is to run on a particular core. By setting the thread affinity, we can bind a thread to a particular core, which can be beneficial to the performance of numerical kernels. This also allows us to ensure our threads are running on the same socket on machines with multiple sockets. This may be desirable for numerical algorithms, especially if there is effective utilization of a cache shared between the cores on a socket (e.g., L3 cache on Nehalem). When setting the thread affinity, we set the affinity for each thread to a different core on the same socket.

3   Numerical Experiments

We implemented a level set triangular solve prototype that solves triangular systems of ten levels, with the same number of rows for each level. For this prototype, the rows in a level are distributed in a block fashion to different threads that will perform the computation on those rows. This simple set up allows us to easily control the factor of data locality. We experiment on a range of different size matrices and run our experiments on one, two, and four threads to study the scalability of the algorithm variants.

We have performed these experiments on two different multi-core systems. The first system is an Intel Nehalem system running Linux with a two socket motherboard with 2.93 GHz quad-core Intel Xeon processors for a total of 8 cores. Intel's Turbo Boost Technology is turned off on this system, so two threads should run at the same clock speed as one thread. The second system is an AMD Istanbul system running Linux with a two socket motherboard with 2.6 GHz six-core AMD Opteron processors for a total of 12 cores.

3.1   Barriers

First, we compare the results for the different types of barriers. Figures 3 and 4 show results for the triangular solves on the good data locality matrices of various sizes when the thread affinity is set. For the Nehalem system, we show results for 2, 4, and 8 threads. For the Istanbul system, we show results for 2, 6, and 12 threads. Parallel speedups are presented for both the active and passive barrier variants.

For both the Nehalem and the Istanbul systems, it is clear that the active barrier is necessary to obtain good scalability, especially for the smaller sized matrices and runs with many threads. Figure 5 shows a runtime comparison of the two implementations using different barriers for two of the matrices (bad data locality and thread affinity on) for 1, 2, and 4 threads. It is clear from both the Nehalem and Istanbul plots that having an active barrier is important for scalability.

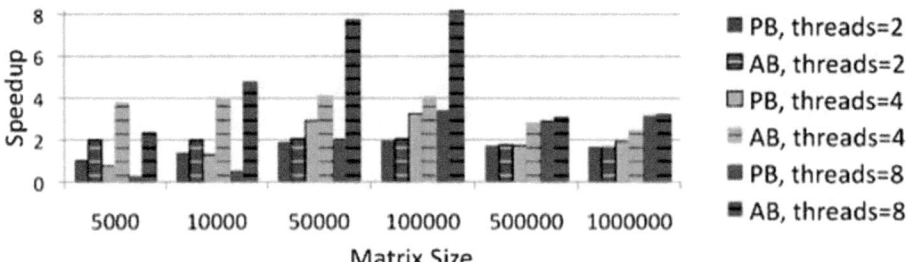

Fig. 3. Effects of different barriers on Nehalem system: thread affinity on, good data locality. Speedups for passive barrier (PB) and active barrier (AB) for 2, 4, and 8 threads.

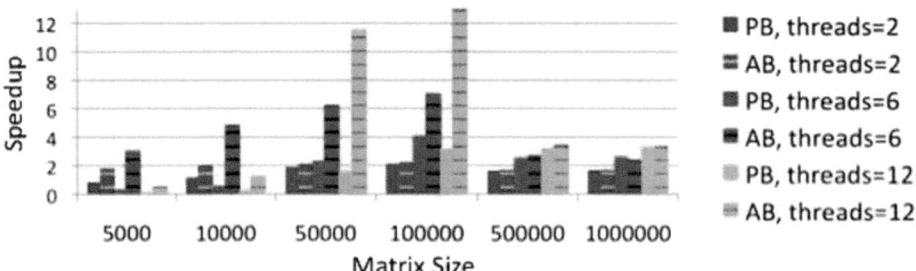

Fig. 4. Effects of different barriers on Istanbul system: thread affinity on, good data locality. Speedups for passive barrier (PB) and active barrier (AB) for 2, 6, and 12 threads.

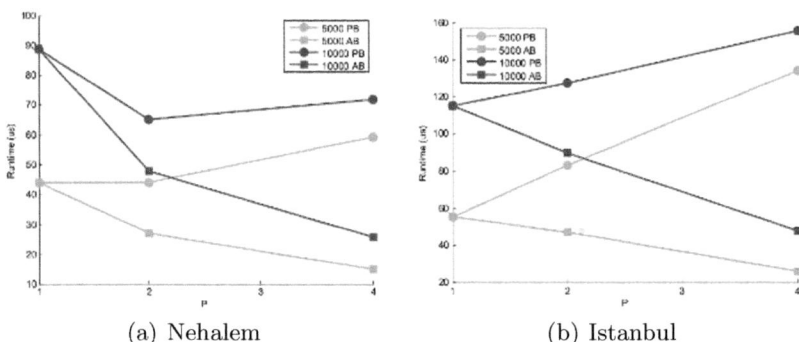

(a) Nehalem   (b) Istanbul

Fig. 5. Effects of different barriers: sizes 5000 and 10000, thread affinity on, bad data locality. Runtimes for passive barrier (PB) and active barrier (AB).

3.2   Thread Affinity

Next, we examine the effects of thread affinity on the scalability of our triangular solve algorithm. Figures 6 and 7 show results for the triangular solves on the good data locality matrices of various sizes with the active barrier. For both the Nehalem and the Istanbul systems, it is clear that the thread affinity is not as important of a factor in scalability as the barrier type. However, for some of the smaller data sizes, setting the thread affinity does seem to improve the scalability. Figure 8 shows the runtimes for two matrices (bad data locality and active barrier) with thread affinity on or off for 1, 2, and 4 threads. It seems that

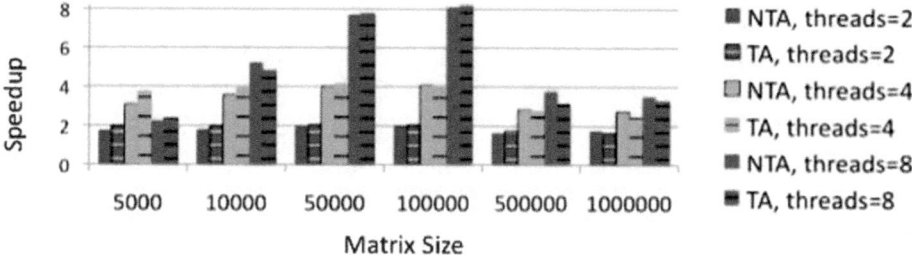

Fig. 6. Effects of binding threads to cores by setting thread affinity on Nehalem system: active barrier, good data locality. Speedups for algorithm when setting thread affinity (TA) and not setting thread affinity (NTA) for 2, 4, and 8 threads.

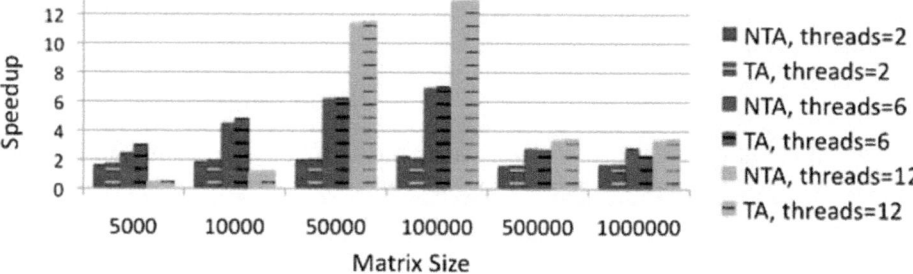

Fig. 7. Effects of binding threads to cores by setting thread affinity on Istanbul system: active barrier, good data locality. Speedups for algorithm when setting thread affinity (TA) and not setting thread affinity (NTA) for 2, 6, and 12 threads.

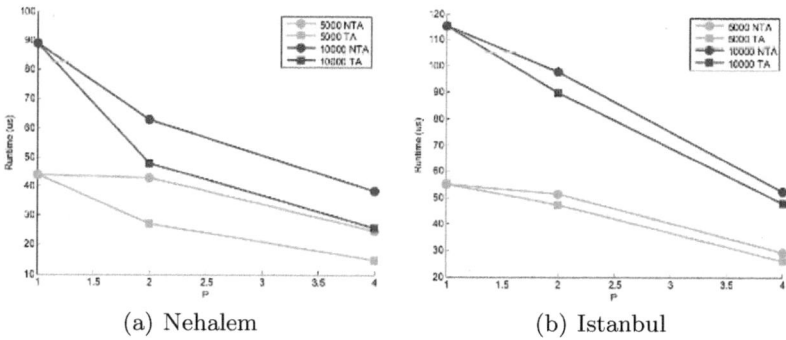

(a) Nehalem   (b) Istanbul

Fig. 8. Effects of setting thread affinity: sizes 5000 and 10000, active barrier, bad data locality. Runtimes for setting thread afinity (TA) and not setting thread affinity (NTA).

thread affinity is somewhat important for these problem sizes, especially on the Nehalem system.

3.3   Data Locality

Finally, we examine the impact of data locality on the scalability of the triangular solves. Figure 9 shows a comparison between the results of the two different types of matrices (one with good locality and the other with bad data locality) of size 50000 and 100000 rows. We see basically no difference for these types of matrices for Nehalem and only a very slight difference for Istanbul. The results for these two sizes was typical of what we observed overall.

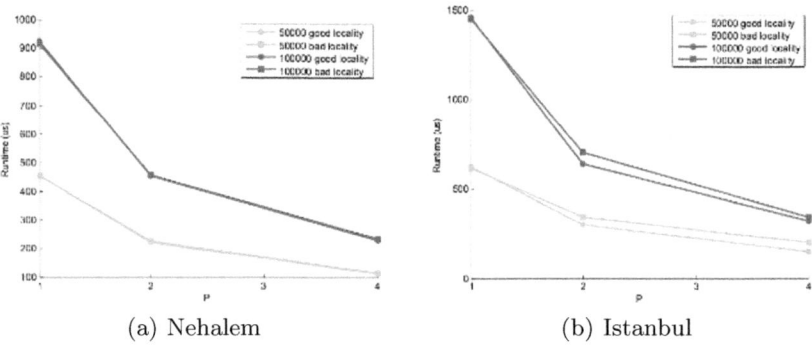

(a) Nehalem   (b) Istanbul

Fig. 9. Effects of data locality: active barrier, thread affinity on. Runtimes for good and bad data locality matrices (sizes 50000 and 100000).

3.4   More Realistic Problems

In the previous subsections, we solved triangular systems for a very specific set of matrices. These matrices were designed to have a specific structure that allowed us to study the importance of data locality in a very simple environment. These matrices were sufficient to get a good handle of the factors affecting performance in an ideal scenario. In this subsection, we study the impact of barrier type and thread affinity in more realistic situation, solving triangular systems resulting from four symmetric matrices obtained from the University of Florida Sparse Matrix Collection [8]. These four matrices are shown in Table 1 with their respective number of rows, number of nonzeros, and application areas. We generalized the prototype solver used in the previous subsections to calculate level sets, permute the matrices, and solve the triangular system for any lower triangular system.

We take the lower triangular part of the matrices shown in Table 1 to be our lower triangular matrices (i.e., zero fill incomplete factorizations). The fourth column of Table 1 gives the average number of rows per level for the level-sets determined from the lower triangular part of these matrices. We picked these four matrices deliberately to cover a range for this statistic. As we did with the

Table 1. Symmetric Matrix Info

Name	N	nnz	N/nlevels	application area
asic680ks	682,712	2,329,176	13932.9	circuit simulation
cage12	130,228	2,032,536	1973.2	DNA electrophoresis
pkustk04	55,590	4,218,660	149.4	structural engineering
bcsstk32	44,609	2,014,701	15.1	structural engineering

simple matrices of the previous subsections, we compare the results for the active
and passive barrier variants. Figures 10 and 11 show results for the triangular
solves when the thread affinity is set, comparing the two barrier types.

Again we see for both the Nehalem and the Istanbul systems that the active
barrier is necessary to obtain good scalability. The difference is particularly striking for the larger numbers of threads. As expected, the solves scaled better when
the matrices had large numbers of rows per level. In particular, the **asic680ks**
and **cage12** matrices, which had the largest numbers of rows per level, scaled
very well (especially on the Istanbul architecture). However, the **bcsstk32** matrix, which had approximately 15.1 rows per level, actually required more runtime as the number of threads increased. This is not too surprising since with an

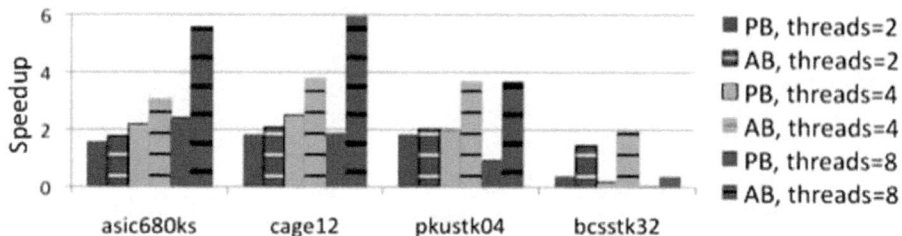

Fig. 10. Application matrices. Effects of different barriers on Nehalem system: thread
affinity on. Speedups for passive barrier (PB) and active barrier (AB) for 2, 4, and 8
threads.

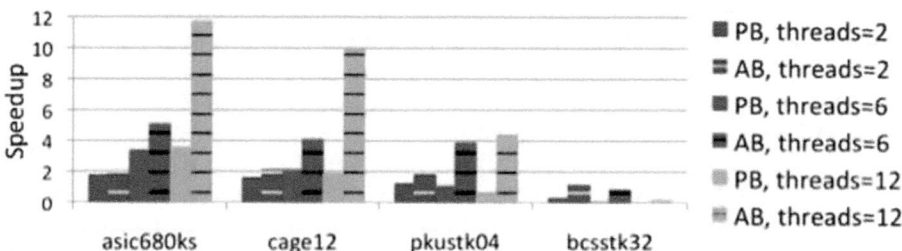

Fig. 11. Application matrices. Effects of different barriers on Istanbul system: thread
affinity on. Speedups for passive barrier (PB) and active barrier (AB) for 2, 6, and 12
threads.

average of 15.1 rows per level, many levels would not have one row per thread, let alone provide enough work to amortize the cost of the synchronization step.

Again, we examine the effects of thread affinity on the scalability of our triangular solve algorithm. Figures 12 and 13 show results for the triangular solves for these more realistic matrices with the active barrier. For both the Nehalem and the Istanbul systems, it is clear that the thread affinity is not as important of a factor in scalability as the barrier type. For several problems, we see a slight increase in speedup when thread affinity is on. However, there are several counter examples where the speedup slightly decreases.

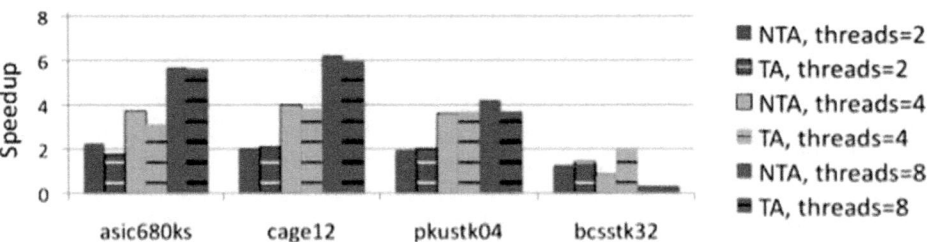

Fig. 12. Application matrices. Effects of binding threads to cores by setting thread affinity on Nehalem system: active barrier. Speedups for algorithm when setting thread affinity (TA) and not setting thread affinity (NTA) for 2, 4, and 8 threads.

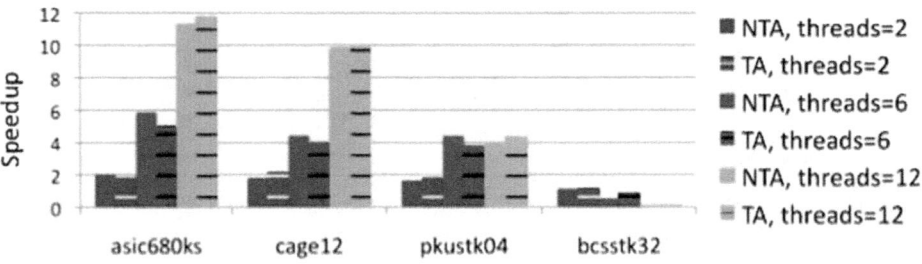

Fig. 13. Application matrices. Effects of binding threads to cores by setting thread affinity on Istanbul system: active barrier. Speedups for algorithm when setting thread affinity (TA) and not setting thread affinity (NTA) for 2, 6, and 12 threads.

4   Summary and Conclusions

In pursuit of more scalable solvers that scale to hundreds of thousands of computational cores on multi-core architectures, we are researching hybrid MPI/threaded algorithms that should lower iterations counts by reducing the number of MPI tasks (and subdomains). An essential part of these algorithms are scalable threaded numerical kernels such as the triangular solver on which we focused. We examined three different factors that affect the performance of the threaded level-set triangular solver. Of these three factors, the barrier type was shown to have the most impact, with an active barrier greatly increasing the parallel performance when compared to a more passive barrier. Although it is not always possible (e.g., if additional computation takes place on the same cores concurrently), we advocate using as aggressive of a barrier as possible in this type of algorithm. Our results showed that binding the threads to processor cores had less impact than the barrier type. However, it did improve the performance for some cases and may be a reasonable approach to take for multithreaded numerical kernels where the number of active threads is not more than the number of computational cores. With an active barrier and thread binding to cores, we were able to achieve excellent parallel performance for a range of matrix sizes for the ideal matrices as well as three of the four more realistic matrices that we studied.

We also examined the impact of data locality on the scalability of the triangular solves, comparing matrices with good and bad data locality. It is unclear from our results whether data locality is an important factor in the parallel performance. It is possible that our bad data locality matrices do not have poor enough data locality to see a very large effect. It is also possible that our matrices are too sparse and that we would see more of an effect for denser matrices. But perhaps the memory systems are too fast for the locality of the data in these sparse matrix triangular solves to greatly impact the scalability of the algorithm. If the data locality becomes an issue more general classes of triangular matrices, we would need to explore ordering techniques to mitigate this problem.

Of more importance was the sparsity structure of the matrices and how this sparsity translated into level-sets. This was apparent in our study that utilized symmetric matrices obtained from various application areas. We saw a strong correlation between parallel performance of our multithreaded triangular solver and the average number of rows per level in the level-set permuted matrices. The matrix obtained from the **bcsstk32** matrix resulted in only 15.1 rows per level. The solution of these system actually slowed down as threads were added. The lower triangular part of a tridiagonal matrix would be the worse case with a linear DAG and only 1 row per level. Clearly for these types of matrices, the level-set method is not scalable. Perhaps a feasible approach is to calculate this statistic in the DAG analysis phase to determine whether or not to use the level-set algorithm.

Acknowledgments

We thank Chris Baker, Cédric Chevalier, Karen Devine, Doug Doerfler, Paul Lin, and Kevin Pedretti for their input and many helpful discussions. This work was funded as part of the Extreme-scale Algorithms and Software Institute (EASI) by the Department of Energy, Office of Science. Sandia is a multiprogram laboratory operated by Sandia Corporation, a Lockheed Martin Company, for the United States Department of Energy's National Nuclear Security Administration, under contract DE-AC-94AL85000.

References

1. Lin, P., Shadid, J., Sala, M., Tuminaro, R., Hennigan, G., Hoekstra, R.: Performance of a parallel algebraic multilevel preconditioner for stabilized finite element semiconductor device modeling. Journal of Computational Physics 228(17), 6250–6267 (2009)
2. Hennigan, G., Hoekstra, R., Castro, J., Fixel, D., Shadid, J.: Simulation of neutron radiation damage in silicon semiconductor devices. Technical Report SAND2007-7157, Sandia National Laboratories (2007)
3. Lin, P.T., Shadid, J.N.: Performance of an MPI-only semiconductor device simulator on a quad socket/quad core InfiniBand platform. Technical Report SAND2009-0179, Sandia National Laboratories (2009)
4. Li, X.S., Shao, M., Yamazaki, I., Ng, E.G.: Factorization-based sparse solvers and preconditioners. Journal of Physics: Conference Series 180(1), 012015 (2009)
5. Saltz, J.H.: Aggregation methods for solving sparse triangular systems on multiprocessors. SIAM Journal on Scientific and Statistical Computing 11(1), 123–144 (1990)
6. Rothberg, E., Gupta, A.: Parallel iccg on a hierarchical memory multiprocessor – addressing the triangular solve bottleneck. Parallel Computing 18(7), 719–741 (1992)
7. Mayer, J.: Parallel algorithms for solving linear systems with sparse triangular matrices. Computing 86(4), 291–312 (2009)
8. Davis, T.A.: The University of Florida Sparse Matrix Collection (1994), Matrices found at http://www.cise.ufl.edu/research/sparse/matrices/

Performance and Numerical Accuracy Evaluation of Heterogeneous Multicore Systems for Krylov Orthogonal Basis Computation

Jérôme Dubois[1,2], Christophe Calvin[1], and Serge Petiton[2]

[1] Commissariat l'Energie Atomique,
CEA-Saclay/DEN/DANS/DM2S/SERMA/LLPR
F-91191 Gif-sur-Yvette Cedex, France
Tel.: +33 169 081 563; Fax: +33 169 084 572
[2] Université de Lille 1, Laboratoire d'Informatique Fondamentale de Lille
F-59650 Villeneuve d'Ascq Cedex, France
Tel.: +33 328 767 339; Fax: +33 328 778 537

Abstract. We study the numerical behavior of heterogeneous systems such as CPU with GPU or IBM Cell processors for some orthogonalization processes. We focus on the influence of the different floating arithmetic handling of these accelerators with Gram-Schmidt orthogonalization using single and double precision. We observe for dense matrices a loss of at worst 1 digit for CUDA-enabled GPUs as well as a speed-up of 20x, and 2 digits for the Cell processor for a 7x speed-up. For sparse matrices, the result between CPU and GPU is very close and the speed-up is 10x. We conclude that the Cell processor is a good accelerator for double precision because of its full IEEE compliance, and not sufficient for single precision applications. The GPU speed-up is better than Cell and the decent IEEE support delivers results close to the CPU ones for both precisions.

Keywords: parallel and distributed computing, numerical algorithms for CS&E, performance analysis.

1 Introduction

In the scientific computing domain, many subproblems require the computation of an orthogonal basis. The purpose of this technique is to compute an orthogonal basis spanning some linear subspace. The orthogonality of this basis is critical for problems such as solving systems with the GMRES method[16], or computing eigenvalues with the QR method or the Arnoldi process[4]. Depending on the orthogonalization process, the quality of the basis may be impacted due to numerical rounding error. The behavior of this error is predictable among modern mainstream processors essentially because of the IEEE floating arithmetic norm[13] and it is generally taken for granted that mainstream processors are fully IEEE compliant. Consequently the error for elementary floating calculations should be the same for any fully IEEE compliant processors : 10^{-8} in Single Precision (or SP) and 10^{-16} for Double Precision (or DP).

J.M.L.M. Palma et al. (Eds.): VECPAR 2010, LNCS 6449, pp. 45–57, 2011.
© Springer-Verlag Berlin Heidelberg 2011

Emerging computing architectures do not always fully respect this IEEE norm. It is the case of first and second[1] generation NVidia CUDA-enabled GPUs and the STI Cell processor. The upcoming Fermi/GT4xx architecture from NVidia will fully support the recent IEEE 754-2008, respecting a more recent standard than modern processors. In this aspect, the study described in this paper still applies for this upcoming hardware. The two accelerated architectures offer very high achievable computing power compared to classical multicore CPUs, for less Watts / GFLOPs. The peak computing power of the Cell is 200 GFLOPs in single precision(SP) and 100 GFLOPs in double precision(DP). For the best 2009 scientific CUDA-enabled GPU, it is 933 GFLOPs for SP and almost 100 GFLOPs for DP. Also, the memory bandwidth of GPUs varies from dozens of GBytes/s to almost 200 GBytes/s. It is more than any mainstream CPU-based system can achieve.

With the purpose of orthogonalizing faster using GPUs or Cell processors, we want in this paper to focus on the influence of the non fully IEEE-compliance of these architectures compared to a fully IEEE compliant CPU in SP and DP. We first test GPU and Cell with dense matrices by using some well known generated matrices from MatrixMarket, and apply the orthogonalization process using manufacturer BLAS [1] routines. We then analyze the results in terms of performance and accuracy and test the sparse case for the GPU.

This paper will be organized as follows. In section 1 we will explain the different orthogonalization algorithms that will be used. In section 2 we will describe the orthogonalization process. Section 3 will focus on the NVidia CUDA-enabled GPU and the IBM Cell processor to explain the hardware and IEEE differences with CPUs. Section 4 will present the implementation of the different orthogonalization processes. Finally we will discuss in section 5 the results for dense and sparse matrices in terms of quality of the orthogonal basis and performances.

2   Orthogonalization Process

Several variants of the orthogonalization process exist [11] : Householder reflections, Givens rotations and the Gram-Schmidt process are the most common ones. The Gram-Schmidt process is declined in different versions : classical(CGS), modified(MGS) and classical with reorthogonalization(CGSr) for instance. These algorithms are described in table 1. The classical version is the most simple and easily parallelizable as seen in [17] : it is faster, as it can take advantage of BLAS[2] 2 operations instead of BLAS 1 in the modified algorithm. But the main drawback of CGS is the possibly high loss of orthogonality within the computed basis due to round-off error [9]. The MGS algorithm tries to correct this, and in fact it manages to provide a more accurate version of the Gram-Schmidt process. Mathematically both are the same, but the MGS provides less parallelism. The CGSr process provides an even more accurate version,

---

[1] First generation is hardware 1.0 and 1.1 for G80/90 GPUs, and second generation is Hardware 1.2 and 1.3 for GT200 GPUs.

[2] Basic Linear Algebra Subroutines.

Table 1. Classical Gram-Schmidt(CGS), Modified G-S(MGS), CGS with reorthogonalization(CGSr)

CGS	MGS	CGSr
1. for i = 1 : m	1. for i = 1 : m	1. for i = 1 : m
2. $v_{i+1} = Av_i$	2. $v_{i+1} = Av_i$	2. $v_{i+1} = Av_i$
3. $H_{1:i,i} = (v_{i+1}, v_{1:i})$	3. for j = 1 : i	3. $H_{1:i,i} = (v_{i+1}, v_{1:i})$
4. $v_{i+1} = v_{i+1} - H_{1:i,i}.v_{1:i}$	4. $H_{j,i} = (v_{i+1}, v_j)$	4. $v_{i+1} = v_{i+1} - H_{1:i,i}.v_{1:i}$
5. $H_{i,i+1} = |v_{i+1}|_2$	5. $v_{i+1} = v_{i+1} - H_{j,i}.v_j$	5. $C_{1:i,i} = (v_{i+1}, v_{1:i})$
6. $v_{i+1} = \frac{v_{i+1}}{H_{i,i+1}}$	6. end	6. $v_{i+1} = v_{i+1} - C_{1:i,i}.v_{1:i}$
7. end	7. $H_{i,i+1} = |v_{i+1}|_2$	7. $H_{1:i,i} += C_{1:i,i}$
	8. $v_{i+1} = \frac{v_{i+1}}{H_{i,i+1}}$	8. $H_{i,i+1} = |v_{i+1}|_2$
	9. end	9. $v_{i+1} = \frac{v_{i+1}}{H_{i,i+1}}$
		10. end

by reorthogonalizing each computed vector. In practice it is not necessary to reorthogonalize at each iteration, and so the cost of this algorithm using selective reorthogonalization is close to CGS while being more accurate.

Accuracy of the orthogonal basis is a crucial matter when we apply the orthogonalization process for the Arnoldi Iteration. The orthogonality has an impact on the computed eigenvalues [4], as the constructed orthogonal basis is involved in the projection of the Ritz vectors to obtain the eigenvectors.

3 Accelerators Programming

In this section we will expose the architecture as well as the programming paradigm of the GPUs and the Cell processor.

3.1 Nvidia CUDA-Enabled GPUs

For several years, GPU processing power has kept steadily increasing, outperforming the peak computing power of the best multicore processors [15]. Because of this, researchers started to study the use of GPUs for scientific computations, and one of the most successful attempt was Brook [5], a scientific language to exploit GPUs.

Hardware and language. In 2006 Nvidia, a major mainstream graphics card manufacturer, released a first version of CUDA[3], its programming language for Nvidia G80 and above series. CUDA aims at providing a comprehensive paradigm able to exploit the massive parallelism of a GPU. This language can not be used with other manufacturer's graphics cards. The memory is faster than classical central memory and the access pattern must be regular to achieve great performance just like for a vector machine.

---

[3] Compute Unified Device Architecture.

IEEE compliance. The hardware IEEE floating point norm is handled slightly differently than on fully IEEE architectures. Here are some examples of these differences for SP only, taken from the programming manual [15]:

- Addition and multiplication are often combined into a single multiply-add instruction (FMAD), which truncates the intermediate result of the multiplication;
- Division is implemented via the reciprocal in a non-standard-compliant way;
- Square root is implemented via the reciprocal square root in a non- standard-compliant way;
- For addition and multiplication, only round-to-nearest-even and round-towards-zero are supported via static rounding modes; directed rounding towards +/- infinity is not supported;

There are more details about IEEE support of Nvidia GPUs in the programming manual [15].

3.2   STI Cell Processor

The Cell processor is the product of the joint between Sony, Toshiba and IBM. The objective of the Cell is to provide an efficient novel multicore design for multimedia applications. As for video cards, its original field is far from the scientific computing field. Despite this, it has been improved to fully handle IEEE double precision arithmetic. The fastest supercomputer in the top500 from June 2009 [14] is the Roadrunner from IBM, delivering more than one PetaFLOPs.

Hardware and language. The Cell processor embeds 9 cores : one classical PowerPC core, called PPE[4], and 8 computing cores called SPEs[5]. As a result, the basic paradigm to program the Cell is as follows : one initializes the application using the PPE, and then spawns threads on the SPEs to do or help computations.

IEEE compliance. The PPE is fully IEEE 754 compliant as it is a PowerPC processor. In single precision, SPEs only implement round-towards-zero rounding mode as opposed to the standard round-to-even mode. It can impact the calculations as seen in [8] where 2 to 3 bits of accuracy are lost. The data format follows the IEEE standard 754 definition, but single precision results are not fully compliant with this standard (different overflow and underflow behavior, support only for truncation rounding mode, different denormal results) [2]. The programmer should be aware that, in some cases, the computation results will not be identical to IEEE Standard 754 ones. For double precision, the Cell processor is fully IEEE-compliant.

4   Optimizations

In this section, we will see the implementation and optimizations applied to the orthogonalization processes for each hardware : reference CPU, CUDA-enabled GPUs and the STI Cell processor.

---

[4] PowerPC Processing Element.
[5] Synergistic Processing Element.

4.1   BLAS Operations

We saw in table 1 the different versions of the Gram-Schmidt process. For each of them, the same basic BLAS operations apply.

- Matrix vector multiplication : gemv
- Vector dot product : dot
- Scaled vector addition : axpy
- Scaling of a vector : scal
- *optionally* : vector norm 2 computation, nrm2. It is the square root of the dot product of the vector.

If we take CGS as an example, then we have :

1. for i $= 1 : m$
2. $v_{i+1} = Av_i \leftarrow$ one gemv operation
3. $H_{1:i,i} = (v_{i+1}, v_{1:i}) \leftarrow$ several dot products
4. $v_{i+1} = v_{i+1} - H_{1:i,i}.v_{1:i} \leftarrow$ several axpy operations
5. $H_{i,i+1} = |v_{i+1}|_2 \leftarrow$ one norm2 computation
6. $v_{i+1} = \frac{v_{i+1}}{H_{i,i+1}} \leftarrow$ one scaling of a vector
7. end.

Same basic kernels are being used for the other versions of the GS orthogonalization. Consequently, we may use optimized BLAS operations to take advantage of the different hardware in an optimized and portable manner.

4.2   CPU

The implementation of the dense orthogonalization process on classical CPU follows the standard algorithm described in table 1 and uses ATLAS[6] subroutines [18]. ATLAS is a superset of BLAS library, that adapts some parameters to the architecture by probing its caches and performances according to the number of cores used, at library compilation time. In our case all the operations are BLAS 1 or 2. ATLAS implementations of the BLAS 1 and 2 operations use only one thread, and so does our application.

Concerning the sparse case, we follow the standard CSR matrix vector product, mixed with ATLAS BLAS 1 routines to comply with the described GS algorithms.

4.3   GPU

Implementation on the GPU for the dense case is also close to the original algorithms and uses mainly CUBLAS [6], except that we have to handle the memory operations between host and device memory : allocation and transfers.

For sparse matrices, we use the optimized sparse matrix vector multiply from NVidia, which can handle different matrix format : COO[7], CSR[8], ELL[9], DIA[10],

---

[6] Automatically Tuned Linear Algebra Subroutines.
[7] COOrdinate.
[8] Compressed Sparse Row.
[9] ELLPACK.
[10] Diagonal.

and hybrid (DIA+ELL). See [3] for more details about the implementation and the hybrid format. Provided this optimized Matrix-Vector product, we use CUBLAS routines to exploit the GPU power in a portable way, following the steps described in table 1.

4.4   Cell Broadband Engine

As for the GPU hardware, the global algorithm was respected, but an accelerated implementation on the Cell processor implied more programming effort. After data initialization on the PPE, the computing part is distributed among the SPEs equally. There is a BLAS interface used locally on each SPE. This way we have an optimized BLAS operation using all the SPE power. The matrix-vector product uses a rowwise computing pattern, using the EIB to stream data between the processing elements. The locally computed part of the vector is used for the vector operations, an data exchanges may be necessary between SPEs.

5   Experimentation

This section will present the results in terms of precision and performance, each time for both the dense and sparse matrices. The hardware characteristics are described in table 2.

Table 2. Hardware used for experimentation. Frequency is in GHertz and memory in GBytes. The bandwidth is the peak memory bandwidth and actual is the benchmarked bandwidth. Both are expressed in GBytes/s. The peak power is expressed in GFLOPs.

Hardware	Frequency	cores	memory	bandwidth / actual	Peak Power (DP/SP)
Xeon e5440	2.83	4	4	5.33 / 5.0	45.6 / 91.2
Tesla c1060	1.296	320	4	100 / 75	75 / 933
Cell QS22	3.2	8	16	25.6 / 25.6	100 / 200

5.1   Hardware Precision

Experimentation methodology. The precision of the GS process for one sub-space size is calculated using the largest absolute dot product between vectors of the orthogonal basis:

$$\text{Orthogonal error}: \max |< v_i, v_j >|, \ i \neq j \text{ and } v_i, v_j \in V \text{ basis.}$$

Precisely, this would be the worst orthogonality between two vectors of the basis. Also, the accumulator of the dot products uses the same precision as the rest of the calculations : if the precision is SP, then the accumulator is SP too. Same applies for DP. This way we do not add another parameter to the calculations.

Dense matrices. We tested CGS and CGSr orthogonalization with dense matrices. CGS is known to be very sensitive to machine precision [12], which usually

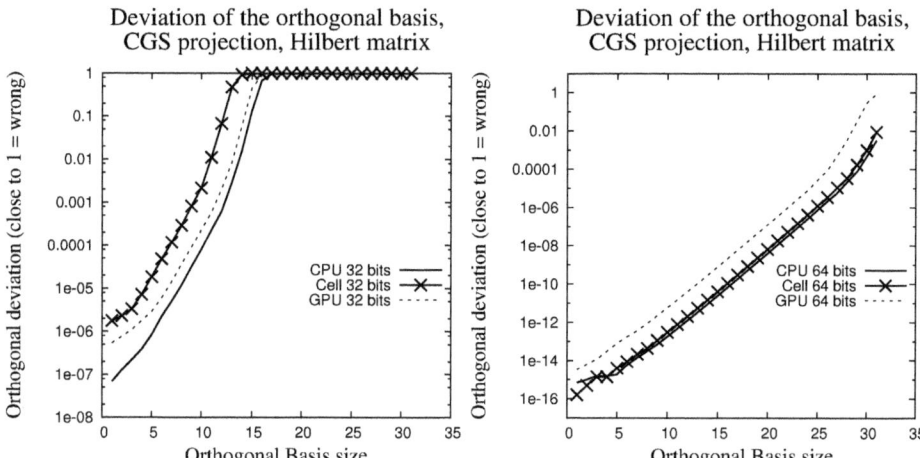

Fig. 1. Precision of the CGS orthogonalization of a 10240 square Hilbert matrix, using different hybrid architectures. The left figure shows the single precisiob computations, and the right one double precision. The X axis represents the subspace size starting from one, andthe Y axis the accuracy of the basis, e.g. the worst orthogonality b"tween two vectors of V.

is IEEE-754 compliant. This standard does not define how the machine has to implement floating point operations, but mainly the global behavior and characteristics [10]. On figure 1, we clearly see the difference between 32 bits and 64 bits precision, and the use of an accelerator or not.

Some statistical indicators were extracted from figures 1 and 2, to try to characterize each device in table 3. As we can see in this table, the Cell is the least accurate device in single precision, and the GPU is less accurate than the CPU. The reason behind is that every computation for the Cell, including division, was done on the non fully IEEE-compliant SPE. Furthermore the Cell is the least IEEE compliant device in SP.

When switching to double precision, then each accelerator tends to be very close to the reference CPU in terms of precision. But the interesting thing with the Cell processor is its full double precision IEEE compliance : the result differs from the fully double precision IEEE compliant CPU. It is due to the multi SIMD core nature of the Cell, which will handle computations in a different order than in the CPU case, here used as a single core processor. Some tests conducted on basic dot products parallelized on the Cell provided different results wether we use each core of the Cell as a scalar processor or a SIMD unit. The difference was within the order of the machine precision and it could explain the different Gram-Schmidt orthogonalization behavior.

Sparse matrices. We chose two test matrices from the eigenvalues field : Lin and Andrews matrices from the sparse matrices collection of the University of Florida[7]. We experimented the sparse CGSr with the GPU against the CPU.

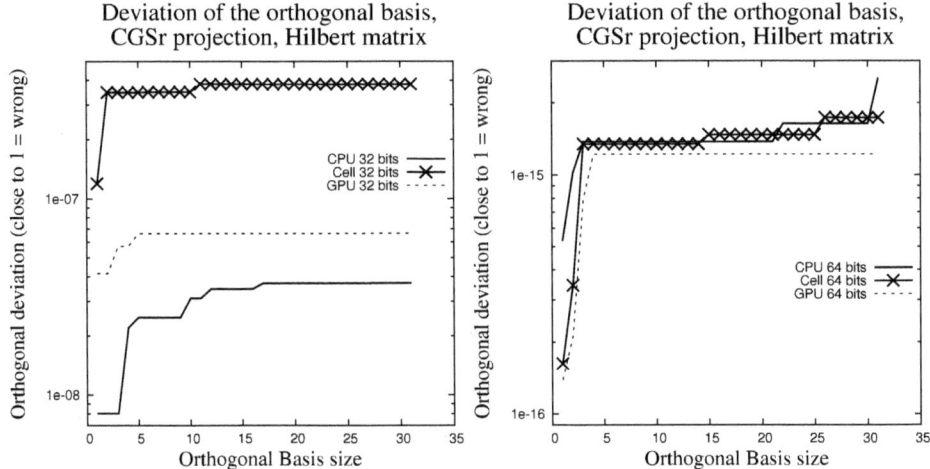

Fig. 2. Precision of the dense CGS process with reorthogonalization on a 10240 square Hilbert matrix, using different hybrid architectures. The left figure shows single precision accuracy results and the right one double precision accuracy results. X axis bears the subspace size, and the Y axis the accuracy of the orthogonal basis, e.g. the worst orthogonality between two vectors of V.

Table 3. Precision factor achieved for the dense CGS and CGSr orthogonalization. Factor = $\text{Accelerator}_{precision}$ / $\text{CPU}_{precision}$.

Hardware	Precision	CGS		CGSr	
		Max. Err	Median Err.	Max. Err	Median Err.
Cell	32 bits	170x	22.74x	43x	11x
	64 bits	3x	1.64x	1.07x	0.98x
GPU	32 bits	8x	2.98x	7.1x	1.91x
	64 bits	30x	31.66x	0.9x	0.89x

We did not test the Cell because the achievable speed-up would have implied a great programming effort. The trends in figure 3 and 4 are different than the dense case. By looking at the matrix H, which is built using dot products of V, we see very close results for the first columns. Then some noise appears, which gives 31x higher elements than the CPU-computed result. Because the coefficients of the H matrix are used throughout the orthogonalization process, there is a quick propagation of this noise through each vector operation.

Also, and due to the datasets, the resulting vectors tend to contain numbers relatively close to the machine precision. The GPU truncates these, which explains the apparent better accuracy of the GPU when we seek the "worst" dot product of the V vectors : many zeros imply a lower dot product, and so a lower error.

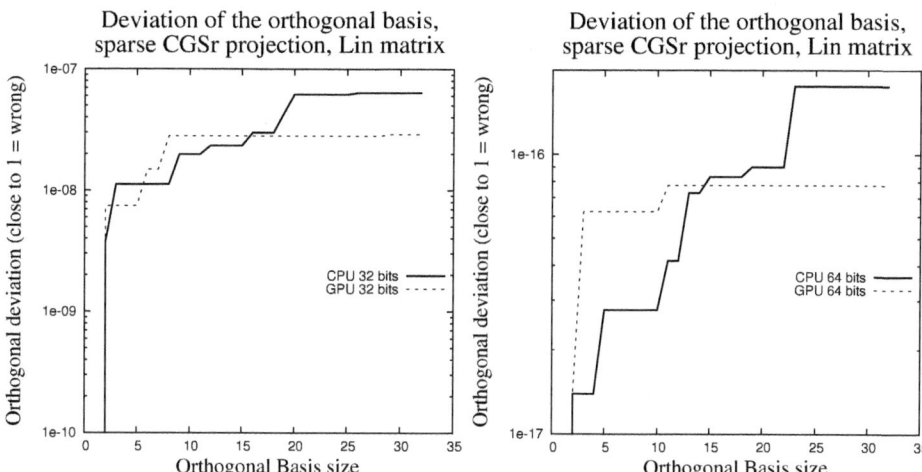

Fig. 3. Precision of the sparse CGS process with reorthognalization on sparse Lin matrix, using different hybrid architectures. Top figure shows single precision results and bottom figure double precision. This matrix is square with a square size of 256k. The total number of non zeros is 1.8m numbers and the filling ratio is 0.0027%. The subspace size starts from 1 and increases along the X axis, and the y axis shows the error of the orthogonal basis, e.g. the worst orthogonality between two vectors of V.

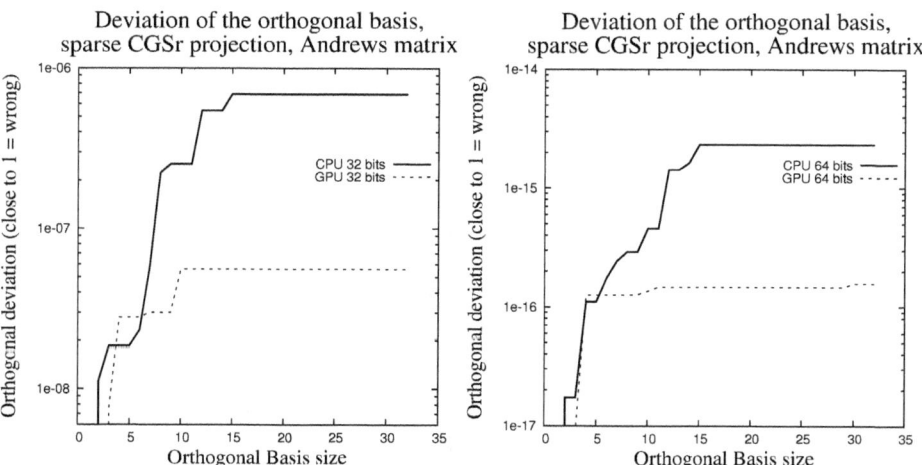

Fig. 4. Precision of the sparse CGSr process on sparse Andrews matrix, using different hybrid architectures. Top figure represents the single precision accuracy results and the lower one double precision. This matrix is square with 60k elements per dimension, and has 760k non zeros and the filling ratio is 0.021%. X axis represents the subspace size and y axis the precision of the orthogonal basis, e.g. the worst orthogonality between two vectors of V.

5.2   Performance Achieved

We present in this section the performance achieved for mainstream x86 multi-core, Nvidia GPUs and the Cell processor with dense matrices, as well as performance achieved for sparse matrices with x86 multicore and GPU hardware.

Expected Results. The dominant computing operation is the memory-bandwidth bound matrix vector product, which for a matrix of nz elements does $2nz$ floating operations implying a 2 factor between computation and memory used. Say we have Z GTransfer/s, where the size of one transfer is a double precision element. Then we may do 2Z computations. So the GFLOPs for each architecture will be 2 times the GT/s. From the performance and memory speed shown in table 2 we can deduce the maximum performance expected for these accelerators, summarized in table 4.

Table 4. Maximum peak performance for the Gram-Schmidt process in GFlops

Precision	CPU	GPU	Cell
32 bits	2.667	37	12.8
64 bits	1.333	18	6.4

Dense matrices. In practice, the observed results differ from the theoretical ones, due to data transfers which cannot utilize the whole available bandwidth, be it for the CPU or the accelerators. The fastest implementation is the GPU's one, with a speed-up of 15x in SP and 21x in DP compared to our CPU. The second one is the Cell with a speed-up of 6x compared to the CPU in SP and DP. Table 5 shows the performance results in more details. The efficiency of our solution for the GPU is of 55% in SP and 82% in DP. Concerning the Cell, it is 70% and 66%. So, even with the use of high level BLAS routines, we were able to achieve a decent portion of the accelerators' available bandwidth.

Sparse matrices. For the sparse case, we focus on the CUDA-enabled Tesla GPU, and compare it to the CPU. Here, the performance highly depends on the pattern of the matrices : Lin and Andrews in our case. Table 5 shows the actual performance of the sparse CGSr. As we can see, the GPU is the fastest, with a varying factor of 6x to 7x times faster. By using the DIA format, then the performance increases with the GPU, speeding-up by a factor of 9x to 10x. The results with the sparse Andrews matrix are similar, except that the performance is lower for GPU hardware 2.5 GFLOPs in DP and 3 GFlops in SP for the CSR format. Using the Hybrid format from Nvidia, the GPU performs at 3.17 GFLOPs in DP and 4.23 GFLOPs in SP for a speed-up of 5.3x in DP and 3.25x in SP compared to the CPU.

Table 5. Execution time and performance for the CGSr process for a dense Hilbert matrix of size 10240x10240 and a sparse Lin matrix of size 256kx256k and 1.8m nnz, with a subspace size of 32. Time is expressed in seconds and performance in GFLOPs.

	Dense			Sparse		
	CPU	GPU	Cell	CPU	GPU(CSR)	GPU(Best)
Time(sec)				Time(sec)		
32 bits	5.01	0.33	0.75	1.02	0.156	0.118(DIA)
64 bits	10.12	0.46	1.6	1.8	0.242	0.188(DIA)
Performance(GFLOPs)				Performance(GFLOPs)		
32 bits	1.36	20.5	9.02	1.195	7.82	10.32(DIA)
64 bits	0.69	14.7	4.22	0.65	5.05	6.50(DIA)

6 Synthesis

We presented implementations of the dense and sparse Classical Gram-Schmidt process with and without reorthogonalization using high level manufacturer BLAS routines for the STI Cell processor and the CUDA-enabled GPUs.

Performance. These implementations outperformed their CPU counterpart with a factor of 20x for the GPU and 7x for the Cell processor with dense matrices, and 10x for sparse matrices using the GPU.

Floating point arithmetic. The Cell is less IEEE compliant than the GPU in SP and fully IEEE compliant in DP, providing the potential same accuracy as a classical processor. In DP, the GPU is close to be fully IEEE compliant, with differences only in exceptions handling and static rounding modes. Consequently, here is the expected ranking of these architectures in terms of precision:

- $SP_{accuracy}$: $Cell_{acc.} \leq GPU_{acc.} \leq CPU_{acc.}$
- $DP_{accuracy}$: $GPU_{acc.} \leq Cell_{acc.} \leq CPU_{acc.}$

Precision achieved. Concerning the precision with dense matrices, using the Cell implies an actual precision of 10^{-6}. For the GPU, the precision is around 10^{-7}, close to the CPU. In DP, accelerators' results were very close to the CPU ones. We saw varying results in DP due to a different parallel order of execution of the Cell and GPU. For the precision with sparse matrices, the results are less clear, as the GPU seems to give a better orthogonal basis than the CPU, even if it is due to cancellation of some elements of the V basis. Still, it is reliable as the results are close to the reference CPU.

7 Conclusion

The Cell is not IEEE-compliant enough for sensitive SP computation and may only be used mixed with or in full IEEE DP mode if precision is a concern. It provides a good speed-up but at a great programming effort. The GPU has a very good potential because it tends to become more and more IEEE-compliant and it remains backward compatible. Thus, the upcoming Fermi NVidia scientific GPU will fully support the IEEE-norm for both SP and DP, implying a better precision than previous GPUs. It will also integrate L1 and L2 caches improving performances for irregular data patterns like sparse matrices. But because of the parallel nature of these architectures, one has to take into account the parallel execution of the operations, which may have an impact on the precision as with the Cell and the Xeon or maybe the 2 caches of the Fermi. Furthermore, we saw it is possible to accelerate a memory bandwidth bound algorithm such as the Gram-Schmidt process with CUDA-GPUs. Finally, if high-level libraries such as CUBLAS are used then the code may be tested with the next-generation GPUs. These accelerators or new emerging ones may be the key to reach the post-petascale era, certainly mixing precision on levels we know now -32 and 64 bits- but also extended precision : 128 bits. It is thus possible to experiment the same benchmarks to measure the promising improvement in accuracy and performance of the Fermi GPUs as well as new accelerators.

Acknowledgements

We thank the CINES[11] research center from Montpellier for providing an access to the IBM QS21 Bladeserver powered by Cell processors as well as an access to Nvidia Tesla servers, and we also thank the IBM research center of Montpelier, which gave us access to a QS22 Double Precision enhanced Cell.

References

1. An updated set of basic linear algebra subprograms (blas). ACM Trans. Math. Softw. 28(2), 135–151 (2002)
2. Arevalo, A., Matinata, R.M., (Raj)Pandian, M., Peri, E., Ruby, K., Thomas, F., Almond, C.: Architecture overview and its impact on programming. In: Programming the Cell Broadband Engine Architecture: Examples and Best Practices, ch. 4.61. IBM (2008)
3. Bell, N., Garland, M.: Implementing sparse matrix-vector multiplication on throughput-oriented processors. In: SC 2009: Proceedings of the 2009 ACM/IEEE Conference on Supercomputing. ACM, New York (2009)
4. Braconnier, T., Langlois, P., Rioual, J.C.: The influence of orthogonality on the arnoldi method. Linear Algebra and its Applications 309(1-3), 307–323 (2000)
5. Buck, I., Foley, T., Horn, D., Sugerman, J., Fatahalian, K., Houston, M., Hanrahan, P.: Brook for gpus: stream computing on graphics hardware. ACM Trans. Graph. 23(3), 777–786 (2004)
6. NVidia Corporation. Nvidia: Cublas library. Technical report. Whitepaper. Part of CUDA Toolkit
7. Duff, I.S., Grimes, R.G., Lewis, J.G.: Sparse matrix test problems. ACM Trans. Math. Softw. 15(1), 1–14 (1989)

---

[11] Centre Informatique National de l'Enseignement Supérieur.

8. Frigo, M., Johnson, S.G.: Fftw on the cell processor, `http://www.fftw.org/cell/`
9. Giraud, L., Langou, J., Rozložník, M., van den Eshof, J.: Rounding error analysis of the classical Gram-Schmidt orthogonalization process. Numerische Mathematik 101(1), 87–100 (2005)
10. Goldberg, D.: What every computer scientist should know about floating-point arithmetic. ACM Computing Surveys (1991)
11. Golub, G.H., Van Loan, C.F.: Matrix Computations (Johns Hopkins Studies in Mathematical Sciences). The Johns Hopkins University Press, Baltimore (1996)
12. Hernandez, V., Roman, J.E., Tomas, A.: Parallel arnoldi eigensolvers with enhanced scalability via global communications rearrangement. Parallel Comput. 33(7-8), 521–540 (2007)
13. IEEE: IEEE standard for binary floating-point arithmetic. ACM SIGPLAN Notices 22(2), 9–25 (1985)
14. Meuer, H., Strohmaier, E., Dongarra, J., Simon, H.: Architecture share over time, `http://www.top500.org/overtime/list/32/archtype`
15. NVIDIA. NVIDIA CUDA Programming Guide 2.0 (2008)
16. Rozlozník, M., Strakos, Z., Tuma, M.: On the role of orthogonality in the gmres method. In: Král, J., Bartosek, M., Jeffery, K. (eds.) SOFSEM 1996. LNCS, vol. 1175, pp. 409–416. Springer, Heidelberg (1996)
17. Takuya, Y., Daisuke, T., Taisuke, B., Mitsuhisa, S.: Parallel implementation of classical gram-schmidt orthogonalization using matrix multiplication. IPSJ SIG Technical Reports (63(HPC-106)), 31–36 (2006)
18. Clint Whaley, R., Petitet, A., Dongarra, J.J.: Automated empirical optimizations of software and the atlas project. Parallel Computing 27, 2001 (2001)

An Error Correction Solver for Linear Systems: Evaluation of Mixed Precision Implementations

Hartwig Anzt, Vincent Heuveline, and Björn Rocker

Karlsruhe Institute of Technology (KIT)
Institute for Applied and Numerical Mathematics 4
Fritz-Erler-Str. 23
76133 Karlsruhe, Germany
hartwig.anzt@kit.edu, vincent.heuveline@kit.edu, bjoern.rocker@kit.edu

Abstract. This paper proposes an error correction method for solving linear systems of equations and the evaluation of an implementation using mixed precision techniques.

While different technologies are available, graphic processing units (GPUs) have been established as particularly powerful coprocessors in recent years. For this reason, our error correction approach is focused on a CUDA implementation executing the error correction solver on the GPU.

Benchmarks are performed both for artificially created matrices with preset characteristics as well as matrices obtained from finite element discretizations of fluid flow problems.

Keywords: Mixed Precision, Iterative Refinement Method, Computational Fluid Dynamics (CFD), Large Sparse Linear Systems, Hardware-aware Computing, GPGPU.

1 Introduction

The development of modern technology is characterized by simulations, that often are no longer performed through physical experiments, but through mathematical modeling and numerical simulation. In many cases, for example in computational fluid dynamics (CFD), massive computation power is needed, in order to handle large systems of linear equations.

Often iterative solvers are chosen for the solving process, since they can exploit the sparse structure of the affiliated matrix to compute an approximation of a certain accuracy usually faster than a direct solver.

The computational complexity of this problem depends on the characteristics of the linear system, the properties of the used linear solver and the floating point format. The floating point format determines not only the execution time when performing computations, but also the occurring rounding errors. A more complex floating point format usually leads to higher accuracy and higher computational effort.

Today, most hardware architectures are configured for the IEEE 754 [1] standard containing single precision and double precision as the main floating point formats. As their names indicate, the double precision format has twice the size

J.M.L.M. Palma et al. (Eds.): VECPAR 2010, LNCS 6449, pp. 58–70, 2011.

of the single precision format, leading to a factor of two in computational cost while offering a higher precision.

In many cases single precision floating point accuracy is not sufficient for scientific computation. The question arises, whether the whole algorithm has to be performed in the double precision format, or whether one can gain speed by computing parts of it in single precision and other parts in double precision, and still obtain double precision accuracy for the final result. One approach is to modify the algorithm of an error correction method such that the inner error correction solver uses a lower format than the working precision in the outer loop. As the final accuracy only depends on the stopping criterion of the refinement solver, the solution approximation is not affected. Still, it can be expected that the mixed precision approach performs faster than a plain solver in high precision, since the cheaper error correction solver in the low precision format may overcompensate the additional computations and typecasts.

2 Mixed Precision Error Correction Methods

2.1 Mathematical Background

Error correction methods have been known for more than 100 years, and have finally become of interest with the rise of computer systems in the middle of the last century. The core idea is to use the residual of a computed solution as the right-hand side to solve a correction equation.

The motivation for the error correction method can be obtained from Newton's method. Newton developed a method for finding successively better approximations to the zeros of a function $f(\cdot)$ by updating the solution approximation x_i through

$$x_{i+1} = x_i - (\nabla f(x_i))^{-1} f(x_i). \tag{1}$$

We now apply Newton's method (1) to the function $f(x) = b - Ax$ with $\nabla f(x) = -A$. By defining the residual $r_i := b - Ax_i$, we obtain

$$
\begin{aligned}
x_{i+1} &= x_i - (\nabla f(x_i))^{-1} f(x_i) \\
&= x_i + A^{-1}(b - Ax_i) \\
&= x_i + A^{-1} r_i.
\end{aligned}
$$

Denoting the solution update with $c_i := A^{-1} r_i$, we can design an algorithm.

1: initial guess as starting vector: x_0
2: compute initial residual: $r_0 = b - Ax_0$
3: **while** ($\| Ax_i - b \|_2 > \varepsilon \| r_0 \|$) **do**
4:     $r_i = b - Ax_i$
5:     solve: $Ac_i = r_i$
6:     update solution: $x_{i+1} = x_i + c_i$
7: **end while**

Algorithm 1. Error Correction Method

Here x_0 is an initial guess. In each iteration, the inner correction solver searches for a c_i, such that $Ac_i = r_i$ with r_i being the residual of the solution approximation x_i. Then, the approximation of the solution x_i is updated to $x_{i+1} = x_i + c_i$.

2.2 Mixed Precision Approach

The underlying idea of mixed precision error correction methods is to use different precision formats within the algorithm of the error correction method, updating the solution approximation in high precision, but computing the error correction term in lower precision. This approach was also suggested by [8],[12], [10], and [9].

Hence, one regards the inner correction solver as a black box, computing a solution update in lower precision. The term high precision refers to the precision format that is necessary to display the accuracy of the final solution and we can obtain the following algorithm where $.^{high}$ denotes the high precision value and $.^{low}$ denotes the value in low precision. The conversion between the formats will be left abstract throughout this paper. Because especially the conversion of the matrix A is expensive, it should be stored in both precision formats, high and low precision. This leads to the drawback of a higher memory need. In the case of using hybrid hardware, A should be stored in the local memory of the hardware devices in the respectively used format. E.g. using a hybrid CPU-GPU system, the matrix A^{high} should be stored in the CPU memory and A^{low} should be stored in the GPU memory.

Using the mixed precision approach to the error correction method, we have to be aware of the fact that the residual error bound of the error correction solver may not exceed the accuracy of the lower precision format. Furthermore, each error correction produced by the inner solver in lower precision cannot exceed the data range of the lower precision format. This means that the smallest possible error correction is the smallest number ϵ_{low}, that can be represented in the lower precision. Thus, we can not guarantee an accuracy of the final solution exceeding ϵ_{low} either. This can become a problem when working with very small numbers, because then the solution correction terms can not be denoted in low precision. However, in most cases, the problem can be avoided by converting the original values to a lower order of magnitude.

Using the displayed algorithm we obtain a mixed precision solver. If the final accuracy does not exceed the smallest number ϵ_{low} that can be represented in the lower precision, it gives exactly the same solution approximation as if the solver was performed in the high precision format. Theoretically, any precision can be chosen, but in most cases it is comfortable to use the IEEE 754 standard formats.

The computation of the correction loop $A^{low}c^{low} = r^{low}$ can be performed with a direct solver, or again with an iterative method. This implies that it is even possible to cascade a number of error correction solvers using decreasing precision.

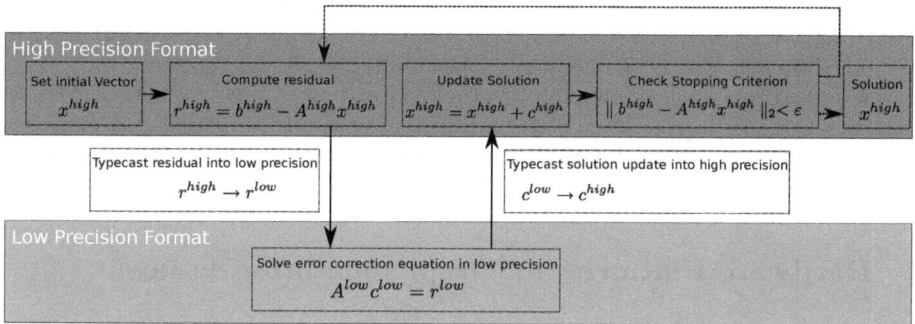

Fig. 1. Visualizing the mixed precision approach to an error correction solver

In the case of an iterative solver as error correction solver, especially the iterative approaches to the Krylov subspace methods are of interest, since these provide an approximation of the residual error iteratively in every computation loop. Hence, one is able to set a certain relative residual stopping criterion for the iterative error correction solver. Possible Krylov subspace solvers include the CG algorithm, GMRES, BiCGStab etc. (see e.g. [2], [3]). The mixed precision error correction method based on a certain error correction solver poses the same demands to the linear problem, as the within used Krylov subspace solver.

In the case of a direct error correction solver, the solution update usually has a quality depending on the condition number of the system and the lower precision format [4]. Hence, the solution improvement normally depends on the specific case, but is generally high. Despite the fact that direct methods are for many cases computationally expensive and have a high memory consumption, they are of interest as error correction solver, since some of them own pleasant characteristics:

Using for example the LU solver as error correction solver, the LU decomposition has to be computed only in the first error correction loop. In the following loops, the stored decomposition can be used to perform the forward and backward substitution. Since these substitutions imply only low computational effort, they can, depending on the hardware structure, even be performed in the high precision format. This leads to accuracy advantages and economizes the algorithm by omitting the computationally expensive typecasts of the residual and the solution update.

It should be mentioned, that the solution update of the error correction solver is usually not optimal for the outer system, since the discretization of the problem in the lower precision format contains rounding errors, and it therefore solves a perturbed problem. When comparing the algorithm of an error correction solver to a plain solver, it is obvious, that the error correction method has more computations to execute. Each outer loop consists of the computation of the residual error term, a typecast, a vector update, the scaling process, the inner solver for the correction term, the reconversion of the data and the solution update. The computation of the residual error itself consists of a matrix-vector multiplication, a vector addition and a scalar product. Using a hybrid architecture, the converted data additionally has to be transmitted between the devices.

The mixed precision refinement approach to a certain solver is superior to the plain solver in high precision, if the additional computations and typecasts are overcompensated by the cheaper inner correction solver using a lower precision format.

3   Hardware Platform and Implementation Issues

The utilized TESLA-System is equipped with one NVIDIA TESLA S1070[1]. Both host nodes, each connected via a PCIe 2.0 x16 to the S1070, are equipped with two Intel Xeon 5450 CPUs. The Intel MKL in version 10.1.1.019, the Intel compiler in version 11.0.074 and the CUDA in version 2.0 are used.

The InstitutsCluster[2] (IC1) is located at the Karlsruhe Institute of Technology (KIT) and consists of 200 computing nodes each equipped with two Intel quad-core EM64T Xeon 5355 processors, owning 16 GB of main memory. Peak performance of one node is about 85,3 GFlops. For a detailed performance evaluation see [17]. On the software side, the Intel CMKL in version 10.1.2.024 and the Intel compiler in version 10.1.022 are used.

The implementation of the GMRES algorithm (taken from [2]), and the surrounding error correction method is based on the elementary kernels of the Intel CMKL and the CUBLAS library.

4   Numerical Experiments

4.1   Test Configurations

To be able to compare the performance of different implementation of the GMRES-(10) solver, we perform tests with different linear systems. In this work 10 denotes the restart parameter for the GMRES.

All solvers use the relative residual stopping criterion $\varepsilon = 10^{-10} \parallel r_0 \parallel_2$. Due to the iterative residual computation in the case of the plain GMRES-(10) solvers, the mixed GMRES-(10) solvers based on the mixed precision error correction method usually iterate to a better approximation since they compute the residual error explicitly, but as the difference is generally small, the solvers are comparable. In case of the mixed precision GMRES-(10) on the TESLA-System, the error correction solver is performed on one of the four available GPUs, while the solution update is led to the CPU of the same system. This is done to be able to handle larger problems since the amount of memory on the GPU is limited to 4 GB. Our hardware platform is therefore similar to a system equipped with one TESLA C1060, but in the following we denote the results with S1070.

[1] http://www.nvidia.com
[2] http://www.scc.kit.edu

On the one hand, we use matrices with a preset condition number, preset sparsities, and increase the dimension. Depending on the sparsity, the matrices are stored in the matrix array storage format (MAS) or the compressed row storage format (CRS).

M1 The first test matrix, is a dense matrix that is generated with the DLATMR-routine out of the lapack-library [18]. As parameter set, we choose all entries smaller 1, and a condition number of 3. One drawback is that we cannot set positive definiteness in the routine itself. To ensure this property, we set the diagonal entries to be $10 \cdot n$, where n is the dimension of the matrix. By doing so, we lose the control of the condition number, but it is bounded by the former choice.

M2 We also choose the second artificial test matrix to be a dense matrix, but this time with a higher condition number. As it is not easy to control the condition number of a dense matrix, we choose the term $W = 2 \cdot 10^3 \cdot n + n$ on the diagonal. The term on the upper and lower second diagonal is $V = 10^3 \cdot n$, and the rest of the matrix is filled with random double precision numbers between 0 and 1. These entries are the only entries we cannot control, and in one row, they can at most sum up to $(n-3) \cdot (1-\varepsilon)$, but they can also sum up to $(n-3) \cdot \varepsilon$, with $\varepsilon > 0$. Since the random numbers are for large dimension evenly distributed, we assume that they sum up to $0.5 \cdot (n-3)$.

M3 The third test case is a sparse matrix similar to a 5-point stencil. The difference is the the term $H = 4 + 10^{-3}$ instead of $H = 4$ on the diagonal. Furthermore, the second and fifth upper and lower diagonal is filled with -1. The term 10^{-3} on the diagonal is used to control the condition number.

Table 1. Structure plots and properties of the artificial test-matrices

M1	M2	M3
$\begin{pmatrix} 10\cdot n & * & \cdots & \cdots & * \\ * & 10\cdot n & & & \vdots \\ \vdots & & 10\cdot n & & \vdots \\ \vdots & & & & * \\ * & \cdots & \cdots & * & 10\cdot n \end{pmatrix}$	$\begin{pmatrix} W & V & * & \cdots & * \\ V & W & V & & \vdots \\ * & V & W & & * \\ \vdots & & & & V \\ * & \cdots & * & V & W \end{pmatrix}$	$\begin{pmatrix} H & -1 & 0 & \cdots & -1 & 0 & \cdots & 0 \\ -1 & H & -1 & & & & & \vdots \\ 0 & & H & & & & & 0 \\ & & & & & & & 0 & -1 \\ -1 & & & & & & & \vdots \\ 0 & & & & & & & 0 \\ \vdots & & & & & & H & -1 \\ 0 & \cdots & 0 & -1 & \cdots & 0 & -1 & H \end{pmatrix}$
problem: artificial matrix	problem: artificial matrix	problem: artificial matrix
problem size: variable	problem size: variable	problem size: variable
sparsity: $nnz = n^2$	sparsity: $nnz = n^2$	sparsity: $nnz \approx 5n$
cond. number: $\kappa < 3$	cond. number: $\kappa \approx 8 \cdot 10^3$	cond. number: $\kappa \approx 8 \cdot 10^3$
storage format: MAS	storage format: MAS	storage format: CRS

On the other hand, we additionally use linear problems that were obtained from discretization in the area of CFD. The three systems of linear equations CFD1, CFD2 and CFD3 are affiliated with the 2D modeling of a Venturi Nozzle in different discretization fineness. The distinct number of supporting points leads to different matrix characteristics concerning the dimension, the number of non-zeros, and the condition number.

Table 2. Sparsity plots and properties of the CFD test-matrices

CFD1	CFD2	CFD3
problem: 2D fluid flow	problem: 2D fluid flow	problem: 2D fluid flow
problem size: $n = 395009$	problem size: $n = 634453$	problem size: $n = 1019967$
sparsity: $nnz = 3544321$	sparsity: $nnz = 5700633$	sparsity: $nnz = 9182401$
storage format: CRS	storage format: CRS	storage format: CRS

4.2   Numerical Results

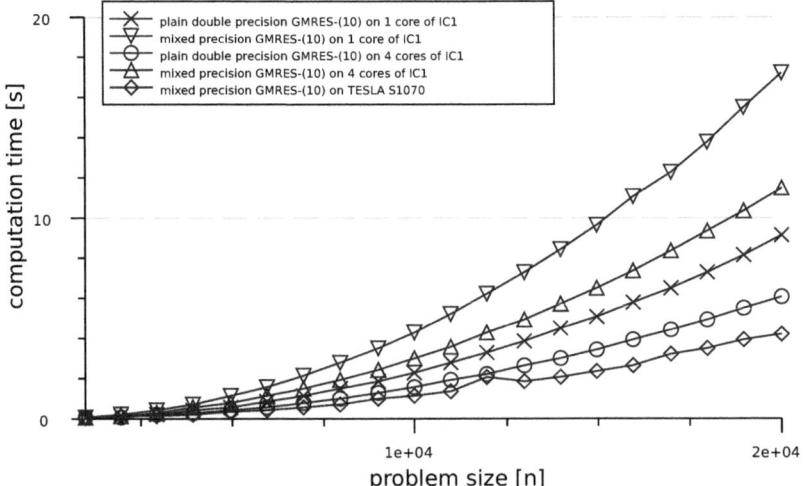

Fig. 2. Test case M1, relative residual stopping criterion $\varepsilon = 10^{-10}$

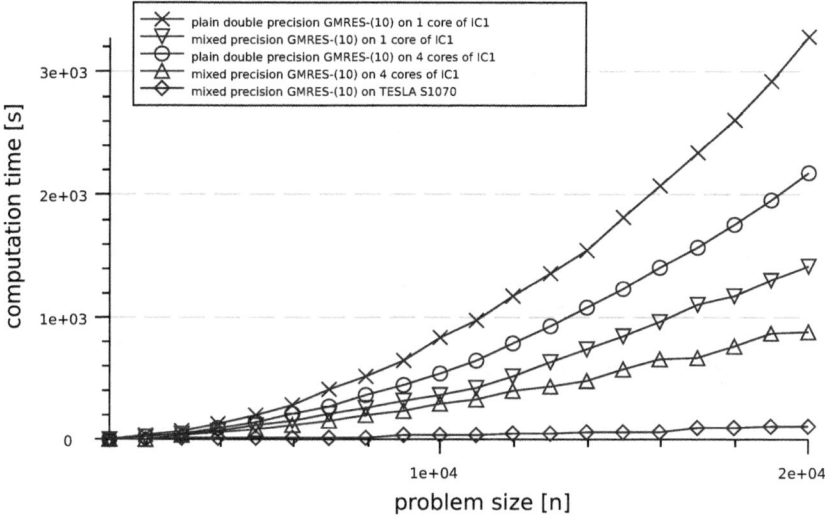

Fig. 3. Test case M2, relative residual stopping criterion $\varepsilon = 10^{-10}$

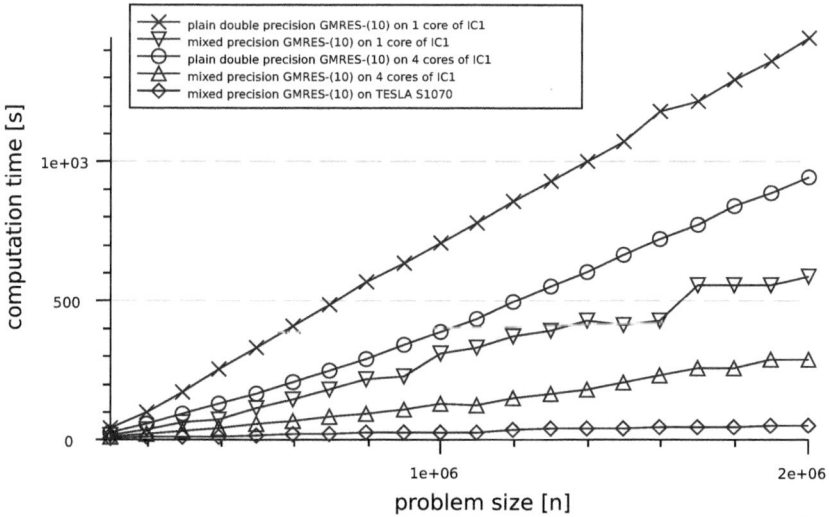

Fig. 4. Test case M3, relative residual stopping criterion $\varepsilon = 10^{-10}$

CFD1

solver type	computation time [s]
plain double GMRES-(10) on 1 core of IC1	3146.61
mixed GMRES-(10) on 1 core of IC1	1378.56
plain double GMRES-(10) on 4 cores of IC1	1656.53
mixed GMRES-(10) on 4 cores of IC1	712.83
mixed GMRES-(10) on TESLA S1070	438.13

CFD2

solver type	computation time [s]
plain double GMRES-(10) on 1 core of IC1	13204.70
mixed GMRES-(10) on 1 core of IC1	5924.32
plain double GMRES-(10) on 4 cores of IC1	6843.66
mixed GMRES-(10) on 4 cores of IC1	3495.09
mixed GMRES-(10) on TESLA S1070	2092.84

CFD3

solver type	computation time [s]
plain double GMRES-(10) on 1 core of IC1	60214.50
mixed GMRES-(10) on 1 core of IC1	41927.40
plain double GMRES-(10) on 4 cores of IC1	32875.10
mixed GMRES-(10) on 4 cores of IC1	19317.00
mixed GMRES-(10) on TESLA S1070	10316.70

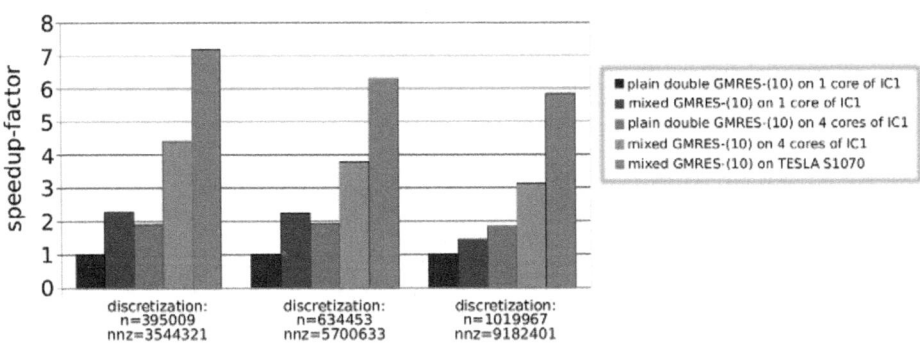

Fig. 5. Speedup of the different solvers for the CFD simulation of the Venturi Nozzle in different discretization fineness with $\varepsilon = 10^{-10}$ and $\varepsilon_{\text{inner}} = 0.1$

4.3   Result Interpretation

In the first test (Fig. 2), the low condition number leads to a good convergence rate of GMRES-(10) and there are only few iterations necessary to obtain a solution approximation fulfilling the stopping criterion. The additional computational cost of the mixed precision iterative refinement approach is large compared to the computational cost of the pure double solver. Therefore is the mixed precision GMRES-(10) neither for the sequential, nor for the parallel case able to compete with the plain double precision GMRES-(10). The TESLA S1070-implementation of the mixed GMRES-(10) outperforms the solvers on the IC1 due to the larger number of cores and the excellent single precision performance of the GPU, that can be exploited by the inner error correction solver. It should be mentioned, that the factors between the computation time of the different solver types are independent of the dimension n of the linear system that is solved.

The difference of the second test (Fig. 3) to the first test case (Fig. 2) is the fairly high condition number of $\kappa \approx 8 \cdot 10^3$ of the linear system. Due to the high number of iterations the linear solvers have to perform, the overhead for the mixed precision method is considerably small. Therefore, also on the IC1, the additional costs can be overcompensated by the speedup gained by performing the inner solver in a lower precision format. Both, for the parallel and the sequential case we gain a factor of about two using the mixed precision iterative refinement approach instead of the plain double precision GMRES-(10). Using a lower precision format leads to a shorter execution time when performing elementary computations on the one hand, and to a more efficient use of the memory bandwidth on the other hand. The memory space needed to store one single precision floating point number is half the size that is needed for one double precision floating point number. Usually, the memory bandwidth is the limiting factor of the computational power of a system. Using a lower precision format, the processors have shorter waiting time for the data, and the system gains a higher efficiency. Since this argument applies to all memory levels, the speedup using single precision for the GMRES-(10) can even exceed the factor 2 that characterizes the speedup of a general purpose CPU when switching from double to single precision computations.

The speedup factor gained by performing the mixed GMRES-(10) on the TESLA S1070 is almost 15 with respect to the sequential plain double GMRES-(10) on the IC1. Again we can observe, that the speedup factors between the different solvers on the different Hardware platforms remain constant, independent of the problem size.

For the third test case (Fig. 4), again an artificial test matrix is used with a condition number of $\kappa \approx 8 \cdot 10^3$. In difference to the former test cases, we now apply the solvers to sparse linear systems where the matrices are stored in the CRS format. The low number of nonzero entries leads no longer to a computational cost that is quadratically increasing with the dimension, but linearly. Furthermore is the total computational effort lower compared to the tests with matrix structure M2. Despite some perturbations, that can be explained by rounding effects and the use of different cache levels, we can still observe that the quotients between the solver types remain the same, independently of the dimension of the linear system. Again, both for the sequential and the parallel case, the mixed precision GMRES-(10) on the IC1 outperform the plain double implementations due to the fact, that the additional computational cost of the iterative refinement scheme is overcompensated by the speedup gained through the execution of the inner solver in single precision. The implementation on the TESLA S1070 can additionally exploit the excellent single precision performance of the highly parallelized GPU and overcompensate the additional data transfer costs. Furthermore approximate the speedups gained by using the mixed precision GMRES-(10) the speedups of test case M2.

The tests with the matrices CFD1, CFD2 and CFD3 show that the mixed precision iterative refinement approach is also beneficial when applying solvers to real world problems (Fig. 5). The mixed GMRES-(10) solvers outperform the sequential plain double GMRES-(10) implementations for all test problems, both in the sequential and the parallel case. The reason is again the fact, that the additional computational cost of the iterative refinement approach is overcompensated by the cheaper inner solver using a lower precision format.

Using hybrid hardware, the mixed GMRES-(10) on the TESLA S1070 even generates speedups up to 7 with respect to the plain double implementation on the IC1. It can be observed, that this factor decreases for increasing dimension. The reason is, that for large data amounts, the connection between the host CPU and the GPU slows the mixed GMRES-(10) down.

5   Conclusions and Future Work

Numerical tests have shown the high potential of using different precision formats within the proposed error correction solver.

While the obtained algorithm is flexible in terms of choosing the inner correction solver, it is robust in terms of numerical stability. The possibility of performing the error correction solver on a coprocessor increases the potential of mixed precision methods, as they can be implemented efficiently on hybrid systems. Performing the error correction solver of an error correction method in a lower format leads to an overall increase in performance for a large number of problems.

On a CPU, performing the error correction method in mixed precision, one often achieves a speedup factor of two compared to the plain solver in double precision.

When using hybrid hardware, consisting of coprocessors specialized on low precision performance, even higher speedup factors can be expected. In the numerical experiments for the FEM discretizations of the Venturi Nozzle we achieved speedups of more than seven for our CUDA implementation.

Still, a very ill-conditioned problem can lead to a high number of additional outer iterations necessary to correct the rounding errors, that arise from the use of a lower precision format in the error correction solver. In the worst case, the inner solver will not converge. Due to the fact that we are usually not able to determine a priori whether the mixed precision method is superior for a specific problem, an optimized implementation of the solver would execute the first solution update of the mixed precision error correction method and determine, depending on the improvement of the solution approximation, whether it should continue in the mixed precision mode or whether it should use the plain solver in high precision. The next step beyond this strategy of changing between single and double precision is to use techniques around adaptive precision, where the precision is adjusted according to the convergence in the inner solver. FPGAs and related technologies may provide the capabilities for such algorithms.

For an efficient implementation of the mixed precision error correction techniques in a solver suite, some additional work is necessary, especially concerning the use of preconditioners. This may not only increase the stability of the solver, but also its performance. In such an environment, the mixed precision error correction methods form powerful solvers for FEM simulations and beyond.

References

1. Microprocessor Standards Committee of the IEEE Computer Society: IEEE Standard for Floating-Point Arithmetic (2008)
2. Saad, Y.: Iterative Methods for Sparse Linear Systems. Society for Industrial and Applied Mathematics, Philadelphia (2003)
3. van der Vorst, H.A.: Iterative Krylov Methods for Large Linear Systems. Cambridge University Press, Cambridge (2003)
4. Higham, N.J.: Accuracy and Stability of Numerical Algorithms. Society for Industrial and Applied Mathematics, Philadelphia (1996)
5. Dongarra, J.J., Duff, I.S., Sorensen, D.C., van der Vorst, H.A.: Numerical Linear Algebra for High-Performance Computers. Society for Industrial and Applied Mathematics, Philadelphia (1998)
6. Demmel, J.W.: Applied Numerical Linear Algebra. Society for Industrial and Applied Mathematics, Philadelphia (1997)
7. Bai, Z., Demmel, J., Dongarra, J.J., Ruhe, A., van der Vorst, H.: Templates for the Solution of Algebraic Eigenvalue Problems. Society for Industrial and Applied Mathematics, Philadelphia (2000)
8. Göddeke, D., Strzodka, R., Turek, S.: Performance and accuracy of hardware–oriented native–, emulated– and mixed–precision solvers in FEM simulations. Fakultät für Mathematik, TU Dortmund (2007)
9. Buttari, A., Dongarra, J., Langou, J., Langou, J., Luszcek, P., Kurzak, J.: Mixed Precision Iterative Refinement Techniques for the Solution of Dense Linear Systems (2007)
10. Baboulin, M., Buttari, A., Dongarra, J., Langou, J., Langou, J., Luszcek, P., Kurzak, J., Tomov, S.: Accelerating Scientific Computations with Mixed Precision Algorithms (2008)
11. Strzodka, R., Göddeke, D.: Pipelined Mixed Precision Algorithms on FPGAs for Fast and Accurate PDE Solvers from Low Precision Components. In: IEEE Proceedings on Field–Programmable Custom Computing Machines (FCCM 2006). IEEE Computer Society Press, Los Alamitos (2006)
12. Göddeke, D., Strzodka, R.: Performance and accuracy of hardware–oriented native–, emulated– and mixed–precision solvers in FEM simulations (Part 2: Double Precision GPUs). Fakultät für Mathematik, TU Dortmund (2008)
13. NVIDIA CUDA Compute Unified Device Architecture Programming Guide. NVIDIA Corporation. edit. 2.3.1 (2009)

14. NVIDIA CUDA CUBLAS Library Programming Guide. NVIDIA Corporation. edit. 1.0 (2007)
15. Technical Brief NVIDIA GeForce GTX 200 GPU Architectural Overview. NVIDIA Corporation (2008)
16. InstitutsCluster User Guide. Universität Karlsruhe (TH), Steinbuch Centre for Computing. edit. 0.92 (2008)
17. Heuveline, V., Rocker, B., Ronnas, S.: Numerical Simulation on the SiCortex Supercomputer Platform: a Preliminary Evaluation. EMCL Preprint Series (2009)
18. BLAS LAPACK User's Guide. Fujitsu (2003)

Multifrontal Computations on GPUs
and Their Multi-core Hosts

Robert F. Lucas[1], Gene Wagenbreth[1], Dan M. Davis[1], and Roger Grimes[2]

[1] Information Sciences Institute, University of Southern California
4676 Admiralty Way, Suite 1001
Marina del Rey, California 90230
{rflucas,genew,ddavis}@isi.edu
[2] Livermore Software Technology Corporation
7374 Las Positas Rd
Livermore, California 94551
grimes@lstc.com

Abstract. The use of GPUs to accelerate the factoring of large sparse symmetric matrices shows the potential of yielding important benefits to a large group of widely used applications. This paper examines how a multifrontal sparse solver performs when exploiting both the GPU and its multi-core host. It demonstrates that the GPU can dramatically accelerate the solver relative to one host CPU. Furthermore, the solver can profitably exploit both the GPU to factor its larger frontal matrices and multiple threads on the host to handle the smaller frontal matrices.

Keywords: GPU acceleration, GPGPU, multifrontal algorithms, MCAE.

1 Introduction

Solving the system of linear equations Ax = b, where A is both large and sparse, is a computational bottleneck in many scientific and engineering applications. Therefore, over the past forty years, a tremendous amount of research has gone into this problem, exploring both direct and iterative methods [1]. This paper focuses on a subset of this large space of numerical algorithms, factoring large sparse symmetric indefinite matrices. Such problems often arise in Mechanical Computer Aided Engineering (MCAE) applications. For decades, researchers have sought to exploit novel computing systems to accelerate the performance of sparse matrix factorization algorithms. This paper continues that trend, exploring whether or not one can accelerate the factorization of large sparse matrices, which is already parallelized on a modern multi-core microprocessor, by additionally exploiting graphics processing units (GPUs).

The GPU is a very attractive candidate as an accelerator to ameliorate a computational bottleneck such as sparse matrix factorization. Unlike previous generations of accelerators, such as those designed by Floating Point Systems [2] for the relatively small market of scientific and engineering applications, current GPUs are designed to improve the end-user experience in mass-market arenas such as gaming. Together

J.M.L.M. Palma et al. (Eds.): VECPAR 2010, LNCS 6449, pp. 71–82, 2011.

with other niche chips, such as Sony, Toshiba, and IBM's (STI) Cell [3], they are a new generation of devices whose market share is growing rapidly, independently of science and engineering. The extremely high peak floating point performance of these new commodity components begs the question as to whether or not they can be exploited to increase the throughput and/or reduce the cost of applications beyond the markets for which they are targeted. The quest to explore broader use of GPUs is often called GPGPU, which stands for General Purpose computation on GPUs [4].

There are many algorithms for factoring large sparse linear systems. The multifrontal method [5] is particularly attractive, as it transforms the sparse matrix factorization into a hierarchy of dense matrix factorizations. Multifrontal codes can effectively exploit the memory hierarchies of cache-based microprocessors, routinely going out-of-core to disk as needed. With the right data structures, the vast majority of the floating point operations can be performed with calls to highly tuned BLAS3 routines, such as the SGEMM matrix-matrix multiplication routine [6], and near peak throughput is expected. Not surprisingly, all of the major commercial MCAE applications use multifrontal solvers.

Recent GPGPU work has demonstrated that dense, single-precision linear algebra computations, *e.g.,* SGEMM, can achieve very high levels of performance on GPUs [7][8][9]. This in turn led to early efforts to exploit GPUs in multifrontal linear solvers by investigators at USC [10], ANSYS [11], and AAI [12]. These early efforts compared the performance of early model NVIDIA G80 GPUs to that of single CPU hosts. In the work reported herein, we extend the previous work and report on the performance of a multifrontal linear solver exploiting both state-of-the-art NVIDIA Tesla GPUs as well as shared memory concurrency on the dual-socket, quad-core Intel Nehalem host microprocessor.

The remainder of the paper is organized as follows. The next section provides a brief overview of the multifrontal method and illustrates how it turns a sparse problem into a tree of dense ones. This is followed by a brief overview of the NVIDIA Tesla C1060 GPU used in the earlier phase of this experiment. We discuss both the unique nature of its architecture as well as its CUDA programming language. Section 4 presents our strategy for factoring individual frontal matrices on the GPU and provides performance results on the GPU. Section 5 presents the impact on the overall performance of the multifrontal sparse solver of utilizing both shared memory parallelism and the GPU. Finally, we summarize the results of our experiment and suggest directions for future research.

2   Overview of a Multifrontal Sparse Solver

Figure 1 depicts the non-zero structure of a small sparse matrix. Coefficients that are initially non-zero are represented by an 'x', while those that fill-in during factorization are represented by a '*'. Choosing an optimal order in which to eliminate these equations is in general an NP-complete problem, so heuristics, such as METIS [13], are used to try to reduce the storage and operations necessary. The multifrontal method treats the factorization of the sparse matrix as a hierarchy of dense sub-problems. Figure 2 depicts the multifrontal view of the matrix in Fig.1. The directed acyclic graph of the order in which the equations are eliminated is called the elimination tree. When each equation is eliminated, a small dense matrix called the frontal matrix is assembled. In Figure 2, the numbers to the left of each frontal matrix are its row indices. Frontal matrix assembly proceeds in the following fashion: the frontal matrix is cleared, it is loaded with the initial values from the pivot column (and row if it's asymmetric), then any updates generated when factoring the pivot equation's children in the elimination tree are accumulated. Once the frontal matrix has been assembled, the variable is eliminated. Its Schur complement (the shaded area in Fig.2) is computed as the outer product of the pivot row and pivot column from the frontal matrix. Finally, the pivot equation's factor (a column of L) is stored and its Schur complement placed where it can be retrieved when needed for the assembly of its parent's frontal matrix. If a post-order traversal of the elimination tree is used, the Schur complement matrix can be placed on a stack of real values.

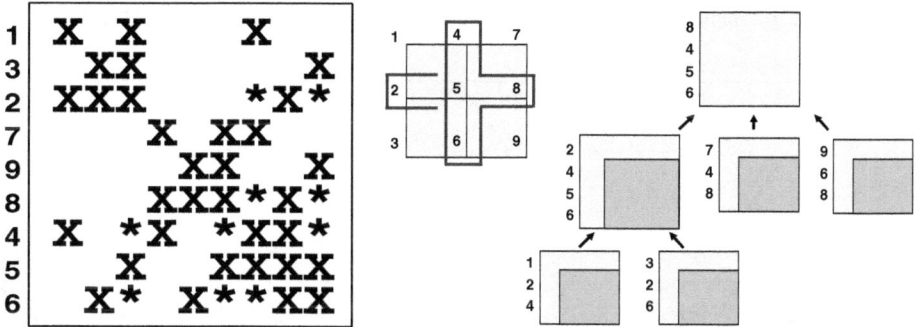

Fig. 1. Sparse matrix with symmetric non-zero structure   **Fig. 2.** Multifrontal view of sparse matrix from Fig.1

The cost of assembling frontal matrices is reduced by exploiting supernodes. A supernode is a group of equations whose non-zero structures in the factored matrix are indistinguishable. For example, zeros filled-in during the factorization of the matrix in Fig.1 turn its last four equations into a supernode. The cost of assembling one frontal matrix for the entire supernode is amortized over the factorization of all the constituent equations, reducing the multifrontal method's overhead. Furthermore, when multiple equations are eliminated from within the same frontal matrix, their Schur complement can be computed very efficiently as the product of two dense matrices.

Fig.3 depicts a finite element grid generated by the LS-DYNA MCAE code (www.lstc.com). The matrix for the grid in Fig. 3 is relatively small, having only 235,962 equations. Matrices with two orders-of-magnitude more equations are routinely factored today. Factoring such large problems can take many hours, a time that is painfully apparent to the scientists and engineers waiting for the solution.

Figure 4 illustrates the elimination tree for the matrix corresponding to the grid in Fig 3, as ordered by METIS. This particular elimination tree has 12,268 relaxed [14] supernodes in it. There are thousands of leaves and one root. The leaves are relatively small, O(10) equations being eliminated from O(100). The supernodes near the root are much bigger. Hundreds of equations are eliminated from over a thousand. Because dense factor operations scale as order N^3, approximately two-dozen supernodes at the top of the tree contain half of the total factor operations.

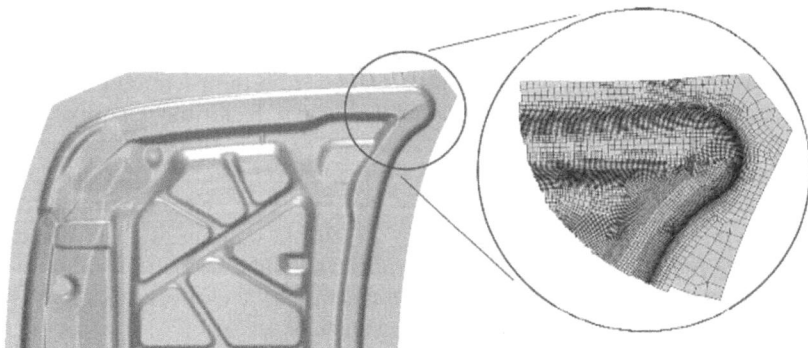

Fig. 3. Example of an MCAE Finite Element Problem and Grid (courtesy LSTC)

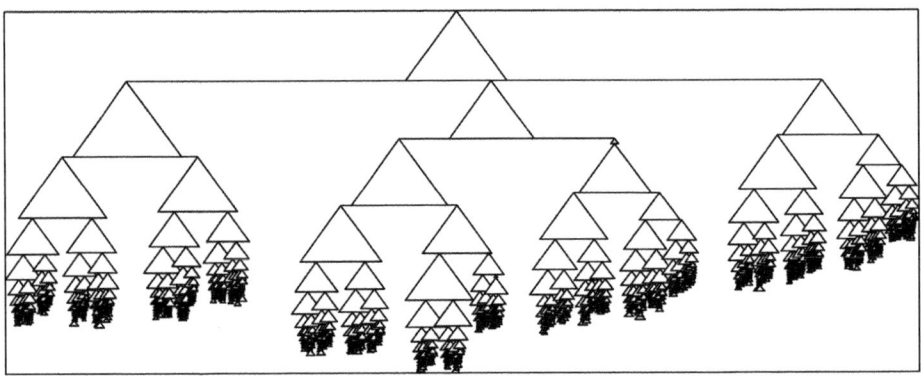

Fig. 4. Supernodal elimination tree for problem in Figure 3 (courtesy Cleve Ashcraft)

The multifrontal code discussed in this paper has two strategies for exploiting shared-memory, multithreaded concurrency. The frontal matrices at the leaves of the elimination tree can all be assembled and factored independently. At the lower levels in the tree, there can be thousands of such leaves, dwarfing the number of processors, and hence each supernode is assigned to an individual processor. This leads to a breadth-first traversal of the elimination tree, and a real stack can no longer be used to manage the storage of the update matrices [15]. Near the top of the elimination tree, the number of supernodes drops to less than the number of processors. Fortunately, for the finite element matrices considered in this work, these few remaining supernodes are large, and a right-looking code can be sped up by dividing the matrix into panels and assigning them to different processors.

The objective of the work reported here is to attempt to use GPUs as inexpensive accelerators to factor the large supernodes near the root of the elimination tree, while processing the smaller supernodes near the bottom of the tree by exploiting shared-memory concurrency on the multicore host. This should lead to a significant increase in the throughput of sparse matrix factorization compared to a single CPU. The next section gives a brief description of the NVIDIA Tesla C1060 and its CUDA programming language, highlighting just those features used in this work to factor individual frontal matrices.

3 Graphics Processing Units

The NVIDIA Tesla GPU architecture consists of a set of multiprocessors. Each of the C1060's thirty multiprocessors has eight Single Instruction, Multiple Data (SIMD) processors. This GPU supports single precision (32 bit) IEEE 754 [16] formatted floating-point operations. It also supports double precision, but at a significantly lower performance. Each SIMD processor can perform two single precision multiplies and one add at every clock cycle. The clock rate on the C1060 card is 1.3 GHz. Therefore, the peak performance is:

$$1.3 \text{ GHz} * 3 \text{ results/cycle} * 8 \text{ SIMD/mp} * 30 \text{ mp} = 936 \text{ GFlops/s}$$

The ratio of multiplies to adds in matrix factorization is one, so for a linear solver, the effective peak performance is 624 GFlop/s. In practice, the NVIDIA CuBLAS SGEMM routine delivers just over half of that performance.

Memory on the Tesla GPU is organized into device memory, shared memory and local memory. Device memory is large (4 GBytes), is shared by all multiprocessors, is accessible from both host and GPU, and has high latency (over 100 clock cycles). Each multiprocessor has a small (16 KBytes) shared memory that is accessible by all of its SIMD processors. Shared memory is divided into banks and, if accessed so as to avoid bank conflicts, has a one cycle latency. Shared memory should be thought of a user-managed cache or buffer between device memory and the SIMD processors. Local memory is allocated for each thread. It is small and can be used for loop variables and temporary scalars, much as registers would be used. The constant memory and texture memory were not used in this effort.

In our experience, there are two primary issues that must be addressed to use the GPU efficiently:

- code must use many threads, without conditionals, operating on separate data to keep the SIMD processors busy
- code must divide data into small sets, which can be cached in the shared memory. Once in shared memory, data must be used in many operations (10 – 100) to mask the time spent transferring between shared and device memory.

It is not yet feasible to convert a large code to execute on the GPU. Instead, compute-bound subsets of the code should be identified that use a large percentage of the execution time. Only those subsets should be converted to run on the GPU. Their input data is transferred from the host to the GPU's device memory before initiating computation on the GPU. After the GPU computation is complete, the results are transferred back to the host from the GPU's device memory.

To facilitate general-purpose computations on their GPU, NVIDIA developed the Compute Unified Device Architecture (CUDA) programming language [17]. CUDA is a minimal extension of the C language and is loosely type-checked by the NVIDIA compiler (and preprocessor), nvcc, which translates CUDA programs (.cu) into C programs. These are then compiled with the gcc compiler and linked as an NVIDIA provided library. Within a CUDA program, all functions have qualifiers to assist the compiler with identifying whether the function belongs on the host of the GPU. For variables, the types have qualifiers to indicate where the variable lives, *e.g.*, __device__ or __shared__. CUDA does not support recursion, static variables, functions with arbitrary numbers of arguments, or aggregate data types.

4   Algorithm for Factoring Individual Frontal Matrices on the GPU

In earlier work, we determined that, in order to get meaningful performance using the GPU, we had to both maximize use of the NVIDIA supplied SGEMM arithmetic kernel and minimize data transferred between the host and the GPU. We decided to adopt the following strategy for factoring individual frontal matrices on the GPU:

- Download the factor panel of a frontal matrix to the GPU. Store symmetric data in a square matrix, rather than a compressed triangular. This wastes storage, but is easy to implement.
- Use a left-looking factorization, proceeding over panels from left to right:
  - Update a panel with SGEMM
  - Factor the diagonal block of the panel
  - Eliminate the off-diagonal entries from the panel
- Update the Schur complement of this frontal matrix with SGEMM
- Return the entire frontal matrix to the host, converting back from square to triangular storage
- Return an error if the pivot threshold was exceeded or a diagonal entry was zero

Table 1. Log of time spent factoring a model frontal matrix

Method Name	GPU msec	%GPU time
Copy data to and from GPU	201.0	32.9%
Factor 32x32 diagonal blocks	42.6	7.0%
Eliminate off diagonal panels	37.0	6.1%
Update with SGEMM	330.6	54.1%
Total time	611.4	100.0%

The time log for factoring a large, simulated frontal matrix with the fully optimized CUDA factorization code is in Table 1. This timing was taken when the GPU was eliminating 3072 equations from 4096. Approximately half of the execution time on the GPU is spent in SGEMM. Eliminating off-diagonals and factoring diagonal blocks takes only 13% of the time. The remaining third of the time is spent realigning the matrices and copying data to and from the host. A further 0.029 seconds are spent on the host, and not reflected in Table 1. The computation rate for the entire dense symmetric factorization is 163 GFlops/s. In contrast, four cores of the Intel Xeon Nehalem host achieve 29 GFlop/s when factoring the same sized frontal matrix and using the same 32-column panel width. Performance results using the GPU to factor a variety of model frontal matrices is presented in Table 2. These range in the number of equations eliminated from the frontal matrix (size) as well as the number of equations left in the frontal matrix, *i.e.*, its external degree (degree). As expected, the larger the frontal matrix gets, the more operations one has to perform to factor it, and the higher the performance of the GPU.

Table 2. Performance of the C1060 GPU frontal matrix factorization kernel

Size	Degree	Secs	GFlop/s
1024	1024	0.048	51.9
1536	1024	0.079	66.3
2048	1024	0.117	79.7
512	2048	0.045	60.2
1024	2048	0.079	86.5
1536	2048	0.123	101.3
2048	2048	0.179	112.2
512	3072	0.076	74.7
1024	3072	0.128	103.9
1536	3072	0.188	122.4
2048	3072	0.258	136.0
512	4096	0.116	84.0
1024	4096	0.185	118.3
1536	4096	0.267	137.3
2048	4096	0.361	150.9

5   Performance of the Accelerated Multifrontal Solver

In this section we examine the performance impact of the GPU on overall multifrontal sparse matrix factorization. We will use a matrix extracted from the LS-DYNA MCAE application. It is derived from a three dimensional problem composed of three cylinders nested within each other, and connected with constraints. The rank of this symmetric matrix is 760320 and its diagonal and lower triangle contain 29213357 non-zero entries. After reordering with Metis, it takes 7.104E+12 operations to factor the matrix. The resulting factored matrix contains 1.28E+09 entries.

Figure 5 plots the time it takes to factor the matrix, as a function of the number of cores employed, both with and without the GPU. The dual socket Nehalem host sustains 10.3 GFlop/s when using one core, and 59.7 GFlop/s when using all eight. When the GPU is employed, it performs 6.57E+12 operations, 92% of the total, and sustains 98.1 GFlop/s in doing so. The overall performance with the GPU improves to 61.2 GFlop/s when one host core is used, and 79.8 GFlop/s with all eight. For perspective, reordering and symbolic factorization take 7.9 seconds, permuting the input matrix takes 2.64 seconds, and the triangular solvers take 1.51 seconds.

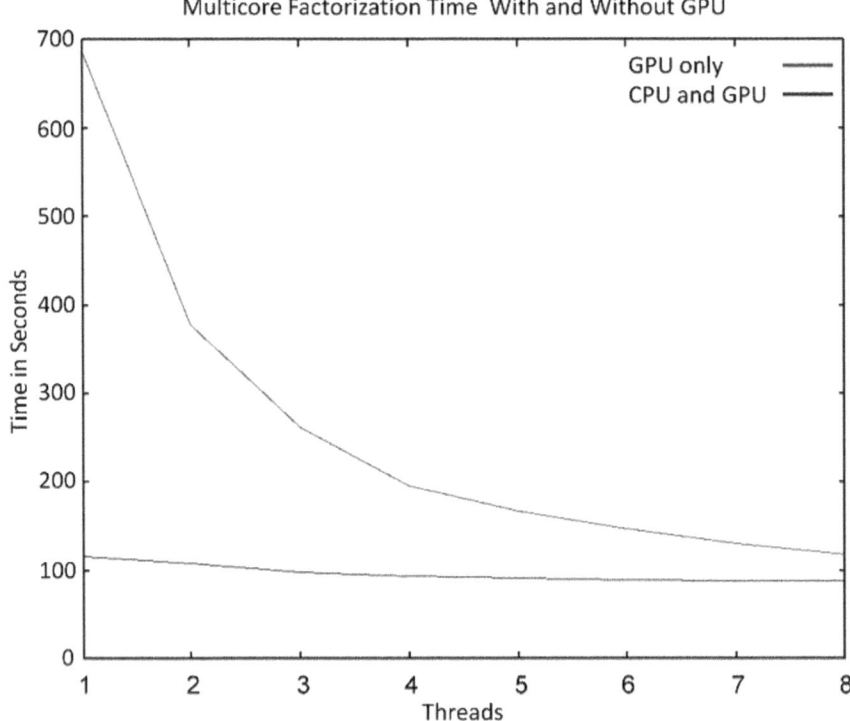

Fig. 5. Multicore factorization time, with and without the GPU

To understand why there seems to be so little speedup when the GPU-enhanced solver goes from one core to eight, consider Figure 6. It displays the number of supernodes per level in the elimination tree for the three cylinder matrix, along with the number of operations required to factor the supernodes at each level. Notice that the vast majority of the operations are in the top few levels of the tree, and these are processed by the GPU.

Figure 7 plots the performance achieved by the multicore host when factoring the supernodes at each level of the tree. Note, near the leaves, performance is nowhere near the peak. This is true even for one core, as the supernodes are too small to facilitate peak SGEMM performance. As multiple cores are used, relatively little speedup is observed, which is likely due to the relatively low ratio of floating point operations to memory loads and stores for these small supernodes, leaving them memory bound on the multicore processor.

The performance study described above was performed in single precision, using an NVIDIA Tesla C1060. In most implicit MCAE applications, double precision is preferred. Therefore, it was not until the Tesla C2050 (Fermi) was available that it made send to try to integrate the GPU-enhanced multifrontal code into an application such as LS-DYNA. The results of this integration as presented in Figure 8, which depicts the time for LS-DYNA to run a static analysis of the same three nested cylinders. LS-DYNA itself is compiled to use eight-byte integers and double precision

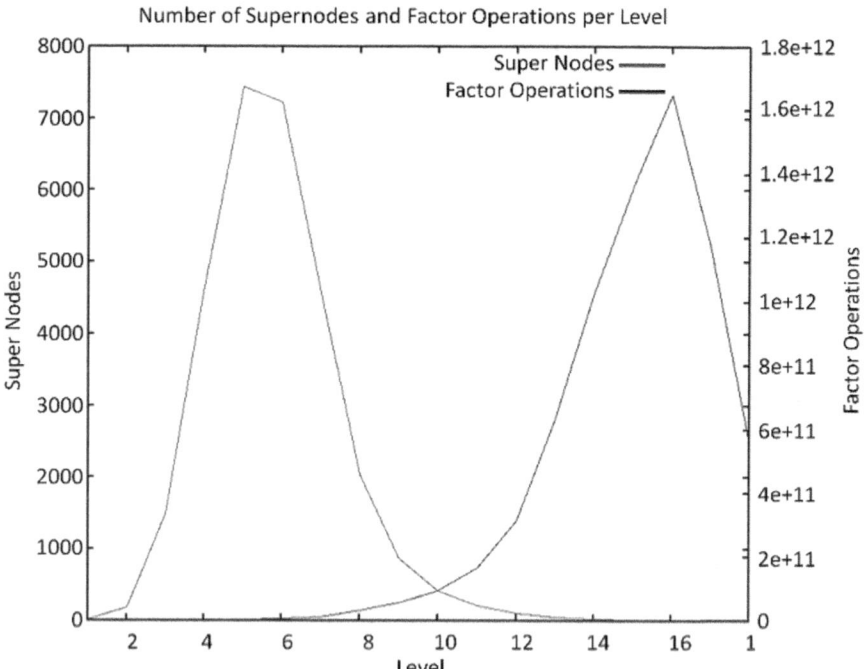

Fig. 6. Number of supernodes and factor operations per level in the tree

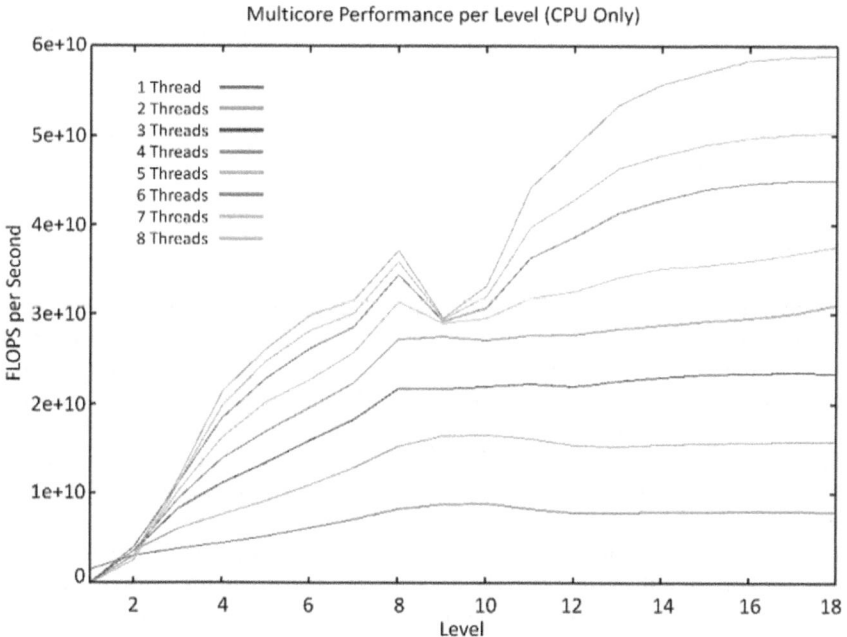

Fig. 7. Multicore performance per level in the elimination tree

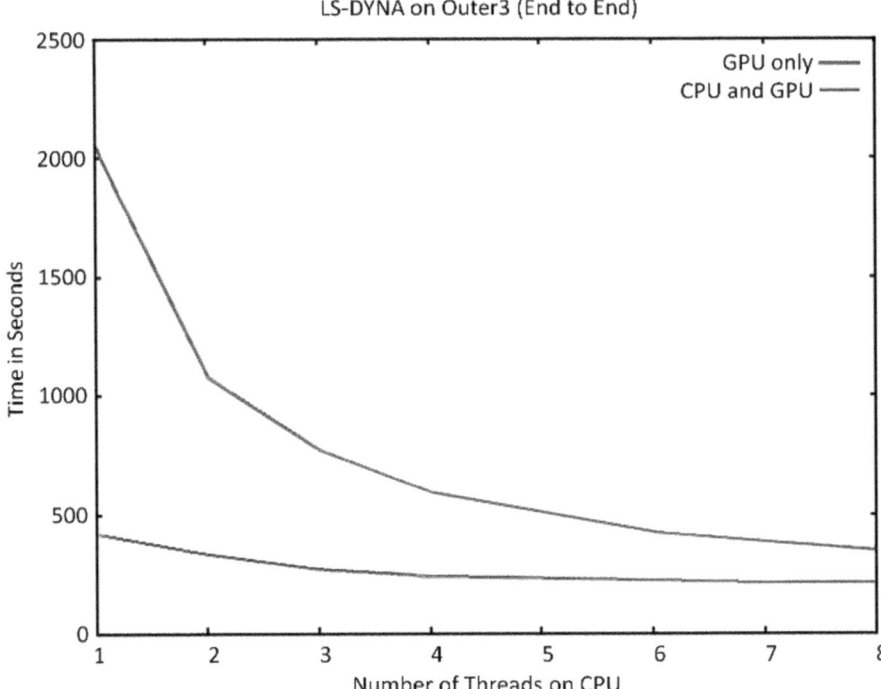

Fig. 8. LS-DYNA run time, with and without the GPU

floating point operations. The frontal matrix factorization kernel on the C2050 GPU uses 32-bit integers and double precision floating point numbers. Because sparse matrix factorization dominates the run time in this example, the shape of the curves depicting the run time of LS-DYNA on multiple cores, with and without the C2050 GPU, is very similar to those of the solver in Fig. 5.

6 Summary

This paper has demonstrated that a GPU can in fact be used to significantly accelerate the throughput of a multi-frontal sparse symmetric factorization code, even when exploiting shared memory concurrency on the host multicore microprocessor. We have demonstrated factorization speed-up of 5.91 relative to one core on the host, and 1.34 when using eight cores. This was done by designing and implementing a symmetric factorization algorithm for the NVIDIA C1060 in the CUDA language and then offloading a small number of large frontal matrices, containing over 90% the total factor operations, to the GPU.

In addition, we have demonstrated that with suitable support for double precision arithmetic, as is found in the NVIDIA C2050, the GPU-enhanced solver can accelerate implicit MCAE analysis. We note that both ANSYS and Simulia have recently reported similar results, having integrated solver kernels developed by Acceleware in to their MCAE codes, ANSYS and ABAQUS. However, more work needs to be done before the use of GPUs will be common for such applications. The GPU frontal matrix factorization code implemented for this experiment should be revisited to make it more efficient in its use of memory on the GPU. It should be modified to implement pivoting so that indefinite problems can be factored entirely on the GPU. Further, it should be extended to work on frontal matrices that are bigger than the relatively small device memory on the GPU, much as the multifrontal code goes out-of-core when the size of a sparse matrix exceeds the memory of the host processor.

Finally, if one GPU helps, why not more? Researchers have been implementing parallel multifrontal codes for over two decades [18]. In fact, the multifrontal code used in these experiments has both OpenMP and MPI constructs. Therefore exploiting multiple GPUs is not an unreasonable thing to consider. However, when one considers that one would have to simultaneously overcome both the overhead of accessing the GPU as well as the costs associated with communicating amongst multiple processors; it may be very challenging to efficiently factor one frontal matrix with multiple GPUs.

Acknowledgement

We would like thank Norbert Juffa, Stan Posey, and Peng Wang of NVIDIA for their encouragement and support for this work. This has included guidance in performance optimization as well as access to the latest NVIDIA GPUs. The work presented herein derives from earlier work supported by the United States Joint Forces Command via a subcontract from the Lockheed Martin Corporation and SimIS, Inc., as well as research sponsored by the Air Force Research Laboratory under agreement numbers F30602-02-C-0213 and FA8750-05-2-0204.

References

[1] Heath, M., Ng, E., Peyton, B.: Parallel algorithms for sparse linear systems. Society for Industrial and Applied Mathematics Review 33, 420–460 (1991)

[2] Charlesworth, A., Gustafson, J.: Introducing Replicated VLSI to Supercomputing: the FPS-164/MAX Scientific Computer. IEEE Computer 19(3), 10–23 (1986)

[3] Pham, D.C., Aipperspach, T., Boerstler, D., Bolliger, M., Chaudhry, R., Cox, D., Harvey, P., Harvey, P.M., Hofstee, H.P., Johns, C., Kahle, J., Kameyama, A., Keaty, J., Masubuchi, Y., Pham, M., Pille, J., Posluszny, S., Riley, M., Stasiak, D.L., Suzuoki, M., Takahashi, O., Warnock, J., Weitzel, S., Wendel, D., Yazawa, K.: Overview of the Architecture, Circuit Design, and Physical Implementation of a First-Generation Cell Processor. IEEE Journal of Solid State Circuits 41(1) (January 2006)

[4] Lastra, A., Lin, M., Minocha, D.: ACM Workshop on General Purpose Computations on Graphics Processors (2004)

[5] Duff, I., Reid, J.: The Multifrontal Solution of Indefinite Sparse Symmetric Linear Systems. ACM Transactions on Mathematical Software 9, 302–335 (1983)

[6] Dongarra, J.J., Du Croz, J., Hammarling, S., Duff, I.S.: A Set of Level 3 Basic Linear Algebra Subprograms. ACM Transactions on Mathematical Software 16(1), 1–17 (1990)

[7] Scott Larson, E., McAllister, D.: Fast matrix multiplies using graphics hardware. In: Proceedings of the 2001 ACM/IEEE Conference on Supercomputing, p. 55. ACM Press, New York (2001)

[8] Fatahalian, K., Sugarman, J., Hanrahan, P.: Understanding the Efficiency of GPU Algorithms for Matrix-Matrix Multiplication. In: Proceedings of the ACM Sigraph/Eurographics Conference on Graphics Hardware. Eurographics Association, pp. 133–138 (2004)

[9] Govindaraju, N., Manocha, D.: Cache-Efficient Numerical Algorithms Using Graphics Hardware, University of North Carolina Technical Report (2007)

[10] Lucas, R.F.: GPU-Enhanced Linear Solver Results. In: The Proceedings of Parallel Processing for Scientific Computing. SIAM, Philadelphia (2008)

[11] Private communication with Gene Poole, ANSYS Inc., at SCI2008, Austin, TX (November 2008)

[12] http://cqse.ntu.edu.tw/cqse/download_file/DPierce_20090116.pdf

[13] Karypis, G., Kumar, V.: A fast and high quality multilevel scheme for partitioning irregular graphs. In: Haridi, S., Ali, K., Magnusson, P. (eds.) Euro-Par 1995. LNCS, vol. 966, pp. 113–122. Springer, Heidelberg (1995)

[14] Ashcraft, C., Grimes, R.: The Influence of Relaxed Supernode Partitions on the Multifrontal Method. ACM Transactions in Mathematical Software 15, 291–309 (1989)

[15] Ashcraft, C., Lucas, R.: A Stackless Multifrontal Method. In: Tenth SIAM Conference on Parallel Processing for Scientific Computing (March 2001)

[16] Arnold, M.G., Bailey, T.A., Cowles, J.R., Winkel, M.D.: Applying Features of IEEE 754 to Sign/Logarithm Arithmetic. IEEE Transactions on Computers 41(8), 1040–1050 (1992)

[17] Buck, I.: GPU Computing: Programming a Massively Parallel Processor. In: International Symposium on Code Generation and Optimization, San Jose, California

[18] Duff, I.: Parallel Implementation of Multifrontal Schemes. Parallel Computing 3, 193–204 (1986)

Accelerating GPU Kernels for Dense Linear Algebra*

Rajib Nath, Stanimire Tomov, and Jack Dongarra**

Department of Electrical Engineering and Computer Science,
University of Tennessee, Knoxville
{rnath1,tomov,dongarra}@eecs.utk.edu

Abstract. Implementations of the Basic Linear Algebra Subprograms (BLAS) interface are major building block of dense linear algebra (DLA) libraries, and therefore have to be highly optimized. We present some techniques and implementations that significantly accelerate the corresponding routines from currently available libraries for GPUs. In particular, *Pointer Redirecting* – a set of GPU specific optimization techniques – allows us to easily remove performance oscillations associated with problem dimensions not divisible by fixed blocking sizes. For example, applied to the matrix-matrix multiplication routines, depending on the hardware configuration and routine parameters, this can lead to two times faster algorithms. Similarly, the matrix-vector multiplication can be accelerated more than two times in both single and double precision arithmetic. Additionally, GPU specific acceleration techniques are applied to develop new kernels (e.g. syrk, symv) that are up to 20× faster than the currently available kernels. We present these kernels and also show their acceleration effect to higher level dense linear algebra routines. The accelerated kernels are now freely available through the MAGMA BLAS library.

Keywords: BLAS, GEMM , GPUs.

1 Introduction

Implementations of the BLAS interface are major building block of dense linear algebra libraries, and therefore have to be highly optimized. This is true for GPU computing as well, especially after the introduction of shared memory in modern GPUs. This is important because it enabled fast Level 3 BLAS implementations for GPUs [2,1,4], which in turn made possible the development of DLA for GPUs to be based on BLAS for GPUs [1,3]. Earlier attempts (before the introductions of shared memory) could not rely on memory reuse, only on the GPU's high bandwidth, and as a result were slower than the corresponding CPU implementations.

---

* Candidate to the Best Student Paper Award.
** Research reported here was partially supported by the National Science Foundation, NVIDIA, and Microsoft Research.

J.M.L.M. Palma et al. (Eds.): VECPAR 2010, LNCS 6449, pp. 83–92, 2011.

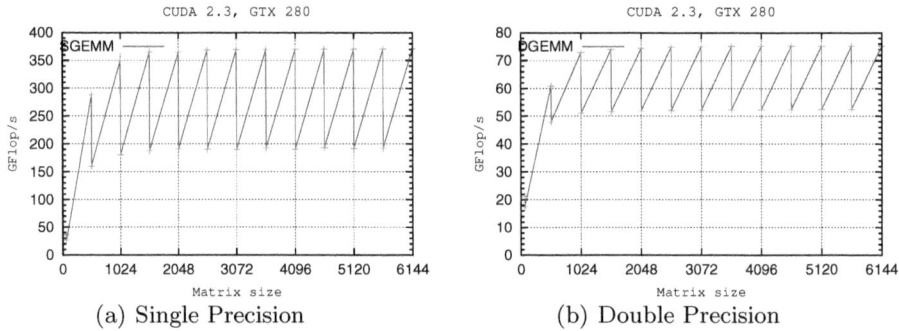

(a) Single Precision   (b) Double Precision

Fig. 1. GEMM Performance on Square Matrices

Despite the current success in developing highly optimized BLAS for GPUs [2,1,4], the area is still new and presents numerous cases/opportunities for improvements. This paper addresses several very important kernels, namely the matrix-matrix multiplication that are crucial for the performance throughout DLA, and matrix-vector multiplication that are crucial for the performance of linear solvers and two-sided matrix factorizations (and hence eigen-solvers). The new implementations are included in the recently released and freely available *Matrix Algebra for GPU and Multicore Architectures* (MAGMA) version 0.2 BLAS Library [3].

The rest of the paper is organized as follows. Section 2 gives some performance results of current kernels and points out our optimization targets. Section 3 presents the *Pointer Redirecting* techniques and their use to accelerate the xAXPY, xGEMV, and xGEMM routines. Section 4 summarizes the results on accelerating selected MAGMA BLAS kernels. Next, in Section 5 we give the performance results for the new kernels. Finally, Section 6 summarizes this work and describes on-going efforts.

2   Performance of Current BLAS for GPUs

One current BLAS library for GPUs is NVIDIA's CUBLAS [2]. Figure 1(a) shows the performance of the single precision matrix-matrix multiplication routine (SGEMM) for a discrete set of matrix dimensions. Figure 1(b) shows similar data but for double precision arithmetic. Note that at some dimensions the performance is much higher than at other dimensions, e.g. taken at odd numbers like 65, 129, etc. These performance dips, that actually happen in the majority of matrix dimensions are one of our acceleration targets. The reason for these dips is very likely related to an implementation that has even inner-blocking size to match various hardware parameters and considerations to get high performance. The performance graphs illustrate a quite high performance loss for the cases when the matrix dimension is obviously not a multiple of the inner blocking size. In particular, the performance gap is more than 24 GFlops/s in double precision (around.34% of the peak performance), and is worse for single precision.

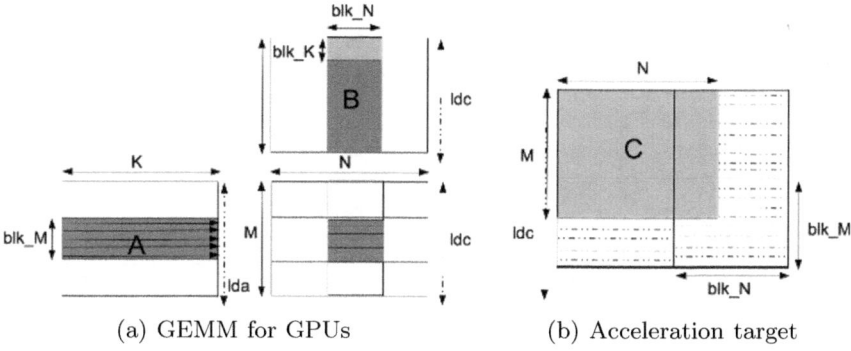

(a) GEMM for GPUs   (b) Acceleration target

Fig. 2. The algorithmic view of GEMM for GPUs

There are ways around to work with these BLAS routines and still get high performance in high level algorithms. One possible solution is to force the user to allocate and work with matrices multiple of the blocking size. This though leads to memory waste. Sometimes it is a burden to the user if the application is already written, and in general is obviously not a good solution. Another solution is padding with 0s to fit the blocking factor, do the computation and keep this transparent to the user. This approach has the overhead of copying data back and forth, and possibly some extra computation. A third approach is to rewrite the kernels in such a way that there are no extra computations, no data movement or any other overheads. This rewriting though is difficult and time consuming, especially taken into account different GPU specifics as related to data coalescing, data parallel computation, computation symmetry, and memory bank layout.

3   Pointer Redirecting

The matrix-matrix multiplication (xGEMM; e.g. $C = AB$) algorithm for GPUs is schematically represented in Figure 2(a). Matrix C is divided into blocks of size $blk_M \times blk_N$ and each block is assigned to a block of $nthd_x \times nthd_y$ threads. Each thread inside a thread block computes a row of sub matrix $blk_M \times blk_N$. Each thread accesses corresponding row of matrix A as shown by an arrow and uses the sub-matrix $K \times blk_N$ of matrix B for computing the final result. As the portion of matrix B needed by each thread inside a thread block is the same, they load a sub-matrix of matrix B of size $blk_N \times blk_K$ from global memory to shared memory in a coalesced way, synchronize themselves, do the computation and repeat until the computation is over. All these happen in a series of synchronized steps. With an optimal selection of $blk_M, blk_N, blk_K, nthd_X, nthd_Y$, we can get the best kernel for the matrix sizes that are divisible by blocking factors, i.e. $M\%blk_M = 0$, $N\%blk_N = 0$, $K\%blk_K = 0$.

The question is how to deal with matrix dimensions that are not divisible by the blocking factor. Whatever solution we choose, we have to keep it transparent

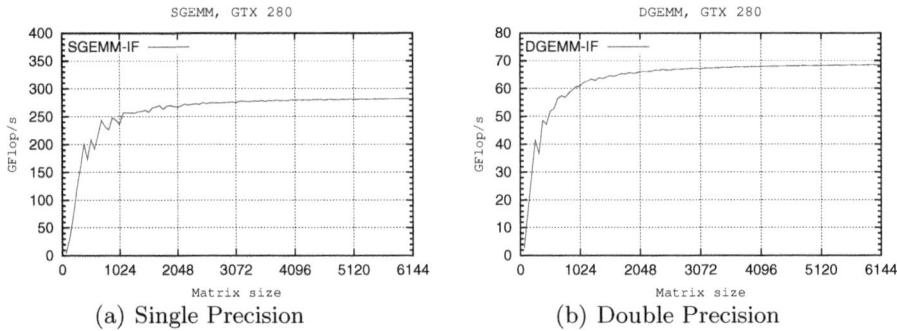

(a) Single Precision   (b) Double Precision

Fig. 3. GEMM Implementation with Conditional Statement in Inner Loop

to the user while maintaining highest flexibility. The goal is to allow reasonable overhead (if needed) and to achieve high performance in general cases. We show in Figure 2(b) matrix C of a xGEMM operation $(C = \alpha C + \beta Op(A)Op(B))$ where dimensions M and N are not divisible by the blocking factor. The matrix has only one full block. We can do the computation for the full block and do the other partial blocks by loading data and doing computation selectively. This will introduce several if-else statements in the kernel which will prevent the threads inside a thread-block to run in parallel. Figure 3 shows the performance of one such implementation. Note that GPUs run all the threads inside a thread block in parallel as long as they execute the same instruction on different data. If the threads ever execute different instruction, their processing would become temporary sequential until they start executing the same instructions again.

Another approach is to let the unnecessary threads do similar work so that the whole thread block can run in data parallel mode. In Figure 2(b) the dashed blue lines correspond to unnecessary flops that are done by respective thread. It is not clear yet which data they will operate on, but it also does not matter because the computation will be discarded. Lets take a look at the scenario where all the threads assume that the matrix fits into the block and do the work in a natural way until updating matrix C. In Figure 4, the shaded region corresponds to original matrix and the outmost rectangle corresponds to the largest matrix that best fits in terms of blocking factor. We are going to make $\lceil \frac{M}{dim_M} \rceil \times \lceil \frac{N}{dim_N} \rceil$ number of grids and allow threads at the partial block to compute the same way as it is done in a full block. It is evident that memory accesses inside the shaded region in Figure 4, denoted by white diamond, are always valid. Memory accesses denoted by red diamonds are always invalid. Memory accesses represented by green diamond could be valid or illegal. As we can see in the Figure 4, the leftmost green diamond could be an element from the next column, e.g. when $lda \leq blk_M \times \lceil \frac{M}{blk_M} \rceil$. It could be an element in the same column when $lda > blk_M \times \lceil \frac{M}{blk_M} \rceil$, or it could be invalid memory reference.

In Figure 5(Left), the blue lines in last row and last column are last valid memory reference irrespective of any values of $lda, M, N, K, blk_M, blk_N, nthd_X, nthd_Y$.

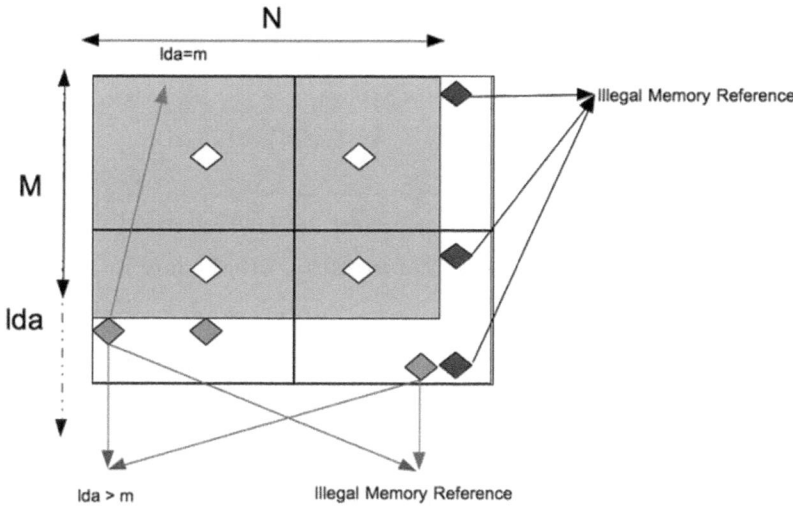

Fig. 4. Possible Illegal Memory Reference in Matrix Multiply

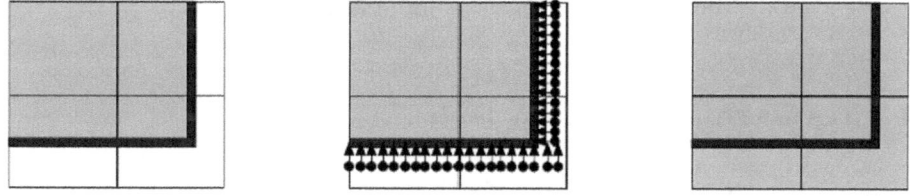

Fig. 5. (Left) Last Valid Access (Middle) Pointer Redirecting (Right) Mirroring

If some thread needs to access some memory location beyond this last row/column, we are going to force him reference to this last row/column by adjusting the pointer. These threads will be doing unnecessary computation, we don't care from where this data is coming from. All we care is that together they make best use of memory bandwidth and layout, access data in a coalesced manner. Figure 5(Middle) depicts the complete scenario how the memory is referenced. As a result the matrix will have some virtual row where rows beyond the last row are replication of last row and columns beyond the last column are replication of last column. It is shown in Figure 5.

Let's see how it fits into xGEMM's(Op(A) = Op(B) =Non-Transposed) context in terms of accessing matrix A. As in Figure 6(a), thread t1, t2, t3, t4 will be accessing valid memory location. And all the threads beyond thread t4, e.g. thread t5, t6 will be accessing same memory thread t4 is accessing. As a result no separate memory read operation will be issued and no latency will be experienced for this extra load. If we look at Figure 6(b), $blk_K \times blk_N$ data of matrix

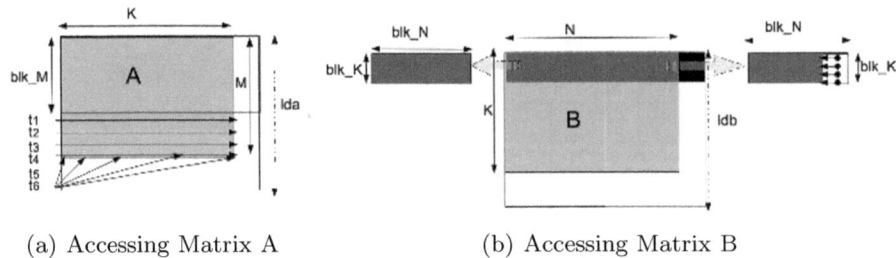

(a) Accessing Matrix A   (b) Accessing Matrix B

Fig. 6. Algorithmic view of GEMM for GPUs with Pointer Redirecting

B are brought into shared memory by $nthd_X \times nthd_Y$ threads in a coalesced manner. The left $blk_K \times blk_N$ block is necessary as we can see. But the right $blk_K \times blk_N$ is partially needed. The black portions are unnecessary memory access. As discussed before, it will access the last row or column that is needed instead of accessing invalid memory. This will still be done in a coalesced way and it is accessing less memory now. Some memory are accessed more than once, which doesn't hamper performance. This a simple solution to the problem with little overhead that doesn't break the pattern of coalesced memory access. Note that we will not be doing any extra computation in K dimension, so we don't need to zeroing out values to keep the computation valid.

4   MAGMA BLAS Kernels

MAGMA BLAS includes a subset of CUDA BLAS that are crucial for the performance of MAGMA routines. The pointer redirecting technique were applied to most of the kernels. Here we mention a few of the new kernels and their use in high level MAGMA routines.

xGEMM: Various kernels were developed as an extension to the approach previously presented in [4]. The extensions include more parameters to explore xGEMM's design space to find best performing versions in an auto-tuning approach. The new algorithms are of extreme importance for both one-sided and two-sided matrix factorizations as they are in general based on xGEMMs involving rectangular matrices, and these are the cases that we managed to accelerate most significantly.

xGEMV: Similarly to xGEMM, various implementations were developed and parametrized to prepare them for auto-tuning based acceleration. Different implementations are performing best in different settings. xGEMVs are currently used in MAGMA's mixed-precision iterative refinement solvers and the Hessenberg reduction algorithm.

xSYMV: Similarly to xGEMM and xGEMV, various implementations were developed. xSYMV is used similarly to when xGEMV is used with the difference when symmetric matrices are involved. This is again the mixed-precision iterative refinement solvers and the reduction to three diagonal form.

xTRSM: Algorithms that trade off parallelism and numerical stability, especially in algorithms related to triangular solvers, have been known and studied before, but now are getting extremely relevant with the emerging highly parallel architectures, like the GPUs. We use an approach where diagonal blocks of the matrix are explicitly inverted and used in a block algorithm. Multiple kernels, including kernels where the inverses are computed on the CPU or GPU, with various block sizes (e.g., recursively increasing it from 32), are developed.

xSYRK: A block index reordering technique is used to initiate and limit the computation only to blocks that are on the diagonal or in the lower (correspondingly upper) triangular part of the matrix. In addition, all the threads in a diagonal block are responsible to compute redundantly half of the block in a data parallel fashion in order to avoid expensive conditional statements that would have been necessary otherwise. Some threads also load unnecessary data so that data is fetched from global memory in a coalesced manner. These routines are used in both some one-sided and two-sided matrix factorization algorithms.

5 Performance

For the unnecessary computation there will be some overhead. Figure 7 shows the percentage of extra flops needed for different dimensions of matrix with parameters $blk_M = 64, blk_N = 16, blk_K = 16, nthd_X = 16, nthd_Y = 4$ for different matrix sizes. The overhead is scaled to 100 for visibility. Figure 9 and Figure 8 shows the performance results for GEMM in single and double precision respectively. In double precision we are seeing an improvement of 24 GFlops/s and in single precision it is like 170 GFlops/s. As we have discussed before other than small dimensions the improvement is significant The zig-zag patterns in performance graph resembles the blocking factor of the kernel.

As we have discussed before, if the matrices are in CPU memory one can use padding, e.g., as in [5]. We have to allocated a bigger dimension of matrix in GPU memory, put zeroes in the extra elements, then transfer the data from CPU to GPU and then call the Kernel. Figure 10 shows the performance comparison when data is in CPU memory. It is evident that for small matrix size our implementation is better and for higher dimension they are very identical. We note that the pointer redirecting approach does not use extra memory, does not require a memory copy if non padded matrix is given on the GPU memory, and finally does not require initialization of the padded elements.

Finally Figure 11 gives an illustration on the effect of optimized BLAS on high level routines. We see similar results throughout MAGMA algorithms. Table 1 shows the performance of the one-sided QR factorization using CUBLAS and MAGMA BLAS for matrix sizes not divisible by the kernel's block size. The pointer redirecting approach brings 20% to 50% performance improvement over CUBLAS in this case.

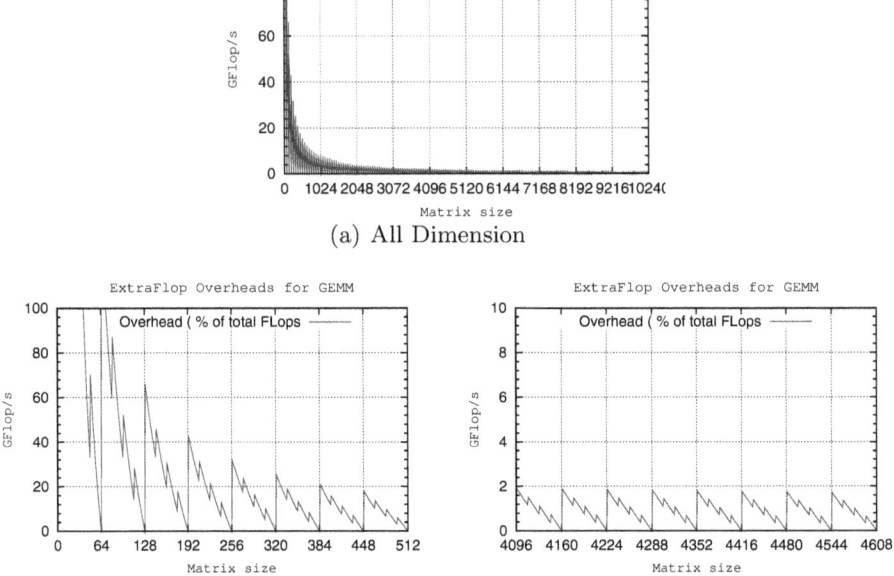

(a) All Dimension

(b) Small Dimension   (c) Large Dimension

Fig. 7. Flops overhead in xGEMM

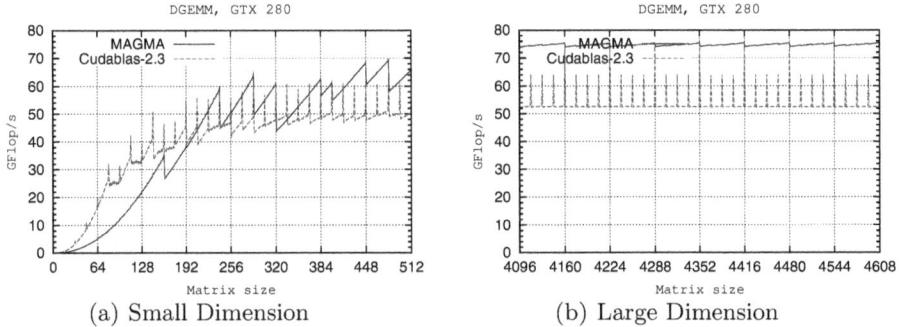

(a) Small Dimension   (b) Large Dimension

Fig. 8. Performance dGEMM

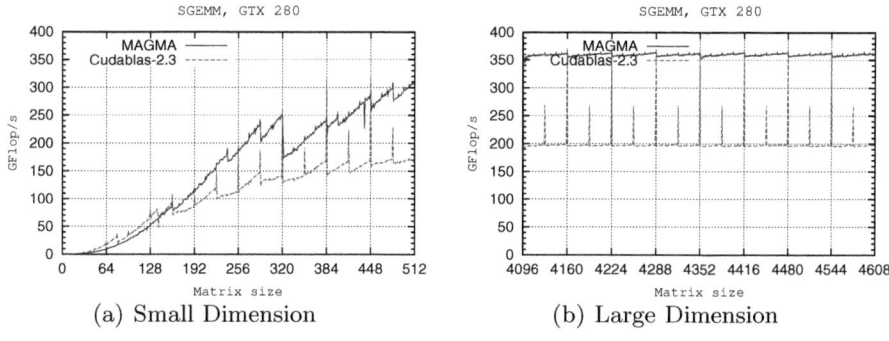

Fig. 9. Performance sGEMM

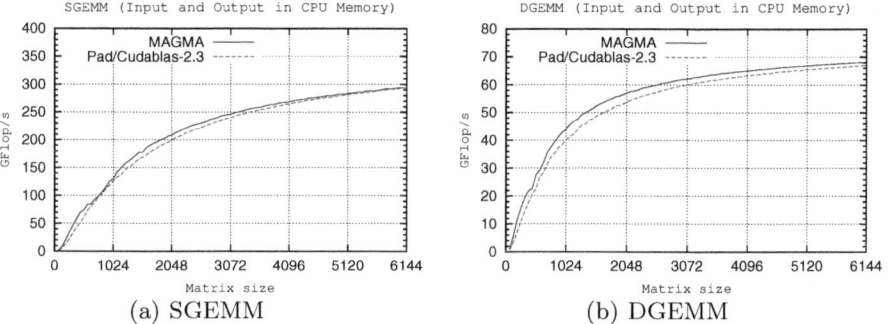

Fig. 10. Performance xGEMM with Padding (Data In/Out in CPU Memory)

Table 1. Performance comparison between MAGMA BLAS with pointer redirecting and CUBLAS for the QR factorization in single precision arithmetic

Matrix Size	CUBLAS	MAGMA BLAS
1001	47.65	46.01
2001	109.69	110.11
3001	142.15	172.66
4001	154.88	206.34
5001	166.79	226.43
6001	169.03	224.23
7001	175.45	246.75
8001	177.13	251.73
9001	179.11	269.99
10001	180.45	262.90

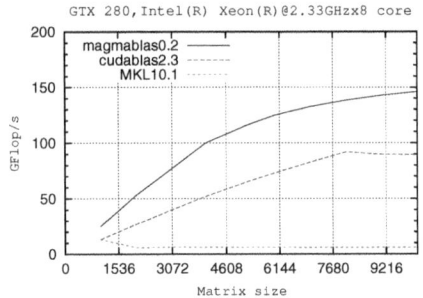

Fig. 11. Effect of optimized SGEMV on the Hessenberg reduction

6   Conclusions and On-going Work

We presented techniques to accelerate GPU BLAS kernels that are crucial for the performance of DLA algorithms. Performance results, demonstrating significant kernels acceleration and the effect of this acceleration on high level DLA, were also presented. On-going work includes the extension of these techniques to more routines, and their inclusion in the MAGMA BLAS library.

References

1. Volkov, V., Demmel, J.: Benchmarking gpus to tune dense linear algebra. In: Proc. of SC 2008, Piscataway, NJ, USA, pp. 1–11 (2008)
2. CUDA CUBLAS Library, http://developer.download.nvidia.com
3. Tomov, S., Nath, R., Du, P., Dongarra, J.: MAGMA version 0.2 Users' Guide (November 2009), http://icl.cs.utk.edu/magma
4. Li, Y., Dongarra, J., Tomov, S.: A note on auto-tuning GEMM for gPUs. In: Allen, G., Nabrzyski, J., Seidel, E., van Albada, G.D., Dongarra, J., Sloot, P.M.A. (eds.) ICCS 2009. LNCS, vol. 5544, pp. 884–892. Springer, Heidelberg (2009)
5. Barrachina, S., Castillo, M., Igual, F., Mayo, R., Quintana-Orti, E.: Evaluation and Tuning of the Level 3 CUBLAS for Graphics Processors. In: PDSEC 2008 (2008)

A Scalable High Performant Cholesky Factorization for Multicore with GPU Accelerators

Hatem Ltaief, Stanimire Tomov, Rajib Nath, Peng Du, and Jack Dongarra*

Department of Electrical Engineering and Computer Science,
University of Tennessee, Knoxville
{ltaief,tomov,rnath1,du,dongarra}@eecs.utk.edu

Abstract. We present a Cholesky factorization for multicore with GPU accelerators systems. The challenges in developing scalable high performance algorithms for these emerging systems stem from their heterogeneity, massive parallelism, and the huge gap between the GPUs' compute power *vs* the CPU-GPU communication speed. We show an approach that is largely based on software infrastructures that have already been developed for homogeneous multicores and hybrid GPU-based computing. This results in a scalable hybrid Cholesky factorization of unprecedented performance. In particular, using NVIDIA's Tesla S1070 (4 C1060 GPUs, each with 30 cores @1.44 GHz) connected to two dual-core AMD Opteron @1.8GHz processors, we reach up to 1.163 TFlop/s in single and up to 275 GFlop/s in double precision arithmetic. Compared with the performance of the embarrassingly parallel xGEMM over four GPUs, where no communication between GPUs are involved, our algorithm still runs at 73% and 84% for single and double precision arithmetic respectively.

1 Introduction

When processor clock speeds flatlined in 2004, after more than fifteen years of exponential increases, the era of routine and near automatic performance improvements that the HPC application community had previously enjoyed came to an abrupt end. CPU designs moved to multicores and are currently going through a renaissance due to the need for new approaches to manage the exponentially increasing (a) appetite for power of conventional system designs, and (b) gap between compute and communication speeds.

Compute Unified Device Architecture (CUDA) [1] based multicore platforms stand out among a confluence of trends because of their low power consumption and, at the same time, high compute power and bandwidth. Indeed, as power consumption is typically proportional to the cube of the frequency, accelerators using GPUs have a clear advantage against current homogeneous multicores,

---

* Research reported here was partially supported by the National Science Foundation, NVIDIA, and Microsoft Research.

J.M.L.M. Palma et al. (Eds.): VECPAR 2010, LNCS 6449, pp. 93–101, 2011.

as their compute power is derived from many cores that are of low frequency. Initial GPU experiences across academia, industry, and national research laboratories have provided a long list of success stories for specific applications and algorithms, often reporting speedups on the order of 10 to 100× compared to current x86-based homogeneous multicore systems [2]. The area of dense linear algebra (DLA) is no exception as evident from previous work on a single core with a single GPU accelerator [3,4], as well as BLAS for GPUs (see the CUBLAS library [5]). Despite the current success stories involving hybrid GPU-based systems, the large scale enabling of those architectures for computational science would still depend on the successful development of fundamental numerical libraries for using the CPU-GPU in a hybrid manner. Major issues in terms of developing new algorithms, programmability, reliability, and user productivity have to be addressed. Our work is a contribution to the development of these libraries in the area of dense linear algebra and will be included in the *Matrix Algebra for GPU and Multicore Architectures* (MAGMA) Library [9]. Designed to be similar to LAPACK in functionality, data storage, and interface, the MAGMA library will allow scientists to effortlessly port their LAPACK-relying software components and to take advantage of the new hybrid architectures.

The challenges in developing scalable high performance algorithms for multicore with GPU accelerators systems stem from their heterogeneity, massive parallelism, and the huge gap between the GPUs' compute power *vs* the CPU-GPU communication speed. We show an approach that is largely based on software infrastructures that have already been developed – namely, the *Parallel Linear Algebra for Scalable Multicore Architectures* (PLASMA) [6] and MAGMA libraries. On one hand, the tile algorithm concepts from PLASMA allow the computation to be split into tiles along with a scheduling mechanism to efficiently balance the work-load between GPUs. On the other hand, MAGMA kernels are used to efficiently handle heterogeneity and parallelism on a single tile. Thus, the new algorithm features two levels of nested parallelism. A coarse-grained parallelism is provided by splitting the computation into tiles for concurrent execution between GPUs (following PLASMA's framework). A fine-grained parallelism is further provided by splitting the work-load within a tile for high efficiency computing on GPUs but also, in certain cases, to benefit from hybrid computations by using both GPUs and CPUs (following MAGMA's framework). Furthermore, to address the challenges related to the huge gap between the GPUs' compute power *vs* the CPU-GPU communication speed, we developed a mechanism to minimize the communications overhead by trading off the amount of memory allocated on GPUs. This is crucial for obtaining high performance and scalability on multicore with GPU accelerators systems. Indeed, although the computing power of order 1 TFlop/s is concentrated in the GPUs, communications between them are still performed using the CPUs as a gateway, which only offers a shared connection on the order of 1 GB/s. As a result, by reusing the core concepts of our existing software infrastructures along with data persistence optimizations, the new hybrid Cholesky factorization not only achieves unprecedented high performance but also, scales while the number of GPUs increases.

The paper is organized as follows. Section 2 introduces the principles of the new technique, which permits the overall algorithm to scale on multiple GPUs. It also gives implementation details about various Cholesky versions using different levels of optimizations. Section 3 presents the performance results of those different versions. Section 4 describes the on-going work in this area and finally, Section 5 summarizes this work.

2   Cholesky Factorization on Multicore+MultiGPUs

In this section, we describe our new technique to efficiently perform the Cholesky factorization on a multicore system enhanced with multiple GPUs.

2.1   Principles and Methodology

This section represents our main twofold contribution.

First, the idea is to extend the runtime environment (RTE) of PLASMA, namely the static scheduler [7], to additionally handle computation on GPUs. Instead of assigning tasks to a single CPU, the static scheduler is now able to assign tasks to a CPU+GPU couple. Each CPU host is dedicated to a particular GPU device to offload back and forth data. PLASMA's RTE ensures dependencies are satisfied before a host can actually trigger the computation on its corresponding device. Moreover, there are four kernels to compute the Cholesky factorization and they need to be redefined (from PLASMA). Three of them – xTRSM, xSYRK and xGEMM – can be efficiently executed on the GPU using CUBLAS or the MAGMA BLAS libraries. In particular, we developed and used optimized xTRSM and xSYRK (currently included in MAGMA BLAS). But most importantly, the novelty here is to replace the xPOTRF LAPACK kernel by the corresponding hybrid kernel from MAGMA. High performance on this kernel is achieved by allowing both host and device to factorize the diagonal tile together in a hybrid manner. This is paramount to improve the kernel because the diagonal tiles are located in the critical path of the algorithm.

Second, we developed a data persistence strategy that optimizes the number of transfers between the CPU hosts and GPU devices, and vice versa. Indeed, the host is still the only gateway to any transfers occurring between devices which appears to be a definite bottleneck if communications are not handled cautiously. To bypass this issue, the static scheduler gives us the opportunity to precisely keep track of the location of any particular data tile during runtime. One of the major benefits of such a scheduler is that each processing CPU+GPU couple knows ahead of time its workload and can determine where a data tile resides. Therefore, many assumptions can be taken before the actual computation in order to limit the amount of data transfers to be performed.

The next sections present incremental implementations of the new tile Cholesky factorization on multicore with GPU accelerators systems. The last implementation is the most optimized version containing both contributions explained above.

2.2   Implementations Details

We describe four different implementations of the tile Cholesky factorization designed for hybrid systems. Each version introduces a new level of optimizations and simultaneously includes the previous ones. Each GPU device is dedicated to a particular CPU host, and this principle holds for all versions described below.

2.3   Memory Optimal

This version of the tile Cholesky factorization is very basic in the sense that the static scheduler from PLASMA is reused out of the box. The scheduler gives the green light to execute a particular task after all required dependencies have been satisfied. Then, three steps occur in this following order. First, the core working on that task triggers the computation on its corresponding GPU by offloading the necessary data. Second, the GPU performs the current computation. Third, the specific core requests the freshly computed data back from the GPU. Those three steps are repeated for all kernels except for the diagonal factorization kernel, i.e., xPOTRF, where no data transfers are needed since the computation is only done by the host. This version only requires, at most, the size of three data tiles to be allocated on the GPU (due to the xGEMM kernel). However, the amount of communication involved is tremendous as for each kernel call (except xPOTRF) , two data transfers are needed (steps one and three).

2.4   Data Persistence Optimizations

In this implementation, the amount of communications is significantly decreased by trading off the amount of memory allocated on GPUs. To understand how this works, it is important to mention that each data tile located on the left side of the current panel being factorized corresponds to the final output, i.e., they are not transient data tiles. And this is obviously due to the nature of the left-looking Cholesky factorization. Therefore, the idea is to keep in GPU's memory any data tile loaded for a specific kernel while processing the panel, in order to be eventually reused by the same GPU for subsequent kernels. After applying all operations on a specific data tile located on the panel, each GPU device uploads back to its CPU host the final data tile to ensure data consistency between hosts/devices for the next operations. As a matter of fact, another progress table has been implemented to determine whether a particular data tile is already present in the device's memory or actually needs to be uploaded from host's memory. This technique requires, at most, the amount of half the matrix to be stored in GPU's memory. Besides optimizing the number of data transfers between hosts and devices, we also try to introduce asynchronous communications to overlap communications by computations (using the `cudaMemcpy2DAsync` function and pinned CPU memory allocation).

2.5   Hybrid xPOTRF Kernel

The implementation of this version is straightforward. The xPOTRF kernel has been replaced by the hybrid xPOTRF MAGMA kernel, where both host and device compute the factorization of the diagonal tile.

2.6   xSYRK and xTRSM Kernel Optimizations

This version integrates new implementations of the BLAS xSYRK and xTRSM routines, which are highly optimized for GPU computing as explained below.

xSYRK: A block index reordering technique is used to initiate and limit the computation only to blocks that are on the diagonal or in the lower (correspondingly upper) triangular part of the matrix (since the resulting matrix is symmetric). Thus, no redundant computations are performed for blocks off of the diagonal. Only the threads that would compute diagonal blocks are let to compute redundantly half of the block in a data parallel fashion in order to avoid expensive conditional statements that would have been necessary otherwise.

xTRSM: Similarly to [3,8], we explicitly invert blocks of size 32×32 on the diagonal of the matrix and use them in blocked xTRSM algorithms. The inverses are computed simultaneously, using one GPU kernel, so that the critical path of the blocked xTRSM can be greatly reduced by doing it in parallel (as a matrix-matrix multiplication). We have implemented multiple kernels but this performed best for the tile sizes used in the Cholesky factorization (see Section 3.2) and our particular hardware configuration.

3   Experimental Results

3.1   Environment Setup

The experiments have been performed on a dual-socket dual-core host machine based on an AMD Opteron processor operating at 1.8 GHz. The NVIDIA S1070 graphical card is composed of four GPUs C1060 with two PCI Express connectors driving two GPUs each. Each GPU has 1.5 GB GDDR-3 of memory and 30 processing cores each, operating at 1.44 GHz. Each processing core has eight SIMD functional units and each functional unit can issue three floating point operations per cycle (1 mul-add + 1 mul = 3 flops). The single precision theoretical peak performance of the S1070 card is then $30 \times 8 \times 3 \times 1.44 \times 4 = 4.14$ Tflop/s. However, only two flops per cycle can be used for general purpose computations in our dense linear algebra algorithm (1 mul-add per cycle). So, in our case, the single precision peak performance drops to $2/3 \times 4.14 = 2.76$ Tflop/s. The double precision peak is computed similarly with the only difference being that there is only one SIMD functional unit per core, i.e., the peak will be $30 \times 1 \times 2 \times 1.44 \times 4 = 345$ Gflop/s. The host machine is running Linux 2.6.18 and provides GCC Compilers 4.1.2 together with the CUDA 2.3 library. All the experiments presented below focus on asymptotic performance and have been conducted on the maximum amount of cores and GPUs available on the machine, i.e., four cores and four GPUs.

3.2   Tuning

The performance of the new factorization strongly depends on tunable execution parameters, most notably various block sizes for the two levels of nested parallelism in the algorithm, i.e., the outer and inner block sizes. These parameters are usually computed from an auto-tuning procedure (e.g., established at installation time) but for now, manual tuning based on empirical data is used to determine their "optimal" values. The selection of the tile size (the outer blocking size) is determined by the performance of xGEMM. The goal is to determine from which tile size the performance of xGEMM on a single GPU starts to asymptotically flatten. Several values in that region were tested to finally select the best performing ones, namely $b_s = 576$ in single and $b_d = 832$ in double precision arithmetic. The selection of the inner blocking sizes for the splitting occurring within the hybrid kernels (i.e., MAGMA's xPOTRF) and the GPU kernels (i.e., MAGMA BLAS's xSYRK, xTRSM, xGEMM) is done similarly, based on empirical data for problem sizes around 500 and 800 for single and double precision arithmetic, respectively [10].

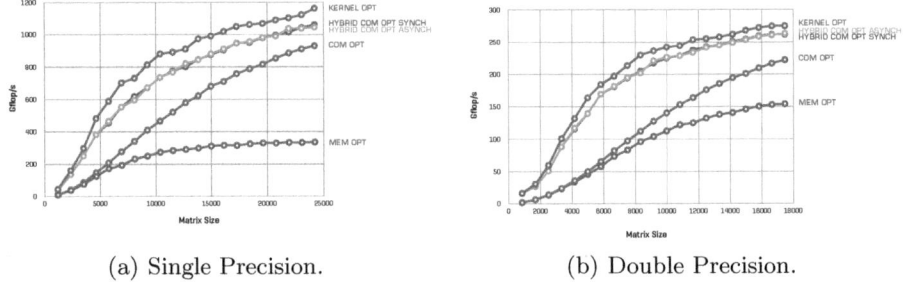

(a) Single Precision.   (b) Double Precision.

Fig. 1. Performance comparisons of various implementations

3.3   Performance Results

Figure 1 shows the incremental performance in single and double precision arithmetic of the tile hybrid Cholesky factorization using the entire system resources, i.e. four CPUs and four GPUs. Each curve represents one version of the Cholesky factorization. The memory optimal version is very expensive due to the high number of data transfers occurring between hosts and devices. The communication optimal or data persistence techniques trigger a considerable boost in the overall performance, especially for single precision arithmetic. The integration of the hybrid kernel (i.e., MAGMA's xPOTRF) to accelerate the execution of tasks located on the critical path improves further the performance. To our surprise, we did not see any improvements between the synchronous and the asynchronous version. Most probably this feature is not yet handled efficiently at the level of the driver. Finally, the additional optimizations performed on the other MAGMA BLAS kernels (i.e., MAGMA BLAS's xSYRK, xTRSM, xGEMM) make the Cholesky factorization reach up to 1.163 Tflop/s for single and 275 Gflop/s for double precision arithmetic. Compared with the performance of the embarrassingly parallel xGEMM over four GPUs, i.e. $400 \times 4 = 1.6$ Tflop/s for single precision (58% of the theoretical peak of the NVIDIA card) and $82 \times 4 = 328$ Gflop/s for double precision arithmetic (95% of the theoretical peak of the NVIDIA card), our algorithm runs correspondingly at 73% and 84%. Figure 2 highlights the scalable speed-up of the tile hybrid Cholesky factorization using four CPUs - four GPUs in single and double precision arithmetics. The performance doubles as the number of CPU-GPU couples doubles.

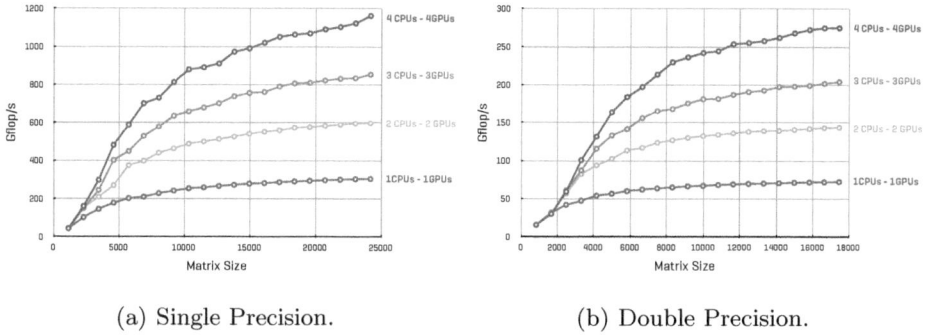

(a) Single Precision.   (b) Double Precision.

Fig. 2. Speed up of the tile hybrid Cholesky factorization

4   Related Work

Several authors have presented work on multiGPU algorithms for dense linear algebra. Volkov and Demmel [3] presented an LU factorization for two GPUs (NVIDIA GTX 280) running at up to 538 GFlop/s in single precision. The algorithm uses 1-D block cyclic partitioning of the matrix between the GPUs and achieves 74% improvement *vs* using just one GPU. Although extremely impressive, it is not clear if this approach will scale for more GPUs, especially by taking into account that the CPU work and the CPU-GPU bandwidth will not scale (and actually will remain the same with more GPUs added).

Closer in spirit to our work is [11]. The authors present a Cholesky factorization and its performance on a Tesla S1070 (as we do) and a host that is much more powerful than ours (two Intel Xeon Quad-Core E5440 @2.83 GHz). It is interesting to compare with this work because the authors, similarly to us, split the matrix into tiles and schedule the corresponding tasks using a dynamic scheduling. Certain optimizations techniques are applied but the best performance obtained is only close to our memory optimal version, which is running three times slower compared to our best version. The algorithm presented in here performs better for a set of reasons, namely the data persistence optimization techniques along with the efficiency of our static scheduler, the integration of the hybrid kernel, and the overall optimizations of the other GPU kernels.

5   Summary and Future Work

This paper shows how to redesign the Cholesky factorization to greatly enhance its performance in the context of multicore with GPU accelerators systems. It initially achieves up to 20 GFlop/s in single and up to 10 GFlop/s in double precision arithmetic by using only two dual-core 1.8 GHz AMD Opteron processors. Adding four GPUs and redesigning the algorithm accelerates the computation up to 65× and 27× for single and double precision arithmetic respectively. This acceleration is due to a design that enables efficient cooperation between the four Opteron cores and the four NVIDIA GPUs (30 cores per GPU, @1.44 GHz per core). By reusing concepts developed in the PLASMA and MAGMA libraries along with data persistence techniques, we achieve an astounding performance of $1,163$ TFlop/s in single and 275 GFlop/s in double precision arithmetic. Compared with the performance of the embarrassingly parallel xGEMM over four GPUs, where no communication between GPUs are involved, our algorithm still runs at 73% and 84% for single and double precision arithmetic respectively. Although this paper focused only on the Cholesky factorization, a full high-performance linear solver is possible [12]. This hybrid algorithm will eventually be included in the future release of MAGMA. Future work includes the extension to LU and QR factorizations.

References

1. NVIDIA CUDA Compute Unified Device Architecture - Programming Guide (2007), http://developer.download.nvidia.com
2. NVIDIA CUDA ZONE, http://www.nvidia.com/object/cuda_home.html
3. Volkov, V., Demmel, J.: Benchmarking GPUs to tune dense linear algebra. In: Proc. of SC 2008, Piscataway, NJ, USA, pp. 1–11 (2008)
4. Tomov, S., Dongarra, J.: Accelerating the reduction to upper Hessenberg form through hybrid GPU-based computing. LAPACK Working Note 219 (May 2009)
5. CUDA CUBLAS Library, http://developer.download.nvidia.com
6. Agullo, E., Dongarra, J., Hadri, B., Kurzak, J., Langou, J., Langou, J., Ltaief, H., Luszczek, P., YarKhan, A.: PLASMA version 2.0 user guide (2009), http://icl.cs.utk.edu/plasma
7. Kurzak, J., Buttari, A., Dongarra, J.J.: Solving systems of linear equations on the CELL processor using Cholesky factorization. IEEE Transactions on Parallel and Distributed Systems 19(9), 1–11 (2008)
8. Baboulin, M., Dongarra, J., Tomov, S.: Some issues in dense linear algebra for multicore and special purpose architectures. LAPACK Working Note 200 (May 2008)
9. Tomov, S., Nath, R., Du, P., Dongarra, J.: MAGMA version 0.2 User Guide (November 2009), http://icl.cs.utk.edu/magma

10. Li, Y., Dongarra, J., Tomov, S.: A note on auto-tuning GEMM for gPUs. In: Allen, G., Nabrzyski, J., Seidel, E., van Albada, G.D., Dongarra, J., Sloot, P.M.A. (eds.) ICCS 2009. LNCS, vol. 5544, pp. 884–892. Springer, Heidelberg (2009)
11. Ayguadé, E., Badia, R., Igual, F., Labarta, J., Mayo, R., Quintana-Ortí, E.: An extension of the starSs programming model for platforms with multiple gPUs. In: Sips, H., Epema, D., Lin, H.-X. (eds.) Euro-Par 2009. LNCS, vol. 5704, pp. 851–862. Springer, Heidelberg (2009)
12. Tomov, S., Nath, R., Ltaief, H., Dongarra, J.: Dense Linear Algebra Solvers for Multicore with GPU Accelerators. In: Proceedings of IPDPS 2010, Atlanta, GA (April 2010)

On the Performance of an Algebraic Multigrid Solver on Multicore Clusters*

Allison H. Baker, Martin Schulz, and Ulrike M. Yang

Center for Applied Scientific Computing
Lawrence Livermore National Laboratory
P.O. Box 808, L-560, Livermore, CA 94551, USA
{abaker,schulzm,umyang}@llnl.gov

Abstract. Algebraic multigrid (AMG) solvers have proven to be extremely efficient on distributed-memory architectures. However, when executed on modern multicore cluster architectures, we face new challenges that can significantly harm AMG's performance. We discuss our experiences on such an architecture and present a set of techniques that help users to overcome the associated problems, including thread and process pinning and correct memory associations. We have implemented most of the techniques in a MultiCore SUPport library (MCSup), which helps to map OpenMP applications to multicore machines. We present results using both an MPI-only and a hybrid MPI/OpenMP model.

1 Motivation

Solving large sparse systems of linear equations is required by many scientific applications, and the AMG solver in *hypre* [14], called BoomerAMG [13], is an essential component of simulation codes at Livermore National Laboratory (LLNL) and elsewhere. The implementation of BoomerAMG focuses primarily on distributed memory issues, such as effective coarse grain parallelism and minimal inter-processor communication, and, as a result, BoomerAMG demonstrates good weak scalability on distributed memory machines, as demonstrated for weak scaling on BG/L using 125,000 processors [11].

Multicore clusters, however, present new challenges for libraries such as *hypre*, caused by the new node architectures: multiple processors each with multiple cores, sharing caches at different levels, multiple memory controllers with affinities to a subset of the cores, as well as non-uniform main memory access times. In order to overcome these new challenges, the OS and runtime system must map the application to the available cores in a way that reduces scheduling conflicts, avoids resource contention, and minimizes memory access times. Additionally, algorithms need to have good data locality at the micro and macro level, few synchronization conflicts, and increased fine-grain parallelism [4]. Unfortunately,

---

* This work was performed under the auspices of the U.S. Department of Energy by Lawrence Livermore National Laboratory under contract DE-AC52-07NA27344 (LLNL-CONF-429864).

J.M.L.M. Palma et al. (Eds.): VECPAR 2010, LNCS 6449, pp. 102–115, 2011.

sparse linear solvers for structured, semi-structured and unstructured grids do not naturally exhibit these desired properties. Krylov solvers, such as GM-RES and conjugate gradient (CG), comprise basic linear algebra kernels: sparse matrix-vector products, inner products, and basic vector operations. Multigrid methods additionally include more complicated kernels: smoothers, coarsening algorithms, and the generation of interpolation, restriction and coarse grid operators. Various recent efforts have addressed performance issues of some of these kernels for multicore architectures. While good results have been achieved for dense matrix kernels [1,21,5], obtaining good performance for sparse matrix kernels is a much bigger challenge [20,19]. In addition, efforts have been made to develop cache-aware implementations of multigrid smoothers [9,15], which, while not originally aimed at multicore computers, have inspired further research for such architectures [18,12].

Little attention has been paid to effective core utilization and to the use of OpenMP in AMG in general, and in BoomerAMG in particular. However, with rising numbers of cores per node, the traditional MPI-only model is expected to be insufficient, both due to limited off-node bandwidth that cannot support ever-increasing numbers of endpoints, and due to the decreasing memory per core ratio, which limits the amount of work that can be accomplished in each coarse grain MPI task. Consequently, hybrid programming models, in which a subset of or all cores on a node will have to operate through a shared memory programming model (like OpenMP), will become commonplace.

In this paper we present a comprehensive performance study of AMG on a large multicore cluster at LLNL and present solutions to overcome the observed performance bottlenecks. In particular, we make the following contributions:

- A performance study of AMG on a large multicore cluster with 4-socket, 16-core nodes using MPI, OpenMP, and hybrid programming;
- Scheduling strategies for highly asynchronous codes on multicore platforms;
- A MultiCore SUPport (MCSup) library that provides efficient support for mapping an OpenMP program onto the underlying architecture;
- A demonstration that the performance of AMG on the coarsest grid levels can have a significant effect on scalability.

Our results show that both the MPI and the OpenMP version suffer from severe performance penalties when executed on our multicore target architecture without optimizations. To avoid the observed bottlenecks we must pin MPI tasks to processors and provide a correct association of memory to cores in OpenMP applications. Further, a hybrid approach shows promising results, since it is capable of exploiting the scaling sweet spots of both programming models.

2   The Algebraic Multigrid (AMG) Solver

Multigrid methods are popular for large-scale scientific computing because of their algorithmically scalability: they solve a sparse linear system with n unknowns with $O(n)$ computations. Multigrid methods obtain the $O(n)$ optimality by utilizing a sequence of smaller linear systems, which are less expensive to compute on, and by capitalizing on the ability of inexpensive smoothers (e.g., Gauss-Seidel) to resolve high-frequency errors on each grid level. In particular, because multigrid is an iterative method, it begins with an estimate to the solution on the fine grid. Then at each level of the grid, a smoother is applied, and the improved guess is transferred to a smaller, or coarser, grid. On the coarser grid, the smoother is applied again, and the process continues. On the coarsest level, a small linear system is solved, and then the solution is transferred back up to the fine grid via interpolation operators. Good convergence relies on the smoothers and the coarse-grid correction process working together in a complimentary manner.

AMG is a particular multigrid method that does not require an explicit grid geometry. Instead, coarsening and interpolation processes are determined entirely based on matrix entries. This attribute makes the method flexible, as often actual grid information may not be available or may be highly unstructured. However, the flexibility comes at a cost: AMG is a rather complex algorithm.

We use subscripts to indicate the AMG level numbers for the matrices and superscripts for the vectors, where 1 denotes the finest level, so that $A_1 = A$ is the matrix of the original linear system to be solved, and m denotes the coarsest level. AMG requires the following components: grid operators $A_1, \ldots, A_m$, interpolation operators P_k, restriction operators R_k (here we use $R_k = (P_k)^T$), and smoothers S_k, where $k = 1, 2, \ldots m - 1$. These components of AMG are determined in a first step, known as the *setup phase*. During the setup phase, on each level k, the variables to be kept for the next coarser level are determined using a coarsening algorithm, P_k and R_k are defined, and the coarse grid operator is computed: $A_{k+1} = R_k A_k P_k$.

Once the setup phase is completed, the *solve phase*, a recursively defined cycle, can be performed as follows, where $f^{(1)} = f$ is the right-hand side of the linear system to be solved and $u^{(1)}$ is an initial guess for u:

> **Algorithm:** $MGV(A_k, R_k, P_k, S_k, u^{(k)}, f^{(k)})$.
>   If $k = m$, solve $A_m u^{(m)} = f^{(m)}$.
>   Otherwise:
>   Apply smoother S_k μ_1 times to $A_k u^{(k)} = f^{(k)}$.
>   Perform coarse grid correction:
>   Set $r^{(k)} = f^{(k)} - A_k u^{(k)}$.
>   Set $r^{(k+1)} = R_k r^{(k)}$.
>   Set $e^{(k+1)} = 0$.
>   Apply $MGV(A_{k+1}, R_{k+1}, P_{k+1}, S_{k+1}, e^{(k+1)}, r^{(k+1)})$.
>   Interpolate $e^{(k)} = P_k e^{(k+1)}$.
>   Correct the solution by $u^{(k)} \leftarrow u^{(k)} + e^{(k)}$.
>   Apply smoother S_k μ_2 times to $A_k u^{(k)} = f^{(k)}$.

The algorithm above describes a V(μ_1, μ_2)-cycle; other more complex cycles such as W-cycles are described in [3].

Determining appropriate coarse grids is non-trivial, particularly in parallel, where processor boundaries require careful treatment (see, e.g.,[6]). In addition, interpolation operators often require a fair amount of communication to determine processor neighbors (and neighbors of neighbors) [7]. The setup phase time is non-trivial and may cost as much as multiple iterations in the solve phase. The solve phase performs the multilevel iterations (often referred to as cycles). These iterations consist primarily of applying the smoother, restricting the error to the coarse-grid, and interpolating the error to the fine grid. These operations are all matrix-vector multiplications (MatVecs) or MatVec-like, in the case of the smoother. An overview of AMG can be found in [11,17,3].

For the results in this paper, we used a modification of the BoomerAMG code in the *hypre* software library. We chose one of our best performing options: HMIS coarsening [8], one level of aggressive coarsening with multipass interpolation [17], and extended+i(4) interpolation [7] on the remaining levels. Since AMG is generally used as a preconditioner, we investigate it as a preconditioner for GMRES(10).

The results in this paper focus on the solve phase (since this can be completely threaded), though we will also present some total times (setup + solve times). Note that because AMG is a fairly complex algorithm, each individual component (e.g., coarsening, interpolation, and smoothing) affects the convergence rate. In particular, the parallel coarsening algorithms and the hybrid Gauss-Seidel parallel smoother, which uses sequential Gauss-Seidel within each task and delayed updates across cores, are dependent on the number of tasks, and the partitioning of the domain. Since the number of iterations can vary based on the experimental setup, we rely on average cycle times (instead of the total solve time) to ensure a fair comparison.

BoomerAMG uses a parallel matrix data structure. Matrices are distributed across cores in contiguous block of rows. On each core, the matrix block is split into two parts, each of which are stored in compressed sparse row (CSR) format. The first part contains the coefficients local to the core, whereas the second part contains the remaining coefficients. The data structure also contains a mapping that maps the local indices of the off-core part to global indices as well as information needed for communication. A complete description of the data structure can be found in [10].

Our test problem is a 3D Laplace problem with a seven-point stencil generated by finite differences, on the unit cube, with $100 \times 100 \times 100$ grid points per node. Note that the focus of this paper is a performance study of AMG on a multicore cluster, and not a convergence study, which would require a variety of more difficult test problems. This test problem, albeit simple from a mathematical point of view, is sufficient for its intended purpose. While the matrix on the finest level has only a seven-point stencil, stencil sizes as well as the overall density of the matrix increase on the coarser levels. We therefore encounter various scenarios that can reveal performance issues, which would also be present in more complex test problems.

3   The Hera Multicore Cluster

We conduct our experiments on Hera, a multicore cluster installed at LLNL with 864 nodes interconnected by Infiniband. Each node consists of four AMD Quad-core (8356) 2.3 GHz processors. Each core has its own L1 and L2 cache, but four cores share a 2 MB L3 cache. Each processor provides its own memory controller and is attached to a fourth of the 32 GB memory per node. Despite this separation, a core can access any memory location: accesses to memory locations served by the memory controller on the same processor are satisfied directly, while accesses through other memory controllers are forwarded through the Hypertransport links connecting the four processors. This leads to non-uniform memory access (NUMA) times depending on the location of the memory.

Each node runs CHAOS 4, a high-performance computing Linux variant based on Redhat Enterprise Linux. All codes are compiled using Intel's C and OpenMP/C compiler (Version 11.1). We rely on MVAPICH over IB as our MPI implementation and use SLURM [16] as the underlying resource manager. Further, we use SLURM in combination with an optional affinity plugin, which uses Linux's NUMA control capabilities to control the location of processes on sets of cores. The impact of these settings are discussed in Section 4.

4   Using an MPI-Only Model with AMG

As mentioned in Section 1, the BoomerAMG solver is highly scalable on the Blue Gene class of machines using an MPI-only programming model. However, running the AMG solver on the Hera cluster using one MPI task for each of the 16 cores per node yields dramatically different results (Figure 1). Here the problem size is increased in proportion to the number of cores (using $50\times50\times25$ grid points per core), and BG/L shows nearly perfect weak scalability with almost constant execution times for any number of nodes for both total times and cycle times. On Hera, despite having significantly faster cores, overall scalability is severely degraded, and execution times are drastically longer for large jobs.

To investigate this observation further we first study the impact of affinity settings on the AMG performance, which we influence using the before mentioned affinity plugin loaded as part of the SLURM resource manager. The black line in Figure 2 shows the performance of the AMG solve phase for a single cycle on 1, 64, and 216 nodes with varying numbers of MPI tasks per node without affinity optimizations (Aff=16/16 meaning that each of the 16 tasks has equal access to all 16 cores). The problem uses $100\times100 \times100$ grid points per node. Within a node we partition the domain into cuboids so that communication between cores is minimized, e.g., for 10 MPI tasks the subdomain per core consists of $100\times50\times20$ grid points, whereas for 11 MPI tasks the subdomains are of size $100\times100\times10$ or $100\times100\times9$, leading to decreased performance for the larger prime numbers. From these graphs we can make two observations: the performance generally increases for up to six MPI tasks per node; adding more tasks is counterproductive. Second, this effect is growing with the number of

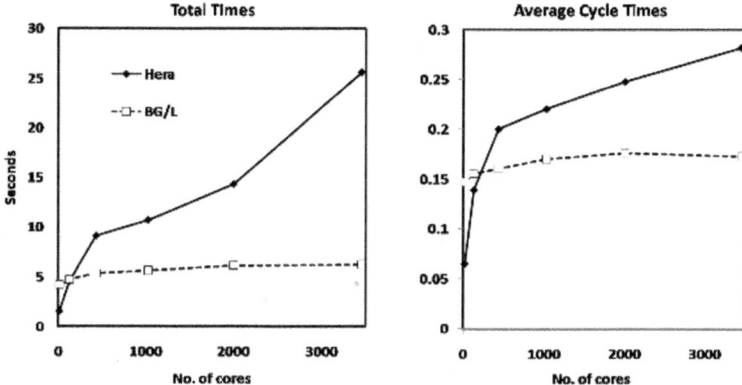

Fig. 1. Total times, including setup and solve times, (left) and average times per iteration (right) for AMG-GMRES(10) using MPI only on BG/L and Hera. Note that the setup phase scales much worse on Hera than the solve phase.

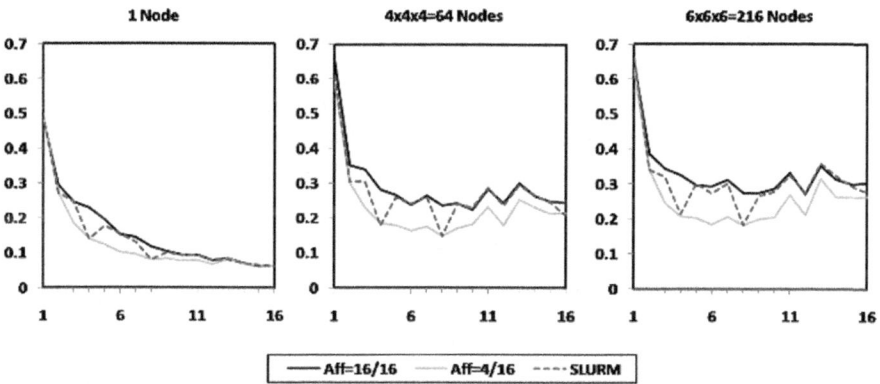

Fig. 2. Average times in seconds per AMG-GMRES(10) cycle for varying numbers of MPI tasks per node

nodes. While for a single node, the performance only stagnates, the solve time increases for large node counts. These effects are caused by a combination of local memory pressure and increased pressure on the internode communication network.

Additionally, the performance of AMG is impacted by affinity settings: while the setting discussed so far (Aff=16/16) provides the OS with the largest flexibility for scheduling the tasks, it also means that a process can migrate between cores and with that also between processors. Since the node architecture based on the AMD Opteron chip uses separate memory controllers for each processor, this means that a process, after it has been migrated to a different processor, must satisfy all its memory requests by issuing remote memory accesses. The consequence is a drastic loss in performance. However, if the set of cores that an MPI task can be executed on is fixed to only those within a processor, then we leave the OS with the flexibility to schedule among multiple cores, yet eliminate cross-processor migrations. This choice results in significantly improved performance (gray, solid line marked Aff=4/16). Additional experiments have further shown that restricting the affinity further to a fixed core for each MPI task is ineffective and leads to poor performance similar to Aff=16/16.

It should be noted that SLURM is already capable of applying this optimization for selected numbers of tasks, as indicated by the black dashed line in Figure 2, but a solution across all configurations still requires manual intervention. Note that for the remaining results in this paper optimal affinity settings were applied (either manually using command line arguments for SLURM's affinity plugin or automatically by SLURM itself).

5   Replacing On-Node MPI with OpenMP

The above observations clearly show that an MPI-only programming model is not sufficient for machines with wide multicore nodes, such as our experimental platform. Further, the observed trends indicate that this problem will likely get more severe with increasing numbers of cores. With machines on the horizon for the next few years that offer even more cores per node as well as more nodes, solving the observed problems is becoming critical. Therefore, we study the performance of BoomerAMG on the Hera cluster using OpenMP and MPI.

5.1   The OpenMP Implementation

Here we describe in more detail the OpenMP implementation within Boomer-AMG. OpenMP is generally employed at the loop level. In particular for m OpenMP threads, each loop is divided into m parts of approximately equal size. For most of the basic matrix and vector operations, such as the MatVec or dot product, the OpenMP implementation is straight-forward. However, the use of OpenMP within the highly sequential Gauss-Seidel smoother requires an algorithm change. Here we use the same technique as in the MPI implementation, i.e., we use sequential Gauss-Seidel within each OpenMP thread and delayed updates for those points belonging to other OpenMP threads. In addition, because the parallel matrix data structure essentially consists of two matrices in CSR storage format, the OpenMP implementation of the multiplication of the transpose of the matrix with a vector is less efficient than the corresponding MPI implementation; it requires a temporary vector to store the partial matrix-vector product within each OpenMP thread and the subsequent summation of these vectors.

Overall, the AMG solve phase, including GMRES, is completely threaded, whereas in the setup phase, only the generation of the coarse grid operator (a triple matrix product) has been threaded. Both coarsening and interpolation do not contain any OpenMP statements.

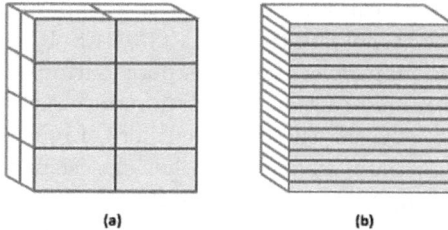

Fig. 3. Two partitionings of a cube into 16 subdomains on a single node of Hera. The partitioning on the left is optimal, and the partitioning on the right is the partitioning used for OpenMP.

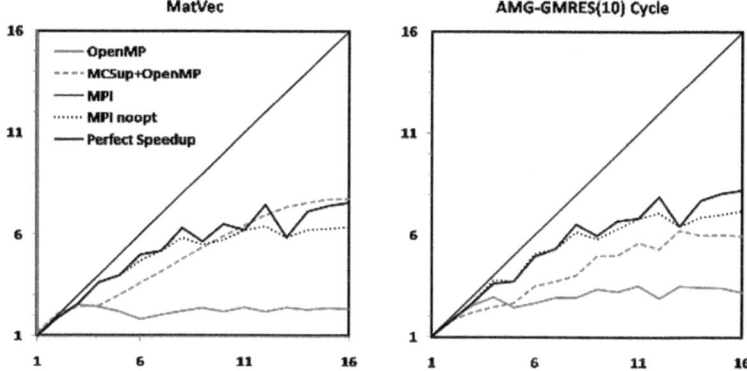

Fig. 4. Speedup for the MatVec kernel and a cycle of AMG-GMRES(10) on a single node of Hera

Note that, in general, the partitioning used for the MPI implementation is not identical to that of the OpenMP implementation. Whereas we attempt to optimize the MPI implementation to minimize communication (see Figure 3(a)), for OpenMP the domain of the MPI task is sliced into m parts due to the loop-level parallelism, leading to a less optimal partitioning (see Figure 3(b)). Therefore, Figure 4 (discussed in Section 5.2) also contains timings for MPI using the less-optimal partitioning (Figure 3(b)), denoted 'MPI noopt', which allows a comparison of MPI and OpenMP with the same partitioning.

5.2 Optimizing Memory Behavior with MCSup

The most time intensive kernels, the sparse MatVec and the smoother, account for 60% and 30%, respectively, of the solve time. Since these two kernels are similar in terms of implementation and performance behavior, we focus our investigation on the MatVec kernel. The behavior of the MatVec kernel closely matches the performance of the full AMG cycle on a single node. Figure 4 shows the initial performance of the OpenMP version compared to MPI in terms of speedup for the MatVec kernel and the AMG-GMRES(10) cycle on a single node of Hera (16 cores). The main reason for this poor performance lies in the code's memory behavior and its interaction with the underlying system architecture.

On NUMA systems, such as the one used here, Linux's default policy is to allocate new memory to the memory controller closest to the executing thread. In the case of the MPI application, each rank is a separate process and hence allocates its own memory to the same processor. In the OpenMP case, though, all memory gets allocated and initialized by the master thread and hence is pushed onto a single processor. Consequently, this setup leads to long memory access times, since most accesses will be remote, as well as memory contention on the memory controller responsible for all pages. Additionally, the fine-grain nature of threads make it more likely for the OS to migrate them, leading to unpredictable access times.

Note that in this situation even a first-touch policy, implemented by some NUMA-aware OS and OS extensions, would be insufficient. Under such a policy, a memory page would be allocated on a memory close to the core that first uses (typically writes) to it, rather than to the core that is used to allocate it. However, in our case, memory is often also initialized by the master thread, which still leads to the same locality problems. Further, AMG's underlying library *hypre* frequently allocates and deallocates memory to avoid memory leakage across library routine invocations. This causes the heap manager to reuse previously allocated memory for subsequent allocations. Since this memory has already been used/touched before, its location is now fixed and a first touch policy is no longer effective.

To overcome these issues, we developed MCSup (MultiCore SUPport), an OpenMP add-on library capable of automatically co-locating threads with the memory they are using. It performs this in three steps: first MCSup probes the memory and core structure of the node and determines the number of cores and memory controllers. Additionally, it determines the maximal concurrency used by the OpenMP environment and identifies all available threads. In the second step, it pins each thread to a processor to avoid later migrations of threads between processors, which would cause unpredictable remote memory accesses.

For the third and final step, it provides the user with new memory allocation routines that they can use to indicate which memory regions will be accessed globally and in what pattern. MCSup then ensures that the memory is distributed across the node in a way that memory is located locally to the threads most using it. This is implemented using Linux's NUMAlib, a set of low-level routines that provide fine-grain control over page and thread placements.

5.3   Optimized OpenMP Performance

Using the new memory and thread scheme implemented by MCSup greatly improves the performance of the OpenMP version of our code, as shown in Figure 4. The performance of the 16 OpenMP thread MatVec kernel improved by a factor of 3.5, resulting in comparable single node performance for OpenMP and MPI.

Note that when using the same partitioning the OpenMP+MCSup version of the MatVec kernel shows superior performance than the MPI version for 8 or more threads. Also the performance of the AMG-GMRES(10) cycle improves significantly. However, in this case using MPI tasks instead of threads still results in better performance on a single node. The slower performance is primarily caused by the less efficient OpenMP version of the multiplication of the transpose of the matrix with a vector.

6   Mixed Programming Model

Due to the apparent shortcomings of both MPI- and OpenMP-only programming approaches, we next investigate the use of a hybrid approach allowing us to utilize the scaling sweet spots for both programming paradigms and present early results. Since we want to use all cores, we explore all combinations with m MPI processes and n OpenMP threads per process with $m * n = 16$ within a node. MPI is used across nodes. Figure 5 shows total times and average cycle times for various combinations of MPI with OpenMP. Note, that since the setup phase of AMG is only partially threaded, total times for combinations with large number of OpenMP threads such as OpenMP or MCSup are expected to be worse, but they outperform the MPI-only version for 125 and 216 nodes. While MCSup outperforms native OpenMP, its total times are generally worse than the hybrid tests. However when looking at the cycle times, its overall performance is comparable to using 8 MPI tasks with 2 OpenMP threads (Mix 8×2) or 2 MPI tasks with 8 OpenMP threads (Mix 2×8) on 27 or more nodes. Mix 2×8 does not use MCSup, since this mode is not yet supported, and therefore shows a similar, albeit much reduced, memory contention than OpenMP. In general, the best performance is obtained for Mix 4×4, which indicates that using a single MPI task per socket with 4 OpenMP threads is the best strategy.

7   Investigating the MPI-Only Performance Degradation

Conventional wisdom for multigrid is that the largest amount of work and, consequently, the most time is spent on the finest level. This also coincides with our previous experience on closely coupled large-scale machines such as Blue Gene/L, and hence we expected that the performance and scalability of a version of the AMG preconditioner restricted to just two levels is similar to that of the multilevel version. However, our experiments on the Hera cluster show a different result.

The left plot on the bottom of Figure 5 illustrates that on two levels the MPI-only version performs as well as Mix 8×2 and Mix 4×4, which indicates that the performance degradation within AMG for the MPI-only model occurs on one or more of the lower levels. The right plots in Figure 5 confirm that, while the MPI-only version shows good scalable performance on two levels, its overall time is increasing much more rapidly than the other versions with increasing numbers of levels. While both OpenMP and MCSup do not appear to be significantly

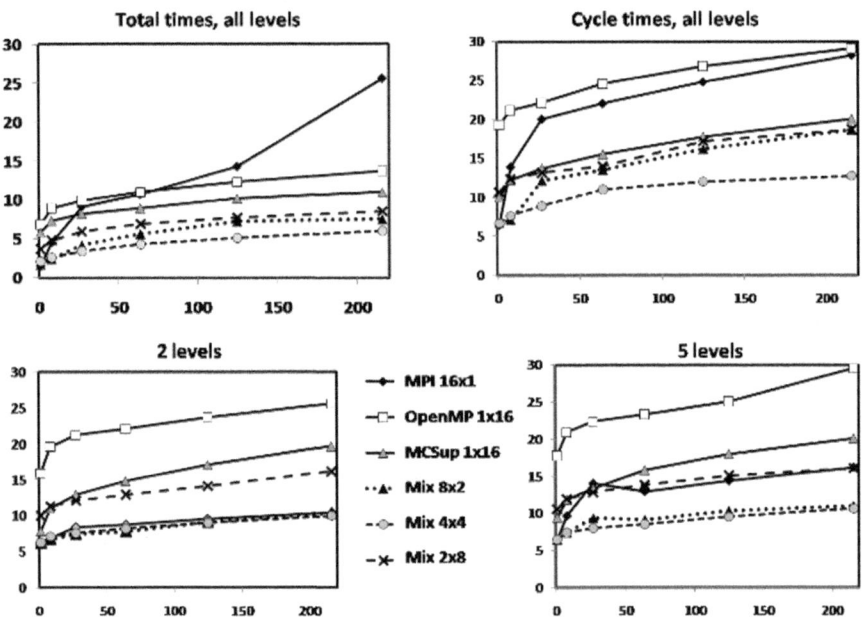

Fig. 5. Total times (setup + solve phase) in seconds of AMG-GMRES(10) (top left) and times in seconds for 100 AMG-GMRES(10) cycles (top right) using all levels (7 to 9) of AMG. Times for 100 cycles using two (bottom left) or five (bottom right) levels only. '$m \times n$' denotes m MPI tasks and n OpenMP threads per node.

affected by varying the number of levels, performance for the variants that use more than one MPI task per node decreases (the Mix 4×4 case is least affected). We note that while we have only shown the degradation in MPI-only performance with increasing numbers of levels for the solve phase, the effect is even more pronounced in the setup phase.

To understand the performance degradation for the MPI-only version on coarser levels, we must first consider the difference in the work done at the finer and coarser levels. In general, on the fine grid the matrix stencils are smaller (our test problem is a seven-point stencil on the finest grid), and the matrices are sparser. Neighbor processors, with which communication is necessary, are generally fewer and "closer" in terms of process ranks and messages passed between processors are larger in size. As the grid is coarsened, processors own fewer rows in the coarse grid matrices, eventually owning as little as a single row or even no rows at all on the coarsest grids.[1] On the coarsest levels there is very little computational work to be done, and the messages that are sent are generally small. However, because there are few processes left, the neighbor

---

[1] When all levels are generated, the AMG algorithm coarsens such that the coarsest matrix has fewer than nine rows.

processes may be farther away in terms of process ranks. The mid-range levels are a mix of all effects and are difficult to categorize. All processors will remain at the mid-levels, but the stencil is likely bigger, which increases the number of neighbors. Figure 6 shows the total communication volume (setup and solve phase) collected with TAU/ParaProf [2] in terms of number of messages sent between pairs of processes on 128 cores (8 nodes) of Hera using the MPI-only version of AMG. From left to right in the figure, the number of AMG levels is restricted to 4, 6, and 8 (all) levels, respectively. Note that the data in these plots is cumulative, e.g., the middle 6-level plot contains the data from the left 4-level plot, plus the communication totals from levels 5 and 6. The fine grid size for this problem is 8,000,000 unknowns. The coarsest grid size with the 4, 6, and 8 levels is 13643, 212, and 3 unknowns, respectively.

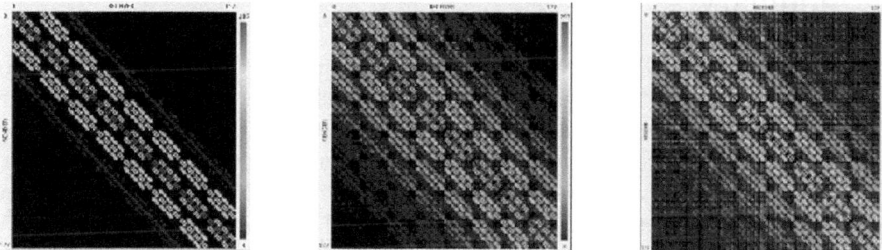

Fig. 6. Communication matrices indicating the total number of communications between pairs of 128 cores on 8 nodes. The x-axis indicates the id of the receiving MPI task, the the y-axis indicates the id of the sender. Areas of black indicate zero messages between cores. From left to right, results are shown for restricting AMG to 4, 6, and 8 (all) levels, respectively.

These figures show a clear difference in the communication structure in different refinement levels. For 4 levels we see a very regular neighborhood communication pattern with very little additional communication off the diagonal (black areas on the top/right and bottom/left). However, on the coarser levels, the communication added by the additional levels becomes more arbitrary and long-distance, and on the right-most plot with 8 levels of refinement, the communication has degraded to almost random communication. Since our resource manager SLURM generally assigns process ranks that are close together to be physically closer on the machine (i.e., processes 0-15 are one a node, processes 16-31 are on the next node, etc.), we benefit from regular communication patterns like we see in the finer levels. The more random communication in coarser levels, however, will cause physically more distant communication as well as the use of significantly more connection pairs, which need to be initialized. The underlying Infiniband network used on Hera is not well suited for this kind of communication due to its fat tree topology and higher cost to establish connection pairs. The latter is of particular concern in this case since these connections are short lived and only used to exchange very little communication and hence the setup overhead can no longer be fully amortized.

When comparing the setup and the solve phase, we notice that the solve phase is less impacted by the performance degradation. While the setup phase touches each level only once, the solve phase visits each level at least twice (except the coarsest) in each iteration. This enables some reuse of communication pairs and helps to amortize the associated overhead.

We note that on more closely coupled machines, such as Blue Gene/L with a more even network topology and faster communication setup mechanisms, we don't see this degradation. Further the degradation is less in the hybrid OpenMP case, since fewer MPI tasks, and with that communication end points as well as communication pairs, are involved.

8   Summary

Although the *hypre* AMG solver scales well on distributed-memory architectures, obtaining comparable performance on multicore clusters is challenging. Here we described some of the issues we encountered in adapting our code for multicore architectures and make several suggestions for improving performance. In particular, we greatly improved OpenMP performance by pinning threads to specific cores and allocating memory that the thread will access on that same core. We also demonstrated that a mixed model of OpenMP threads and MPI tasks on each node results in superior performance. However, many open questions remain, particularly those specific to the AMG algorithm. In the future, we plan to more closely examine kernels specific to the setup phase and include OpenMP threads in those that have not been threaded yet. We will also explore the use of new data structures.

References

1. Asanovic, K., Bodik, R., Catanzaro, B., Gebis, J., Husbands, P., Keutzer, K., Patterson, D., Plishker, W., Shalf, J., Williams, S., Yelick, K.: The landscape of parallel computing research: A view from Berkeley (2006)
2. Bell, R., Malony, A.D., Shende, S.S.: *ParaProf*: A Portable, Extensible, and Scalable Tool for Parallel Performance Profile Analysis. In: Kosch, H., Böszörményi, L., Hellwagner, H. (eds.) Euro-Par 2003. LNCS, vol. 2790, pp. 17–26. Springer, Heidelberg (2003)
3. Briggs, W., Henson, V., McCormick, S.: A Multigrid Tutorial, 2nd edn. SIAM, Philadelphia (2000)
4. Buttari, A., Dongarra, J., Kurzak, J., Langou, J., Luszczek, P., Tomov, S.: The impact of multicore on math software. In: Kågström, B., Elmroth, E., Dongarra, J., Waśniewski, J. (eds.) PARA 2006. LNCS, vol. 4699, pp. 1–10. Springer, Heidelberg (2007)
5. Buttari, A., Lusczek, P., Kurzak, J., Dongarra, J., Bosilca, G.: A rough guide to scientific computing on the PlayStation 3 (2007)

6. Chow, E., Falgout, R., Hu, J., Tuminaro, R., Yang, U.: A survey of parallelization techniques for multigrid solvers. In: Heroux, M., Raghavan, P., Simon, H. (eds.) Parallel Processing for Scientific Computing. SIAM Series on Software, Environments, and Tools (2006)

7. De Sterck, H., Falgout, R.D., Nolting, J., Yang, U.M.: Distance-two interpolation for parallel algebraic multigrid. Num. Lin. Alg. Appl. 15, 115–139 (2008)

8. De Sterck, H., Yang, U.M., Heys, J.: Reducing complexity in algebraic multigrid preconditioners. SIMAX 27, 1019–1039 (2006)

9. Douglas, C., Hu, J., Kowarschik, M., Ruede, U., Weiss, C.: Cache optimization for structures and unstructured grid multigrid. Electronic Transactions on Numerical Analysis 10, 21–40 (2000)

10. Falgout, R., Jones, J., Yang, U.M.: Pursuing scalability for hypre's conceptual interfaces. ACM ToMS 31, 326–350 (2005)

11. Falgout, R.D.: An introduction to algebraic multigrid. Computing in Science and Eng. 8(6), 24–33 (2006)

12. Garcia, C., Prieto, M., Setoain, J., Tirado, F.: Enhancing the performance of multigrid smoothers in simultaneous multithreading architectures. In: Daydé, M., Palma, J.M.L.M., Coutinho, Á.L.G.A., Pacitti, E., Lopes, J.C. (eds.) VECPAR 2006. LNCS, vol. 4395, pp. 439–451. Springer, Heidelberg (2007)

13. Henson, V.E., Yang, U.M.: BoomerAMG: a parallel algebraic multigrid solver and preconditioner. Applied Numerical Mathematics 41, 155–177 (2002)

14. *hypre*. High performance preconditioners,
    http://www.llnl.gov/CASC/linear_solvers/

15. Kowarschik, M., Christadler, I., Rüde, U.: Towards cache-optimized multigrid using patch-adaptive relaxation. In: Dongarra, J., Madsen, K., Waśniewski, J. (eds.) PARA 2004. LNCS, vol. 3732, pp. 901–910. Springer, Heidelberg (2006)

16. Lawrence Livermore National Laboratory. SLURM: Simple Linux Utility for Resource Management (June 2005), http://www.llnl.gov/linux/slurm/

17. Stüben, K.: An introduction to algebraic multigrid. In: Trottenberg, U., Oosterlee, C., Schüller, A. (eds.) Multigrid, pp. 413–532. Academic Press, London (2001)

18. Wallin, D., Loef, H., Hagersten, E., Holmgren, S.: Multigrid and Gauss-Seidel smoothers revisited: Parallelization on chip multiprocessors. In: Proceedings of ICS 2006, pp. 145–155 (2006)

19. Williams, S., Oliker, L., Vuduc, R., Shalf, J., Yelick, K.: Optimization of sparse matrix-vector multiplication on emerging multicore platforms. Parallel Computing 35, 178–194 (2009)

20. Williams, S., Oliker, L., Vuduc, R., Shalf, J., Yelick, K., Demmel, J.: Optimization of sparse matrix-vector multiplication on emerging multicore platforms. In: Proceedings of IEEE/ACM Supercomputing 2007 (2007)

21. Williams, S., Shalf, J., Oliker, L., Kamil, S., Husbands, P., Yelick, K.: Scientific computing kernels on the cell processor. International Journal of Parallel Programming (2007)

An Hybrid Approach for the Parallelization of a Block Iterative Algorithm

Carlos Balsa[1], Ronan Guivarch[2], Daniel Ruiz[2], and Mohamed Zenadi[2]

[1] CEsA–FEUP, Porto, Portugal
balsa@ipb.pt
[2] Université de Toulouse, INP (ENSEEIHT), IRIT
{guivarch,ruiz,mzenadi}@enseeiht.fr

Abstract. The Cimmino method is a row projection method in which the original linear system is divided into subsystems. At every iteration, it computes one projection per subsystem and uses these projections to construct an approximation to the solution of the linear system.

The usual parallelization strategy in block algorithms is to distribute the different blocks on the available processors. In this paper, we follow another approach where we do not perform explicitly this block distribution to processors within the code, but let the multi-frontal sparse solver MUMPS handle the data distribution and parallelism. The data coming from the subsystems defined by the block partition in the Block Cimmino method are gathered in an unique block diagonal sparse matrix which is analysed, distributed and factorized in parallel by MUMPS. Our target is to define a methodology for parallelism based only on the functionalities provided by general sparse solver libraries and how efficient this way of doing can be.

1   Introduction

The Cimmino method is a row projection method in which the original linear system is divided into subsystems. At each iteration, it computes one projection per subsystem and uses these projections to construct an approximation to the solution of the linear system. The Block-CG method can also be used within the Block Cimmino Iteration to accelerate its convergence. Therefore, we present an implementation of a parallel distributed Block Cimmino method where the Cimmino iteration matrix is used as a preconditioner for the Block-CG.

In this work we propose to investigate a non usual methodology for parallelization of the Block Cimmino method where the direct solver package MUMPS (MUltifrontal Massively Parallel sparse direct Solver [1,2]) is incorporated in order to schedule and perform most of the parallel tasks. This library offers to the user the facilities to call the different phases of the solver (analysis, factorization, solution) while taking care of the distribution of data and processes.

The outline of the paper is the following: the Block Cimmino Algorithm is described in section 2 and parallelization strategies are exposed in section 3. More details about our strategy are given in section 4. We finish by a presentation of

J.M.L.M. Palma et al. (Eds.): VECPAR 2010, LNCS 6449, pp. 116–128, 2011.

some preliminary numerical results in section 5 that give us some hints for future
improvements and developments.

2   Block Cimmino Algorithm

The Block Cimmino method is a generalization of the Cimmino method [3].
Basically, we partition the linear system of equations:

$$\mathbf{A}x = b, \tag{1}$$

where $\mathbf{A}$ is a $m \times n$ matrix, into l subsystems, with $l \leq m$, such that:

$$\begin{pmatrix} \mathbf{A}_1 \\ \mathbf{A}_2 \\ \vdots \\ \mathbf{A}_l \end{pmatrix} x = \begin{pmatrix} b_1 \\ b_2 \\ \vdots \\ b_l \end{pmatrix} \tag{2}$$

The block method [4] computes a set of l row projections, and a combination
of these projections is used to build the next approximation to the solution of
the linear system. Now, we formulate the Block Cimmino iteration as:

$$\delta_i^{(k)} = \mathbf{A}_i^+ b_i - \mathbf{P}_{\mathcal{R}(A_i^T)} x^{(k)} \tag{3}$$

$$= \mathbf{A}_i^+ \left(b_i - \mathbf{A}_i x^{(k)} \right)$$

$$x^{(k+1)} = x^{(k)} + \nu \sum_{i=1}^{l} \delta_i^{(k)} \tag{4}$$

In (3), the matrix $\mathbf{A}_i^+$ refers to the classical pseudo-inverse of $\mathbf{A}_i$ defined as:
$\mathbf{A}_i^+ = \mathbf{A}_i^T \left(\mathbf{A}_i \mathbf{A}_i^T \right)^{-1}$.

However, the Block Cimmino method will converge for any other pseudo-
inverse of $\mathbf{A}_i$ and in our parallel implementation we use a generalized pseudo-
inverse [6], $\mathbf{A}_{i\ \mathbf{G}^{-1}}^- = \mathbf{G}^{-1} \mathbf{A}_i^T \left(\mathbf{A}_i \mathbf{G}^{-1} \mathbf{A}_i^T \right)^{-1}$, were $\mathbf{G}$ is some symmetric and
positive definite matrix. The $\mathbf{P}_{\mathcal{R}(A_i^T)}$ is an orthogonal projector onto the range
of $\mathbf{A}_i^T$.

We use the augmented systems approach, as in [7] and [8], for solving the
subsystems (3)

$$\begin{bmatrix} \mathbf{G} & \mathbf{A}_i^T \\ \mathbf{A}_i & 0 \end{bmatrix} \begin{bmatrix} u_i \\ v_i \end{bmatrix} = \begin{bmatrix} 0 \\ b_i - \mathbf{A}_i x \end{bmatrix} \tag{5}$$

with solution:

$$v_i = - \left(\mathbf{A}_i \mathbf{G}^{-1} \mathbf{A}_i^T \right)^{-1} r_i$$

$$u_i = \mathbf{A}_{i\ \mathbf{G}^{-1}}^- (b_i - \mathbf{A}_i x) \tag{6}$$

$$= \delta_i$$

The Block Cimmino method is a linear stationary iterative method with a symmetrizable iteration matrix [9]. When A has full rank, the symmetrized iteration matrix is symmetric positive definite (SPD). We can accelerate its rate of convergence with the use of Block Conjugate Gradient method (Block-CG) [10,11].

3   Parallelization Strategy

The Block-Cimmino algorithm generates a set of fully independent blocks that are the basis for a first level of natural parallelism. As opposed to usual parallelization strategies that we will recall shortly in the next subsection, and that consist in distributing in some way each block onto processors, while having more blocks in general than processors to manage better load balancing, we introduce an other strategy (detailed in subsection 3.2) that resumes in gathering the blocks into a larger but single sparse linear system with a block diagonal structure, and handle all the levels of parallelism at the same time.

3.1   Manual Parallelism Description

As described in [12], a first, and natural strategy would be to distribute the blocks evenly on the available processors.

Blocks' distribution. It could be difficult to achieve a good load balancing since the distribution of the augmented systems blocks must be closely related to the number of processors. From a technical point we can consider the following issues:

- if there is a high number of processors, we can create the same number of blocks, but the speed of convergence of such block iterative algorithms is often slowed down when the number of blocks increases;
- if the blocks are too small, the cost of communication can become prohibitive relatively to the computations.

   In general, it's better to have more blocks than processors to have more degree of freedom for better load-balancing.

Parallelism exploitation. In a problem with different physical properties having blocks with different dimensions is a common thing, which may overload the processors supposed to handle them. Moreover the other processors will stall once they finish their work as there is a synchronization after the solve part for the projectors sum.

   Furthermore, we may not have a fully and unrestricted choice in the block partitioning of our matrix. Some problems involving different physical properties require a particular partitioning that have to be respected otherwise it may disturb the convergence rate.

3.2   Automatic Parallelism with MUMPS

In order to get rid of the problems noticed in the manual block distribution, we chose to use the direct sparse solver MUMPS (**MU**ltifrontal **M**assively **P**arallel sparse direct **S**olver)[1], which will handle the block distribution and the parallelism.

In our approach we create the block diagonal sparse matrix from the data coming from the subsystems defined by the block partition (2) in the Block Cimmino method. The matrix is then given to MUMPS for an in-depth analysis which will permit a fine grain parallelism handling while respecting the matrix structure and augmented systems' properties. The matrix is then distributed following the mapping generated by this analysis, and afterward factorized in parallel. At each Block-CG iteration, MUMPS solves in parallel the system involved during the preconditioning step.

The added advantage of this way of doing, is that the sparse linear solver will handle (without any extra development for us) all the levels of parallelism available, and in particular those coming from sparsity structure and BLAS3 kernels on top of the block partitioning. This also gives us more degrees of freedom when partitioning the matrix, with the possibility to define less blocks than processors but larger ones. This may help to increase the speed of convergence of the method while still keeping a good parallelism since the three levels of parallelism are managed together.

The analysis and the factorization –part of the preprocessing step– are handled by MUMPS. However, during the solve step –as the solve is done in parallel– we have to gather the distributed results onto the master to perform the remaining Block-CG operations (daxpy, ddot, residual computation) in sequential. The next achievement to reach our goal is to perform most of these operations in parallel by keeping the results distributed and performing the Block-CG operations in parallel. To achieve this, we need to exploit MUMPS-embedded functionalities for data management and communications (distributed solution) and implement others (distributed right-hand sides).

4   Strategy Details

Our approach inherits from the direct methods the three main steps which are the *Analysis*, the *Factorization* and the *Solution*. For the sake of simplicity we consider two main steps by merging the Analysis and the Factorization steps into the *Preprocessing* step. Meanwhile, Block-Cimmino and Block-CG will be assimilated into the *Solve* step.

4.1   Preprocessing

During this step most of the Block-Cimmino preprocessing is done. As described in Fig.1, we go through the following processes:

120   C. Balsa et al.

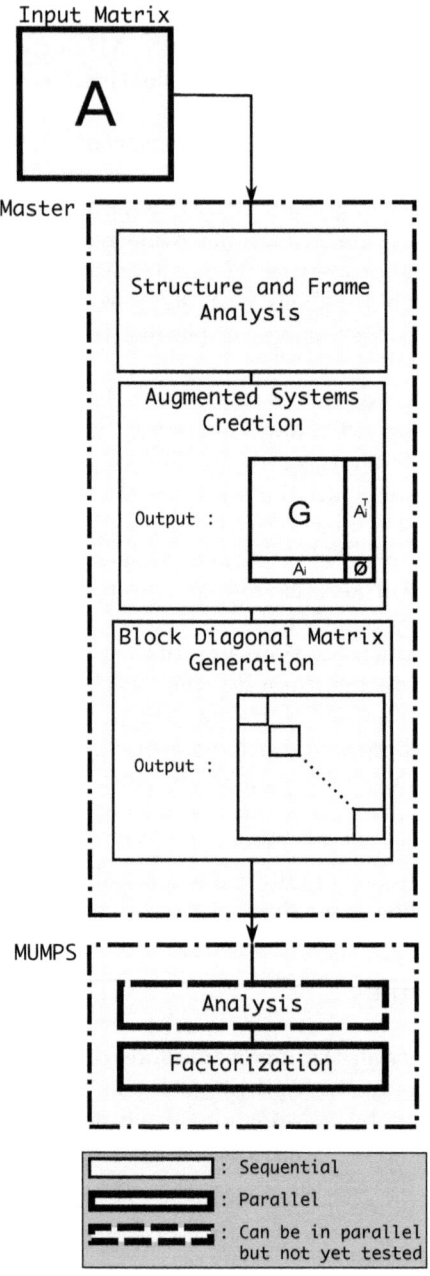

Fig. 1. Preprocessing

Structure and Frame analysis. Using the matrix structure and the block decomposition information we start by analyzing the dependence between the blocks by finding the non-zero values on the common columns. This information helps us to find the parts of the projections to sum during the Block-CG iterations.

Augmented Systems and Block Diagonal Matrix creation. The new blocks' structure is used to generated the augmented systems as described in (5). We then combine these augmented systems into a block diagonal matrix.

The Analysis. MUMPS handles during this step the following steps:

- structure analysis and Scaling;
- graph ordering using METIS;
- mapping the nodes over the available processors by respecting matrix block diagonal structure. MUMPS will recognize it as a forest and thus maps the nodes over the processors according to the number of nodes and their weight;
- distribute the matrix over the processors.

The Factorization. The distributed matrix is LDL^T factorized in parallel. The factors of the block diagonal matrix are kept in place and distributed onto the processors for future access during the solves in the Block-CG iterations.

4.2   Solve: The Block-CG Acceleration

The Block Cimmino method is a linear stationary iterative method with a symmetrizable iteration matrix [9].

$$E_{RJ}\, x \;=\; h_{RJ} \tag{7}$$

$$\text{With}: E_{RJ} \;=\; \sum_{i=1}^{l} A_i^T (A_i A_i^T)^{-1} A_i$$
$$\text{and}: h_{RJ} \;=\; \sum_{i=1}^{l} A_i^T (A_i A_i^T)^{-1} b_i$$

One way to solve (7) in parallel is to use a distributed QR factorization which will easily help to calculate E_{RJ}, an implementation is under current development and released under the name of QR-MUMPS: http://mumps.enseeiht.fr/doc/ud_2010/qr_talk.pdf. The other approach –our case– is using augmented systems described in Section 2.

This system has two main properties:

- when A is consistent our system is consistent and symmetric positive semi-definite;
- when A has full rank our system is symmetric positive definite.

The second property made our choice to use Block-CG as acceleration. This method [10,11] simultaneously searches for the next approximation to the system's solution in a given number of Krylov subspaces, and this number is given by the *block size* of the Block-CG method. The Block-CG method converges in a finite number of iterations in absence of roundoff errors. The straightforward implementation of the Block-CG loop is described in Algorithm 1. We can see that this implies matrix-matrix operations with better granularity than in the classical CG implementation.

It also involves in step 6 the solution of augmented systems with several right-hand sides (RHS) inducing thus some extra parallelization in the solution phase. We also mention that, for reasons of numerical stability, this straightforward implementation needs some slight modifications as described in [5].

Algorithm 1. Block-CG acceleration algorithm

1: $X^{(0)} \leftarrow arbitrary$
2: $\tilde{R}^{(0)} \leftarrow H - EX^{(0)}$
3: $\tilde{P}^{(0)} \leftarrow \tilde{R}^{(0)}$
4: $k \leftarrow 0$
5: **loop**
6:     $\Omega^{(k)} \leftarrow EP^{(k)}$
7:     $\beta^{(k)} \leftarrow (\tilde{P}^{(k)T} G \Omega^{(k)})^{-1} (\tilde{R}^{(k)T} G \tilde{R}^{(k)})$
8:     $X^{(k+1)} \leftarrow X^{(k)} + \tilde{P}^{(k)} \beta^{(k)}$
9:     **if** Converged **then**
10:        *exit loop*
11:    **end if**
12:    $\tilde{R}^{(k+1)} \leftarrow \tilde{R}^{(k)} - \Omega^{(k)} \beta^{(k)}$
13:    $\alpha^{(k)} \leftarrow (\tilde{R}^{(k)T} G \tilde{R}^{(k)})^{-1} (\tilde{R}^{(k+1)T} G \tilde{R}^{(k+1)})$
14:    $\tilde{P}^{(k+1)} \leftarrow \tilde{R}^{(k+1)} + \tilde{R}^{(k)} \alpha^{(k+1)}$
15:    $k \leftarrow k + 1$
16: **end loop**

We use MUMPS during the operations involved in the lines 2 and 6 of Algorithm 1. These operations are equivalent to solve the subsystems (8) in parallel, and then centralize the solutions to combine the projectors. The solve process can be presented as in the Fig.2.

$$\begin{bmatrix} \mathbf{G} & \mathbf{A}_i^T \\ \mathbf{A}_i & 0 \end{bmatrix} \begin{bmatrix} u_i \\ v_i \end{bmatrix} = \begin{bmatrix} 0 \\ b_i - \mathbf{A}_i x \end{bmatrix} \tag{8}$$

The main Block-CG loop (run on the master) prepares the RHS for each augmented system and generates a single RHS –to accelerate the convergence rate of Block-CG we can generate multiple RHS– corresponding to the system to be solved with the block diagonal matrix generated during the preprocessing. The RHS is then sent to MUMPS to solve the system.

MUMPS distributes the RHS according to the way the augmented systems were mapped on the available processors during the *Preprocessing* step. Then each subsystem (8) is solved in parallel with no association between the number of subsystems and the number of processors. Once solved the solutions are sent back to the master to be combined and used in the rest of Block-CG iteration.

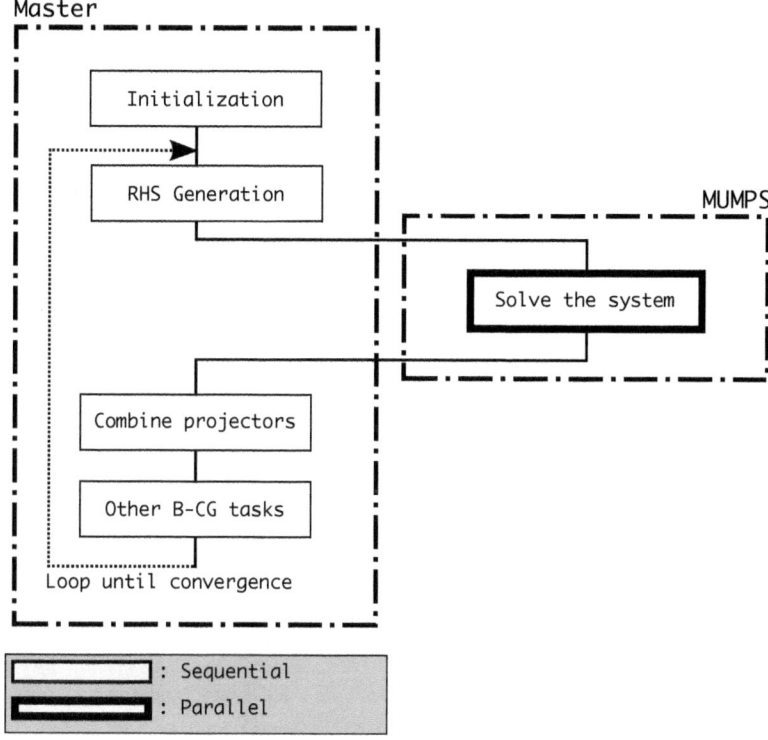

Fig. 2. Block-CG acceleration loop

5 Numerical Results

The numerical experiments were carried out on the Hyperion supercomputer (http://www.calmip.cict.fr/spip/spip.php?rubrique90). Hyperion is the latest supercomputer of the CICT (Centre Interuniversitaire de Calcul de Toulouse) and was ranked 223rd in November's Top500.org ranking. With its 352 bi-Intel "Nehalem" EP quad-core nodes it can develop a peak of 33TFlops. Each node has 4.5GB memory dedicated for each of the cores with an overall of 32GB fully available memory on the node. Our testing method respects the following:

- each MPI-Process is mapped on a single Quad-Core CPU with a dedicated memory access;
- we run each test case –relative to the number of processors– with two different numbers of partitions;
- all partitions have same number of rows;
- we use the Block-CG acceleration with 16 RHS;
- the stopping criterion is $\dfrac{\|b - Ax\|}{\|A\|\|\tilde{x}\| + \|b\|} <= 10^{-14}$.

The results we present are for a symmetric, indefinite and cyclic band diagonal matrix. The matrix is 3 dimensional with $N_i \times N_j \times N_k = 96 \times 128 \times 128$ for a total of 1,572,864 nodes. The problem results from the pressure equation discretization in wind energy generation and is provided by CEsA-FEUP[1].

5.1   Factorization Step

Usually the first problematic part is the factorization in which the structure of the matrix determines how fast it will be performed. From the results we got in Fig.3, we can see that we benefit directly from the fact that MUMPS handles the three levels of parallelism.

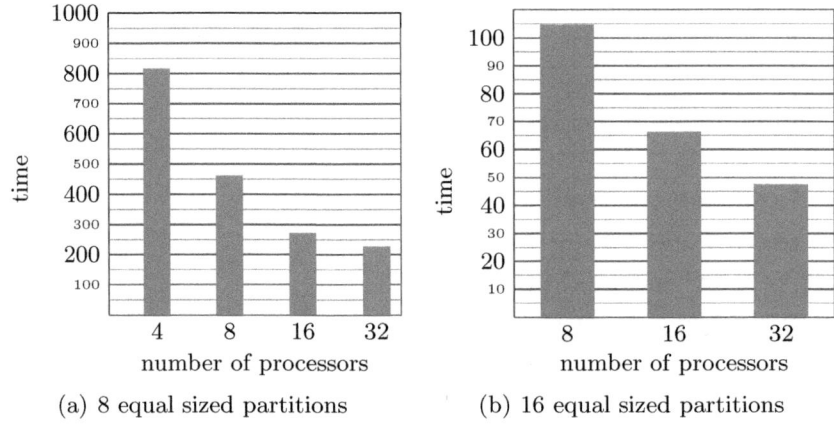

(a) 8 equal sized partitions   (b) 16 equal sized partitions

Fig. 3. Factorization Step

Table 1. Number of operations during the node elimination

8 partitions	16 partitions
1.162D+12	6.924D+10

From the factorization results we retain the following information:

– The more partitions we have the faster the factorization goes in general. This is mainly due to the narrow band matrix given to MUMPS resulting from the small diagonal block size. We see this effect also from the number of operations to be done during the factorization (Table 1). Still comparing gains in Fig.3 when increasing the number of processors,we can observe that a larger number of partitions implies a larger block diagonal matrix with smaller blocks on the diagonal. Thus there is less potential for parallelism within the factorization of each of these diagonal blocks.

[1] http://paginas.fe.up.pt/~cesa/

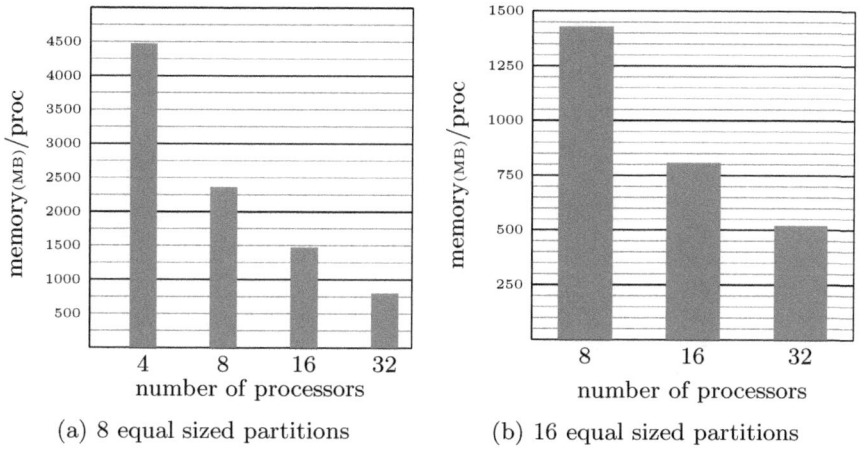

(a) 8 equal sized partitions   (b) 16 equal sized partitions

Fig. 4. Memory usage per processor during factorization

- The good locality of data –result of a good mapping during the analysis– gives a fast distributed computing.
- The gains are lower on higher number of processors as they do not get enough work compared to the communications that keep increasing.
- The memory usage (Fig.4) decreases when the number of partitions or processors increases. This is a desirable property to expect for very large 3D PDE problems in particular.

5.2   Solve Step

In Fig.5 the different parts refer to:

- **Sequential** is the time spent in the Block-CG iterations and the summation of projections.
- **Communication** is the time spent in the scatter and gather of RHS and solution vectors.
- **Parallel Computing** is the time spent in MUMPS. It refers to the solution in parallel of the augmented systems and thus computing the projections.

For the solution phase we have:

- MUMPS handles well the solution of the augmented systems. This behaviour is shown by the good scalability of the *Parallel Computing* phase.
- Communication becomes slightly more important when we increase the number of processors due to an increase of the number of messages between processors.
- Increasing the number of partitions makes the communication longer. Indeed the block diagonal matrix given to MUMPS becomes larger and the length of vectors to gather and scatter is larger.

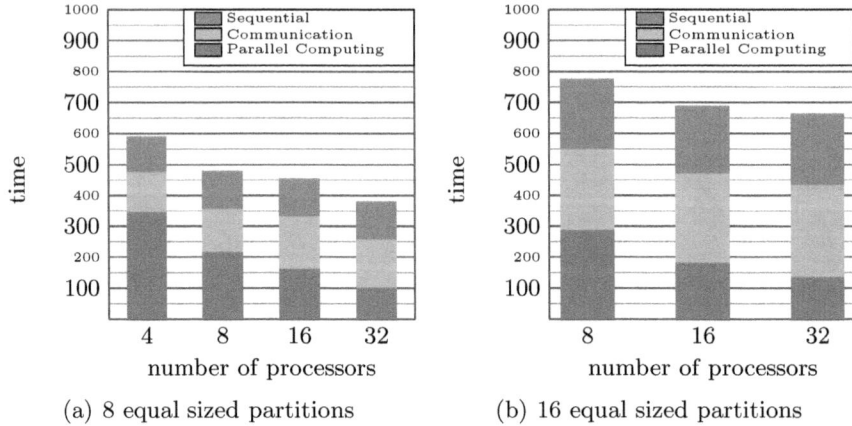

Fig. 5. Solve Step

– The overall time is longer with increasing number of partitions as the block-iterative methods converge slower with more partitions (98 iterations for the 8-block partition, 156 for the 16-block partition).

If we consider this overall time, the gains when we increase the number of processor are slight. These results can be explained and softened if we look at Table 2 that shows the details about the time spent in each step in a solve call. We can see the large percentage taken by communications and which increases proportionally to the number of processors and partitions.

Implementing in parallel the rest of Block-CG's operations (daxpy, ddot, etc.) will help to get rid of the sequential part that represents about 30% of the overall time. At the same time, we will reduce the communications as the summation of the projections will be done in parallel too.

Table 2. Time spent on each part of one solve call (seconds)

		Scatter	Gather	Computing
	8 processors	0.92	1.27	3.23
8-block partition	16 processors	1.02	1.46	2.53
	32 processors	1.10	1.35	1.84
	8 processors	1.01	1.42	2.81
16-block partition	16 processors	1.50	1.51	1.90
	32 processors	1.24	1.53	1.42

In previous work [14], the solution time was increasing with the increase of the number of processor, and even appeared to be heavily long in the case of the 16-block partition. It was due to the fact that the mapping was not taking into account the forest structure in the dependency tree. With the current fine tuning, MUMPS notices the block diagonal structure of the matrix and maps the data with respect to that and gains can be obtained when increasing the number of processors.

6 Ongoing Work

The current results show the good parallelism performance and scalability. It helped us to identify the bottlenecks and problems for future improvements.

To implement the final targeted parallel version, we have to use the basic embedded functionalities already available in the MUMPS package, such as residual computation used in iterative refinement for instance, and design new features whenever necessary.

We have also to define a user interface to address directly these functionalities within the parallel iterative solver in order to appropriately exploit the data distribution already established and handled by MUMPS. This will help us especially during the parallelisation of Block-CG's sequential parts and get rid of the actual but useless gather and scatter communications.

We plan also to experiment the potential of the final solver on matrices coming from real problems coming from an industrial software for the simulation of atmospheric circulation above mountain in the field of wind energy production [13].

Acknowledgment. This particular work has strongly benefited from sustained interactions with the members of the MUMPS project development team. Indeed, the strategy for parallelism and the particular sparse structure exploited in this approach was needing fine tuning as well as some tiny adaptations of the MUMPS package to achieve encouraging results. We are therefore very grateful to all the MUMPS project members, and in particular to P. Amestoy, J-Y. L'Excellent, F-H. Rouet, and B. Uçar for the various fruitful exchanges. We also wish to thank K. Kaya from CERFACS for the very useful suggestions concerning pre-ordering strategies.

References

1. Amestoy, P., Buttari, A., Combes, P., Guermouche, A., L'Excellent, J.Y., Pralet, S., Ucar, B.: Multifrontal massively parallel solver - user's guide of the version 4.9.4 (2009)
2. Amestoy, P., Duff, I., L'Excellent, J.Y., Koster, J.: A fully asynchronous multifrontal solver using distributed dynamic scheduling. SIAM Journal on Matrix Analysis and Applications 23, 15–41 (2001)
3. Cimmino, G.: Calcolo approssimato per le soluzioni dei sistemi di equaziono lineari. In: Ricerca Sci. II, vol. 9(I), pp. 326–333 (1938)
4. Arioli, M., Duff, I.S., Noailles, J., Ruiz, D.: A block projection method for sparse matrices. SIAM Journal on Scientific and Statistical Computing, 47–70 (1992)
5. Arioli, M., Duff, I.S., Ruiz, D., Sadkane, M.: Block Lanczos techniques for accelerating the Block Cimmino method. SIAM Journal on Scientific and Statistical Computing, 1478–1511 (1995)

6. Campbell, S.L., Meyer, J.C.D.: Generalized inverses of linear transformations. Pitman, London (1979)
7. Bartels, R.H., Golub, G.H., Saunders, M.A.: Numerical techniques in mathematical programming. In: Rosen, J.B., Mangasarian, O.L., Ritter, K. (eds.) Nonlinear Programming. Academic Press, New York (1970)
8. Hachtel, G.D.: Extended applications of the sparse tableau approach - finite elements and least squares. In: Spillers, W.R. (ed.) Basic question of design theory. North Holland, Amsterdam (1974)
9. Hageman, L.A., Young, D.M.: Applied Iterative Methods. Academic Press, London (1981)
10. O'Leary, D.P.: The block conjugate gradient algorithm and related methods. Linear Algebra Appl. 29, 293–322 (1980)
11. Arioli, M., Ruiz, D.: Block conjugate gradient with subspace iteration for solving linear systems. In: Second IMACS International Symposium on Iterative Methods in Linear Algebra, Blagoevgrad, Bulgaria, pp. 64–79 (1995)
12. Arioli, M., Drummond, A., Duff, I.S., Ruiz, D.: A parallel scheduler for block iterative solvers in heterogeneous computing environments. In: Proceedings of the Seventh SIAM Conference on Parallel Processing for Scientific Computing, pp. 460–465. SIAM, Philadelphia (1995)
13. Silva Lopes, A., Palma, J.M.L.M., Castro, F.A.: Simulation of the Askervein flow. Part 2: Large-eddy simulations. Boundary-Layer Meteorology 125, 85–108 (2007)
14. Balsa, C., Guivarch, R., Raimundo, J., Ruiz, D.: MUMPS Based Approach to Parallelize the Block Cimmino Algorithm. In: International Meeting High Performance Computing for Computational Science (VECPAR), Toulouse (2008), http://vecpar.fe.up.pt/2008/papers/45.php

Towards an Efficient Tile Matrix Inversion of Symmetric Positive Definite Matrices on Multicore Architectures

Emmanuel Agullo[1], Henricus Bouwmeester[2], Jack Dongarra[1], Jakub Kurzak[1], Julien Langou[2], and Lee Rosenberg[2]

[1] Dpt. of Electrical Engineering and Computer Science, University of Tennessee, 1122 Volunteer Blvd, Claxton Building, Knoxville, TN 37996-3450, USA
{Emmanuel.Agullo,Jack.Dongarra,Jakub.Kurzak}@eecs.utk.edu
[2] Dpt. of Mathematical and Statistical Sciences, University of Colorado Denver, Campus Box 170, P.O. Box 173364, Denver, Colorado 80217-3364, USA, Research was supported by the National Science Foundation grant no. NSF CCF-811520.
{Henricus.Bouwmeester,Lee.Rosenberg}@email.ucdenver.edu,
Julien.Langou@ucdenver.edu

Abstract. The algorithms in the current sequential numerical linear algebra libraries (*e.g.* LAPACK) do not parallelize well on multicore architectures. A new family of algorithms, the *tile algorithms*, has recently been introduced. Previous research has shown that it is possible to write efficient and scalable tile algorithms for performing a Cholesky factorization, a (pseudo) LU factorization, a QR factorization, and computing the inverse of a symmetric positive definite matrix. In this extended abstract, we revisit the computation of the inverse of a symmetric positive definite matrix. We observe that, using a dynamic task scheduler, it is relatively painless to translate existing LAPACK code to obtain a ready-to-be-executed tile algorithm. However we demonstrate that, for some variants, non trivial compiler techniques (array renaming, loop reversal and pipelining) need then to be applied to further increase the parallelism of the application. We present preliminary experimental results.

1 Introduction

The appropriate direct method to compute the solution of a symmetric positive definite system of linear equations consists of computing the Cholesky factorization of that matrix and then solving the underlying triangular systems. It is not recommended to use the inverse of the matrix in this case. However some applications need to explicitly form the inverse of the matrix. A canonical example is the computation of the variance-covariance matrix in statistics. Higham [15, p.260,§3] lists more such applications.

With their advent, multicore architectures [21] induce the need for algorithms and libraries that fully exploit their capacities. A class of such algorithms – called tile algorithms [8,9] – has been developed for one-sided dense factorizations

J.M.L.M. Palma et al. (Eds.): VECPAR 2010, LNCS 6449, pp. 129–138, 2011.

(Cholesky, LU and QR) and made available as part of the Parallel Linear Algebra Software for Multicore Architectures (PLASMA) library [2]. In this paper, we study this class of algorithms to the case of the (symmetric positive definite) matrix inversion. An identical study has already been performed in 2008 [11], and the associated software is present in the libflame library [23].

Besides constituting an important functionality for a library such as PLASMA, the study of the matrix inversion on multicore architectures represents a challenging algorithmic problem. Indeed, first, contrary to standalone one-sided factorizations that have been studied so far, the matrix inversion exhibits many anti-dependences [4] (Write After Read). Those anti-dependences can be a bottleneck for parallel processing, which is critical on multicore architectures. It is thus essential to investigate (and adapt) well known techniques used in compilation such as using temporary copies of data to remove anti-dependences to enhance the degree of parallelism of the matrix inversion. This technique is known as *array renaming* [4] (or *array privatization* [14]). Second, *loop reversal* [4] is to be investigated. Third, the matrix inversion consists of three successive steps (first of which is the Cholesky decomposition). In terms of scheduling, it thus represents an opportunity to study the effects of *pipelining* [4] those steps on performance.

The current version of PLASMA (version 2.1) is scheduled statically. Initially developed for the IBM Cell processor [17], this static scheduling relies on POSIX threads and simple synchronization mechanisms. It has been designed to maximize data reuse and load balancing between cores, allowing for very high performance [3] on today's multicore architectures. However, in the case of the matrix inversion, the design of an ad-hoc static scheduling is a time consuming task and raises load balancing issues that are much more difficult to address than for a stand-alone Cholesky decomposition, in particular when dealing with the pipelining of multiple steps. Furthermore, the growth of the number of cores and the more complex memory hierarchies make executions less deterministic. In this paper, we rely on an experimental in-house dynamic scheduler named QUARK [16]. This scheduler is based on the idea of expressing an algorithm through its sequential representation and unfolding it at runtime using data hazards (Read after Write, Write after Read, Write after Write) as constraints for parallel scheduling. The concept is rather old and has been validated by a few successful projects. We could as well have used schedulers from the Jade project from Stanford University [20], from the SMPSs project from the Barcelona Supercomputer Center [18], or from the SuperMatrix framework [10].

Our discussions are illustrated with experiments conducted on a dual-socket quad-core machine based on an Intel Xeon EMT64 processor operating at 2.26 GHz. Composed of 8 cores, the theoretical peak is equal to 9.0 Gflop/s per core or 72.3 Gflop/s for the whole machine. The machine is running Mac OS X 10.6.2 and is shipped with the Apple vecLib v126.0 multithreaded BLAS [1] and LAPACK vendor library. We have installed reference LAPACK [5] v3.2.1, reference ScaLAPACK [7] v1.8.0, and libflame v3935 [23].

The rest of the paper is organized as follows. In Section 2, we explain a possible algorithmic variant for matrix inversion based on tile algorithms; we explain how we articulated it with our dynamic scheduler to take advantage of multicore architectures and we compare its performance against state-of-the-art libraries. In Section 3, we investigate the impact on parallelism and performance of different well known techniques used in compilation: loop reversal, array renaming and pipelining. We conclude and present future work directions in Section 4.

2   Tile in-place Matrix Inversion

Tile algorithms are a class of Linear Algebra algorithms that allow for fine granularity parallelism and asynchronous scheduling, enabling high performance on multicore architectures [3,8,9,11,19]. The matrix of order n is split into $t \times t$ square submatrices of order b ($n = b \times t$). Such a submatrix is of small granularity (we fixed $b = 200$ in this paper) and is called a *tile*. So far, tile algorithms have been essentially used to implement one-sided factorizations [3,8,9,11,19].

Algorithm 1 extends this class of algorithms to the case of the matrix inversion. As in state-of-the-art libraries (LAPACK, ScaLAPACK), the matrix inversion is performed *in-place*, *i.e.*, the data structure initially containing matrix A is directly updated as the algorithm is progressing, without using any significant temporary extra-storage; eventually, A^{-1} substitutes A. Algorithm 1 is composed of three steps. Step 1 is a Tile Cholesky Factorization computing the Cholesky factor L (lower triangular matrix satisfying $A = LL^T$). This step was studied in [9]. Step 2 computes L^{-1} by inverting L. Step 3 finally computes the inverse matrix $A^{-1} = L^{-1^T}L^{-1}$.

A more detailed description is beyond the scope of this extended abstract and is not essential to the understanding of the rest of the paper. However, we want to point out that the stability analysis of the block (or tile) triangular inversion is quite subtle and one should not replace too hastily "TRSM-then-TRTRI" by "TRTRI-then-TRMM" See [13] for a comprehensive explanation.

Each step is composed of multiple fine granularity tasks (since operating on tiles). These tasks are part of the BLAS (SYRK, GEMM, TRSM, TRMM) and LAPACK (POTRF, TRTRI, LAUUM) standards. Indeed, from a high level point of view, an operation based on tile algorithms can be represented as a Directed Acyclic Graph (DAG) [12] where nodes represent the fine granularity tasks in which the operation can be decomposed and the edges represent the dependences among them. For instance, Figure 1(a) represents the DAG of Step 3 of Algorithm 1.

Algorithm 1 is based on the variants used in LAPACK 3.2.1. Bientinesi, Gunter and van de Geijn [6] discuss the merit of algorithmic variations in the case of the computation of the inverse of a symmetric positive definite matrix. Although of definite interest, this is not the focus of this extended abstract. We will use the same variant enumerations as in [6]. With these notations, Algorithm 1 is called 242: variant 2 for POTRF, variant 4 for TRTRI and variant 2 for LAUUM. Variant 4 of TRTRI is identical to variant 1 of TRTRI but it starts

---

Algorithm 1. Tile In-place Cholesky Inversion (lower format). Matrix A is the on-going updated matrix (in-place algorithm).

Input: A, Symmetric Positive Definite matrix in tile storage ($t \times t$ tiles).
Result: A^{-1}, stored in-place in A.
Step 1: Tile Cholesky Factorization (compute L such that $A = LL^T$);
  Variant 2;
for $j = 0$ **to** $t - 1$ **do**
  for $k = 0$ **to** $j - 1$ **do**
    $A_{j,j} \leftarrow A_{j,j} - A_{j,k} * A_{j,k}^T$ (SYRK(j,k)) ;
  $A_{j,j} \leftarrow CHOL(A_{j,j})$ (POTRF(j)) ;
  for $i = j + 1$ **to** $t - 1$ **do**
    for $k = 0$ **to** $j - 1$ **do**
      $A_{i,j} \leftarrow A_{i,j} - A_{i,k} * A_{j,k}^T$ (GEMM(i,j,k)) ;
  for $i = j + 1$ **to** $t - 1$ **do**
    $A_{i,j} \leftarrow A_{i,j} / A_{j,j}^T$ (TRSM(i,j)) ;

Step 2: Tile Triangular Inversion of L (compute L^{-1})
  Variant 4;
for $j = t - 1$ **to** 0 **do**
  for $i = t - 1$ **to** $j + 1$ **do**
    $A_{i,j} \leftarrow A_{i,i} * A_{i,j}$ (TRMM(i,j)) ;
    for $k = i - 1$ **to** $j + 1$ **do**
      $A_{i,j} \leftarrow A_{i,j} + A_{i,k} * A_{k,j}$ (GEMM(i,j,k)) ;
    $A_{i,j} \leftarrow -A_{i,j} / A_{j,j}$ (TRSM(i,j)) ;
  $A_{j,j} \leftarrow TRINV(A_{j,j})$ (TRTRI(j)) ;

Step 3: Tile Product of Lower Triangular Matrices (compute $A^{-1} = L^{-1^T} L^{-1}$)
  Variant 2;
for $i = 0$ **to** $t - 1$ **do**
  for $j = 0$ **to** $i - 1$ **do**
    $A_{i,j} \leftarrow A_{i,i}^T * A_{i,j}$ (TRMM(i,j)) ;
  $A_{i,i} \leftarrow A_{i,i}^T * A_{i,i}$ (LAUUM(i)) ;
  for $j = 0$ **to** $i - 1$ **do**
    for $k = i + 1$ **to** $t - 1$ **do**
      $A_{i,j} \leftarrow A_{i,j} + A_{k,i}^T * A_{k,j}$ (GEMM(i,j,k)) ;
  for $k = i + 1$ **to** $t - 1$ **do**
    $A_{i,i} \leftarrow A_{i,i} + A_{k,i}^T * A_{k,i}$ (SYRK(i,k)) ;

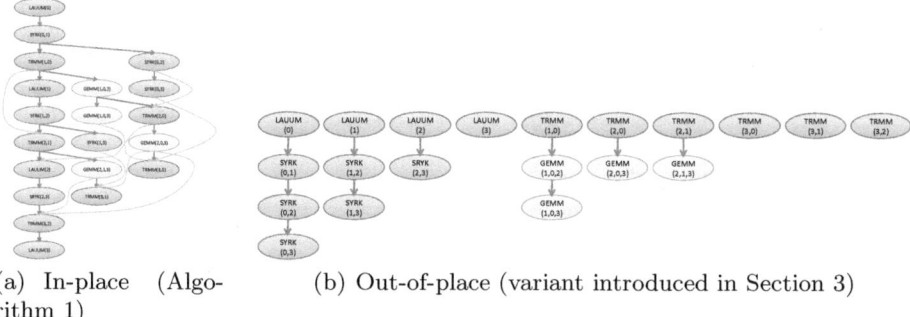

(a) In-place   (Algo-   (b) Out-of-place (variant introduced in Section 3)
rithm 1)

Fig. 1. DAGs of Step 3 of the Tile Cholesky Inversion ($t = 4$)

from the bottom-right corner and ends at the top left corner. (Variant 1 is the reverse.)

We will see in the experimental section, Section 3, that this choice of variants is not optimal, however it gives an interesting case study.

We have implemented Algorithm 1 using our dynamic scheduler QUARK introduced in Section 1. Figure 2 shows its performance against state-of-the-art libraries and the vendor library on the machine described in Section 1. For a matrix of small size while keeping the tile size reasonably large, it is difficult to extract parallelism and have full use of all the cores [3,8,9,19]. We indeed observe a limited scalability ($N = 1000$, Figure 2(a)). However, tile algorithms (Algorithm 1) still benefit from a higher degree of parallelism than blocked algorithms [3,8,9,19]. Therefore Algorithm 1 (in place) consistently achieves a significantly better performance than vecLib, ScaLAPACK and LAPACK libraries. With the tile size kept constant, a larger matrix size ($N = 4000$, Figure 2(b)) allows for a better use of parallelism. In this case, an optimized implementation of a blocked algorithm (vecLib) competes well against tile algorithms (in place) on few cores (left part of Figure 2(a)). However, only tile algorithms scale to a larger number of cores (rightmost part of Figure 2(b)) due to a higher degree of parallelism. In other words, the tile Cholesky inversion achieves a better *strong scalability* than the blocked versions, similarly to what had been observed for the factorization step [3,8,9,19].

We see that the performance of the 242 variant (green and yellow lines) is mediocre and a more appropriate variant would be 331 (red and purple lines) or even better 312 (dashed purple). The reason for this is that the combinantion of variants in the 242 variant does not lend itself well to interleaving. Variant 2 of POTRF starts from the top-left corner and ends bottom-right, then variant 4 of TRTRI starts from the bottom-right corner and ends at the top-left corner being followed by variant 2 of LAUUM which starts from the top-left corner and ends at the bottom-right. Due to the progression of each step within this combination, it is very difficult to interleave the tasks. More appropriately, a variant of TRTRI which progresses from top-left to bottom-right would afford a more aggressive interleaving of POTRF, TRTRI, and LAUUM. Variants 331 and 312 provide

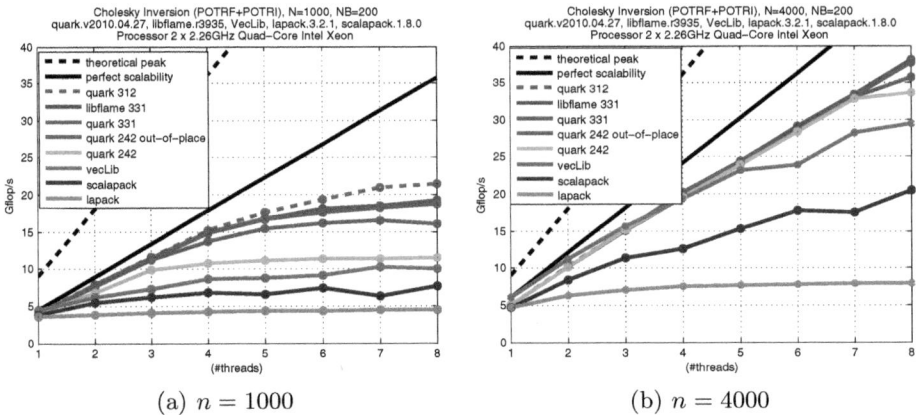

Fig. 2. Scalability of Algorithm 1 (in place) and its out-of-place variant introduced in Section 3, using our dynamic scheduler against libflame, vecLib, ScaLAPACK and LAPACK libraries

this combination. This corroborates the observation of Bientinesi, Gunter and van de Geijn: the 331 variant (e.g.) allows for a "*one-sweep*" algorithm [6].

We note that we obtain similar performance as the libflame libraries when we use the same 331 algorithmic variant. (See red and plain purple curves.) The main difference in this case being the schedulers (QUARK vs Supermatrix) which are performing equally for these experiments. We did not try to tune the parameters of either of these schedulers.

The best algorithmic combination of variants for our experimental conditions was 312 as observed in the plot with the dashed purple curve.

3   Algorithmic Study

In the previous section, we compared the performance of the tile Cholesky inversion against state-the-art libraries. In this section, we focus on tile Cholesky inversion and we discuss the impact of several possible optimizations of Algorithm 1 on its performance.

Array renaming (removing anti-dependences). The dependence between SYRK(0,1) and TRMM(1,0) in the DAG of Step 3 of Algorithm 1 (Figure 1(a)) represents the constraint that the SYRK operation (l. 28 of Algorithm 1) needs to read $A_{k,i} = A_{1,0}$ before TRMM (l. 22) can overwrite $A_{i,j} = A_{1,0}$. This anti-dependence (Write After Read) can be removed by use of a temporary copy of $A_{1,0}$. Similarly, all the SYRK-TRMM anti-dependences, as well as TRMM-LAUMM and GEMM-TRMM anti-dependences can be removed. We have designed a variant of Algorithm 1 that removes all the anti-dependences by usage of a large working array (this technique is called *array renaming* [4] in compilation [4]). The subsequent DAG (Figure 1(b)) is split into multiple pieces

Table 1. Length of the critical path as a function of the number of tiles t

	In-place case	Out-of-place case
Step 1	$3t - 2$ (up)	$3t - 2$ (up)
Step 2	$3t - 2$ (down)	$2t - 1$ (down)
Step 3	$3t - 2$ (up)	t (up)

(Figure 1(b)), leading to a shorter critical path (Table 1). We also implemented the out-of-place algorithm, within the framework of our dynamic scheduler. Figure 2(a) shows that our dynamic scheduler exploits this higher degree of parallelism to achieve a higher strong scalability even on small matrices ($N = 1000$). For a larger matrix (Figure 2(b)), the in-place algorithm already achieved very good scalability. Therefore, using up to 7 cores, their performance are similar. However, there is not enough parallelism with a 4000×4000 matrix to efficiently use all 8 cores with the in-place algorithm; thus the higher performance of the out-of-place version in this case (leftmost part of Figure 2(b)).

Loop reversal (exploiting commutativity). The most internal loop of each step of Algorithm 1 (l. 8, l. 17 and l. 26) consists in successive commutative GEMM operations. Therefore they can be performed in any order, among which increasing order and decreasing order of the loop index. Their ordering impacts the length of the critical path. Algorithm 1 orders those three loops as increasing (U) for POTRF, decreasing (D) for TRTRI, and increasing (U) for LAUUM. We had manually chosen these respective orders (UDU) because they minimize the critical path of each step (values reported in Table 1). A naive approach would have, for example, been comprised of consistently ordering the loops in increasing order (UUU). In this case (UUU), the critical path of TRTRI would have been equal to $t^2 - 2t + 4$ (in-place) or ($\frac{1}{2}t^2 - \frac{1}{2}t + 2$) (out-of-place) instead of $3t - 2$ (in-place) or $2t - 1$ (out-of-place) for (UDU). Figure 3 shows how loop reversal impacts performance.

This optimization is important for libraries relying on a tile BLAS (e.g. libflame [23]). While any loop ordering is fine for a tile GEMM (e.g.) in term of correctness, the loop ordering has an influence on how fast tiles are freed by the tile GEMM operation. It is therefore critical that tile GEMM (e.g.) has the ability of switching the ordering of the tasks depending on the context. In our case, the optimal loop ordering for the 331 variant of Cholesky inversion is UUU and so the good loop ordering comes "naturally".

Pipelining. Pipelining (interleaving) the multiple steps of the inversion reduces the length of its critical path. For the in-place case, the critical path cannot be reduced since the final task of Step 1 must be completed before the first task of Step 2 can proceed and similarly for Step 2 to Step 3. (This is because we have chosen to study the 242 variant.) For the out-of-place case, it is reduced from $6t - 3$ to $5t - 2$ tasks. We studied the effect of pipelining on the performance of the inversion of a 8000×8000 matrix with an artificially large tile size ($b = 2000$ and $t = 4$). For the out-of-place case, the elapsed time grows from 16.4 to 19.0 seconds (16 % overhead) when pipelining is prevented.

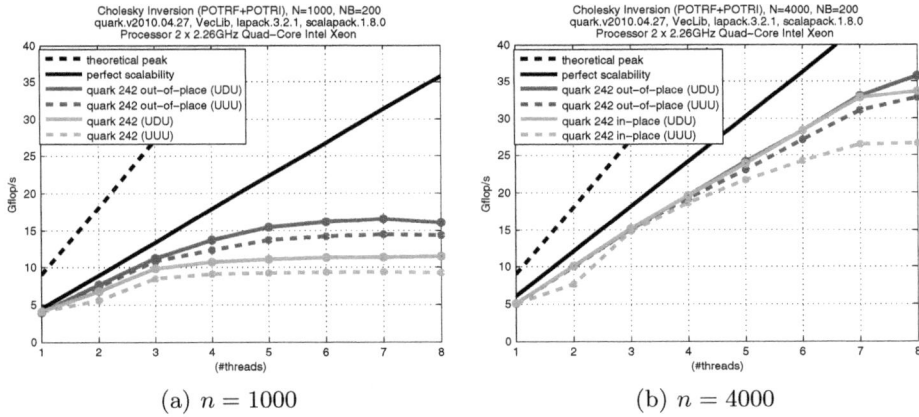

Fig. 3. Impact of loop reversal on performance

4   Conclusion and Future Work

We have studied the problem of the computation of the inverse of a symmetric positive definite matrix on multicore architectures. This problem was already presented by Chan, Van Zee, Bientinesi, Quintana-Ortí, Quintana-Ortí, and van de Geijn in 2008 [11]. We are essentially following the same approach: starting from standard algorithms, we derive tile algorithms whose tasks are then scheduled dynamically.

Our experimental study has shown both an excellent scalability of these algorithms and a significant performance improvement compared to LAPACK and ScaLAPACK based libraries.

In perspective of [11], our contribution is to bring back to the fore well known issues in the domain of compilation. Indeed, we have shown the importance of loop reversal, array renaming and pipelining. The optimization of these are very important in the sense that they influence dramatically the shape of the DAG of tasks that is provided to our dynamic scheduler and consequently determine the degree of parallelism (and scalability) of the application.

The use of a dynamic scheduler allowed an out-of-the-box pipeline of the different steps whereas loop reversal and array renaming required a manual change to the algorithm. The future work directions consist of enabling the scheduler to automatically perform loop reversal and array renaming. We exploited the commutativity of GEMM operations to perform array renaming. Their associativity would furthermore allow them to be processed in parallel (e.g. following a binary tree). Actually, the commutative and associative nature of addition allows one to execute the operations in the fashion of a DOANY loop [22]. The subsequent impact on performance is to be studied. Array renaming requires extra-memory, thus it will be interesting to address the problem of the maximization of performance under memory constraints.

Acknowledgements

The authors would like to thank Anthony Danalis for his insights on compiler techniques, Robert van de Geijn for pointing us to reference [11], (we missed this very relevant reference in the first place!), and Ernie Chan for making the tuning and benchmarking of libflame as easy as possible.

References

1. BLAS: Basic linear algebra subprograms, http://www.netlib.org/blas/
2. Agullo, E., Dongarra, J., Hadri, B., Kurzak, J., Langou, J., Langou, J., Ltaief, H.: PLASMA Users' Guide. Technical report, ICL, UTK (2009)
3. Agullo, E., Hadri, B., Ltaief, H., Dongarrra, J.: Comparative study of one-sided factorizations with multiple software packages on multi-core hardware. In: SC 2009: Proceedings of the Conference on High Performance Computing Networking, Storage and Analysis, pp. 1–12. ACM, New York (2009)
4. Allen, R., Kennedy, K.: Optimizing Compilers for Modern Architectures: A Dependence-based Approach. Morgan Kaufmann, San Francisco (2001)
5. Anderson, E., Bai, Z., Bischof, C., Blackford, L.S., Demmel, J.W., Dongarra, J., Du Croz, J., Greenbaum, A., Hammarling, S., McKenney, A., Sorensen, D.: LAPACK Users' Guide. SIAM, Philadelphia (1992)
6. Bientinesi, P., Gunter, B., van de Geijn, R.: Families of algorithms related to the inversion of a symmetric positive definite matrix. ACM Trans. Math. Softw. 35(1), 1–22 (2008)
7. Blackford, L.S., Choi, J., Cleary, A., D'Azevedo, E., Demmel, J., Dhillon, I., Dongarra, J., Hammarling, S., Henry, G., Petitet, A., Stanley, K., Walker, D., Whaley, R.C.: ScaLAPACK Users' Guide. SIAM, Philadelphia (1997)
8. Buttari, A., Langou, J., Kurzak, J., Dongarra, J.: Parallel tiled QR factorization for multicore architectures. Concurrency Computat.: Pract. Exper. 20(13), 1573–1590 (2008)
9. Buttari, A., Langou, J., Kurzak, J., Dongarra, J.: A class of parallel tiled linear algebra algorithms for multicore architectures. Parallel Computing 35(1), 38–53 (2009)
10. Chan, E.: Runtime data flow scheduling of matrix computations. FLAME Working Note #39. Technical Report TR-09-22, The University of Texas at Austin, Department of Computer Sciences (August 2009)
11. Chan, E., Van Zee, F.G., Bientinesi, P., Quintana-Ortí, E.S., Quintana-Ortí, G., van de Geijn, R.: Supermatrix: a multithreaded runtime scheduling system for algorithms-by-blocks. In: PPoPP 2008: Proceedings of the 13th ACM SIGPLAN Symposium on Principles and practice of parallel programming, pp. 123–132. ACM, New York (2008)
12. Christofides, N.: Graph Theory: An algorithmic Approach (1975)
13. Du Croz, J.J., Higham, N.J.: Stability of methods for matrix inversion. IMA Journal of Numerical Analysis 12, 1–19 (1992)
14. Eigenmann, R., Hoeflinger, J., Padua, D.: On the automatic parallelization of the perfect benchmarks®. IEEE Trans. Parallel Distrib. Syst. 9(1), 5–23 (1998)
15. Higham, N.J.: Accuracy and Stability of Numerical Algorithms, 2nd edn. Society for Industrial and Applied Mathematics, Philadelphia (2002)

16. Kurzak, J., Dongarra, J.: Fully dynamic scheduler for numerical computing on multicore processors. University of Tennessee CS Tech. Report, UT-CS-09-643 (2009)
17. Kurzak, J., Dongarra, J.: QR factorization for the Cell Broadband Engine. Sci. Program. 17(1-2), 31–42 (2009)
18. Perez, J.M., Badia, R.M., Labarta, J.: A dependency-aware task-based programming environment for multi-core architectures. In: Proceedings of IEEE Cluster Computing 2008 (2008)
19. Quintana-Ortí, G., Quintana-Ortí, E.S., van de Geijn, R.A., Van Zee, F.G., Chan, E.: Programming matrix algorithms-by-blocks for thread-level parallelism. ACM Transactions on Mathematical Software 36(3)
20. Rinard, M.C., Scales, D.J., Lam, M.S.: Jade: A high-level, machine-independent language for parallel programming. Computer 6, 28–38 (1993)
21. Sutter, H.: A fundamental turn toward concurrency in software. Dr. Dobb's Journal 30(3) (2005)
22. Wolfe, M.: Doany: Not just another parallel loop. In: Banerjee, U., Gelernter, D., Nicolau, A., Padua, D.A. (eds.) LCPC 1992. LNCS, vol. 757, pp. 421–433. Springer, Heidelberg (1993)
23. Van Zee, F.G.: libflame: The Complete Reference (2009), http://www.lulu.com

A Massively Parallel Dense Symmetric Eigensolver with Communication Splitting Multicasting Algorithm

Takahiro Katagiri[1] and Shoji Itoh[2]

[1] Information Technology Center, The University of Tokyo,
2-11-16 Yayoi, Bunkyo-ku, Tokyo 113-8658, Japan
katagiri@cc.u-tokyo.ac.jp
[2] Advanced Center for Computing and Communication, RIKEN,
2-1 Hirosawa, Wako-shi, Saitama 351-0198, Japan
itosho@riken.jp

Abstract. In this paper, we propose a process grid free algorithm for a massively parallel dense symmetric eigensolver with a communication splitting multicasting algorithm. In this algorithm, a tradeoff exists between speed and memory space to keep the Householder vectors. As a result of a performance evaluation with the T2K Open Supercomputer (U. Tokyo) and the RX200S5, we obtain the performance with 0.86x and 0.95x speed-downs and 1/2 memory space compared to the conventional algorithm for a square process grid. We also show a new algorithm for small-sized matrices in massively parallel processing that takes an appropriately small value of p of the process grid p x q. In this case, the execution time of inverse transformation is negligible.

Keywords: parallel and distributed computing, numerical algorithms for CS&E.

1 Introduction

A parallel dense symmetric eigensolver is a crucial tool for scientific computing. Some applications need few eigenvalues and eigenvectors for a sparse matrix. But other cases need all eigenvalues and all eigenvectors for a dense matrix. For example, all eigenvalues and eigenvectors for a dense matrix are needed in the density functional calculation of the electronic structure of an insulin hexamer [1]. Hundreds of computations of eigenvalues and eigenvectors are required to optimize the structure in some applications. Hence, we need to optimize the eigensolver to match both the dense and small sizes of the matrix in massively parallel processing because of the time restriction of computer services, such as at a supercomputer center.

In addition, current computer architectures are increasing in complexity. Therefore, we need to administrate deep hierarchical caches, non-uniform memory accesses, and increase of the number of cores. Due to the features of current computer architectures, a cache-aware algorithm, that is, a blocking algorithm, is used in many numerical libraries. As an example in eigensolvers, a blocking algorithm for the reduction of dense matrices was proposed by Dongarra *et al.* [2]. After that, to reduce the communication time, a two-step reduction algorithm was proposed by Bischof *et al.* [3].

J.M.L.M. Palma et al. (Eds.): VECPAR 2010, LNCS 6449, pp. 139–150, 2011.
© Springer-Verlag Berlin Heidelberg 2011

These algorithms, however, are not aimed at small matrices. Instead, the target is a huge matrix to obtain high "computational" efficiency for one-time computation. This efficiency does not consider the actual execution limit of computer services. In current massively parallel machines, more than 10,000 cores are implemented. In this massively parallel environment, the conventional approach cannot work well because the actual matrix size that can be solved is very limited. For a dense eigensolver, the computation complexity grows to $O(n^3)$, and hence the computation time increases on the order of 100x of execution time with one core in weak scaling when we use 100,000 cores.

Katagiri et al. [4] proposed a massively parallel algorithm with a communication splitting multicast algorithm. The algorithm established more than a 5x speedup compared to that of ScaLAPACK [4], although the algorithm does not implement a blocking algorithm. The drawback of this algorithm was the restriction of process grid construction. In this paper, we propose an algorithm for a square process grid configuration. The goal of this paper is two-fold.

First, we propose a process grid free algorithm based on [4]. This enables us not only many opportunities to adapt our algorithm, but also another tuning approach.

Second, the process grid free algorithm has a tradeoff between memory space and execution speed. We evaluate the tradeoff by using two kinds of parallel machines. The execution speed with a small-sized matrix is examined in this paper. We focus on the small dimension of 10,000, which represents the real usage for a chemical simulation. The execution time for a small size of 10,000 is shown in the performance evaluation.

This paper is organized as follows. Section 2 explains sequential and parallel algorithms for symmetric eigensolvers. Section 3 proposes a new algorithm with communication-splitting multicasts. Section 4 is a performance evaluation with the T2K Open Supercomputer (U. Tokyo) and the RX200S5. Finally, we summarize our findings in Section 5.

2   Symmetric Dense Eigensolver

To calculate the symmetric standard eigenproblem $Ax = \lambda x$, where $A \in \mathfrak{R}^{n \times n}, \lambda \in \mathfrak{R}^1, x \in \mathfrak{R}^n$, we need to reduce the dense matrix A to a tridiagonal matrix T. This transformation is called "tridiagonalization." After solving the new eigenproblem for T, we obtain an eigenvalue and an eigenvector. The eigenvector, which is y in this example, is not the eigenvector for matrix A. Therefore, we need a transform from y to x, which is the eigenvector of A. This transformation is called "inverse transformation."

Non-blocking Sequential Algorithm

We describe the tridiagonalization processes shown in Figure 1.

In this figure, the notation $A^{(k)}{}_{a:b,c:d}$ indicates the sub-matrix of $A^{(k)}$, which consists of rows from a to b and columns from c to d. Figure 1 includes a dense vector matrix multiplication in line <3>, a dot product in line <4>, a copy in line <5>, and a matrix update in line <6>. To perform inverse transformation, we need a workspace to store the Householder (pivot) vectors $u_1, u_2, \cdots, u_{n-2}$.

Inverse transformation is described in Figure 2. Figure 2 includes a dot product in line <3>, and a matrix update in line <4>.

<1> *do* $k = 1, n-2$

<2> $A^{(k)}{}_{k:n,k} \mapsto (\alpha_k, u_k)$, *where* $\alpha_k \in \Re^1, u_k \in \Re^n$.

<3> $y_k{}^T = \alpha_k u_k{}^T A^{(k)}{}_{k:n,k:n}$

<4> $\mu_k = \alpha_k y_k{}^T u_k$

<5> $x_k = y_k$

<6> $H_k A^{(k)} H_k = A^{(k)} - (x_k - \mu_k u_k) u_k{}^T - u_k y_k{}^T$

<7> *enddo*

Fig. 1. Tridiagonalization Processes

<1> *do* $k = n-2, 1, -1$

<2> *do* $i = k, n$

<3> $\sigma_i = \alpha_k u_k{}^T A^{(k)}{}_{k:n,i}$

<4> $H_k A^{(k)}{}_{k:n,i} = A^{(k)}{}_{k:n,i} - \sigma_i u_k$

<5> *enddo*; *enddo*;

Fig. 2. Inverse Transformation Processes

3   The Communication Splitting Multicasting Algorithm

3.1   The Data Distribution

Let the number of MPI processes be $np = px \times py$, and the process grid be $p \times q$. The process identification is also defined in a 2D manner, that is, (*myidx, myidy*), which ranges from 0 to *p-1* for *myidx* and from 0 to *q-1* for *myidy*.

The symmetric dense matrix A is distributed to each process in a cyclic-cyclic manner with a non-compressed form; thus, it does not use symmetry. In the cyclic-cyclic distribution, indexes of the row and column for matrix A are distributed as follows.

$$\Pi = \{myidx + 1 + (i-1)p\}, \Gamma = \{myidy + 1 + (j-1)q\}, \qquad (1)$$

where,

$$\begin{aligned} i &= 1,2,\cdots, last\ (myidx + 1 + \lfloor n/p \rfloor p, \lfloor n/p \rfloor) \\ j &= 1,2,\cdots, last\ (myidx + 1 + \lfloor n/q \rfloor q, \lfloor n/q \rfloor) \end{aligned}, last\ (a,b) = \begin{cases} b+1\ (if\ a \le n) \\ b\ (if\ a > n) \end{cases}$$

By using the index sets in Equation (1), we can denote the distributed matrix and vectors. For example, $A^{(k)}_{\Pi,k}$ is a vector that consists of a cyclic distribution for the first dimension and an entire k-th column for matrix $A^{(k)}$, and $u_{k\Pi}$ is a vector in which the elements are distributed with the cyclic manner of vector u_k.

3.2   The Square Grid Algorithm for Tridiagonalization

For the parallel algorithm for a square grid [4], that is, the case of $p=q$, tridiagonalization is used, as described in Figure 3.

In Figure 3, all communications can be implemented by using multiple MPI_BCASTs or MPI_ALLREDUCEs on the splitted communicator of MPI. The communication time can be reduced in massively parallel execution, in contrast to the time needed by a conventional algorithm that cannot split the communication. To perform matrix updating in parallel in lines <25>-<29> in Figure 3, we need the partial elements of x_k and u_k. The copies can perform 2x the tridiagonal multicasting operations in lines <5>-<7> to obtain the elements from $u_{k\Pi}$ to $u_{k\Gamma}$, and in lines <12>-<14> to transpose y to x. This enables us to dramatically reduce the communication time for the operation in comparison to that of the conventional algorithm [2].

Figure 4 shows the data distribution of vectors u_k, x_k, and y_k in the square grid algorithm. The data distribution of their duplications is also shown.

3.3   The Process Grid Free Algorithm for Tridiagonalization

The Case of a Rectangle Grid ($p < q$)

In [4], there is no description of rectangle process grid algorithms for the rectangle grid ($p < q$). However, the algorithm can be constructed by exchanging MPI_BCAST in Figure 3 with MPI_ALLREDUCE. We consequently establish a multicasting algorithm, but the communication time increases compared to the case of the square process grid.

The key point of the change is the transpose operations in lines <5>-<7> and <12>-<14> in Figure 3. In the case of $p < q$, there is no data for the elements of y_k with a p cyclic distribution to go with the elements of x_k with a q cyclic distribution. To avoid this situation, we add a copy process of $u_{k\Gamma}=u_{k\Pi}$ stridden (*myidy / px*) with offset (*py / px*), before line <6> in Figure 3 for u_k operation. This is the key implementation technique of this distribution. Figure 5 shows the data distribution of y_k and its multicastings based on this implementation.

The Case of a Rectangle Grid ($p > q$)

There is also no description and no performance evaluation in [4] for the rectangle grid ($p > q$). According to our verification, this algorithm can be described by exchanging MPI_BCASTs with MPI_ALLREDUCE, as in the case of $p < q$. But the stride and offset are changed to (*myidx* / *py*) and (*px* / *py*), respectively. This is also a key implementation technique to establish a grid free algorithm. Figure 6 shows the data distribution of y_k and its multicastings based on the above implementation.

The number of multicasting is reduced to the case of $p < q$, but the number of processes by MPI_ALLREDUCEs is increased. Hence, the best grid configuration depends on the communication performance of the tridiagonalization process.

<1> *do k=1, n-2*
<2>   if (k ∈ Γ) MPI_BCAST ($A_{\Pi,k}^{(k)}$) to Cores sharing rows Π.
<3>   else Receive data with MPI_BCAST ($A_{\Pi,k}^{(k)}$) endif;
<4>   Computation of ($a_k, u_{k\Pi}$) with MPI_ALLREDUCE;
<5>   if (I have diagonal elements of A)
<6>   MPI_BCAST ($u_{k\Pi}$) to Cores sharing columns Γ.
<7>   else Receive data with MPI_BCAST ($u_{k\Gamma}$) endif;
<8>   *do j=k, n*
<9>   if (j ∈ Γ) $y_{k\Pi}^{T} = y_{k\Pi}^{T} + \alpha_k u_k^{T}{}_j A_{\Pi,j}^{(k)}$
<10>   *enddo*
<11>   MPI_ALLREDUCE of $y_{k\Pi}^{T}$ to Cores sharing rows Π.
<12>   if (I have diagonal elements of A)
<13>   MPI_BCAST ($y_{k\Pi}^{T}$) to Cores sharing columns Γ.
<14>   else Receive data with MPI_BCAST ($x_{k\Gamma}$) endif;
<15>   *do j=k, n*
<16> $\mu_k = \alpha_k y_{k\Pi}^{T} u_{k\Pi}$ *enddo*
<17>   MPI_ALLREDUCE of μ_k to Cores sharing rows Π.
<25>   *do j=k, n*
<26>   *do i=k, n*
<27>   if (i ∈ Π .and. j ∈ Γ) then
<28> $A_{i,j}^{(k+1)} = A_{i,j}^{(k+1)} - u_{k_i}(x_k^{T}{}_j - \mu u_k^{T}{}_j) - u_{k_i} y_k^{T}{}_j$ endif;
<29>   *enddo; enddo;*
<30>   if (k ∈ Γ) Γ = Γ − {k} endif
<31>   if (k ∈ Π) Π = Π − {k} endif
<32> *enddo*

Fig. 3. Parallel Tridiagonalization Algorithm with Square Process Grid Proposed in [4]

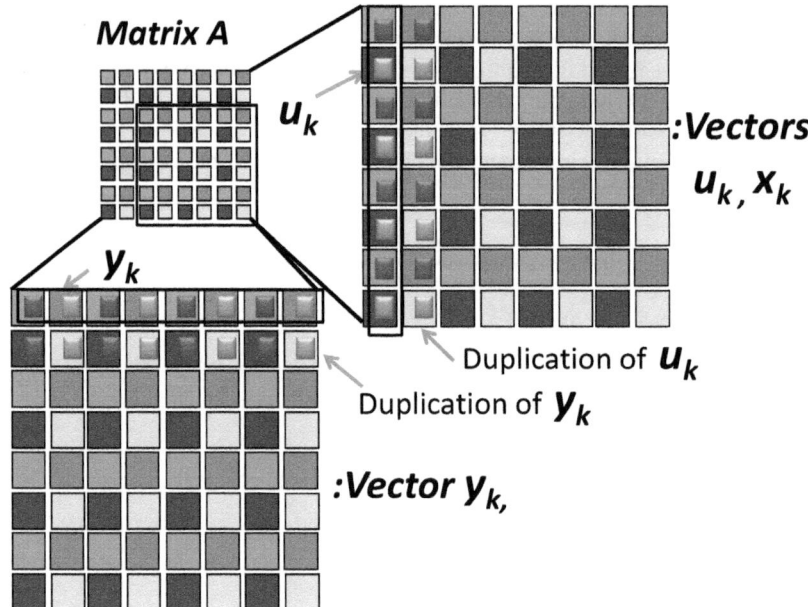

Fig. 4. Data Distribution of Vectors u_k, x_k, and y_k, and Vectors of Their Duplications

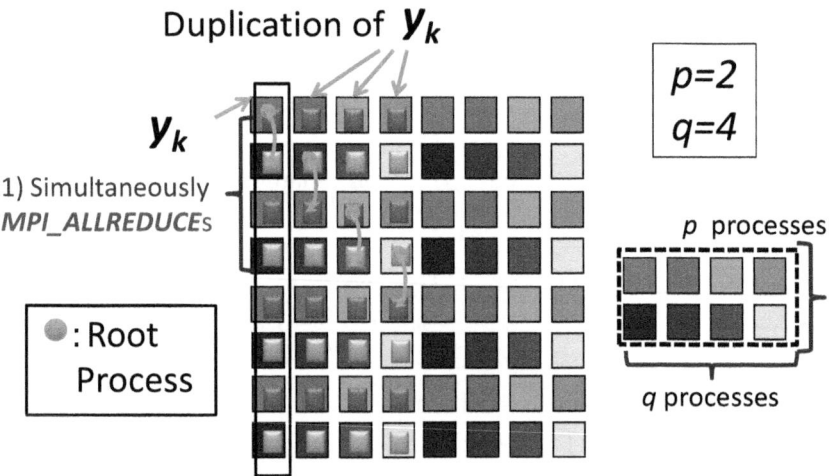

Fig. 5. Data Distribution of Vector y_k and Its Multicastings on the Process Grid Free Algorithm for Tridiagonalization. (the Case of Rectangle Grid $(p < q)$, $p = 2$ and $q = 4$).

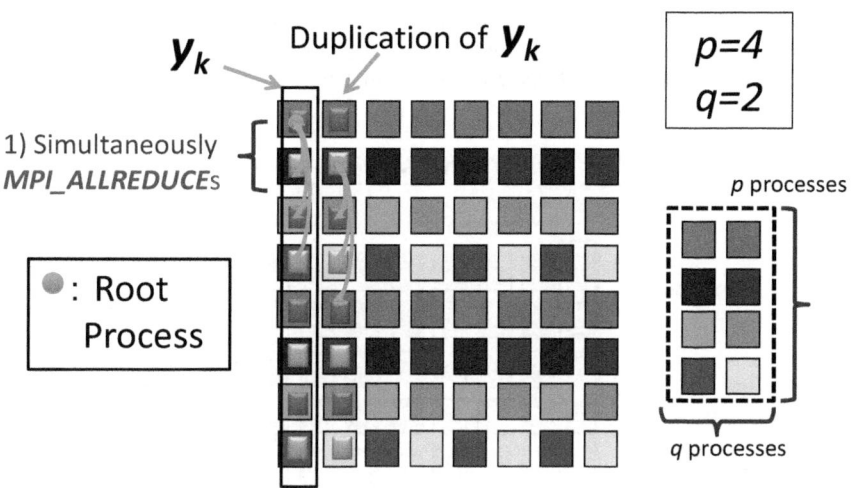

Fig. 6. Data Distribution of Vector y_k and Its Multicastings on the Process Grid Free Algorithm for Tridiagonalization. (the Case of Rectangle Grid ($p > q$), $p = 4$ and $q = 2$).

3.4   The Process Grid Free Algorithm for Inverse Transformation

Figure 7 shows the parallel algorithm for inverse transformation.

<1> *do k=n-2, 1, -1*
<2>   Gather the vector u_k and scalar α_k by using
          p-times of MPI_BCAST for $u_{k\Pi}$ with sharing . columns Γ.
<3>   *do i=kstart, kend*
<4> $\sigma_i = \alpha_k u_k^{T} A^{(k)}{}_{k:n,i}$
<5> $A^{(k)}{}_{k:n,i} = A^{(k)}{}_{k:n,i} - \sigma_i u_{k,}$
<6>   *enddo*
<7> *enddo*

Fig. 7. The Parallel Inverse Transformation Algorithm with Square Grid Proposed in [4]

The algorithm in Figure 7 also can be described with multiple MPI_BCASTs. The process grid affects the execution performance since p is the number of MPI_BCASTs for the p x q grid. The small p seems to perform better, but it depends on network performance. The data distribution of u_k and its multicastings are shown in Figure 8.

However, the memory requirement to store the Householder vector u_k varies according to p. If $p=2$, it needs 2x memory space compared to $p=4$ to keep the Householder (pivot) vectors. Therefore, this algorithm is a tradeoff between execution time and memory space. This algorithm is process grid free. The process grid of the inverse transformation is the same as that of tridiagonalization. From this point of view, the entire performance is determined by the communication performance between tridiagonalization and inverse transformation.

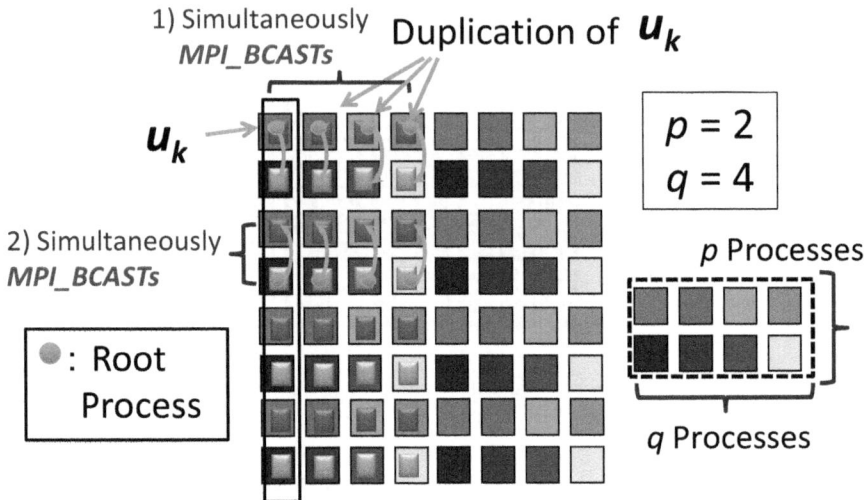

Fig. 8. Data Distribution of u_k and Its Multicastings on the Process Grid Free Algorithm for Inverse Transformation

4 Performance Evaluation

4.1 Machine Environment

We used the T2K Open Supercomputer (TODAI), which is a HITACHI HA8000 installed at the Information Technology Center, The University of Tokyo. Each node contains 4 sockets of the AMD Opteron 8356 (Quad core, 2.3 GHz). The L1 cache is 64 KB/core, the L2 cache is 512 KB/core, and the L3 cache is 2 MB/4 cores. The memory on each node is 32 GB with DDR2-667 MHz. The theoretical peak is 147.2 GFLOPS/node. Inter-node connection is 4 lines of the Myri-10G with a full bisection connection. The inter-node connection attains 5 GB/sec in both directions. We used the HITACHI Fortran90 Compiler version V01-00-/B with option "-opt=ss -noparallel." Users can use a maximum of 64 nodes (1,024 cores) for a personal application in normal service, but a maximum of 256 nodes (4,096 cores) is available for a special service, which can be performed once per month.

We also used the RICC PRIMERGY RX200S5 installed in the Advanced Center for Computing and Communication, RIKEN. Each node contains 2 sockets of the Intel Xeon X5570 (Quad core, 2.93 GHz). The L1 cache is 256 KB/core, the L2 cache is 1 MB/core, and the L3 is 8 MB/4 cores. The memory on the node is 12 GB with DDR3-1333 MHz. The theoretical peak is 93.0 GFLOPS/node. Inter-node connection is one line of the DDR InfiniBand. We used the Fujitsu Fortran90 Compiler version 3.2 with the option "-pc –high." In this experiment, 32 nodes (256 cores) were used.

We used ABCLib_DRSSED version 1.04 [5][6]. No automatic tuning was used in this experiment; hence, the default parameters were set.

4.2 Performance on Different Process Grids

Figure 9 shows the execution time. Table 1 shows the speedups and memory spaces in the cases of the square and rectangle grids on the T2K.

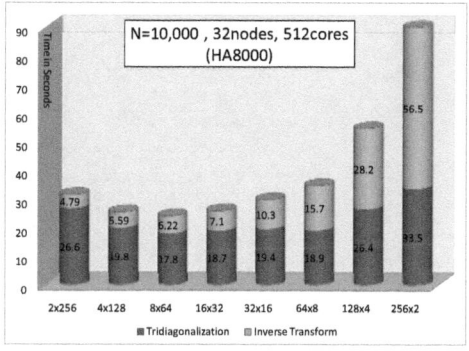
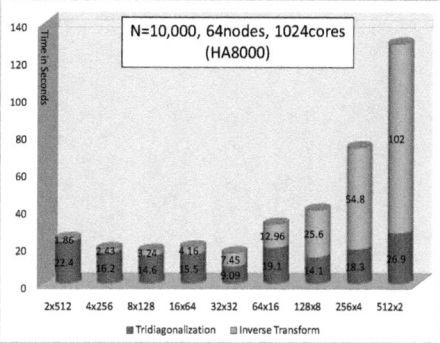

(a) 512 Cores ($p \neq q$)   (b) 1024 Cores ($p = q$)

Fig. 9. Execution Time on Different Process Grids on the T2K

In Figure 9, the execution time of $p > q$ increases because the gathering time for u_k increases according to p. In Table 1, speedup, memory space, speedup per memory (SPM) based on conventional execution are calculated. The conventional executions are 16x32 (512 cores) and 32x32 (1024 cores). If SPM is more than 1.0, it performs with good efficiency with respect to the ratio of speedup based on unit memory space.

Table 1. Speedups and Memory Spaces on the T2K. The Memory Space Is Calculated by the Memory Requirement to Keep Householder Vectors u_k.

(a)   512 Cores ($p \neq q$)   (b) 1024 Cores ($p = q$)

Grid ($p \times q$)	Time [sec.]	Speed UP	Mem.	SPM
16x32	25.8	1x	1x	1
8x64	24.0	1.07x	2x	0.5
4x128	25.3	1.01x	4x	0.2
2x256	31.3	0.82x	8x	0.1
32x16	29.7	0.86x	0.5x	1.7
64x8	24.6	0.74x	0.25x	2.9
128x4	54.6	0.47x	0.125x	3.7
256x2	90.0	0.28x	0.062x	4.5

Grid ($p \times q$)	Time [sec.]	Speed UP	Mem.	SPM
32x32	16.5	1x	1x	1
64x16	32.0	0.51x	0.5x	1.02
128x8	39.7	0.41x	0.25x	1.6
256x4	73.1	0.22x	0.125x	1.7
512x2	128	0.12x	0.062x	1.9

Table 1 shows that the case of $p>q$ has high efficiency with respect to SPM. Especially, the speed-down is only 0.86x, but memory space is reduced to 1/2 in Table 1 (a). In the $p = q$ case, the algorithm of 32x32 is very fast compared to the others. If users accept the 0.41x speed-down, the memory space can be reduced to 1/4.

Figure 10 shows the execution time. Table 2 shows the speedups and memory spaces in the cases of the square and rectangle grids on the RX200S5.

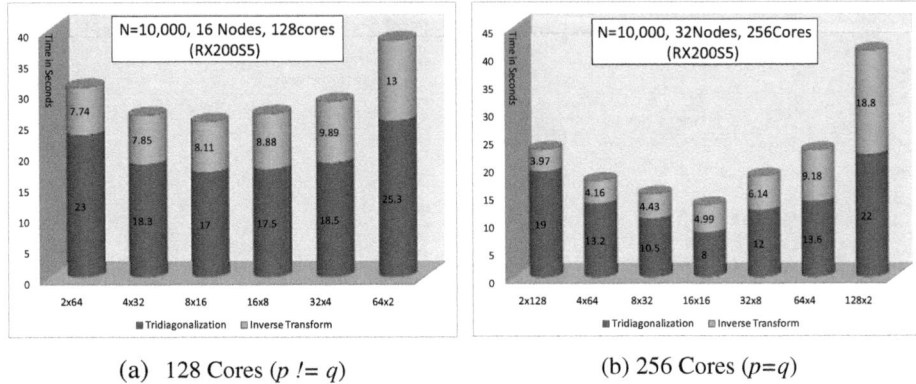

(a) 128 Cores ($p != q$)   (b) 256 Cores ($p=q$)

Fig. 10. Execution Time in Different Processor Grids on the RX200S5

For Table 2 (a), (b), the ratios of SPM are better than those of the T2K. The speed-down is only 0.95x, but the memory space is reduced to 1/2 in Table 2 (a). In Table 2 (a), the speed-down is only 0.71x with 32x8 compared to the case of 16x16.

Table 2. Speedups and Memory Spaces on the RX200S5. The Memory Space Is Calculated by the Memory Requirement for Householder Vectors u_k.

(a)   128 Cores ($p != q$)   (b) 1024 Cores ($p = q$)

Grid (pxq)	Time [sec.]	Speed UP	Mem.	SPM
8x16	25.1	1x	1x	1
4x32	26.1	0.96x	2x	0.48
2x64	30.7	0.81x	4x	0.20
16x8	26.3	0.95x	0.5x	1.9
32x4	28.3	0.88x	0.25x	3.5
64x2	38.3	0.65x	0.125x	5.2

Grid (pxq)	Time [sec.]	Speed UP	Mem.	SPM
16x16	12.9	1x	1x	1
32x8	18.1	0.71x	0.5x	1.4
64x4	22.7	0.56x	0.25x	2.2
128x2	40.8	0.31x	0.125x	2.4

4.3 Execution Performance in a Massively Parallel Environment

Figure 11 shows the execution time with 4,096 cores (256 nodes) on the T2K. In this experiment, 11 kinds of processor-grid configurations were tested.

Figure 11 indicates following interesting phenomena:

1. The conventional square grid, 64x64 (total execution time is 11.7+22.4=34.1 seconds), is not the fastest. In this case, 8x512 (the total execution time is 2.05+26.8=28.8 seconds) is the fastest.
2. The ratio between the time of inverse transformation and the total time decreases when p is reduced. The fastest case of inverse transformation takes only 0.283 seconds of execution in the process grid 2x2048. The ratio of execution time of inverse transformation to the total time is 0.283/59.7=0.47%. This is negligible time.
3. When p increases, the execution time of inverse transformation greatly increases. In addition, it causes a bottleneck due to the increase of processes according to multicastings. In contrast, the time of tridiagonalization only slightly affects the total time compared to the time of inverse transformation. This is due to good load balancing for the heavy computational part of tridiagonalization.

Phenomena 2 and 3 provide another possibility of optimization for symmetric eigensolvers. If we use much memory to store pivot vectors, we take small values of p. Because the heavy computational part is only tridiagonalization in this case, we can tune the routine in a simple manner. Conventional tuning is very complex since there is a tradeoff between tridiagonalization and inverse transformation in the communication, especially in a conventional square process grid. Again, our target is small-sized matrices. There is a room for memory space in our target. Hence, the algorithm with small values of p is a candidate for an efficient parallel algorithm to be considered in the future.

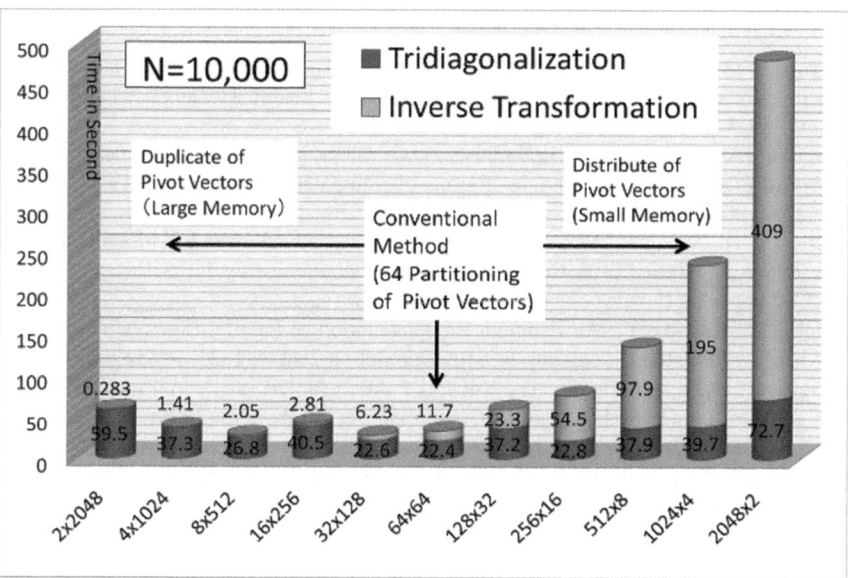

Fig. 11. Execution Time with 4,096 Cores (256 Nodes) on the T2K Using Different Processor Grids

5 Conclusion

In this paper, we propose a process grid free algorithm for a massively parallel dense symmetric eigensolver with a communication splitting multicasting algorithm. A tradeoff exists between speed and memory space in this algorithm. As a result of the performance evaluation with the T2K Open Supercomputer (HITACHI HA8000) and RICC RX200S5, we found that 0.86x and 0.95x speed-downs with 1/2 memory space allows us to keep the Householder vectors.

We showed the possibility of this new algorithm for small-sized matrices on massively parallel processing to take appropriately small values of p of process grid p x q. In this case, the execution time of inverse transformation is negligible.

The blocking parallel algorithm is now being studied in [7] and takes into account the communication reduction for the symmetric dense eigensolver. Implementing a communication-hiding algorithm with previous sending for the next-step Householder vector is important future work for small-sized matrices on massively parallel processing.

Acknowledgments. We thank the RIKEN Cluster of Clusters (RICC) at RIKEN for the computer resources used for the experiment. This work is partially supported by Grant-in-Aid for Scientific Research (B) "Development of the Framework to Support Large-scale Numerical Simulation on Multi-platform," No. 21300017, and Grant-in-Aid for Scientific Research (B) "Development of Auto-tuning Specification Language Towards Manycore and Massively Parallel Processing Era," No. 21300007.

References

1. Inaba, T., Tsunekawa, N., Hirano, T., Yoshihiro, T., Kashiwagi, H., Sato, F.: Density Functional Calculation of the Electronic Structure on Insulin Hexamer. Chemical Physics Letters 434(4-6), 331–335 (2007)
2. Dongarra, J.J., Hammarling, S.J., Sorensen, D.C.: Block Reduction of Matrices to Condensed Forms for Eigenvalue Computations. Journal of Computational and Applied Mathematics 27, 215–227 (1989)
3. Bischof, C.H., Marques, M., Sun, X.: Parallel Bandreduction and Tridiagonalization. In: Proceedings of Sixth SIAM Conference on Parallel Processing for Scientific Computing, pp. 22–24 (1993)
4. Katagiri, T., Kanada, Y.: An Efficient Implementation of Parallel Eigenvalue Computation for Massively Parallel Processing. Parallel Computing 27(14), 1831–1845 (2001)
5. Katagiri, T., Kise, K., Honda, H., Yuba, T.: ABCLib_DRSSED: A Parallel Eigensolver with an Auto-tuning Facility. Parallel Computing 32(3), 231–250 (2006)
6. ABCLib_DRSSED home page, http://www.abc-lib.org/main1.html
7. Imamura, T.: How To Develop The Eigenvalue Solver Which Organizes Beyond Hundred Thousand Cores. IPSJ SIG Notes 2009-HPC-121(19) (2009) (in Japanese)

Global Memory Access Modelling for Efficient Implementation of the Lattice Boltzmann Method on Graphics Processing Units

Christian Obrecht[1], Frédéric Kuznik[1],
Bernard Tourancheau[2], and Jean-Jacques Roux[1]

[1] Centre de Thermique de Lyon
(UMR 5008 CNRS, INSA-Lyon, Université de Lyon)
Bât. Sadi Carnot, 9 rue de la Physique,
69621 Villeurbanne Cedex, France
{christian.obrecht,frederic.kuznik,jean-jacques.roux}@insa-lyon.fr
[2] Laboratoire de l'Informatique du Parallélisme
(UMR 5668 CNRS, ENS de Lyon, INRIA, UCB Lyon 1)
École Normale Supérieure de Lyon, 46 allée d'Italie,
69364 Lyon Cedex 07, France
bernard.tourancheau@ens-lyon.fr

Abstract. In this work, we investigate the global memory access mechanism on recent GPUs. For the purpose of this study, we created specific benchmark programs, which allowed us to explore the scheduling of global memory transactions. Thus, we formulate a model capable of estimating the execution time for a large class of applications. Our main goal is to facilitate optimisation of regular data-parallel applications on GPUs. As an example, we finally describe our CUDA implementations of LBM flow solvers on which our model was able to estimate performance with less than 5% relative error.

Keywords: GPU computing, CUDA, lattice Boltzmann method, CFD.

Introduction

State-of-the-art graphics processing units (GPU) have proven to be extremely efficient on regular data-parallel algorithms [1]. For many of these applications, like lattice Boltzmann method (LBM) fluid flow solvers, the computational cost is entirely hidden by global memory access. The present study intends to give some insight on the global memory access mechanism of the nVidia's GT200 GPU. The obtained results led us to optimisation elements which we used for our implementations of the LBM.

The structure of this paper is as follows. First, we briefly review nVidia's compute unified device architecture (CUDA) technology and the algorithmic aspects of the LBM. Then, we describe our measurement methodology and results. To conclude, we present our CUDA implementations of the LBM.

J.M.L.M. Palma et al. (Eds.): VECPAR 2010, LNCS 6449, pp. 151–161, 2011.
© Springer-Verlag Berlin Heidelberg 2011

1   Compute Unified Device Architecture

CUDA capable GPUs, i.e. the G8x, G9x, and GT200 processors consist in a variable amount of texture processor clusters (TPC) containing two (G8x, G9x) or three (GT200) streaming multiprocessors (SM), texture units and caches [2]. Each SM contains eight scalar processors (SP), two special functions units (SFU), a register file, and shared memory. Registers and shared memory are fast but in rather limited amount, e.g. 64 KB and 16 KB per SM for the GT200. On the other hand, the off-chip global memory is large but suffers from high latency and low throughput compared to registers or shared memory.

The CUDA programming language is an extension to C/C++. Functions intended for GPU execution are named *kernels*, which are invoked on an execution grid specified at runtime. The execution grid is formed of blocks of threads. The blocks may have up to three dimensions, the grid two. During execution, blocks are dispatched to the SMs and split into warps of 32 threads.

CUDA implementations of data intensive applications are usually bound by global memory throughput. Hence, to achieve optimal efficiency, the number of global memory transactions should be minimal. Global memory transactions within a half-warp are coalesced into a single memory access whenever all the requested addresses lie in the same aligned segment of size 32, 64, or 128 bytes. Thus, improving the data access pattern of a CUDA application may dramatically increase performance.

2   Lattice Boltzmann Method

The Lattice Boltzmann Method is a rather innovative approach in computational fluid dynamics [3,4,5]. It is proven to be a valid alternative to the numerical integration of the Navier-Stockes equations. With the LBM, space is usually represented by a regular lattice. The physical behaviour of the simulated fluid is determined by a finite set of *mass fractions* associated to each node. From an algorithmic standpoint, the LBM may be summarised as:

> **for each** time step **do**
>   **for each** lattice node **do**
>   **if** boundary node **then**
>   apply boundary conditions
>   **end if**
>   compute new mass fractions
>   propagate to neighbouring nodes
>   **end for**
> **end for**

The propagation phase follows some specific stencil. Figure 1 illustrates D3Q19, the most commonly used three-dimensional stencil, in which each node is linked to 18 of its 27 immediate neighbours.[1]

---

[1] Taking the stationary mass fraction into account, the number of mass fractions per node amounts to 19, hence D3Q19.

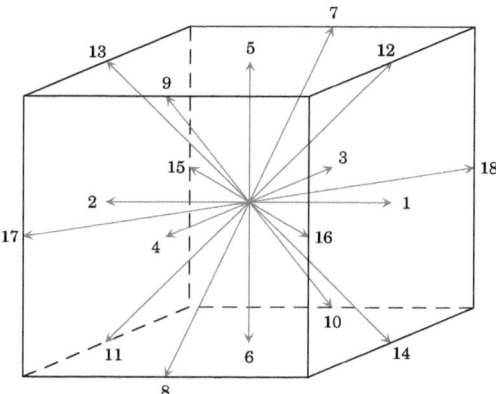

Fig. 1. The D3Q19 stencil

CUDA implementations of the LBM may take advantage of its inherent data parallelism by assigning a thread to each node, the data being stored in global memory. Since there is no efficient global synchronisation barrier, a kernel has to be invoked for each time step [6]. CPU implementations of the LBM usually adopt an array of structures (AoS) data layout, which improves locality of mass fractions belonging to a same node [7]. On the other hand, CUDA implementations benefit from structure of arrays (SoA) data layouts, which allows coalesced global memory accesses [8]. However, this approach is not sufficient to ensure optimal memory transactions, since propagation corresponds to one unit shifts of global memory addresses for the minor spatial dimension. In other words, for most mass fractions, the propagation phase yields misalignments. A way to solve this issue consists in performing propagation partially in shared memory [9]. Yet, as shown in [10], this approach is less efficient than using carefully chosen propagation schemes in global memory.

3   Methodology

To study transactions between global memory and registers, we used kernels performing the following operations:

1. Store time t_0 in a register.
2. Read N words from global memory, with possibly L misalignments.
3. Store time t_1 in a register.
4. Write N words to global memory, with possibly M misalignments.
5. Store time t_2 in a register.
6. Write t_2 to global memory.

Time is accurately determined using the CUDA `clock()` function which gives access to counters that are incremented at each clock cycle. Our observations enabled us to confirm that these counters are per TPC, as described in [11], and not per SM as stated in [2]. Step 6 may influence the timings, but we shall see that it can be neglected under certain circumstances.

The parameters of our measurements are N, L, M, and k, the number of warps concurrently assigned to each SM. Number k is proportional to the occupancy rate α, which is the ratio of active warps to the maximum number of warps supported on one SM. With the GT200, this maximum number being 32, we have: $k = 32\alpha$.

We used a one-dimensional grid and one-dimensional blocks containing one single warp. Since the maximum number of blocks supported on one SM is 8, the occupancy rate is limited to 25%. Nonetheless, this rate is equivalent to the one obtained with actual CUDA applications.

We chose to create a script generating the kernels rather than using runtime parameters and loops, since the layout of the obtained code is closer to the one of actual computation kernels. We processed the CUDA binaries using decuda [12] to check whether the compiler had reliably translated our code. We carried out our measurements on a GeForce GTX 295 graphics board, featuring two GT200 processors.[2]

4   Modelling

At kernel launch, blocks are dispatched to the TPCs one by one up to k blocks per SM [13]. Since the GT200 contains ten TPCs, blocks assigned to the same TPC have identical `blockIdx.x` unit digit. This enables to extract information about the scheduling of global memory access at TPC level. In order to compare the measurements, as the clock registers are peculiar to each TPC [11], we shifted the origin of the time scale to the minimal t_0. We noticed that the obtained timings are coherent on each of the TPCs.

For a number of words read and written $N \le 20$, we observed that:

- Reads and writes are performed in one stage, hence storing of t_2 has no noticeable influence.
- Warps 0 to 8 are launched at once (in a determined but apparently incoherent order).
- Subsequent warps are launched one after the other every ~ 63 clock cycles.

For $N > 20$, reads and writes are performed in two stages. One can infer the following behaviour: if the first n warps in a SM read at least 4,096 words, where $n \in \{4, 5, 6\}$, then the processing of the subsequent warps is postponed. The number of words read by the first n warps being $n \times 32N$, this occurs whenever $n \times N \ge 128$. Hence, $n = 4$ yields $N \ge 32$, $n = 5$ yields $N \ge 26$, and $n = 6$ yields $N \ge 21$.

Time t_0 for the first $3n$ warps of a TPC follow the same pattern as in the first case. We also noticed a slight overlapping of the two stages, all the more

---

[2] In the CUDA environment, the GPUs of the GTX 295 are considered as two distinct devices. It should be noted that our benchmark programs involve only one of those devices.

as storing t_2 should here be taken into account. Nonetheless, the read time for the first warp in the second stage is noticeably larger than for the next ones. Therefore, we may consider, as a first approximation, that the two stages are performed sequentially.

In the targeted applications, the global amount of threads is very large. Moreover, when a set of blocks is assigned to the SMs, the scheduler waits until all blocks are completed before providing new ones. Hence, knowing the average processing time T of k warps per SM allows to estimate the global execution time.

For $N \leq 20$, we have $T = \ell + T_R + T_W$, where ℓ is time t_0 for the last launched warp, T_R is read time, and T_W is write time. Time ℓ only depends on k. For $N > 20$, we have $T = T_0 + \ell' + T'_R + T'_W$, where T_0 is the processing time of the first stage, $\ell'(i) = \ell(i - 3n + 9)$ with $i = 3k - 1$, T'_R and T'_W are read and write times for the second stage.

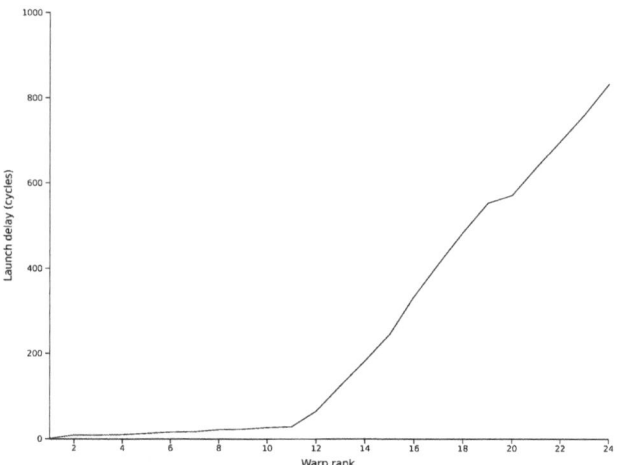

Fig. 2. Launch delay in respect of warp rank

To estimate ℓ, we averaged t_0 over a large number of warps. Figure 2 shows, in increasing order, the obtained times in cycles. Numerically, we have $\ell(i) \approx 0$ for $i \leq 9$ and $\ell(i) \approx 63(i - 10) + 13$ otherwise.

5   Throughput

5.1 $N \leq 20$

Figures 3 and 4 show the distribution of read and write times for 96,000 warps with $N = 19$. The bimodal shape of the read time distribution is due to translation look-aside buffer (TLB) misses [14]. This aspect is reduced when adding misalignments, since the number of transactions increases while the number of misses remains constant. Using the average read time to approximate T is acceptable provided no special care is taken to avoid TLB misses.

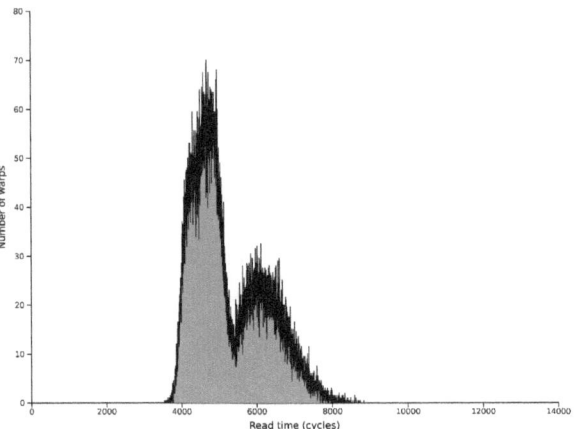

Fig. 3. Read time for $N = 19$

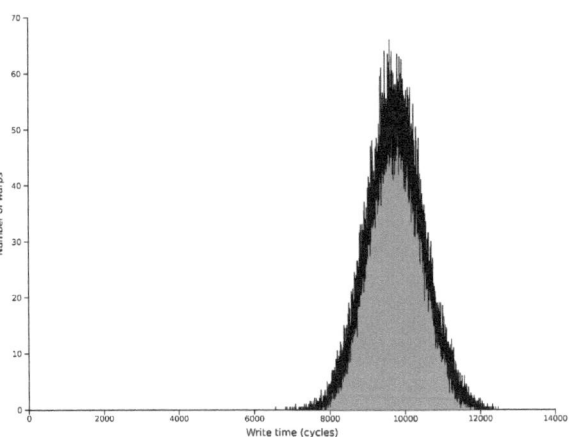

Fig. 4. Write time for $N = 19$

We observed that average read and write times depend linearly of N. Numerically, with $k = 8$, we obtained:

$$T_R \approx 317(N-4) + 440 \qquad T_W \approx 562(N-4) + 1{,}178$$

$$T_{R'} \approx 575(N-4) + 291 \qquad T_{W'} \approx 983(N-4) + 2{,}030$$

where $T_{R'}$ and $T_{W'}$ are read and write times with $L = N$ and $M = N$ misalignments. Hence, we see that writes are more expensive than reads. Likewise, misalignments in writes are more expensive than misalignments in reads.

5.2 $21 \leq N \leq 39$

As shown in figures 5 and 6, T_0, T'_R, and T'_W depend linearly of N in the three intervals $\{21, \ldots 25\}$, $\{26, \ldots 32\}$, and $\{33, \ldots 39\}$. As an example, for the third interval, we obtain:

$$T_0 \approx 565(N - 32) + 15{,}164$$

$$T'_R \approx 112(N - 32) + 2{,}540 \qquad T'_W \approx 126(N - 32) + 3{,}988$$

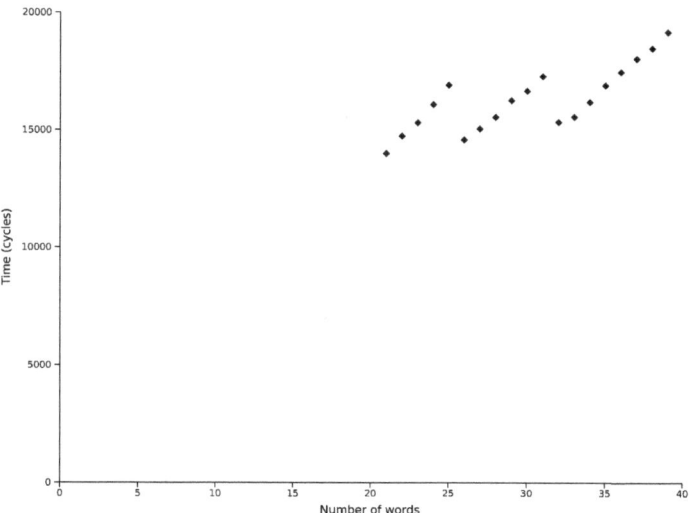

Fig. 5. First stage duration

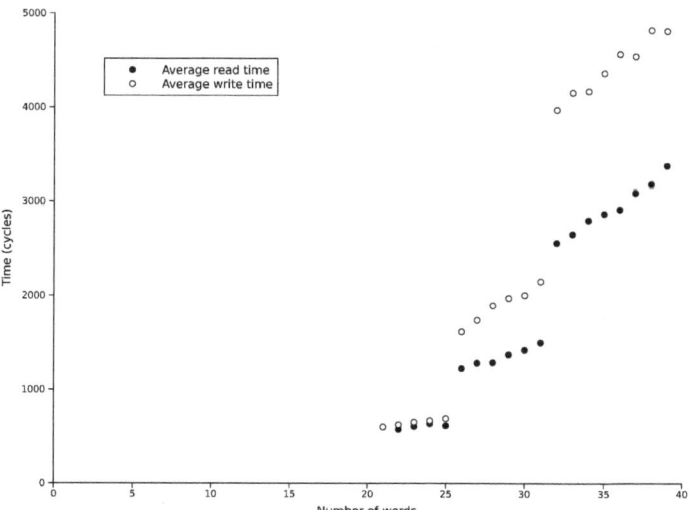

Fig. 6. Timings in second stage

5.3   Complementary Studies

We also investigated the impact of misalignments and occupancy rate on average read and write times. Figures 7 and 8 show obtained results for $N = 19$.

For misaligned reads, we observe that the average write time remains approximatively constant. Read time increases linearly with the number of misalignments until some threshold is reached. From then on, the average read time is maximal. Similar conclusion can be drawn for misaligned writes.

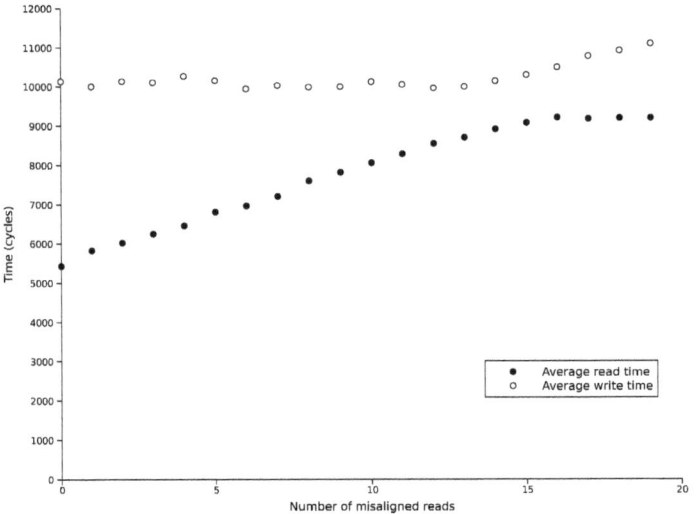

Fig. 7. Misaligned reads

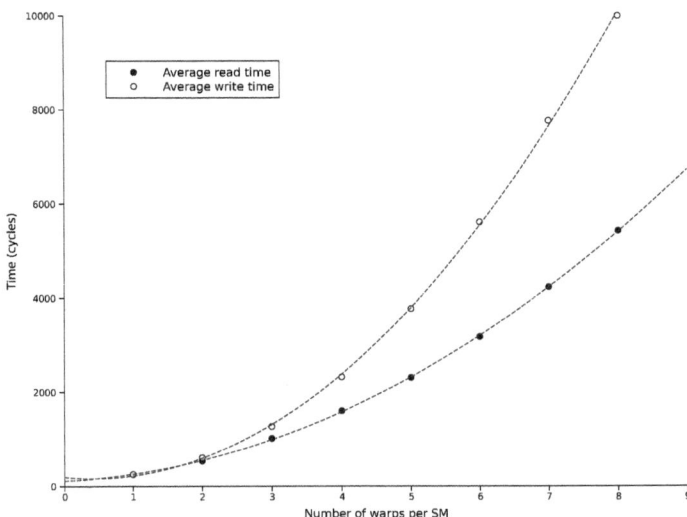

Fig. 8. Occupancy impact

Average read and write times seem to depend quadratically on k. Since the amount of data transferred depends only linearly on k, this leads to think that the scheduling cost of each warp is itself proportional to k.

6   Implementations

We implemented several LBM fluid flow solvers: a D3Q19 LBGK [4], a D3Q19 MRT [5], and a double population thermal model requiring 39 words per node [15]. Our global memory access study lead us to multiple optimisations. For each implementation, we used a SoA like data layout, and a two-dimensional grid of one-dimensional blocks. Since misaligned writes are more expensive than misaligned reads, we experimented several propagation schemes in which misalignments are deferred to the read phase of the next time step. The most efficient appears to be the reversed scheme where propagation is entirely performed at reading, as outlined in figure 9. For the sake of simplicity, the diagram shows a two-dimensional version.

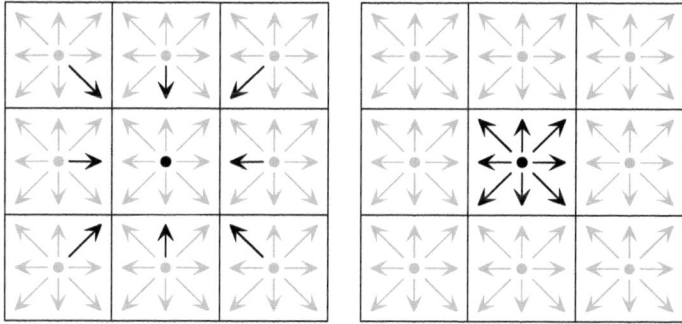

Fig. 9. Reversed propagation scheme

Performance of a LBM based application is usually given in million lattice node updates per second (MLUPS). Our global memory access model enables us to give an estimate of the time T (in clock cycles) required to process k warps per SM. On the GT200, where the number of SMs is 30 and the warp size is 32, k warps per SM amounts to $K = 30 \times k \times 32 = 960k$ threads. Since one thread takes care of one single node, T is therefore the number of clock cycles needed to perform K lattice node updates. Hence, using the global memory frequency F in MHz, the expected performance in MLUPS is: $P = (K/T) \times F$.

With our D3Q19 implementations, for instance, we have $N = 19$ reads and writes, $L = 10$ misaligned reads, no misaligned writes, and 25% occupancy (thus $k = 8$). Using the estimation provided by our measurements, we obtain: $T = \ell + T_R + T_W = 15,594$. Since $K = 7,680$ and $F = 999$ MHz, we have $P = 492$ MLUPS.

To summarize, table 1 gives both the actual and estimated performances for our implementations on a 128^3 lattice. Our estimations appear to be rather accurate, thus validating our model.

Table 1. Performance of LBM implementations (in MLUPS)

Model	Occupancy	Actual	Estimated	Relative error
D3Q19 LBGK	25%	481	492	2.3%
D3Q19 MRT	25%	516	492	4.6%
Thermal LBM	12.5%	195	196	1.0%

Summary and Discussion

In this work, we present an extensive study of the global memory access mechanism between global memory and GPU for the GT200. A description of the scheduling of global memory accesses at hardware level is given. We express a model which allows to estimate the global execution time of a regular data-parallel application on GPU. The cost of individual memory transactions and the impact of misalignments is investigated as well.

We believe our model is applicable to other GPU applications provided certain conditions are met:

- The application should be data-parallel and use a regular data layout in order to ensure steady data throughput.
- The computational cost should be negligible as compared with the cost of global memory reads and writes.
- The kernel should make moderate use of branching in order to avoid branch divergence, which can dramatically impact performance. This would probably not be the case with an application dealing, for instance, with complex boundaries.

On the other hand, our model does not take possible TLB optimisation into account. Hence, some finely tuned applications may slightly outvalue our performance estimation.

The insight provided by our study, turned out to be useful in our attempts to optimize CUDA implementations of the LBM. It may contribute to efficient implementations of other applications on GPU.

Acknowledgement

We thank the reviewers for their helpful comments improving the clarity of our paper.

References

1. Dongarra, J., Moore, S., Peterson, G., Tomov, S., Allred, J., Natoli, V., Richie, D.: Exploring new architectures in accelerating CFD for Air Force applications. In: Proceedings of HPCMP Users Group Conference, Citeseer, pp. 14–17 (2008)
2. nVidia: Compute Unified Device Architecture Programming Guide version 2.3.1 (August 2009)
3. McNamara, G.R., Zanetti, G.: Use of the Boltzmann Equation to Simulate Lattice-Gas Automata. Phys. Rev. Lett. 61, 2332–2335 (1988)
4. Qian, Y.H., d'Humières, D., Lallemand, P.: Lattice BGK models for Navier-Stokes equation. Europhys. Lett. 17(6), 479–484 (1992)
5. d'Humières, D., Ginzburg, I., Krafczyk, M., Lallemand, P., Luo, L.: Multiple-relaxation-time lattice Boltzmann models in three dimensions. Philosophical Transactions: Mathematical, Physical and Engineering Sciences, 437–451 (2002)
6. Ryoo, S., Rodrigues, C.I., Baghsorkhi, S.S., Stone, S.S., Kirk, D.B., Hwu, W.W.: Optimization principles and application performance evaluation of a multithreaded GPU using CUDA. In: Proceedings of the 13th ACM SIGPLAN Symposium on Principles and Practice of Parallel Programming, pp. 73–82. ACM, New York (2008)
7. Pohl, T., Kowarschik, M., Wilke, J., Iglberger, K., Rüde, U.: Optimization and Profiling of the Cache Performance of Parallel Lattice Boltzmann Codes. Parallel Processing Letters 13(4), 549–560 (2003)
8. Kuznik, F., Obrecht, C., Rusaouën, G., Roux, J.J.: LBM Based Flow Simulation Using GPU Computing Processor. Computers and Mathematics with Applications (27) (June 2009)
9. Tölke, J., Krafczyk, M.: TeraFLOP computing on a desktop PC with GPUs for 3D CFD. International Journal of Computational Fluid Dynamics 22(7), 443–456 (2008)
10. Obrecht, C., Kuznik, F., Tourancheau, B., Roux, J.J.: A new approach to the lattice Boltzmann method for graphics processing units. Computers and Mathematics with Applications (in press, 2010)
11. Papadopoulou, M., Sadooghi-Alvandi, M., Wong, H.: Micro-benchmarking the GT200 GPU
12. van der Laan, W.J.: Decuda G80 dissassembler version 0.4 (2007)
13. Collange, S., Defour, D., Tisserand, A.: Power Consumption of GPUs from a Software Perspective. In: Proceedings of the 9th International Conference on Computational Science: Part I, p. 923. Springer, Heidelberg (2009)
14. Volkov, V., Demmel, J.: Benchmarking GPUs to tune dense linear algebra. In: Proceedings of the 2008 ACM/IEEE Conference on Supercomputing. IEEE Press, Piscataway (2008)
15. Peng, Y., Shu, C., Chew, Y.T.: A 3D incompressible thermal lattice Boltzmann model and its application to simulate natural convection in a cubic cavity. Journal of Computational Physics 193(1), 260–274 (2004)

Data Structures and Transformations for Physically Based Simulation on a GPU

Perhaad Mistry[1], Dana Schaa[1], Byunghyun Jang[1], David Kaeli[1],
Albert Dvornik[2], and Dwight Meglan[2]

[1] Department of Electrical and Computer Engineering
Northeastern University, Boston, MA, U.S.A.
{pmistry,dschaa,bjang,kaeli}@ece.neu.edu
[2] Simquest LLC, Boston, MA, USA
{advornik,dmeglan}@simquest.com

Abstract. As general purpose computing on Graphics Processing Units (GPGPU) matures, more complicated scientific applications are being targeted to utilize the data-level parallelism available on a GPU. Implementing physically-based simulation on data-parallel hardware requires preprocessing overhead which affects application performance. We discuss our implementation of physics-based data structures that provide significant performance improvements when used on data-parallel hardware. These data structures allow us to maintain a physics-based abstraction of the underlying data, reduce programmer effort and obtain 6x-8x speedup over previously implemented GPU kernels.

1   Introduction

In any useful surgical simulation system, in order to meet the strict requirements of proper visual and behavioral illusion of reality, the system must solve a number of physics-based problems such as cutting and deformation at interactive speeds [12]. However surgical simulation cannot leverage the tricks that are used in "game physics". Physics engines for soft body simulation concentrate on real time and visually plausible results, while surgical simulation requires numerical stability and accuracy due to the critical nature of the simulation [12].

The present development trend of computational science software libraries is not driven by changes in problem-specific methodology [9], but by the fundamental shift of the underlying hardware towards heterogeneity and parallelism. This is particularly true for data-intensive problems such as finite element analysis. GPUs have become the technology of choice for data-parallel applications due to their potential for impressive speedups and their ability to accelerate a range of general purpose programs [3,5,10].

Our current work involves accelerating the physics simulation library PhysBAM [20] using the Compute Unified Device Architecture (CUDA) on NVIDIA GPUs for a real time surgical simulator. PhysBAM is an object oriented library that works with dynamically generated data structures to simplify the modeling of the underlying physics.

J.M.L.M. Palma et al. (Eds.): VECPAR 2010, LNCS 6449, pp. 162–171, 2011.

Physically-based modeling techniques have been used to properly model time-varying properties such as geometry and topology [21].

Physics simulation algorithms possess inherent data parallelism, however parallelizing such algorithms naively leads to high overhead preprocessing since data parallelism is interspersed throughout the simulation and few compute-intensive "kernels" exist on which optimization efforts can be concentrated. This leads us to search for optimization techniques which can be applied more widely across an application and can improve data parallel performance irrespective of the underlying data structures and algorithms. In this paper, we describe methods to improve data layout and use them to accelerate physical simulation. We present a framework for physically- based simulation that automatically translates dynamic data structures to match the requirements of the GPU memory subsystem.

2 Related Work

A number of prior studies have addressed acceleration of physics simulation and finite element analysis using GPUs [6,7,8,9]. The GPU implementation of FEAST [8] is based on a scalable recursive multi-grid algorithm which prevents us from using it for our surgical simulation implementation due to the real time requirements of our environment and the need to simulate cutting. Farias et al. [16] discuss physically precise deformation and demonstrate very good performance for their particular methodology. Our work attempts to be agnostic to specific algorithms and provides a framework to implement different types of data-parallel physics algorithms that can effectively exploit the resources of a GPU.

The motivation behind our work is to build a simulation engine similar to Bullet [18] that models soft tissue deformation and cutting accurately enough to be applied to surgical simulation. Physics simulation for game and visual realism has been implemented using NVIDIA GPUs in PhysX [2] and is available as a middleware for CUDA capable GPUs. Other physics simulation work for CUDA-based hardware includes [17,18]. For our simulated environment we need to provide accurate soft tissue deformation, so our goal is to more closely couple the physics of the problem with our data parallel implementation.

The implementation described in HONEI [9] is relevant to our work since Dyk et al. also explored the heterogeneity and parallelism between GPUs and CPUs. Our work is different from HONEI in the sense that our work is specific to Physics simulation and the relationship between data-parallel structures and the underlying physics parameters. The data structures provided within HONEI are oriented towards finite element analysis. The Simulation Open Framework Architecture (SOFA) [6] is a framework for surgical simulation, but does not provide the computational infrastructure that will be required for our future work. The Fenics Form Compiler [11] is also related to our work since it deals with the conversion of mathematical expressions into programs that can be executed using low-level linear algebra libraries for general purpose CPUs. Our method is complimentary to this compiler since it is related to data structures and is designed to improve performance for a different computing platform (GPUs).

3   Physically Based Simulation Framework

To describe our data-parallel physics simulation framework, we first discuss how the memory coalescing requirements for NVIDIA GPUs affect the design of our physics-based data structures. We then show how our framework can be used to build structures that exploit the GPU memory sub-system.

3.1   Coalesced Memory Accesses from Arrays of Objects

Physics based data-structures are arrays of dynamically generated objects denoting multiple instances of physical quantities like force, displacement, etc. Within a generic physics simulation engine, due to the lack of a priori knowledge of data layouts, these arrays do not typically reside in contiguous locations in memory. The typical solution for working with arrays of objects on the GPU is to allocate a contiguous block of memory that can hold all the structures, and then copy the complete array of objects into consecutive locations in memory (Figure 1).

The approach of simply contiguously storing data structures will impede performance due to the CUDA memory coalescing rules. Memory coalescing in CUDA is defined as reads or writes by threads to consecutive 4-byte elements in memory. If coalescing is not achieved, accesses are serialized and bandwidth degrades significantly. Figure 1 shows memory accesses when consecutive 3-element data structures try to access memory in CUDA.

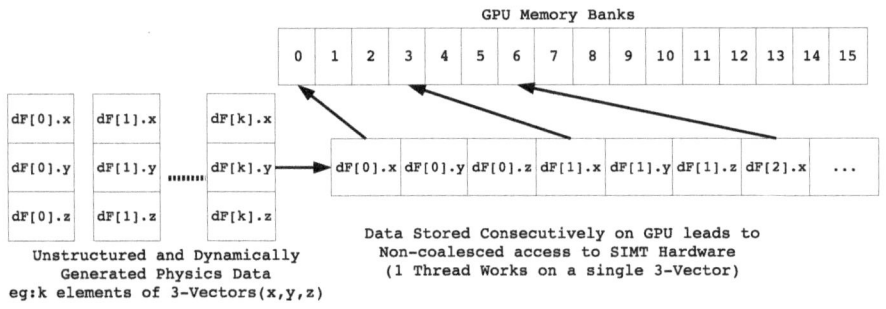

Fig. 1. For data structures stored consecutively, non-coalesced accesses occur

As shown in Figure 2, we need to rearrange the allocated data in linear memory locations to map efficiently to the underlying data-parallel hardware.

3.2   Automated Framework for Physics Data Structures

We have implemented a framework that allows us to create data structures for physics simulation algorithms adhering to the memory coalescing requirements. The motivation

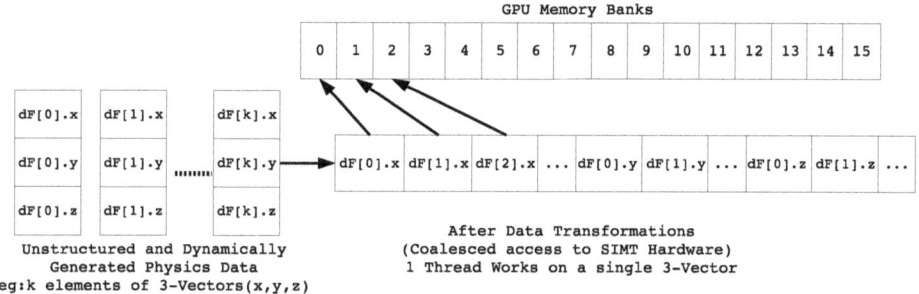

Fig. 2. Data structures that are transformed for coalesced memory access

behind our framework is illustrated by listing some calculations and data structures required for modeling deformation of a silicone cube. The technique used for deformation modeling is based on [14] and entails three main steps.

1. **Singular Value Decomposition(SVD):** The SVD of an array of 3x3 matrices is calculated using approximate methods [14,13].
2. **Stress Derivative:** An array of structures of stress parameters denoting the constitutive model. The results are used to update the stiffness matrices for the iterative solver.
3. **Solving Linear System:** Arrays of 3x3 matrices and 3x3 symmetric matrices for stiffness arrays. This step also requires arrays of vectors for force and displacement.

In our simulator, the data structures and parallel algorithms are closely coupled to the physics theory. The physics based data structures denoted above are only a small subset of the possible dynamically generated arrays which are different for each simulation. Due to the variety of algorithms that could be implemented using PhysBAM [20], little information is known apriori about the data structures and kernels[1] that will be invoked in a simulation. Due to this characteristic of our application, providing a limited set of optimized data structures is not beneficial.

The architecture of our framework called **GPUPhysBAM** (GPU Physics Based Modeling) is shown in Figure 3. The available physics simulation library dynamically generates data whose structure is determined as per the algorithm being simulated. The intermediate layer of GPUPhysBAM will allocate data in the GPU memory while keeping in mind constraints like alignment and ordering for inter-thread access. These constraints have to be satisfied in order to take advantage of the high-bandwidth memory bus between the GPU memory and the GPU's SIMD processors.

Implementation of data structures that can be reused and adapted to different simulation algorithms without sacrificing performance is key since we wish to maintain the generic nature of the simulation library and allow our data-parallel structures to impact a wider range of algorithms.

---

[1] We refer to kernels as code that is parallelized and offloaded to the GPU.

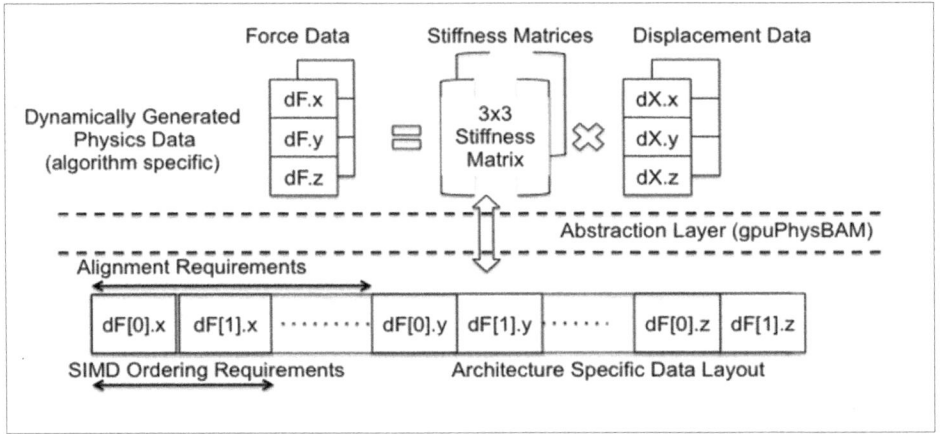

Fig. 3. GPUPhysBAM Architecture

3.3 Data Transformations and Hierarchically Designed Data Structures

Due to the complicated nature of physics based simulation code where both the CPU and the GPU play a significant role in computation, data has to be organized such that computation on both the CPU and the GPU yields optimal performance.

A simple example that illustrates the importance of properly designed data structures for the CPU and GPU is shown in Figure 4. We present a simple program $Y = f(X)$ where X and Y are arrays of 3-element vectors indexed as $X_i[0] - X_i[2]$ and $f()$ is a function that operates on each element of each vector independently. For the CPU layout, function $f()$ was applied as in the loop nest in Figure 4. For the loop in Figure 4, data layout 1 would be optimal. However, as discussed in Section 3.1 such a layout would prevent coalesced reads and writes on the GPU if each iteration of the loop maps to a thread. The optimal layout for CUDA (assuming each iteration of the loop is mapped to 1 thread) is shown in data layout 2, where the access pattern would follow Figure 2. An alignment factor (pitch) is needed, since GPU memory is divided into banks and optimized accesses can only begin at the starting location of the first bank. However, if a large amount of computation is carried out on the CPU as well, we should transform data from layout 2 back to layout 1. These transformations are automated and abstracted using our data structures[2].

The fundamental structure defined within our framework shown in Figure 3 is an **OBJECT ARRAY** which denotes a grouping of similar objects. The term "object" in this context refers to any ordering of data such as arrays, vectors, or matrices. Such structures represent most physics-based simulation data (e.g., force, displacement, stress, etc.) which consist of arrays of vectors or arrays of matrices. Even for the simplistic simulation discussed in Section 3.2, there exist a large number of different data structures. To make our data-parallel simulation framework extensible and independent of any

---

[2] We assume no dependencies occur across elements in each vector and no conflicting compiler optimizations are used.

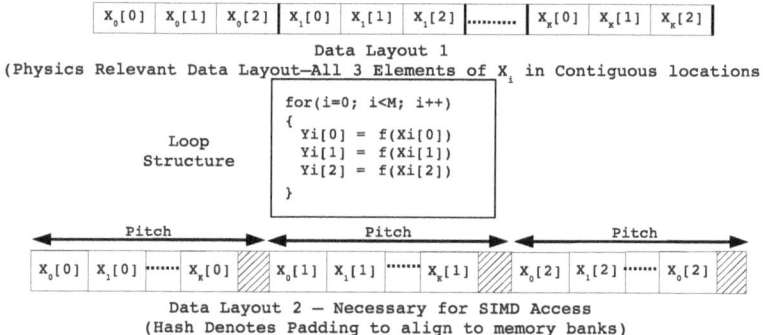

Fig. 4. Optimal data layouts for the CPU and the GPU

particular algorithm, we design a framework for dynamically building data structures such that the data layout generated is always optimal when implemented in CUDA[3].

The Single Instruction Multiple Thread (SIMT) model for CUDA requires well-structured access patterns across threads to coalesce memory accesses. Matching the memory access patterns to the architecture is critical since GPU performance is dependent on exploiting the high bandwidth present between the device memory and the Single Instruction Multiple Data (SIMD) units [1,3,10].

Figure 5 denotes the layout of our data structures used in the implementation of our GPUPhysBAM framework. Our underlying base class behaves as a CPU or GPU memory container. As shown in Figure 5, we abstract out GPU-specific operations in the 2nd level where we store information such as strides, number of elements in an object and number of objects. The 3rd level simply contains indexing functions for each element within an object.

Next, we demonstrate the benefits and extensibility of this approach. Figure 5 shows the implementation of an Object Array of 3x3 symmetric matrices which is commonly used in stiffness matrices. By inheriting from our GPU-generic class and implementing functions ($Get()$ and $Put()$ in SYMMETRIC_MATRIX_ARRAY to access data) **within** each object[4] of the array, we build data-parallel structures that will be efficiently accessed in a SIMD fashion on a GPU. The lower level $getdata()$ and $putdata()$ functions handle the indexing and the pitch calculations that are done to return data when given the object number and the element within the object. Thus, we provide the algorithmically relevant "3x3 Symmetric Matrix Array" while exploiting the wide memory bandwidth and many-core parallelism of a GPU.

A second example (shown in Figure 5) is an Object Array of structs used to describe a constitutive model. The Constitutive Model Array is also created in a similar fashion to the Symmetric Matrix Array. We derive the new class and simply write functions to access the respective value out of 12 elements that make up each object. As shown,

---

[3] Optimal refers to the optimal usage of memory bandwidth which occurs for coalesced data accesses.

[4] In this case one object is 6 elements of a symmetric matrix considered column major.

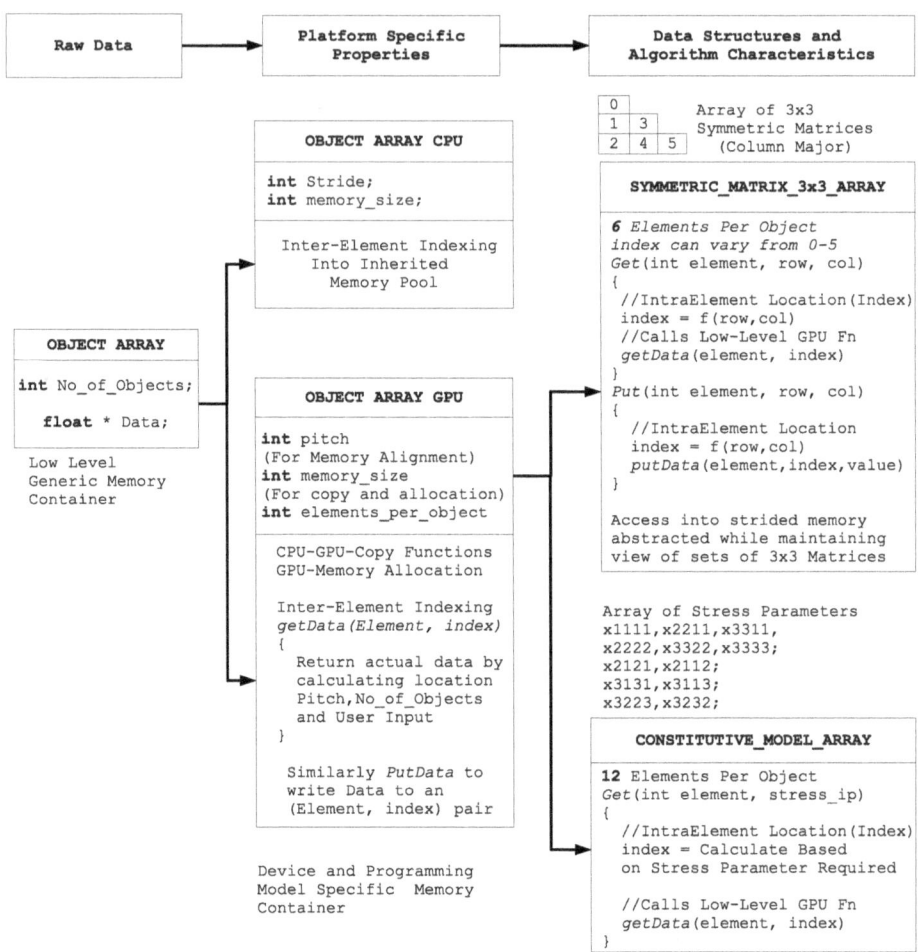

Fig. 5. Hierarchically declared data structures for both the CPU and the GPU

similar to Symmetric Matrices, the lower level base class functions of $getdata()$ and $putdata()$ implement the required indexing.

By using contiguous memory allocated in base classes and controlling indexing using derived objects, we maintain both proper coalescing for the SIMT hardware and the close coupling of the computation to the original physics theory. The utility of our framework lies in the fact that a domain expert could create the appropriate physics based data structures by simply inheriting the base class for the GPU functionality, and based on his/her expertise, simply write functions to access data within each object of the Object Array without knowing how the set of objects are laid out in memory.

4   Performance Results

The datasets used in our work are based on a popular model called the Truth Cube [15]. The Truth Cube serves as a model to validate soft-tissue deformation algorithms by comparing deformation obtained by an analytical method to known mechanical values. We use the Truth Cube to verify our data-structures. Figure 6a shows the Truth Cube in an undeformed position. Figure 6b shows the Truth Cube after performing Quasistatic simulation of deformation.

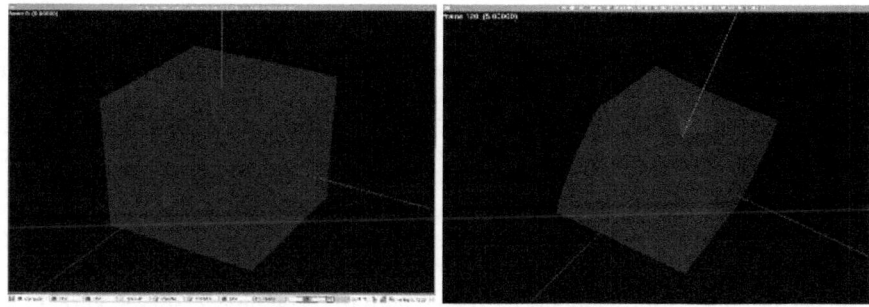

(a) Before Applying Deforming Forces   (b) After Quasistatic Deformation simulation

Fig. 6. Deformation of a Truth Cube

The Quasistatic simulation of the Truth Cube shown in Figure 6 was implemented by using data-structures created within our framework. A range of data structures were required by each benchmark as shown in Table 1. We benchmark 4 physics-relevant kernels from this simulation, which dominate the overall execution time. We compare only the execution time of the GPU kernels. The CPU-GPU I/O and memory transformation overhead does not change across data layouts because the baseline also incurs transformation overhead given that data is dynamically generated and not in contiguous memory. The performance was measured on a system using an NVIDIA GTX-285 GPU, Intel Core 2 Duo with $4GB$ of RAM running Ubuntu 9.04, and CUDA 3.0. Our baseline which is CUDA code implemented without our data structures is naive only with respect to data layout, it is architecturally aware of the GPU. The baseline CUDA code exploits shared memory, textures and uses an optimized thread execution configuration.

The performance improvements of the physics kernels have been denoted in Table 2. The performance improvements are substantial even in computationally intensive kernels like Add Force Differential which use the shared memory of the device to hide most of device memory latency.

"Coalescing Improvement" in Table 2 denotes ratio of *requests* to *actual* memory transactions measured using the CUDA profiler [1]. The increased ratio when using our framework denotes the improvement in memory access efficiency due to coalescing. The improvement in kernel performance when using our framework is due to the reduction in the number of *actual* memory transactions that the GPU memory subsystem processes.

Table 1. Data Structures Built Using GPUPhysBAM For Each Benchmark

Benchmark	Data Structures
Force Calculations	3x3 Matrix, Diagonal Matrices, 1d-vectors
Isotropic Stress Derivative	Constitutive Mode Structs, Diagonal Matrices
Add Force Differential	Symmetric Matrices, 3-Element Vectors
Clamp Particles	Neighbor Lists, 3-Element Vector

Table 2. Performance Results for Different Physics-based Benchmark Kernels

Benchmark	Coalescing Imprrovement		Performance(ms)		Speedup
	Baseline	GPUPhysBAM	Baseline	GPUPhysBAM	
Force Calculations	0.031	0.167	20.240	2.430	8.33x
Isotropic Stress Derivative	0.026	0.222	10.480	1.560	6.72x
Add Force Differential	0.013	0.066	5.900	1.150	5.13x
Clamp Particles	0.053	0.307	12.990	2.143	6.06x

The performance improvements presented here for each physics-based kernel translate to an improvement in application level performance because these kernels constitute the bulk of the computation. For e.g. the *Add Force Differential* kernel contains indirect accesses and is similar to a sparse matrix vector multiplication which is known to consume the bulk of the time spent in the Quasistatic simulation [14,20]. The performance improvements presented in Table 2 are obtained for essentially no increase in programming effort or development time for an application developer because the same physics-derived data structure design API is maintained which allows us to simply insert the improved data structures from underneath the physics simulator.

5   Conclusion

In this work we describe techniques that allow us to implement physics-based simulations efficiently on NVIDIA GPUs. Due to the variety of algorithms that can be implemented using our physics simulator, we focus on implementation techniques and optimizations that are extensible and generic so that they can have impact on a broader class of data-parallel physics simulations. Our framework is extendable to different types of physics-related objects and can also be adapted to other algorithms targeting GPUs. We have used this framework to implement other deformation algorithms based on MultiGrid methods and Backward Euler solvers. Our future work includes supporting more complicated models and evaluating the associated performance enhancements possible.

Acknowledgements

This work was supported by a National Science Foundation Innovation Grant (Award Number EEC-0946463), funding from Simquest International, a grant from AMD and by equipment donations from NVIDIA.

References

1. NVIDIA: NVIDIA CUDA Programming Guide 2.0 (2008),
   `http://www.nvidia.com/cuda`
2. NVIDIA: NVIDIA Physx (2008), `http://www.nvidia.com/physx`
3. Nguyen, H.: Gpu gems 3. Addison-Wesley Professional, Reading (2007)
4. Harris, M.: Optimizing parallel reduction in cuda, NVIDIA Developer Technology (2007)
5. Luebke, D., Harris, M., Govindaraju, N., Lefohn, A., Houston, M., Owens, J., Segal, M., Papakipos, M., Buck, I.: GPGPU: general-purpose computation on graphics hardware. In: Proceedings of the 2006 ACM/IEEE Conference on Supercomputing (2006)
6. Allard, J., Cotin, S., Faure, F., Bensoussan, P.J., Poyer, F., Duriez, C., Delingette, H., Grisoni, L.: Sofa-an open source framework for medical simulation. Studies in Health Technology and Informatics 125, 13
7. Lawlor, O., Chakravorty, S., Wilmarth, T., Choudhury, N., Dooley, I., Zheng, G., Kale, L.: ParFUM: A Parallel Framework for Unstructured Meshes for Scalable Dynamic Physics Applications. Engineering with Computers 22, 215–235 (2006)
8. Turek, S., Becker, C., Kilian, S.: Some concepts of the software package FEAST. In: Hernández, V., Palma, J.M.L.M., Dongarra, J. (eds.) VECPAR 1998. LNCS, vol. 1573, pp. 271–284. Springer, Heidelberg (1999)
9. van Dyk, D., Geveler, M., Mallach, S., Ribbrock, D., Gddeke, D., Gutwenger, C.: HONEI: A collection of libraries for numerical computations targeting multiple processor architectures. Computer Physics Communications 180(12), 2534–2543 (2009)
10. Anderson, J.A., Lorenz, C.D., Travesset, A.: General purpose molecular dynamics simulations fully implemented on graphics processing units. Journal of Computational Physics 227, 5342–5359 (2008)
11. Kirby, R.C., Logg, A.: A compiler for variational forms. ACM Trans. Math. Softw. 32, 417–444
12. Bro-Nielsen, M.: Finite element modeling in surgery simulation. Proceedings of the IEEE 86, 283–291 (1998)
13. Garcia, E.: Information Retrieval Tutorial (2005), `http://www.miislita.com`
14. Teran, J., Sifakis, E., Irving, G., Fedkiw, R.: Robust quasistatic finite elements and flesh simulation. In: SCA 2005: Proceedings of the 2005 ACM SIGGRAPH/Eurographics Symposium on Computer Animation (2005)
15. Kerdok, A.E., Cotin, S.M., Ottensmeyer, M.P., Galea, A.M., Howe, R.D., Dawson, S.L.: Truth cube: Establishing physical standards for soft tissue simulation. Medical Image Analysis 7, 283–291 (2003)
16. de Farias, T.S.M., Almeida, M.W.S., Teixeira, J.M.X., Teichrieb, V., Kelner, J.: A High Performance Massively Parallel Approach for Real Time Deformable Body Physics Simulation. In: 20th International Symposium on Computer Architecture and High Performance Computing 2008. SBAC-PAD 2008, pp. 45–52 (2008)
17. Joselli, M., Clua, E., Montenegro, A.C., Aura Pagliosa, P.: A new physics engine with automatic process distribution between CPU-GPU. In: Sandbox 2008: Proceedings of the 2008 ACM SIGGRAPH Symposium on Video Games, pp. 149–156 (2008)
18. Coumans, E.: Bullet physics library (2009), `http://www.bulletphysics.com`
19. The OpenCL Specification Munshi, A, Khronos OpenCL Working Group (2009)
20. Fedkiw, R., Stam, J., Jensen, H.W.: PhysBAM, `http://physbam.stanford.edu`
21. Melek, Z., Keyser, J.: Multi-representation interaction for physically based modeling. In: SPM 2005: Proceedings of the 2005 ACM Symposium on Solid and Physical Modeling, pp. 187–196. ACM, New York (2005)

Scalability Studies of an Implicit Shallow Water Solver for the Rossby-Haurwitz Problem[*]

Chao Yang[1,2] and Xiao-Chuan Cai[2]

[1] Institute of Software, Chinese Academy of Sciences, Beijing 100190, P.R. China
[2] Department of Computer Science, University of Colorado at Boulder,
Boulder, CO 80309, USA

Abstract. The scalability of a fully implicit global shallow water solver is studied in this paper. In the solver a conservative second-order finite volume scheme is used to discretize the shallow water equations on a cubed-sphere mesh which is free of pole-singularities. Instead of using the popular explicit or semi-implicit methods in climate modeling, we employ a fully implicit method so that the restrictions on the time step size can be greatly relaxed. Newton-Krylov-Schwarz method is then used to solve the nonlinear system of equations at each time step. Within each Newton iteration, the linear Jacobian system is solved by using a Krylov subspace method preconditioned with a Schwarz method. To further improve the scalability of the algorithm, we use multilevel hybrid Schwarz precondi-tioner to suppress the increase of the iteration number as the mesh is re-fined or more processors are used. We show by numerical experiments on the Rossby-Haurwitz problem that the fully implicit solver scales well to thousands of processors on an IBM BlueGene/L supercomputer.

1 Introduction

The numerical simulation of shallow water equations is essential in global cli-mate modeling and is usually used as a testbed for developing and verifying new dynamics cores. In this paper, we study a multilevel fully implicit solver for the shallow water equations on the cubed-sphere mesh. The cubed-sphere mesh is quasi-uniform thus provides better data distribution compared to the popular latitude-longitude mesh and is much easier to implement compared to unstruc-tured mesh. If an explicit time integration is used in the solver, good strong scaling results can be obtained. However, it is well-known that the explicit time step size depends on the mesh size due to the stability condition. As a result, to obtain the solution of a give simulation state, more time steps are needed when the mesh is refined. Therefore the weak scalability of an explicit method is far from ideal. By using a fully implicit time integration, the stability limit on the time step size can be greatly relaxed. And by using a multilevel Schwarz preconditioner, the weak scalability can be further improved.

---

[*] This research was supported in part by DOE under DE-FC-02-06ER25784, in part by NSF under EAR 0934647 and DMS 0913089, in part by NSFC under 10801125 and in part by 863 program of China under 2010AA012300.

J.M.L.M. Palma et al. (Eds.): VECPAR 2010, LNCS 6449, pp. 172–184, 2011.
© Springer-Verlag Berlin Heidelberg 2011

We employ the cubed-sphere mesh of gnomonic type [9] in this study. The mesh is generated by mapping the six faces of an inscribed cube of a sphere to the surface using the gnomonic projection. The six patches are then attached together via interface conditions. Using the local curvilinear coordinates, the shallow water equations have an identical form on each patch:

$$\frac{\partial Q}{\partial t} + \frac{1}{\Lambda}\frac{\partial(\Lambda F)}{\partial x} + \frac{1}{\Lambda}\frac{\partial(\Lambda G)}{\partial y} + S = 0, \quad (x, y) \in [-\pi/4, \pi/4]^2 \qquad (1)$$

with

$$Q = \begin{pmatrix} h \\ hu \\ hv \end{pmatrix}, \ F = \begin{pmatrix} hu \\ huu \\ huv \end{pmatrix}, \ G = \begin{pmatrix} hv \\ huv \\ hvv \end{pmatrix}, \ S = \begin{pmatrix} 0 \\ S_1 \\ S_2 \end{pmatrix}, \qquad (2)$$

where

$$\begin{aligned} S_1 &= \Gamma_{11}^1(huu) + 2\Gamma_{12}^1(huv) + f\Lambda\left(g^{12}hu - g^{11}hv\right), \\ S_2 &= 2\Gamma_{12}^2(huv) + \Gamma_{22}^2(hvv) + f\Lambda\left(g^{22}hu - g^{12}hv\right). \end{aligned} \qquad (3)$$

Here the unknowns are the thickness and velocity of the atmosphere; i.e., h and (u, v). The gravitational constant is g and the Coriolis parameter is f. The coefficients g^{mn}, Λ and Γ_{mn}^ℓ in the equation are only dependent on the curvilinear coordinates, see [15] for details.

2 A Fully Implicit Finite Volume Discretization

Suppose each patch $\mathcal{P}^k$ is covered by a logically rectangular $N \times N$ mesh, which is equally spaced in the computational domain $\{(x, y) \in [-\pi/4, \pi/4]^2\}$ with mesh size $\hbar = \pi/2N$. Patch $\mathcal{P}^k$ is then divided into mesh cells $\mathcal{C}_{ij}^k$ centered at (x_i, y_j), $i, j = 1, \cdots, N$. The approximate solution in cell $\mathcal{C}_{ij}^k$ at time t is defined as

$$Q_{ij}^k \approx \frac{1}{\Lambda_{ij}^k \hbar^2} \int_{y_j - \hbar/2}^{y_j + \hbar/2} \int_{x_i - \hbar/2}^{x_i + \hbar/2} \Lambda(x, y) Q(x, y, t)\, dx dy.$$

Here Λ_{ij}^k is evaluated at the cell center of $\mathcal{C}_{ij}^k$. Since expressions are identical on different patches in most cases, we ignore the superscript k in the sequel.

On each mesh cell we integrate the shallow water equations and use the Green's formula, then we obtain a semi-discrete system

$$\frac{\partial Q_{ij}}{\partial t} + \mathcal{L}(Q_{ij}) = 0, \qquad (4)$$

with

$$\mathcal{L}(Q_{ij}) = \frac{(\Lambda F)_{i+1/2,j} - (\Lambda F)_{i-1/2,j}}{\Lambda_{ij}\hbar} + \frac{(\Lambda G)_{i,j+1/2} - (\Lambda G)_{i,j-1/2}}{\Lambda_{ij}\hbar} + S_{ij}. \qquad (5)$$

The numerical fluxes on cell boundaries are calculated by using the Osher's Riemann solver [7,8]

$$(\Lambda F)_{i\pm 1/2,j} = \Lambda_{i\pm 1/2,j} F\left(Q^*_{i\pm 1/2,j}\right), \quad (\Lambda G)_{i,j\pm 1/2} = \Lambda_{i,j\pm 1/2} G\left(Q^*_{i,j\pm 1/2}\right),$$

where Q^* is evaluated as

$$h^* = \frac{1}{4gg^{11}}\left[\frac{1}{2}\left(u^- - u^+\right) + \sqrt{gg^{11}}\left(\sqrt{h^-} + \sqrt{h^+}\right)\right]^2,$$

$$u^* = \frac{1}{2}\left(u^- + u^+\right) + \sqrt{gg^{11}}\left(\sqrt{h^-} - \sqrt{h^+}\right),$$

$$v^* = \begin{cases} v^- + \left(g^{12}/g^{11}\right)\left(u^* - u^-\right), & \text{if } u^* \geq 0 \\ v^+ + \left(g^{12}/g^{11}\right)\left(u^* - u^+\right), & \text{otherwise,} \end{cases}$$

under the assumption that $|u| < \sqrt{gg^{11}h}$. The calculation of $(\Lambda G)_{i,j+1/2}$ follows a similar scheme, see [14] for more details. The reconstructed states are calculated using a second-order scheme as follows:

$$Q^{\mp}_{i\pm 1/2,j} = Q_{ij} \pm (Q_{i+1,j} - Q_{i-1,j})/4, \quad Q^{\mp}_{i,j\pm 1/2} = Q_{ij} \pm (Q_{i,j+1} - Q_{i,j-1})/4.$$

Only in the construction of the Schwarz preconditioner, we use a piecewise constant method to reconstruct Q by forcing $Q = Q_{ij}$ on $\mathcal{C}_{ij}$, which leads to a first-order scheme [15]. On each patch interface, we extend the mesh with one layer of ghost cells and the numerical fluxes are calculated symmetrically across the interface to insure the numerical conservation of total mass; see [14] for details.

Denote $Q^{(m)}$ as the approximate solution at the m-th time step with a fixed time step size Δt. We use the second-order backward differentiation formula (BDF-2) for time integration

$$\frac{1}{2\Delta t}\left(3Q_{ij}^{(m)} - 4Q_{ij}^{(m-1)} + Q_{ij}^{(m-2)}\right) + \mathcal{L}\left(Q_{ij}^{(m)}\right) = 0. \tag{6}$$

Only at the first time step, a first-order backward Euler (BDF-1) method is used. We also implement an explicit second-order Strong Stability Preserving Runge-Kutta (SSP RK-2) method

$$\overline{Q}^{(m)} = Q^{(m-1)} - \Delta t \mathcal{L}\left(Q^{(m-1)}\right),$$
$$Q^{(m)} = (1/2)\left(Q^{(m-1)} + \overline{Q}^{(m)}\right) - (\Delta t/2)\mathcal{L}\left(\overline{Q}^{(m)}\right) \tag{7}$$

for the purpose of comparison. When using the explicit method, the time step size is adaptively controlled so that the corresponding CFL number is fixed at 0.5. On the other hand, when the fully implicit method is employed, the time step size Δt is no longer constrained by the CFL condition. However, the price to pay by using the fully implicit method is that we have to solve a nonlinear algebraic system for each time step. The development of a scalable algorithm to solve the nonlinear system is the key to the success of the fully implicit solver.

3   An Inexact Newton's Method with Adaptive Stopping Conditions

In the fully implicit solver, a nonlinear system $\mathcal{F}(X) = 0$ arises at each time step. An inexact Newton's method is employed to solve the nonlinear system. In the Newton's method, the approximate solution is updated as

$$X_{n+1} = X_n + \lambda_n S_n, \quad n = 0, 1, \dots \tag{8}$$

Here the steplength λ_n is calculated via linesearch [3] and the initial guess X_0 is chosen as the solution of the previous time step. In the inexact Newton's method, the correction vector S_n is obtained by approximately solving the linear Jacobian system

$$J_n S_n = -\mathcal{F}(X_n). \tag{9}$$

Here $J_n = \mathcal{F}'(X_n)$ is the Jacobian matrix calculated at the current approximate solution X_n. In practice, we solve the right-preconditioned system

$$J_n M^{-1}(M S_n) = -\mathcal{F}(X_n), \tag{10}$$

using a restarted GMRES method until the linear residual $r_n = J_n S_n + \mathcal{F}(X_n)$ satisfies

$$\|r_n\| \le \eta \|\mathcal{F}(X_n)\|. \tag{11}$$

In the implementation, GMRES restarts at every 30 iterations and the nonlinear forcing term is $\eta = 10^{-3}$. A flexible version of GMRES [10] is used for multilevel preconditioners due to the change of the preconditioner during the GMRES iterations. Note that some more flexible methods can be used in choosing the nonlinear forcing term, such as the techniques suggested by Eisenstat and Walker [4].

We stop the Newton iteration (8) when the residual satisfies

$$\|\mathcal{F}(X_{n+1})\| \le \min \left\{ \hat{\varepsilon}_a, \max \left\{ \breve{\varepsilon}_a, \varepsilon_r \|\mathcal{F}(X_0)\| \right\} \right\}. \tag{12}$$

Here $\varepsilon_r = 1.0 \times 10^{-7}$ is the relative tolerance ε_r with safeguard $\hat{\varepsilon}_a = 1.0 \times 10^{-7}$. The absolute tolerance $\breve{\varepsilon}_a$ is initially set to $\breve{\varepsilon}_a^{(0)} = 1.0 \times 10^{-8}$ at the first time step and then adaptively determined by

$$\breve{\varepsilon}_a^{(m)} = \max \left\{ \breve{\varepsilon}_a^{(m-1)}, \|\mathcal{F}(X^{(m-1)})\| \right\}.$$

Here m is the time step index.

4   Some Variants of One-Level and Multilevel Schwarz Preconditioners

The partition of the cubed-sphere mesh is straightforward. Suppose the number of processors is $6p^2$. We decompose each of the six patches of the cubed-sphere

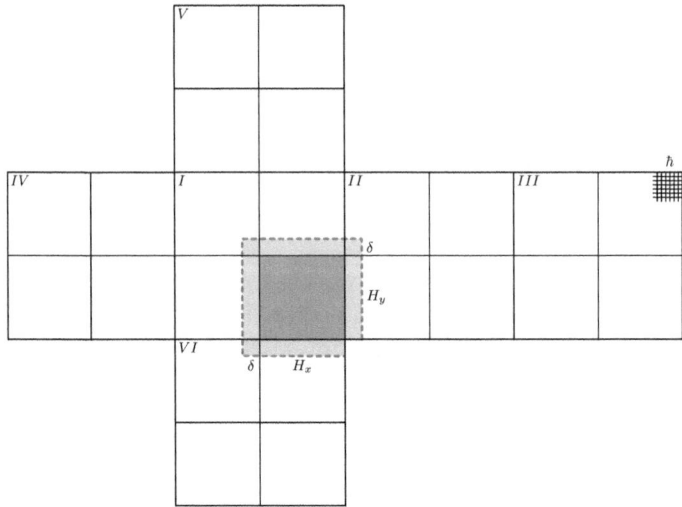

Fig. 1. Partition of the cubed-sphere mesh into subdomains with an overlap δ. Here $p = 2$ and the number of subdomains is $6p^2 = 24$. The solid lines indicate the partition of the domain into non-overlapping pieces of size $H_x \times H_y$, the filled rectangle shows the extended boundary of an overlapping subdomain and the incomplete fine mesh of solid lines illustrates underlying uniform mesh of size $\hbar$.

mesh into $p \times p$ identical non-overlapping subdomains respectively. Each subdomain is then mapped onto one processor. By further extending each subdomain Ω_j to a larger subdomain Ω_j', $j = 1, \cdots, 6p^2$, we get an overlapping decomposition of the cubed-sphere, as shown in Fig 1.

Based on the overlapping domain decomposition, we define the one-level additive Schwarz (AS) preconditioner

$$M_{AS(1)}^{-1} = \sum_{j=1}^{6p^2} (R_j^\delta)^T B_j^{-1} R_j^\delta, \tag{13}$$

and its restricted version (RAS, [2,11])

$$M_{one}^{-1} = \sum_{j=1}^{6p^2} (R_j^0)^T B_j^{-1} R_j^\delta. \tag{14}$$

Here R_j^δ is the restriction operator that maps a vector defined on the whole domain onto subdomain Ω_j'; and $(R_j^\delta)^T$ serves as a prolongation operator that extends a vector defined on subdomain Ω_j' to the whole domain by filling zeros outside Ω_j'. The RAS preconditioner differs from the AS in that the prolongation operator is based on the non-overlapping domain decomposition of the cubed-sphere that reduces communication in the application of the preconditioner. Observations were made [2] that the restricted version of the Schwarz preconditioner also helps in reducing the number of linear iterations in some cases; therefore we will use the RAS preconditioner in this study.

The operator B_j^{-1} represents a subdomain preconditioner that has several different definitions. Denote the Jacobian matrix based on the second-order spatial discretization as J and denote its first-order version as $\tilde{J}$. We list some possible choices of the subdomain preconditioner B_j^{-1}.

(1) The direct inverse of the true subdomain Jacobian matrix based on the LU factorization of $R_j^\delta J (R_j^\delta)^T$.
(2) An approximate inverse of the true subdomain Jacobian matrix based on the incomplete LU factorization of $R_j^\delta J (R_j^\delta)^T$.
(3) The direct inverse of the first-order subdomain Jacobian matrix based on the LU factorization of $R_j^\delta \tilde{J} (R_j^\delta)^T$.
(4) An approximate inverse of the first-order subdomain Jacobian matrix based on the incomplete LU factorization of $R_j^\delta \tilde{J} (R_j^\delta)^T$.

In our implementation, the sparse matrices are stored in a point-block format, and both the LU and ILU factorizations are carried out in the (3×3) point-block format. Compared to factorizations based on other matrix formats, the point-block version helps in keeping the coupling between all physical components of each mesh point which is essential for our fully coupled solver.

If a first-order spatial discretization is used for the fully implicit solution of the shallow water equations, the one-level RAS preconditioner (14) is found to be robust [15]. Some observations made in [13] suggest that the one-level RAS preconditioner constructed based on the low-order scheme is efficient even when the spatial discretization is second-order. Therefore, we only consider the latter two choices of the subdomain preconditioner B_j^{-1}.

When a large number of processors is used to solve a larger problem or larger time step size is considered, it becomes difficult for a one-level method to scale ideally. To obtain better scalability, coarse levels need to be included to the one-level preconditioner. We may compose the one-level additive Schwarz preconditioner B_f with a coarse-level preconditioner B_c either additively

$$M_{two}^{-1} = B_c + B_f, \tag{15}$$

or multiplicatively

$$M_{two}^{-1} = B_c + B_f - B_f J_f B_c, \tag{16}$$

where $B_c = \mathcal{I}_c^f J_c^{-1} \mathcal{I}_f^c$, and $\mathcal{I}_f^c$ and $\mathcal{I}_c^f$ are restriction and prolongation operators mapping between vectors defined on the fine level and the coarse level. A linear system with Jacobian matrix J_c defined on the coarse is solved for each application of the two-level preconditioner (16). Here the linear solver for the coarse level problem does not need to be exact; instead, we may solve it using GMRES with a given tolerance η_c. Note that the coarse level preconditioner can be either one-level or another two-level preconditioner which results in a three-level solver.

5   Numerical Experiments

The numerical experiments are carried out on an IBM BlueGene/L supercomputer with 4096 dual-processor compute nodes. The fully implicit solver is implemented based on the Portable, Extensible Toolkit for Scientific computation (PETSc [1]). We consider the sixth test case provided in [12], which is a four-wave Rossby-Haurwitz problem. Although it is not an analytic solution to the shallow water equations, the Rossby-Haurwitz wave is a good tool as a middle-term test. The height field contour on day 14 using the fully implicit BDF-2 method is given in Fig 2; the result using the explicit RK-2 method is also provided in the same figure for comparison purpose. Our numerical results are in agreement with the reference solutions in [5,6] and although time step sizes are different by two orders of magnitudes, the implicit and explicit results are consistent with each other.

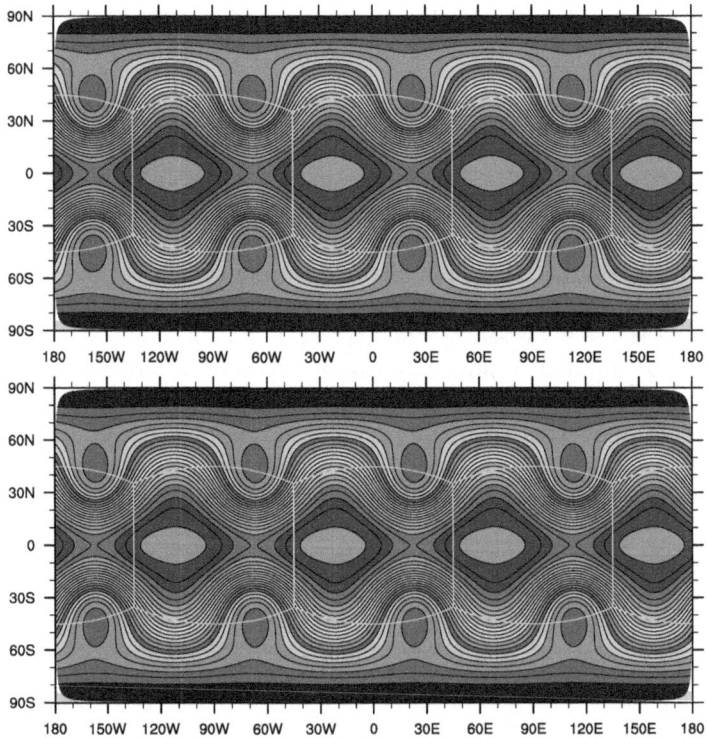

Fig. 2. Height field contours of the Rossby-Haurwitz problem at day 14. The calculations are done on a $128 \times 128 \times 6$ mesh with 96 processors. The top figure is the result using BDF-2 with $\Delta t = 0.1$days (CFL ≈ 55); and the bottom one is obtained using SSP RK-2 with CFL $= 0.5$. The contour levels are from 8100 to 10500m with an interval of 100m. The innermost lines near the equator are at 10500m.

5.1 Numerical Conservation

We then investigate the numerical conservation of the fully implicit solver. Here the normalized conservation error of a specific quantity ρ at time t is measured by $\delta(\rho) = [I(\rho, t) - I(\rho, 0)]/I(\rho, 0)$, where

$$I(\rho, t) = \sum_{k=1}^{6} \sum_{i,j=1}^{N} \left(\Lambda_{ij} \rho_{ij}^k(t) \right). \tag{17}$$

The integral invariants can be the total mass $\delta(h)$, the total energy $\delta(E)$ and the potential enstrophy $\delta(\xi)$, where

$$E = \frac{\Lambda^2 h}{2} (g^{11} v^2 + g^{22} u^2 - 2g^{12} uv),$$

$$\xi = \frac{1}{2\Lambda h} \left\{ \frac{\partial}{\partial x} [\Lambda^2 (g^{11} v - g^{12} u)] - \frac{\partial}{\partial y} [\Lambda^2 (g^{22} u - g^{12} v)] + \Lambda f \right\}^2.$$

The numerical conservation of the three integral invariants are provided in Fig 3. From Fig 3, we have the following observations. When using the explicit method, the mass conservation is accurate to the machine precision and the conservations of the other two integral invariants are also satisfactory. When the fully implicit method is used instead, the numerical conservations are also within acceptable levels; here we believe that the slight discrepancies of the numerical conservation are mainly due to the much larger time step size and the inexact solution of the nonlinear systems.

5.2 Performance Tests

We study the performance of the fully implicit solver by using the preconditioners listed in Table 1. Both the second-order and the first-order Jacobians are evaluated analytically where the latter is used to construct the RAS preconditioners.

Table 1. Preconditioners used in the fully implicit solver. For the three-level preconditioner, the values in the last three entries are provided from the finest level to the coarsest. The $\star$ stands for the direct application of the preconditioner without iteration.

Preconditioner	#Levels	Ratio	η	δ	Subdomain solver
1-level-ILU	1	0.001	-	$2\hbar$	ILU(2)
1-level-LU	1	0.001	-	$2\hbar$	LU
3-level	3	0.001, 0.1, $\star$	1:2	$2\hbar$, 0, $4\hbar$	ILU(2), ILU(2), LU

We first investigate the strong scalability by running a fix-sized problem with an increasing number of processors. In the ideal situation, the execution time should be reduced by a factor of 2 as the number of processors doubles. We

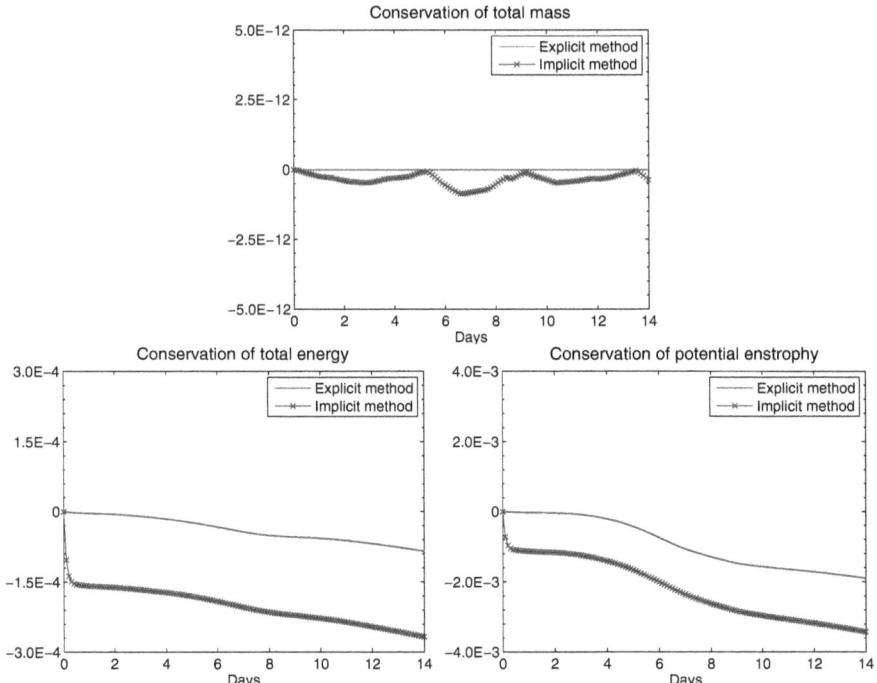

Fig. 3. Numerical conservation of the of the total mass, total energy and potential enstrophy for the Rossby-Haurwitz problem. The calculations are done on a 128×128×6 mesh with 96 processors. Results are compared between BDF-2 with $\Delta t = 0.1$days (CFL ≈ 55) and SSP RK-2 with CFL = 0.5.

run the Rossby-Haurwitz problem on a $1152 \times 1152 \times 6$ mesh (nearly 24 million unknowns) with time step size $\Delta t = 0.1$days for the first 20 time steps. The performance results of the fully implicit solver using the preconditioners listed in Table 1 are provided in Fig 4 and 5. The averaged number of Newton iteration per time step is not provided in the figure because we observe that it is always around 3.0, independent on the number of processors and the preconditioners used inside the linear solver. The results on the averaged number of linear iterations per Newton iteration for each time step are given in Fig 4 from which we find that the number of linear iterations is greatly reduced when using the three-level method instead of one-level versions and more importantly the dependency between the number of linear iterations and the number of processors is successfully removed. The results on the total compute time in the strong scaling tests are provided in Fig 5 in which the explicit results are also given. Fig 5 clearly indicates that the fully implicit solver is orders of magnitudes faster than the explicit solver. The three-level method is about 30% to 50% faster compared to the one-level methods. Suppose the compute speed for 384 processors is ideal, the scalability of the explicit method is nearly ideal and the parallel efficiency of the implicit solver is around 73% to 80% when 7776 processors are used.

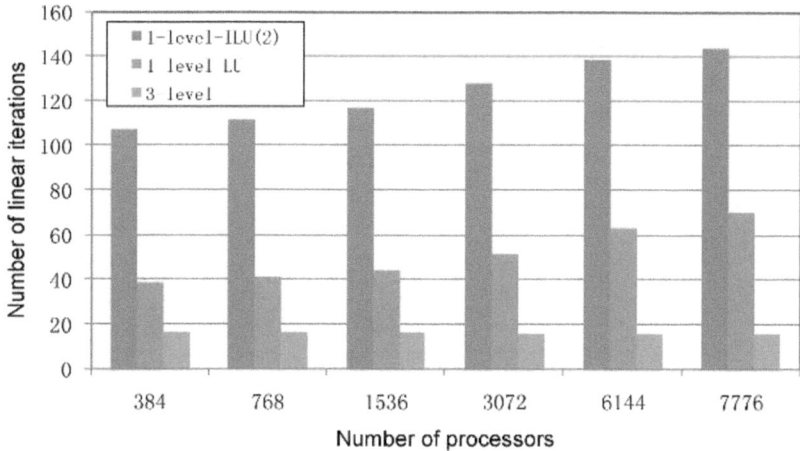

Fig. 4. Strong scaling results on the Rossby-Haurwitz problem, averaged number of linear iterations

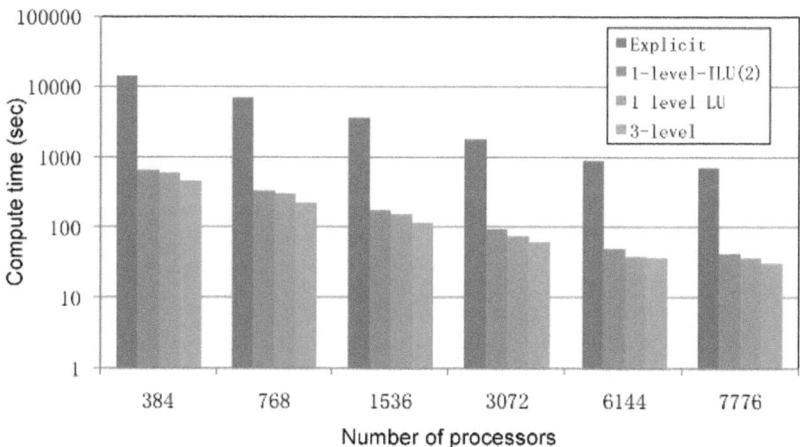

Fig. 5. Strong scaling results on the Rossby-Haurwitz problem, total compute time

We then study the weak scalability of the fully implicit solver. In the weak scaling tests, we fix the per processor mesh size and thus the total problem size increases as more processors are used. As the problem size increases, the CFL number becomes larger, leading to larger condition number of the Jacobian system. In the ideal case, we expect to obtain mesh-independent compute time which is much harder to achieve than the strong scalability. We run the Rossby-Haurwitz problem starting with a $48 \times 48 \times 6$ mesh for six processors and ending up with a $1728 \times 1728 \times 6$ mesh (nearly 54 million unknowns) for 7776 processors. In each test we run the problem for the first 20 time steps also with a fixed time step size $\Delta t = 0.1$days. Analogous to the strong scaling tests, we observe that the

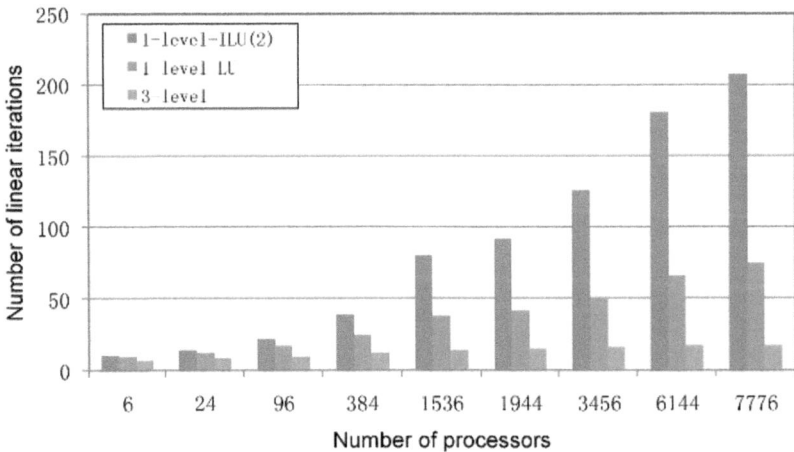

Fig. 6. Weak scaling results on the Rossby-Haurwitz problem, averaged number of linear iterations

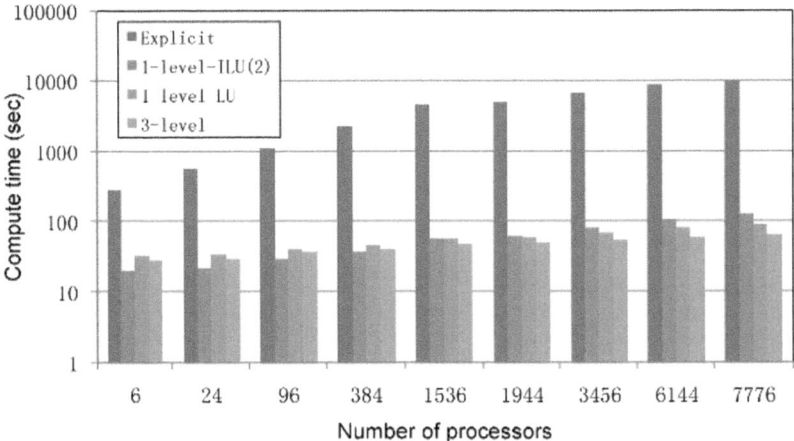

Fig. 7. Weak scaling results on the Rossby-Haurwitz problem, total compute time

number of Newton iterations is insensitive to the number of processors. However the number of linear iterations increases more rapidly as more processors are used, see Fig 6. Although the three-level method provides much better iteration counts, the dependency between the number of linear iterations and the number of processors is not totally removed. As a result, we can not obtain the ideal weak scalability in terms of compute time, see Fig 7. We would like to point out here that if an explicit method is used instead, the weak scalability is far below the ideal situation because the time step size is limited by the CFL condition as the mesh is refined. The results using the explicit RK-2 method is provided in Fig 7 to compare. Although not ideal, with the help of the three-level preconditioner, the compute time only increases by 128% as the number of processor increases from 6 to 7776.

6 Conclusions

The scalability of a global shallow water solver is studied. The solver features a conservative second-order finite volume scheme based on a cubed-sphere mesh and a fully implicit time integration method. Newton-Krylov-Schwarz algorithms are used to solve the nonlinear system arising at each implicit time step. The inexact Newton iteration is controlled by adaptive stopping conditions so that the nonlinear residuals are maintained within a uniform level for all time steps. Multilevel Schwarz preconditioners are studied to improve both strong and weak scalabilities of the fully implicit solver. To show the performance of the algorithm we provide some numerical results for the Rossby-Haurwitz problem carried out on a supercomputer with thousands of processors.

References

1. Balay, S., Buschelman, K., Gropp, W.D., Kaushik, D., Knepley, M., McInnes, L.C., Smith, B.F., Zhang, H.: PETSc Users Manual. Argonne National Laboratory (2010)
2. Cai, X.-C., Sarkis, M.: A restricted additive Schwarz preconditioner for general sparse linear systems. SIAM J. Sci. Comput. 21, 792–797 (1999)
3. Dennis, J.E., Schnabel, R.B.: Numerical Methods for Unconstrained Optimization and Nonlinear Equations. SIAM, Philadelphia (1996)
4. Eisenstat, S.C., Walker, H.F.: Choosing the forcing terms in an inexact Newton method. SIAM J. Sci. Comput. 17, 1064–8275 (1996)
5. Jablonowski, C.: Adaptive Grids in Weather and Climate Modeling. PhD thesis, University of Michigan, Ann Arbor (2004)
6. Jakob-Chien, R., Hack, J.J., Williamson, D.L.: Spectral transform solutions to the shallow water test set. J. Comput. Phys. 119, 164–187 (1995)
7. Osher, S., Chakravarthy, S.: Upwind schemes and boundary conditions with applications to Euler equations in general geometries. J. Comput. Phys. 50, 447–481 (1983)
8. Osher, S., Solomon, F.: Upwind difference schemes for hyperbolic systems of conservation laws. Math. Comp. 38, 339–374 (1982)
9. Ronchi, C., Iacono, R., Paolucci, P.: The cubed sphere: A new method for the solution of partial differential equations in spherical geometry. J. Comput. Phys. 124, 93–114 (1996)
10. Saad, J.: A flexible inner-outer preconditioned GMRES algorithm. SIAM J. Sci. Comput. 14, 461–469 (1993)
11. Toselli, A., Widlund, O.: Domain Decomposition Methods – Algorithms and Theory. Springer, Berlin (2005)
12. Williamson, D.L., Drake, J.B., Hack, J.J., Jakob, R., Swarztrauber, P.N.: A standard test set for numerical approximations to the shallow water equations in spherical geometry. J. Comput. Phys. 102, 211–224 (1992)

13. Yang, C., Cai, X.-C.: Newton-Krylov-Schwarz method for a spherical shallow water model. In: Proceedings of the 19th Intl. Conf. on Domain Decomposition Methods (to appear, 2010)
14. Yang, C., Cai, X.-C.: A parallel well-balanced finite volume method for shallow water equations with topography on the cubed sphere. J. Comput. Appl. Math. (to appear, 2010)
15. Yang, C., Cao, J., Cai, X.-C.: A fully implicit domain decomposition algorithm for shallow water equations on the cubed sphere. SIAM J. Sci. Comput. 32, 418–438 (2010)

Parallel Multigrid Solvers Using OpenMP/MPI Hybrid Programming Models on Multi-Core/Multi-Socket Clusters

Kengo Nakajima

Information Technology Center, The University of Tokyo, 2-11-16 Yayoi,
Bunko-ku, Tokyo 113-8658, Japan
nakajima@cc.u-tokyo.ac.jp

Abstract. OpenMP/MPI hybrid parallel programming models were implemented to 3D finite-volume based simulation code for groundwater flow problems through heterogeneous porous media using parallel iterative solvers with multigrid preconditioning. Performance and robustness of the developed code has been evaluated on the "T2K Open Supercomputer (Tokyo)" and "Cray-XT4" using up to 1,024 cores through both of weak and strong scaling computations. OpenMP/MPI hybrid parallel programming model demonstrated better performance and robustness than flat MPI with large number of cores for ill-conditioned problems with appropriate command lines for NUMA control, first touch data placement and reordering of the data for contiguous "sequential" access to memory.

Keywords: Multigrid, OpenMP/MPI Hybrid, Preconditioning.

1 Introduction

In order to achieve minimal parallelization overheads on SMP (symmetric multiprocessors) and multi-core clusters, a multi-level *hybrid* parallel programming model is often employed. In this method, coarse-grained parallelism is achieved through domain decomposition by message passing among nodes, while fine-grained parallelism is obtained via loop-level parallelism inside each node using compiler-based thread parallelization techniques, such as OpenMP (Fig.1 (a)). Another often used programming model is the single-level *flat MPI* model, in which separate single-threaded MPI processes are executed on each core (Fig.1 (b)). It is well-known that the efficiency of each model depends on hardware performance (CPU speed, communication bandwidth/latency, memory bandwidth/latency, and their balance), application features, and problem size [1].

In the previous work [1], author applied OpenMP/MPI hybrid parallel programming models to finite-element based simulations of linear elasticity problems. The developed code has been tested on the "T2K Open Supercomputer" [2] using up to 512 cores. Performance of OpenMP/MPI hybrid parallel programming model is competitive with that of *flat MPI* using appropriate command lines for NUMA control. Furthermore, reordering of the data for contiguous access to memory with first touch

J.M.L.M. Palma et al. (Eds.): VECPAR 2010, LNCS 6449, pp. 185–199, 2011.
© Springer-Verlag Berlin Heidelberg 2011

data placement provides excellent improvement on performance of OpenMP/MPI hybrid parallel programming models, especially if the problem size for each core is relatively small. Generally speaking, OpenMP/MPI hybrid parallel programming model provides excellent performance for strong scaling cases where problems are *less memory-bound*.

In the present work, OpenMP/MPI hybrid parallel programming models were implemented to 3D finite-volume based simulation code for groundwater flow problems through heterogeneous porous media using parallel iterative solvers with multigrid preconditioning, which was originally developed in [3]. Multigrid is a scalable method and expected to be a promising approach for large-scale computations, but there are no detailed research works where multigrid methods are evaluated on multi-core/multi-socket clusters using OpenMP/MPI hybrid parallel programming models. In this work, developed code has been evaluated on the "T2K Open Super Computer (Todai Combined Cluster) (T2K/Tokyo)" at the University of Tokyo, and "Cray-XT4" at National Energy Research Scientific Computing Center (NERSC) of Lawrence Berkeley National Laboratory [4] using up to 1,024 cores, and performance of *flat MPI* and three kinds of OpenMP/MPI hybrid parallel programming models are evaluated.

The rest of this paper is organized as follows: In section 2, overview of the target hardware ("T2K/Tokyo", "Cray XT4") is provided. In section 3, we outline the details of the present application, and describe the linear solvers and reordering/optimization procedures. In section 4, preliminary results of the computations for both of weak and strong scaling tests are described, while some final remarks are offered in section 5.

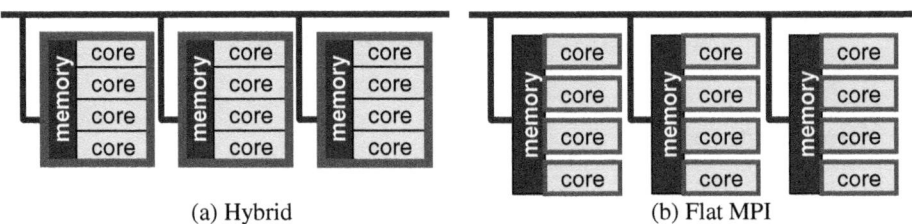

(a) Hybrid   (b) Flat MPI

Fig. 1. Parallel Programming Models

2   Hardware Environment

"T2K Open Super Computer (Todai Combined Cluster) (T2K/Tokyo)" at the University of Tokyo [2] was developed by Hitachi under "T2K Open Supercomputer Alliance" [5]. T2K/Tokyo is an AMD Quad-core Opteron-based combined cluster system with 952 nodes, 15,232 cores and 31TB memory. Total peak performance is 140 TFLOPS. T2K/Tokyo is an integrated system of four clusters. Number of nodes in each cluster is 512, 256, 128 and 56, respectively. Each node includes four "sockets" of AMD Quad-core Opteron processors (2.3GHz), as shown in Fig.2.

Peak performance of each core is 9.2 GFLOPS, and that of each node is 147.2 GFLOPS. Each node is connected via Myrinet-10G network. In the present work, 64 nodes of the system have been evaluated. Because T2K/Tokyo is based on *cache-coherent*

NUMA (cc-NUMA) architecture, careful design of software and data configuration is required for efficient access to local memory.

"Cray XT4 (Franklin)" system at NERSC/LBNL [4] is a large-scale cluster system. A single node of Cray-XT4 corresponds to a single "socket" of AMD Quad-core Opteron processor (2.3 GHz) in Fig.2. Entire system consists of 9,572 nodes, 38,288 cores and 77TB memory. Total peak performance is 352 TFLOPS. Network topology of T2K/Tokyo is multi-stage cross-bar, while that of Cray-XT4 is 3D truss.

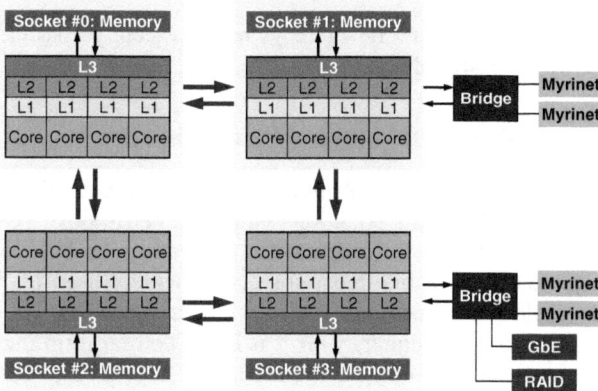

Fig. 2. Overview of a "node" of T2K/Tokyo, each node consists of four sockets of AMD Quad-core Opteron processors (2.3GHz)

3 Implementation and Optimization of Target Application

3.1 Finite-Volume Application

In the present work, groundwater flow problems through heterogeneous porous media (Fig.3) are solved using a parallel finite-volume method (FVM). Problem is described by the following Poisson equation and boundary condition:

$$\nabla \cdot \left(\lambda(x, y, z)\nabla \phi\right) = q, \ \phi = 0 \ at \ z = z_{\max} \tag{1}$$

where ϕ denotes potential of water-head, and $\lambda(x,y,z)$ describes water conductivity. q is value of volumetric flux of each finite-volume cell, and is set to a uniform value (=1.0) in this work. A heterogeneous distribution of water conductivity in each cell is calculated by a sequential Gauss algorithm, which is widely used in the area of geostatistics [6]. The minimum and maximum values of water conductivity are 10^{-5} and 10^5, respectively, with an average value of 1.0. This configuration provides ill-conditioned coefficient matrices whose condition number is approximately 10^{10}. Each cell is a cube, and distribution of cells is structured like finite-difference-type voxels . In this work, entire model is consists of clusters of small models with 128^3 cells. In each small model, distribution pattern is same therefore same pattern of heterogeneity appears periodically. The code is parallelized by domain decomposition using MPI for communications between partitioned domains [3].

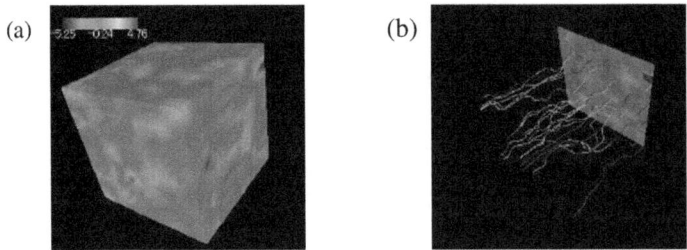

Fig. 3. Example of groundwater flow through heterogeneous porous media, (a) distribution of water conductivity, (b) streamlines

3.2 Iterative Method with Multigrid Preconditioning

The Conjugate Gradient (CG) solver with multigrid preconditioner (MGCG) [3] was applied for solving Poisson's equations with symmetric positive definite (SPD) coefficient matrices. Iterations are repeated until the norm $|r|/|b|$ is less than 10^{-12}. Multigrid is an example of scalable linear solver and widely used for large-scale scientific applications. Relaxation methods such as Gauss-Seidel can efficiently damp high-frequency error, but low-frequency error is left. The multigrid idea is to recognize that this low-frequency error can be accurately and efficiently solved for on a coarser grid [7]. In this work, very simple geometric multigrid with V-cycle, where 8 children form 1 parent cell in isotropic manner for structured finite-difference-type voxels, as shown in Fig.4, has been applied. *Level* of the finest grid is set to 1 and the *level* is numbered from the finest to the coarsest grid, where number of cell is one at each domain (MPI processe). In this work, multigrid operations at each level are done in parallel manner, but the operations at the coarsest levels are executed on a single core by gathering information of entire processes. Total number of cells at the coarsest level is equal to number of domains (MPI processes).

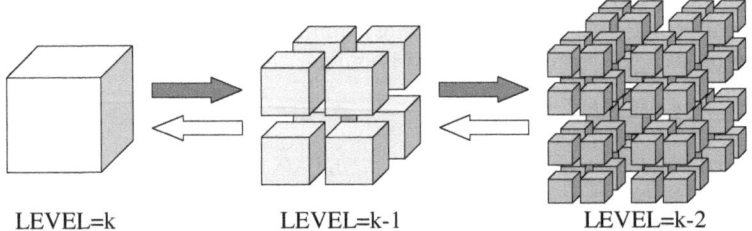

LEVEL=k   LEVEL=k-1   LEVEL=k-2

Fig. 4. Procedure of Geometric Multigrid (8 children = 1 parent)

In multigrid procedure, equations at each level are "relaxed" using smoothing operators, such as Gauss-Seidel iterative solvers. Many types of smoothing operators have been proposed and used [7]. Among those, ILU(0)/IC(0) (Incomplete LU/Cholesky factorization without fill-in's) are widely used. These smoothing operators demonstrate excellent robustness for ill-conditioned problems [3,7,8]. In this study, IC(0) is adopted as a smoothing operator. IC(0) process includes global operations and it is difficult to be parallelized. Block-Jacobi-type localized procedure is possible for distributed parallel operations, but this approach tends to be unstable for ill-conditioned problems. In order to stabilize the localized IC(0) smoothing, additive Schwarz domain decomposition (ASDD) method for overlapped regions [9] has been introduced.

3.3 Procedures for Reordering

The 3D code is parallelized by domain decomposition using MPI for communications between partitioned domains. In the OpenMP/MPI hybrid parallel programming model, multithreading by OpenMP is applied to each partitioned domain. The reordering of elements in each domain allows the construction of local operations without global dependency, in order to achieve optimum parallel performance of IC operations in multigrid processes.

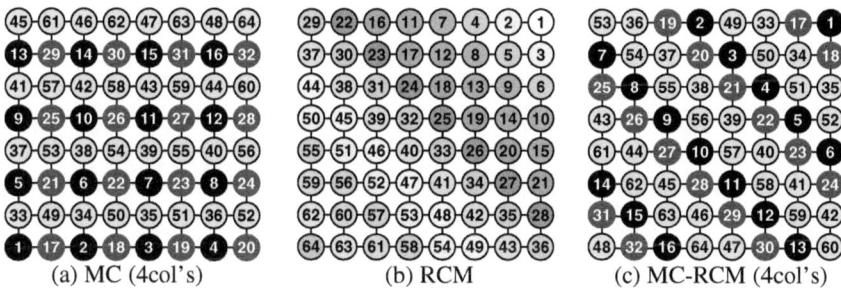

(a) MC (4col's)   (b) RCM   (c) MC-RCM (4col's)

Fig. 5. Various Methods for Coloring

Reverse Cuthill-McKee (RCM) reordering facilitates a faster convergence of iterative solvers with ILU/IC preconditioners than traditional multicolor (MC) reordering, especially for ill-conditioned problems, but leads to irregular numbers of vertices in each level set. The solution to this trade-off is RCM with cyclic-multicoloring (CM-RCM) [11]. In this method, further renumbering in a cyclic manner is applied to vertices that are reordered by RCM, as shown in Fig.5 (c). In CM-RCM, the number of colors should be large enough to ensure that vertices of the same color are independent.

3.4 Procedures for Optimization

In the current work, following three types of optimization procedures have been applied to OpenMP/MPI hybrid parallel programming models:

- Appropriate command lines for NUMA control
- First touch data placement
- Reordering for contiguous "sequential" access to memory,

Same command lines for NUMA control as were used in [1] have been applied in the current work. Detailed information for optimum command lines can be found in [1].

3.4.1 First Touch Data Placement

Minimizing memory access overhead is important for cc-NUMA architecture, such as T2K/Tokyo. In order to reduce memory traffic in the system, it is important to keep the data close to the cores that works with the data. On cc-NUMA architecture, this corresponds to making sure the pages of memory are allocated and owned by the cores that works with the data contained in the page. The most common cc-NUMA page-placement algorithm is the *first touch* algorithm [10], in which the core first referencing a region of memory has the page holding that memory assigned to it. Very common technique in OpenMP programs is to initialize data in parallel using the same loop schedule as will be used later in the computations, as shown in Fig.6.

```
do lev= 1, LEVELtot
    do ic= 1, COLORtot(lev)
!$omp parallel do private(ip,i,j,isL,ieL,isU,ieU)
        do ip= 1, PEsmpTOT
        do i = STACKmc(ip,ic-1,lev)+1, STACKmc(ip,ic,lev)
            RHS(i)= 0.d0; X(i)= 0.d0; D(i)= 0.d0

            isL= indexL(i-1)+1
            ieL= indexL(i)
            do j= isL, ieL
                itemL(j)= 0; AL(j)= 0.d0
            enddo

            isU= indexU(i-1)+1
            ieU= indexU(i)
            do j= isU, ieU
                itemU(j)= 0; AU(j)= 0.d0
            enddo
        enddo
        enddo
!$omp omp end parallel do
    enddo
    enddo
```

Fig. 6. Example of initialization of arrays for *first touch* data placement, where initialization process has been done

3.4.2 Reordering of Data for Contiguous "Sequential" Memory Access
In CM-RCM reordering, initial vector is re-numbered according to color ID, as shown in Fig.5. Elements in each color are distributed to each thread so that load for each thread is balanced.

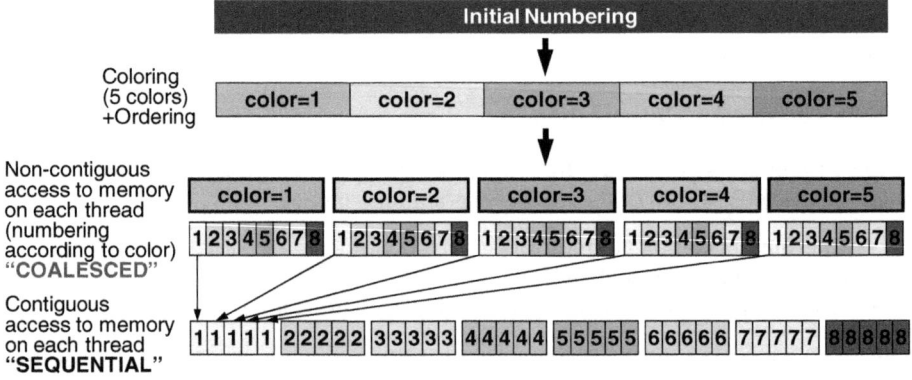

Fig. 7. Data distribution on each thread after further reordering for contiguous "sequential" memory access, number of color: 5, number of thread: 8

Pages of memory are allocated to local memory of each socket through the *first touch* procedure for initialization described in Fig.6. But the problem is that the size of each page is small and addresses of pages in each thread are not contiguous, as shown in Fig.7. This provides inefficient performance of access to memory. In order to provide contiguous address of local pages, further reordering has been applied, as shown in Fig.7.

Thus, each thread can access pages of memory in contiguous manner. This pattern of memory access is called "sequential" and suitable for cc-NUMA architectures with multi-sockets and multi-cores, while original pattern of memory access in CM-RCM is "coalesced" which is rather favorable in GPU computing.

4 Results

4.1 Effect of Coloring and Optimization

Performance of the developed code has been evaluated using between 16 and 1,024 cores of the T2K/Tokyo and Cray-XT4. IC(0) smoothing is applied twice at each level except the coarsest one with a single cycle of ASDD at each smoothing operation. A single V-cycle operation is applied at a preconditioning process of each CG iteration. At the coarsest level, IC(0) smoothing is applied once.

Following three types of OpenMP/MPI hybrid parallel programming models are applied as follows, and results are compared with those of *flat MPI*:

- **Hybrid 4×4 (HB 4×4):** Four OpenMP threads for each of four sockets in Fig.2, four MPI processes in each node, both of T2K/Tokyo and Cray-XT4
- **Hybrid 8×2 (HB 8×2):** Eight OpenMP threads for two pairs of sockets in Fig.2, two MPI processes in each node, only for T2K/Tokyo
- **Hybrid 16×1 (HB 16×1):** Sixteen OpenMP threads for a single node in Fig.2, one MPI process in each node, only for T2K/Tokyo

Because each node of Cray-XT4 has a single socket with four cores, only HB 4×4 has been applied to Cray-XT4 as OpenMP/MPI hybrid cases.

First of all, effect of reordering and optimization for OpenMP/MPI hybrid cases described in the previous chapter has been evaluated using 4 nodes (64 cores) of T2K/Tokyo for *flat MPI* and OpenMP/MPI hybrid parallel programming models. Number of finite-volume cells per each core is 262,144 ($=64^3$), therefore total problem size is 16,777,216. Figure 8 provides relationship between number of iterations for convergence of MGCG solvers with CM-RCM reordering for each parallel programming model and number of colors for CM-RCM reordering. Generally speaking, number of iterations for convergence of iterative solvers with IC/ILU-type preconditioners decreases, as number of colors increases, according to the theory of "incompatible nodes" described in [11]. Convergence of the problem in this work generally follows this theory, as shown in Fig.8.

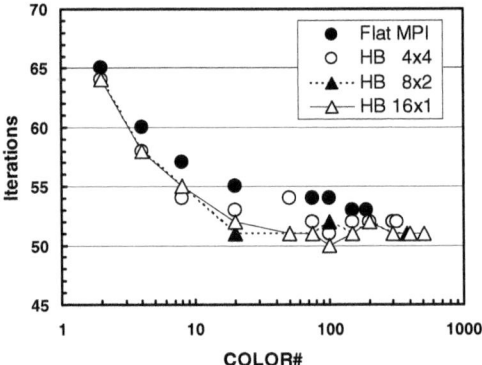

Fig. 8. Performance of MGCG solver with CM-RCM reordering on T2K/Tokyo, 16,777,216 cells, 64 cores, Number of Iterations for Convergence

Each of Fig. 9 (a) and (b) provides the relationship between performance (computation time for linear solvers) and number of colors for each parallel programming model. The procedures for optimization described in the previous chapter for OpenMP/MPI hybrid cases have been already applied. Although number of iterations decreases according to increasing of number of colors, as shown in Fig.8, CM-RCM with only 2 colors (CM-RCM(2)) provides the shortest elapsed computation time of MGCG solvers for each parallel programming model, as shown in Fig.9 (a). In this type of geometry with structured finite-difference-type voxels, 2 colors are enough to ensure that vertices of the same color are independent in CM-RCM procedure. Figure 9 (b) shows computation time for MGCG solvers per iteration, and CM-RCM(2) provides the best computational performance (FLOPS rate). This is because that cache is more efficiently utilized if the number of colors is smaller in CM-RCM ordering for structured finite-difference-type voxels used in the current work. Each of Fig. 10 (a) and (b) shows an example of numbering of cells for 2D structured finite-difference-type voxels with 64 cells using (a) CM-RCM(2) (#1-#32 cells belong to the 1st color, while #33-#64 cells are in the 2nd color) and (b) RCM (with 15 colors), respectively. If forward/backward substitutions (FBS) during ILU operations are considered for cells of #29, #30 and #31 in CM-RCM(2) (Fig.10 (a)), numbering of off-diagonal variables (#59~#64) is contiguous, and diagonal and off-diagonal variables are on separate cache lines. On the contrast, corresponding diagonal and off-diagonal variables could be on a same cache line in RCM with 15 colors (Fig.10 (b)). In this case, the cache line is written back to memory, after one of the diagonal variables is updated in FBS process.

Figure 11 provides computation time of MGCG solver with CM-RCM (2) before and after optimization on T2K/Tokyo. Effect of optimization described in 3. (optimum command lines for NUMA control, first-touch data placement (Fig.6) and reordering of data for contiguous "sequential" memory access (Fig.7)) is significant, especially for HB 8×2 and HB 16×1. In HB 4×4, effect of NUMA control is significant, but effect of first-touch and "sequential" memory access is small, because all data for each process are guaranteed to be on local memory of each socket. *Flat MPI* and optimized OpenMP/MPI Hybrid cases are generally competitive from the viewpoint of computation time.

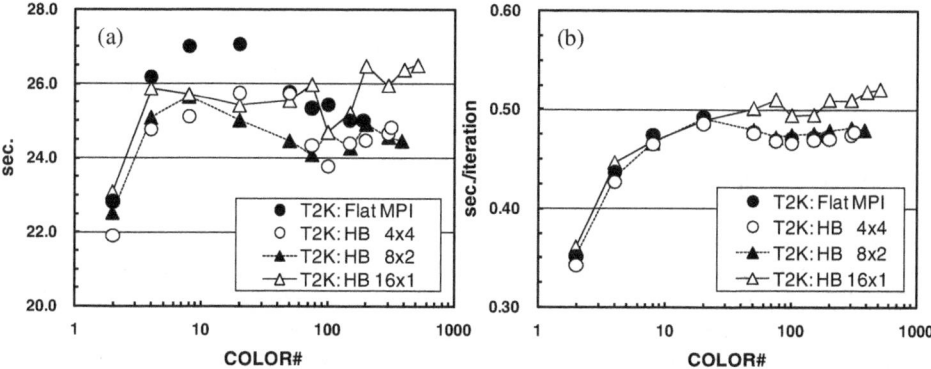

Fig. 9. Performance of MGCG solver with CM-RCM reordering on T2K/Tokyo, 16,777,216 cells, 64 cores (Optimized solvers using first-touch data placement (Fig.6) and reordering of data for contiguous "sequential" memory access (Fig.7)), (a) Elapsed computation time for MGCG solvers, (b) Computation time for MGCG solvers for each iteration

(a)

45	10	39	5	35	2	33	1
17	46	11	40	6	36	3	34
53	18	47	12	41	7	37	4
24	54	19	48	13	42	8	38
59	25	55	20	49	14	43	9
29	60	26	56	21	50	15	44
63	30	61	27	57	22	51	16
32	64	31	62	28	58	23	52

(b)

29	22	16	11	7	4	2	1
37	30	23	17	12	8	5	3
44	38	31	24	18	13	9	6
50	45	39	32	25	19	14	10
55	51	46	40	33	26	20	15
59	56	52	47	41	34	27	21
62	60	57	53	48	42	35	28
64	63	61	58	54	49	43	36

Fig. 10. Examples of numbering of cells for 2D structured finite-difference-type voxels using (a) CM-RCM (2) (#1-#32: 1st color, #33-#64: 2nd color), (b) RCM (with 15 colors)

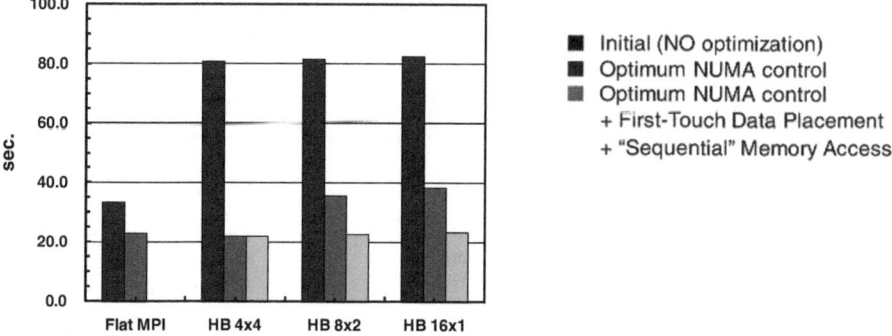

Fig. 11. Performance of MGCG solver with CM-RCM(2) on T2K/Tokyo, 16,777,216 cells, 64 cores (Initial version of solvers, Solvers with optimum command lines for NUMA control, Solvers with additional optimization by first-touch data placement and reordering of data for contiguous "sequential" memory access), Computation time for MGCG solvers

In *flat MPI* cases, time for reordering and set-up of matrices is approximately 1.50 sec. and 6.00 sec., respectively, while computation time for MGCG solvers is about 20 sec. In OpenMP/MPI hybrid cases of the current work, processes for reordering and set-up of matrices are not parallelized yet.

4.2 Weak Scaling

Performance of weak scaling has been evaluated using between 16 and 1,024 cores of the T2K/Tokyo and Cray-XT4. Number of finite-volume cells per each core is 262,144 (=64^3), therefore maximum total problem size is 268,435,456. Figure 12 (a) shows computation time of MGCG solver until convergence, and Fig.12 (b) shows number of iterations for convergence. Number of iterations for convergence of *flat MPI* increases, as number of core is more than 256. On the contrast, number of iterations of OpenMP/MPI hybrid cases stays almost constant between 16 and 1,024 cores for the ill-conditioned problems with condition number of 10^{10}. This feature is more significant, as thread number per process increases. MGCG solvers with OpenMP/MPI hybrid parallel programming model provide excellent scalability even in this type of ill-conditioned problems. Generally speaking, robustness of localized IC(0) preconditioning is getting worse, as number of domains increases in ill-conditioned problems. OpenMP/MPI hybrid is generally more robust than *flat MPI* because number of cells at domain boundaries is relatively fewer.

Performance of Cray-XT4 is generally larger than that of T2K/Tokyo by 40%~50%. Memory performance of Cray-XT4 is about 25% larger than that of T2K/Tokyo according to STREAM benchmarks [12]. Furthermore, cache is utilized more efficiently on Cray-XT4 in multigrid operations especially for coarser level of cells, because no cache coherency is considered on Cray-XT4.

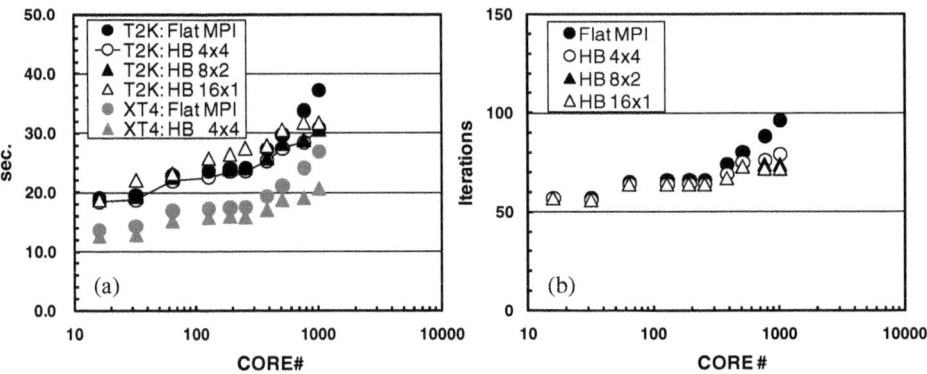

Fig. 12. Performance of MGCG solver with CM-RCM(2) on T2K/Tokyo and Cray-XT4 using up to 1,024 cores, Weak Scaling: 262,144 cells/core, Max. Total Problem Size: 268,435,456 (a) Computation time for MGCG solvers, (b) Number of iterations for convergence

4.3 Strong Scaling

Performance of strong scaling has been evaluated for fixed size of problem with 33,554,432 cells (=512×256×256) using between 16 and 1,024 cores of T2K/Tokyo and Cray-XT4. Figure 13(a) shows relationship between number of cores and number of iterations until convergence for each parallel programming model. Number of iterations for *flat MPI* increases significantly, as number of cores (domains) increases. On the contrast, increasing for hybrid parallel programming model is not so significant. Especially, that of HB 16×1 stays almost constant between 16 and 1,024 cores. Figure 13(b) provides parallel performance of T2K/Tokyo based on the performance of *flat MPI* with 16 cores. At 1,024 cores, parallel performance is approximately 60% of the performance at 16 cores. Decreasing of parallel performance of HB 16×1 is very significant. At 1,024 cores, HB 16×1 is rather slower than *flat MPI* although convergence is much better.

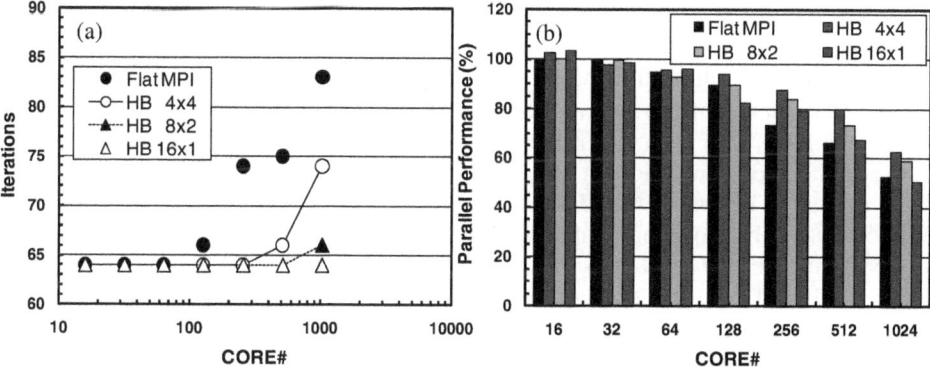

Fig. 13. Performance of MGCG solver with CM-RCM(2) on T2K/Tokyo using up to 1,024 cores, Strong Scaling: 33,554,432 cells (=512×256×256), (a) Number of iterations for convergence, (b) Parallel performance based on the performance of *Flat MPI* with 16 cores

```
!C
!C-- SEND
      do neib= 1, NEIBPETOT
      istart= levEXPORT_index(lev-1,neib) + 1
      iend  = levEXPORT_index(lev  ,neib)
      inum  = iend - istart + 1
!$omp parallel do private (ii)
      do k= istart, iend
         WS(k)= X(EXPORT_ITEM(k))
      enddo
!$omp end parallel do
      call MPI_ISEND (WS(istart), inum, MPI_DOUBLE_PRECISION,        &
   &                  NEIBPE(neib), 0, MPI_COMM_WORLD, req1(neib), ierr)
      enddo
```

Fig. 14. Communications for information exchange at domain boundary (sending process), copies of arrays to/from sending/receiving buffers occur

Communication between partitioned domains at each level occurs in parallel iterative solvers. Information at each domain boundary is exchanged using functions of
MPI for point-to-point communication. In this procedure, copies of arrays to/from
sending/receiving buffers occur, as shown in Fig.14. In the original code using
OpenMP/MPI hybrid parallel programming models, this type of operation for memory copy is parallelized by OpenMP, as shown in Fig.14. But overhead of OpenMP is
significant, if length of loop is short and number of threads is large. If length of loop
is short, operations by a single thread might be faster than those by multi-threading.

In the current work, effect of switching from multi-threading to single-threading at
coarser levels of multigrid procedure has been evaluated. Figure 15 (a) shows results
of HB 16×1 with 1,024 cores (64 nodes) for the strong scaling case. "Communication" part includes processes of memory copies shown in Fig.14. "*Original*" applies
multi-threading by OpenMP at every level of multigrid procedure. "*LEVcri=k*" means
applying multi-threading if *level* of grid is smaller than k. Therefore, single-threading
is applied at every level if "*LEVcri=1*", and multi-threading is applied at only the
finest grid (level=1) if "*LEVcri=2*". Generally speaking, "*LEVcri=2*" provides the
best performance at 1,024 cores for all of HB 4×4, HB 8×2 and HB 16×1, although
effect of switching is not so clear for HB 4×4, as shown in Fig.15 (b). Figure 16
shows effects of this optimization with "*LEVcri=2*" for all OpenMP/MPI hybrid
cases. Performance of HB 8×2 and HB 16×1 are much improved at large number of
cores, and HB 8×2 is even faster than HB 4×4 at 1,024 cores, while performance with
fewer number of cores did not change. Finally, performance of strong scaling has
been evaluated between 16 and 1,024 cores of T2K/Tokyo and Cray-XT4. Each of
Fig.17 (a) and (b) shows parallel speed-up to 1,024 cores with "*LEVcri=2*", based-on
the performance of *flat MPI* with 16 cores of each platform. Performance at 1,024
cores for T2K/Tokyo is 534 (flat MPI), 690 (HB 4×4), 696 (HB 8×2), and 646 (HB
16×1), respectively. Performance of Cray-XT4 is 455 (flat MPI) and 617 (HB 4×4).

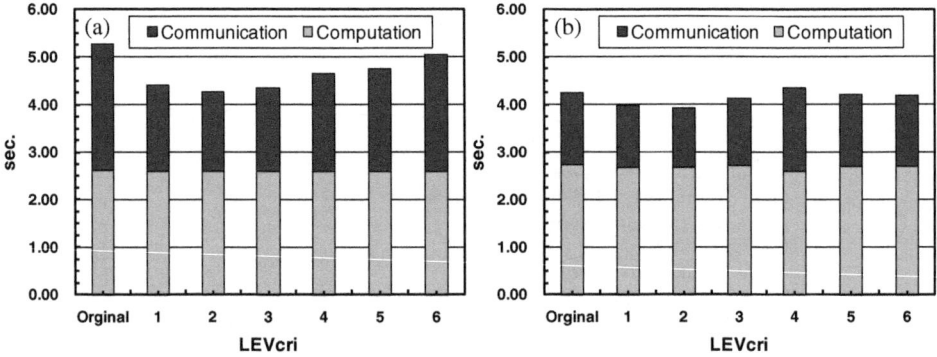

Fig. 15. Effect of switching from multi-threading to single-threading at coarse levels of multigrid procedure in operations of memory copy for communications at domain boundaries using
1,024 cores for strong scaling case with 33,554,432 cells (=512×256×256), "*LEVcri=k*": applying multi-threading if level of grid is smaller than k, (a) HB 16×1, (b) HB 4×4

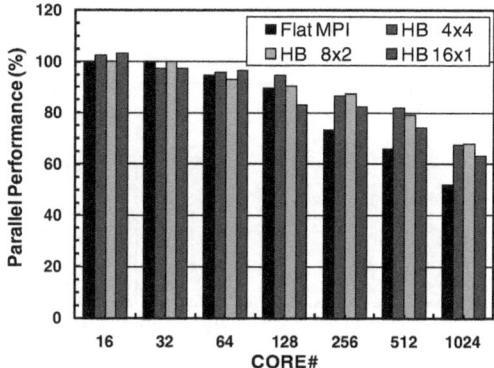

Fig. 16. Performance of MGCG solver with CM-RCM(2) on T2K/Tokyo using up to 1,024 cores, Strong Scaling: 33,554,432 cells (=512×256×256), Parallel performance based on the performance of *Flat MPI* with 16 cores, "*LEVcri=2*" in Fig.15 is applied for OpenMP/MPI hybrid parallel programming models

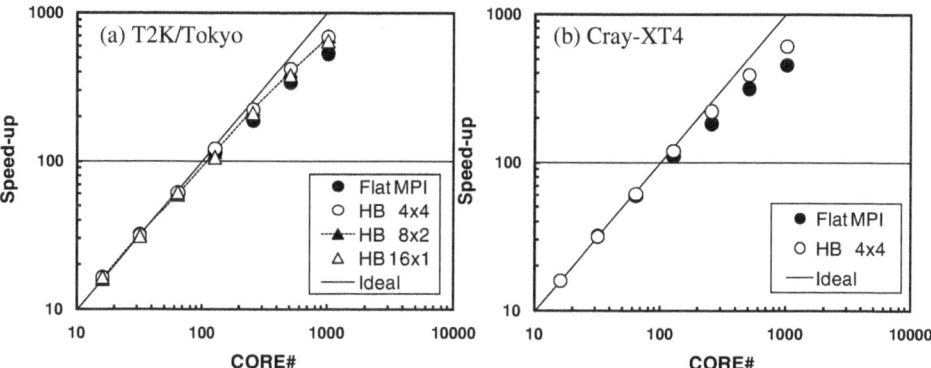

Fig. 17. Performance of MGCG solver with CM-RCM(2) on T2K/Tokyo and Cray-XT4 using up to 1,024 cores, Strong Scaling: 33,554,432 cells (=512×256×256), Speed-up based on the performance of *flat MPI* with 16 cores on each platform, "*LEVcri=2*" in Fig.15 is applied for OpenMP/MPI hybrid parallel programming models

Finally, optimized solver for strong scaling cases with "*LEVcri=2*" has been applied to weak scaling cases, where number of finite-volume cells per each core is $262,144$ ($=64^3$). Figure 18 shows ratio of performance for OpenMP/MPI hybrid cases up to 1,024 cores. Generally speaking, entire performance is not so much changed by optimization for strong scaling cases. Therefore, switching from multi-threading to single-threading at coarser levels for communications at domain boundaries works well for both of weak and strong scaling cases.

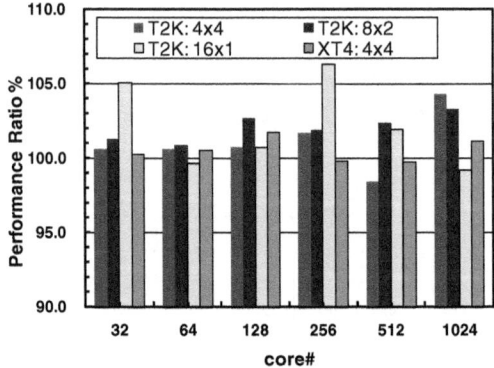

Fig. 18. Performance of MGCG solver with CM-RCM(2) on T2K/Tokyo and Cray-XT4 using up to 1,024 cores, Weak Scaling: 262,144 cells/core, Max. Total Problem Size: 268,435,456, Ratio of performance for the solver with "*LEVcri=2*" to that of original solver in Fig.12

5 Concluding Remarks

OpenMP/MPI hybrid parallel programming models were implemented to 3D finite-volume based simulation code for groundwater flow problems through heterogeneous porous media using parallel iterative solvers with multigrid preconditioning by IC(0) smoothing. Performance and robustness of the developed code has been evaluated on T2K/Tokyo and Cray-XT4 using up to 1,024 cores through both of weak and strong scaling computations. Optimization procedures for OpenMP/MPI hybrid parallel programming models, such as appropriate command lines for NUMA control, first touch data placement and reordering for contiguous "sequential" access to memory, provided excellent improvement of performance on multigrid preconditioners. Furthermore, performance of OpenMP/MPI hybrid at large number of cores in strong scaling is improved by optimization of communication procedure between domains. The developed procedure also provided good performance in weak scaling cases. OpenMP/MPI hybrid demonstrated better performance and robustness than *flat MPI*, especially with large number of cores for ill-conditioned problems, and could be a reasonable choice for large-scale computing on multi-core/multi-socket clusters. Automatic selection of optimum parallel programming models and parameters (e.g. number of colors, switching level for communication) is an interesting area for future works. Furthermore, development of parallel procedures for reordering and more robust procedures for parallel multigrid with HID (Hierarchical Interface Decomposition) [1], and with improvement of solver at the coarsest level, is also ongoing. In the current work, the developed procedures were evaluated under very limited conditions. Various geometries, boundary conditions and problem size will be applied in the future for various types of computer platforms.

References

1. Nakajima, K.: Flat MPI vs. Hybrid: Evaluation of Parallel Programming Models for Pre-conditioned Iterative Solvers on "T2K Open Supercomputer". In: IEEE Proceedings of the 38th International Conference on Parallel Processing (ICPP 2009), pp. 73–80 (2009)

2. Information Technology Center, The University of Tokyo,
   http://www.cc.u-tokyo.ac.jp/
3. Nakajima, K.: Parallel Multilevel Method for Heterogeneous Field. In: IPSJ Proceedings of HPCS 2006, pp. 95–102 (2006) (in Japanese)
4. NERSC, Lawrence Berkeley National Laboratory, http://www.nersc.gov/
5. The T2K Open Supercomputer Alliance,
   http://www.open-supercomputer.org/
6. Deutsch, C.V., Journel, A.G.: GSLIB Geostatistical Software Library and User's Guide, 2nd edn. Oxford University Press, Oxford (1998)
7. Tottemberg, U., Oosterlee, C., Schuller, A.: Multigrid. Academic Press, London (2001)
8. Nakajima, K.: Parallel Multilevel Iterative Linear Solvers with Unstructured Adaptive Grids for Simulations in Earth Science. Concurrency and Computation: Practice and Experience 14-6/7, 484–498 (2002)
9. Smith, B., Bjørstad, P., Gropp, W.: Domain Decomposition, Parallel Multilevel Methods for Elliptic Partial Differential Equations. Cambridge Press, Cambridge (1996)
10. Mattson, T.G., Sanders, B.A., Massingill, B.L.: Patterns for Parallel Programming. Software Patterns Series (SPS). Addison-Wesley, Reading (2005)
11. Washio, T., Maruyama, K., Osoda, T., Shimizu, F., Doi, S.: Efficient implementations of block sparse matrix operations on shared memory vector machines. In: Proceedings of The 4th International Conference on Supercomputing in Nuclear Applications (SNA 2000) (2000)
12. STREAM (Sustainable Memory Bandwidth in High Performance Computers),
   http://www.cs.virginia.edu/stream/

A Parallel Strategy for a Level Set Simulation of Droplets Moving in a Liquid Medium

Oliver Fortmeier and H. Martin Bücker

RWTH Aachen University, Institute for Scientific Computing,
D-52056 Aachen, Germany
{fortmeier,buecker}@sc.rwth-aachen.de

Abstract. The simulation of two-phase flow problems involving two time-dependent spatial regions with different physical properties is computationally hard. The numerical solution of such problems is complicated by the need to represent the movement of the interface. The level set approach is a front-capturing method representing the position of the interface implicitly by the root of a suitably defined function. We describe a parallel adaptive finite element simulation based on the level set approach. For freely sedimenting n-butanol droplets in water, we quantify the parallel performance on a Xeon-based cluster using up to 256 processes.

1   Introduction

Two-phase flow problems play a dominant role in various areas of computational science and engineering. Systems containing different liquids such as oil slicks in coastal waters or liquid-liquid extraction columns are illustrating examples. To develop predictive models for such extraction columns, the study of single droplets is important. At RWTH Aachen University, an interdisciplinary team of researchers from engineering, mathematics and computer science is interested in analyzing the behavior of single droplets in surrounding liquids [1,8,9,15]. These flow problems involve two spatial regions that vary with time. In each of these regions, the physical properties of a material is uniformly distributed in space. The numerical simulation of two-phase flow problems is complicated by the fact that the interface between the two phases needs to be represented for the reconstruction of the interfacial movement. In front-tracking methods, the interface is explicitly represented by computational elements that follow its movement [21]. In contrast, front-capturing methods represent the interface implicitly by suitably defined functions.

The volume of fluid technique [11,13] is a popular front-capturing method representing the interface by a function whose values are interpolated on the underlying mesh. The reconstruction of smooth interfaces, however, requires advanced interpolation techniques. The level set approach [17,18] is another front-capturing method eliminating this drawback. In these methods, the position of the interface is given by the root of a scalar-valued function that splits the computational domain into two regions. An advantage of the level set approach is

J.M.L.M. Palma et al. (Eds.): VECPAR 2010, LNCS 6449, pp. 200–209, 2011.

its elegance and simplicity to handle complicated problems involving breaking or joining regions. While serial level set approaches are widely used, parallel techniques are hardly available in the open literature. An early reference [14] carries out numerical experiments on an Intel/IPSC-860 hypercube but does not focus on parallel computing. A parallel implementation of a hybrid technique that brings together the level set approach and the volume of fluid method is reported in [19].

The approach taken in [22] to parallelize a level set technique for the simulation of two-dimensional dendritic growth is to exploit processor virtualization rather than domain decomposition. Essentially, the program decomposes the computation into a large number of objects called virtual processors which are then mapped to different physical processors. Performance results on a system with Intel Xeon-based dual-core processors are reported demonstrating a good scalability for up to 32 cores, but a degradation of the performance for 64 cores.

The idea behind the recent domain decomposition approach described in [10] is to generate two different grids that can be adapted independently of each other. In addition to the grid on which the flow solution is computed, this approach employs a separate Cartesian grid for tracking the movement of the interface using a level set approach. The approach is shown to scale on a cluster with Intel Xeon-based quad-core processors using up to 2048 cores.

The structure of this note is as follows. In Sect. 2, we sketch a particular two-phase flow problem which is detailed in [1]. This section also includes the underlying mathematical model. The new contributions of this note are given in Sect. 3 and 4. First, the overall strategy to parallelize the finite element method and the level set technique is described. Second, performance results of a new implementation are reported for a freely sedimenting n-butanol droplet in water with up to 256 processes. Concluding remarks are given in Sect. 5.

2   Numerical Simulation of Droplets Sedimenting in Water

Throughout this article, we follow the overall setting reported in [1] in which the solvent extraction standard-test system of n-butanol droplets sedimenting in water is considered [16]. The experimental setup is schematically depicted in Fig. 1(a). The droplets are generated through a nozzle submerged in a cylindrical cell that contains the continuous phase. Upon generation, the droplet starts to accelerate upwards until it reaches its terminal sedimentation velocity which is determined by monitoring the droplet's position by a camera. A numerical simulation is carried out to predict the velocity and the deformation of different droplets over time. In Fig. 1(b), the deformation of a droplet with radius of 2 mm is illustrated. In Fig. 2, the simulated position and velocity are depicted for a droplet of the same radius showing the oscillating behavior [1].

The numerical simulation is based on the following computational model originally introduced in [20]. The incompressible Navier-Stokes equations are employed to model the velocity u and the pressure p of a two-phase flow problem

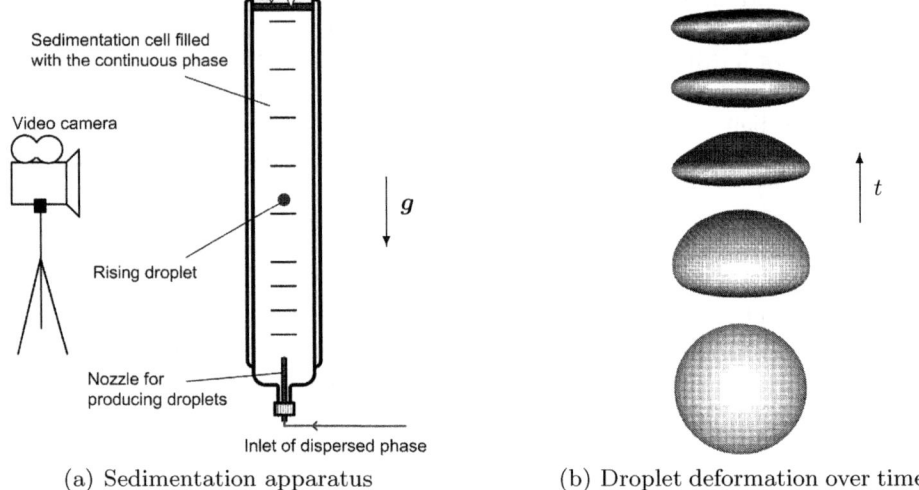

(a) Sedimentation apparatus

(b) Droplet deformation over time

Fig. 1. Experimental setup and simulated shape of a droplet both taken from [1]

in a domain $\Omega \subset \mathbb{R}^3$. The phase interface, denoted by Γ, is described by the root of the scalar level set function $\phi = \phi(\boldsymbol{x}, t)$ where $\boldsymbol{x} \in \Omega$ and t denotes time. The level set function splits the domain Ω into two disjoint subdomains $\Omega_1(t) := \{\boldsymbol{x} \in \Omega \mid \phi(\boldsymbol{x}, t) < 0\}$ and $\Omega_2(t) := \{\boldsymbol{x} \in \Omega \mid \phi(\boldsymbol{x}, t) > 0\}$. Here, the droplet is represented by Ω_1 and the continuous phase by Ω_2. The effect of the surface tension τ is expressed in terms of a localized force at the interface, so-called continuum surface force [3,7]. Combining these approaches leads to the following system of partial differential equations in $\Omega \times [0, T]$:

$$\rho(\phi)\left(\frac{\partial}{\partial t}\boldsymbol{u} + (\boldsymbol{u} \cdot \nabla)\boldsymbol{u}\right) = -\nabla p + \rho(\phi)\boldsymbol{g} + \mathrm{div}\left(\mu(\phi)\mathbf{D}(\boldsymbol{u})\right) + \tau \mathcal{K} \delta_\Gamma \boldsymbol{n}_\Gamma, \quad (1)$$

$$\mathrm{div}\,\boldsymbol{u} = 0, \quad (2)$$

$$\frac{\partial}{\partial t}\phi + \boldsymbol{u} \cdot \nabla\phi = 0 \quad (3)$$

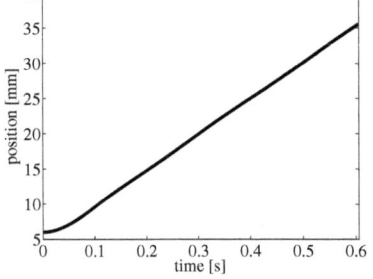

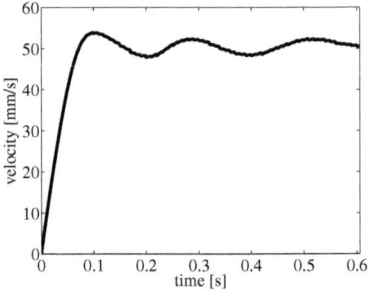

Fig. 2. Position and sedimentation velocity of an n-butanol droplet in water

with appropriate boundary and initial conditions. Here, the density is given by ρ, the viscosity is denoted by μ, the strain tensor is defined by $\mathbf{D}(\boldsymbol{u}) := \nabla\boldsymbol{u}+(\nabla\boldsymbol{u})^T$, and the symbol $\mathcal{K}$ is used for the curvature. Furthermore, the Dirac function with support on Γ is denoted by δ_Γ and $\boldsymbol{n}_\Gamma$ represents the normal vector on Γ. The external gravity force is denoted by $\boldsymbol{g}$.

For the solution of the coupled problem (1)–(3), the finite element solver DROPS [5] is employed. Its main features are as follows: The three-dimensional geometry is discretized by a hierarchical tetrahedral grid accounting for adaptive refinements that change over time. Piecewise quadratic functions are used as finite element functions for the level set and the velocity. The pressure is represented by piecewise linear functions. Various Navier-Stokes, Stokes, and Krylov solvers are implemented. Time integration is based on a linear theta scheme. The level set function and the Navier-Stokes equations are decoupled by a fix-point interaction. A fast-marching variant is used to re-initialize the level set function.

3   Parallel Hierarchy of Triangulations

In DROPS, the hierarchical tetrahedral grid representing the computational domain Ω is decomposed and distributed to the processes. Here, we sketch the data structures used for the parallel refinement algorithm introduced in [6]. Let $\mathcal{T}_0$ denote a coarse triangulation representing the computational domain. A finer triangulation $\mathcal{T}_1$ is obtained by refining some tetrahedra of $\mathcal{T}_0$ by a red/green refinement algorithm [2]. Multiple recursive refinements of tetrahedra lead to a multi-level triangulation $\mathcal{M} = (\mathcal{T}_0, \ldots, \mathcal{T}_{k-1})$, where the triangulation $\mathcal{T}_{k-1}$ represents the finest triangulation. The multi-level triangulation is admissible if the following conditions hold for all levels $l \in \{1, \ldots, k-1\}$:

1. A tetrahedron $T \in \mathcal{T}_l$ is either in $\mathcal{T}_{l-1}$ or is obtained by a refinement of a tetrahedron $T' \in \mathcal{T}_{l-1}$.
2. If $T \in \mathcal{T}_{l-1} \cap \mathcal{T}_l$ then $T \in \mathcal{T}_{l+1}, \ldots, \mathcal{T}_{k-1}$. That is, if the tetrahedron T is not refined then it stays unrefined.

Due to these two conditions, we can assign each tetrahedron T a unique level

$$l(T) := \min \{m \mid T \in \mathcal{T}_m\}.$$

The set of all tetrahedra on level l is denoted by $\mathcal{G}_l$ and is called the hierarchical surplus. These hierarchical surpluses define the hierarchical decomposition $\mathcal{H} = (\mathcal{G}_0, \ldots, \mathcal{G}_{k-1})$. Since each tetrahedron is only located in one hierarchical surplus this decomposition is used to efficiently store all tetrahedra. The equations (1)–(3) are solved on the finest triangulation $\mathcal{T}_{k-1}$. Therefore, the domain decomposition approach is based on distributing the tetrahedra of $\mathcal{T}_{k-1}$ among the processes and leads to a decomposition of the hierarchical surplus $\mathcal{H}$.

In general, repeated local grid modifications lead to large differences in the number of tetrahedra stored on each process. Therefore, a load balancing algorithm is implemented consisting of three steps:

1. The triangulation $\mathcal{T}_{k-1}$ is described by a weighted graph $G = (V, E, \varrho_V, \varrho_E)$. Each node $v \in V$ represents a set of tetrahedra. Each edge $(v, w) \in E$ corresponds to a pair of tetrahedron sets v and w that have at least a face in common. The number of common faces between the tetrahedron sets determines the edge weight $\varrho_E(v, w)$. The vertex weight $\varrho_V(v)$ is given by the number of tetrahedra represented by v. The underlying graph model is discussed in more detail in [4].

2. The weighted graph is partitioned into P parts, where P denotes the number of processes. This leads to a partition of the vertices $V = V_1 \cup \cdots \cup V_P$ where $V_i \cap V_j = \emptyset$ for all $i \neq j$. The library ParMetis [12] partitions the graph in an attempt to minimize the number of edges between V_i and V_j with $i \neq j$ while balancing the sum of the vertex weights in each V_i. In particular, a routine is used that computes a new partition based on a previous one. This is of special interest for solving two-phase problems, since the triangulation $\mathcal{T}_{k-1}$ changes only slightly by performing a time step of the simulation.

3. The final step consists of migrating tetrahedra and its vertices, edges and faces among the processes. After the migration, each process $p = 1, \ldots, P$ stores the tetrahedra represented by the subset V_p. Note that the migration has to take into account not only the tetrahedra in $\mathcal{T}_{k-1}$ but also on other levels so that the complete hierarchy is still admissible.

Since most physical effects occur in the vicinity of the phase interface Γ we refine the grid in this subdomain leading to a high resolution close to Γ. While evolving in simulation time, the location of Γ changes and demands for modifying the finest triangulation. However, this can again imply a large difference in the number of tetrahedra stored on each process. To avoid this imbalance the three load balancing steps are performed each time the grid changes.

DROPS uses piecewise linear and quadratic finite element functions. The corresponding degrees of freedom (DOF) are located at vertices and edges of $\mathcal{T}_{k-1}$. So, the migration algorithm handles not only geometric but also DOF data. Therefore, the non-overlapping decomposition of the tetrahedra in $\mathcal{T}_{k-1}$ leads to a distribution of DOF among the processes. Vertices and edges may be stored by multiple processes if they are located at a process boundary. Hence, the corresponding DOF are stored on multiple processes as well. Linear algebra operations on these DOF require updating mechanisms which include communication between neighboring processes. This communication is overlapped by computation. The systems of linear equations resulting from linearizing and discretizing (1)–(3) are iteratively solved by Krylov subspace methods which involve linear algebra operations such as matrix vector products, inner products, and vector updates. These linear algebra operations are rearranged to minimize the number of communications by re-using updated DOF.

4   Parallel Performance

In this section, we present results of sedimenting n-butanol droplets in water. In [1], droplets of initial radius from 0.5 mm to 2.0 mm are investigated in a

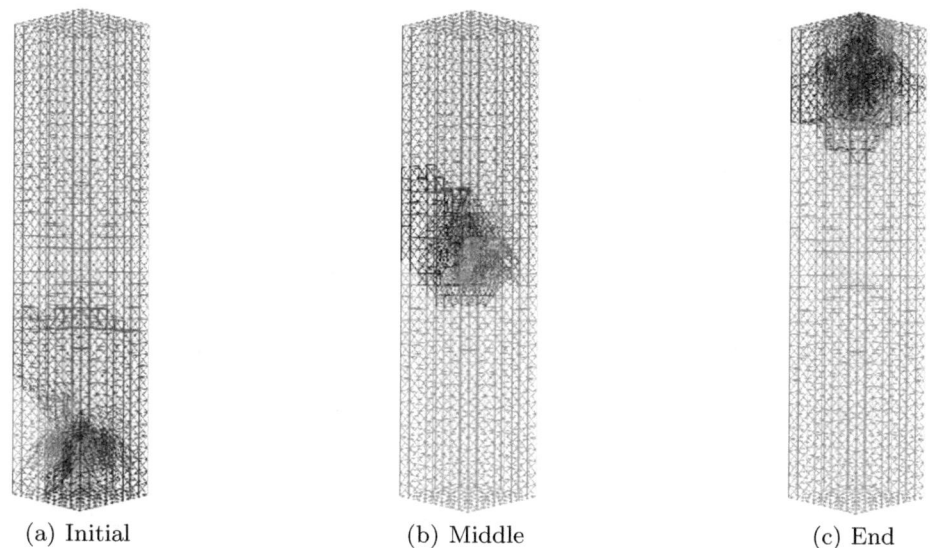

(a) Initial   (b) Middle   (c) End

Fig. 3. Decomposition of a triangulation among 4 processes

brick-shaped domain. However, in this note, we are interested in studying larger problems by simulating droplets of radius 2 mm and 3mm. In Fig. 3(a), the initial distribution of the tetrahedra of the finest triangulation is illustrated. In Fig. 3(b) and 3(c), the distribution of tetrahedra is depicted in the middle and at the end of the simulation time, respectively. These simulations are performed on a triangulation which is 4 or 5 times recursively refined in the vicinity of the phase boundary. That is, each tetrahedron whose barycenter is located at a radius of 0.4 mm around the phase boundary is recursively refined into eight sub-tetrahedra. In Table 1, the problem sizes in terms of number of tetrahedra and DOF are presented.

All experiments are performed on a cluster of Xeon-based quad-core processors (E5450) at the center for computing and communication at RWTH Aachen University. Each node of the cluster consists of two quad-core processors which share one InfiniBand network card for inter-node communication. An Intel implementation of MPI handles the communication between the processes.

To present the performance results, we distinguish between two different placements of MPI processes on the quad-core processors: a compact strategy (*Comp*), and a scatter strategy (*Scat*). In *Comp*, one MPI process is placed on each core

Table 1. Problem size in terms of number of tetrahedra and DOF

Radius [mm]	Refinements	tetrahedra	velocity DOF	pressure DOF	level set DOF
2	4	155 132	525 756	22 485	178 101
2	5	876 776	3 006 030	126 190	1 004 859
3	4	328 472	1 102 782	46 799	370 443

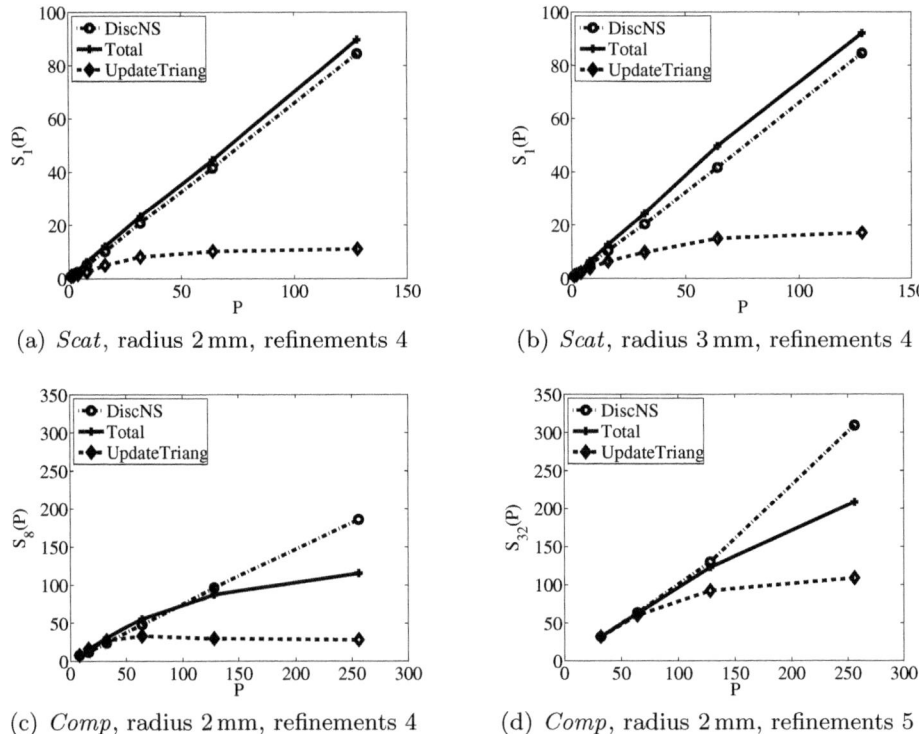

(a) *Scat*, radius 2 mm, refinements 4

(b) *Scat*, radius 3 mm, refinements 4

(c) *Comp*, radius 2 mm, refinements 4

(d) *Comp*, radius 2 mm, refinements 5

Fig. 4. Parallel performance of a single time step

of the processors leading to eight MPI processes per node. In *Scat*, only a single MPI process is located on each processor implying idle cores and two processes per node. The choice of the strategy has a big impact on the time to perform one time step. For example, while simulating a 2 mm drop with 4 levels of refinement, switching from *Scat* to *Comp* for 8 processes increases the runtime by a factor of 3.22. The major reason for this effect seems to be given by the limited memory bandwidth. Since the underlying data structures represent unstructured grids the access pattern to the memory is not consecutive resulting in unstructured memory access. Hence, the cache hierarchy of the processors cannot be exploited. Additionally, the network card is likely to serve two MPI processes more efficiently than eight processes.

In Fig. 4 the speedup of four test cases is shown. In this figure, the speedup for P processes is defined by

$$S_p(P) = \frac{p \cdot T(p)}{T(P)},$$

where $T(P)$ denotes the runtime on P processes and p denotes the smallest number of processes used to perform the corresponding simulation. That is, we assume perfect speedup while using p processes. These plots show the parallel performance of a single time step for the solution of the two-phase flow problem. In Fig. 4(a) and (b), a single process is capable of representing the discrete two-phase flow problem and, thus, in these figures $p = 1$. In Fig. 4(c), at least eight processes are used, i.e., one node is employed whereas in Fig. 4(d) four nodes, $p = 32$, are at least applied to solve the two-phase problem. The time required by a time step is denoted by *Total*. The update of the triangulation including grid modifications and load balancing are referred to as *UpdateTriang*. The discretization of the system (1)–(2) is denoted by *DiscNS* The remaining time required by a time step dominates *Total* and is not shown separately in these plots. The total time scales well up to 128 and 256 processes using the *Scat* or *Comp* strategy, respectively.

The updating does not scale as good as the other parts. In Fig. 4(b), the relative time spent in updating the triangulation w.r.t. the total time is 0.24 % on one process and 1.35 % on 128 processes. In Fig. 4(c), this relative time increases from 0.36 % on 8 processes to 1.47 % on 256 processes. In this scenario of a sedimenting droplet and in typical simulations, the updating is performed about every tenth discrete time step. Therefore, this updating part of the simulation is currently not a major bottleneck.

For the total time, using 128 processes and *Scat* leads to a speedup of $S_1(128) \approx 90$ for both radii and four refinements as presented in Fig. 4(a) and (b). Here, the serial execution time for simulating a 3 mm droplet is decreased from 8801 s to 96 s while using 128 processes. The serial time for updating the triangulation takes 21 s. On 128 processes, modifying the triangulation including the load-balancing steps takes 1.3 s. Figure 4(c) and (d) illustrate that for *Comp* the speedup is given by $S_8(256) \approx 116$ and $S_{32}(256) \approx 209$ when simulating a 2 mm droplet with four and five refinements, respectively. The execution time in Fig. 4(c) decreases from 1713 s on eight processes to 118 s on 256 processes. If simulating a droplet of radius 3 mm with four refinements and *Comp*, the total time of 4508 s on eight processes is reduced to 119 s on 256 processes, whereas the time for updating decreases from 9.6 s to 1.7 s.

5   Concluding Remarks

The strategy to parallelize the three-dimensional computational fluid dynamics software DROPS is presented. This software employs a unique combination of discretization on tetrahedral grids using finite elements, local grid refinement techniques, and level set methods for interface capturing [5]. The parallelization consists of decomposing the computational domain on the finest level of the hierarchy of triangulations. Load balancing is addressed via graph partitioning. In [1], the parallel simulation of a sedimenting n-butanol droplet in water is validated by comparing the numerical results with empirical models as well as with actual experimental measurements. In contrast to [1] where parallel computing is only briefly mentioned, the focus of the present note is on the parallelization strategy and the resulting performance on a cluster of Xeon-based quad-core processors. The reported parallel performance is good, but not excellent. A minor bottleneck in the scalability is shown to be the update of the triangulation.

Moreover, there is room for introducing a more refined graph model for load balancing [4] and more scalable approaches to re-initialize the level set function.

Acknowledgments

We thank our collaborators within SFB 540 "Model-based experimental analysis of kinetic phenomena in fluid multi-phase reactive systems" which is supported by the Deutsche Forschungsgemeinschaft (DFG). The Aachen Institute for Advanced Study in Computational Engineering Science (AICES) provides a stimulating research environment for our work.

References

1. Bertakis, E., Groß, S., Grande, J., Fortmeier, O., Reusken, A., Pfennig, A.: Validated simulation of droplet sedimentation with finite-element and level-set methods. Chemical Engineering Science 65(6), 2037–2051 (2010)
2. Bey, J.: Simplicial grid refinement: On Freudenthal's algorithm and the optimal number of congruence classes. J. Numer. Math. 85(1), 1–29 (2000)
3. Brackbill, J.U., Kothe, D.B., Zemach, C.: A continuum method for modeling surface tension. J. Comput. Phys. 100(2), 335–354 (1992)
4. Fortmeier, O., Henrich, T., Bücker, H.M.: Modeling data distribution for two-phase flow problems by weighted graphs. In: Beigl, M., Cazorla-Almeida, F.J. (eds.) 23rd Workshop on Parallel Sytems and Algorithms, Hannover, Germany, February 12, pp. 31–38. VDE (2010)
5. Groß, S., Reichelt, V., Reusken, A.: A finite element based level set method for two-phase incompressible flows. Comput. Vis. Sci. 9(4), 239–257 (2006)
6. Groß, S., Reusken, A.: Parallel multilevel tetrahedral grid refinement. SIAM J. Sci. Comput. 26(4), 1261–1288 (2005)
7. Groß, S., Reusken, A.: Finite element discretization error analysis of a surface tension force in two-phase incompressible flows. SIAM J. Numer. Anal. 45(4), 1679–1700 (2007)
8. Gross-Hardt, E., Amar, A., Stapf, S., Pfennig, A., Blümich, B.: Flow dynamics inside a single levitated droplet. Ind. Eng. Chem. Res. 1, 416–423 (2006)
9. Gross-Hardt, E., Slusanschi, E., Bücker, H.M., Pfennig, A., Bischof, C.H.: Practical Shape Optimization of a Levitation Device for Single Droplets. Opt. Eng. 9(2), 179–199 (2008)
10. Herrmann, M.: A parallel Eulerian interface tracking/Lagrangian point particle multi-scale coupling procedure. J. Comput. Phys. 229(3), 745–759 (2010)
11. Hirt, C., Nichols, B.: Volume of fluid (VOF) method for the dynamics of free boundaries. J. Comput. Phys. 39(1), 201–225 (1981)
12. Karypis, G., Kumar, V.: A parallel algorithm for multilevel graph partitioning and sparse matrix ordering. J. Parallel Distrib. Comput. 48(1), 71–95 (1998)
13. Li, J., Renardy, Y.: Numerical study of flows of two immiscible liquids at low reynolds number. SIAM Rev. 42(3), 417–439 (2000)
14. Li, X.L.: Study of three-dimensional Rayleigh–Taylor instability in compressible fluids through level set method and parallel computation. Phys. Fluids A-Fluid 5(8), 1904–1913 (1993)

15. Marquardt, W.: Model-based experimental analysis of kinetic phenomena in multiphase reactive systems. Trans. Inst. Chem. Eng. 83(A6), 561–573 (2005)
16. Misek, T., Berger, R., Schröter, J.: Standard test systems for liquid extraction, 2nd edn. Europ. Fed. Chem. Eng. Pub. Ser., Inst. Chem. Eng., Warwickshire, vol. 46 (1985)
17. Osher, S., Sethian, J.A.: Fronts propagating with curvature dependent speed: Algorithms based on Hamilton-Jacobi formulations. J. Comput. Phys. 79(1), 12–49 (1988)
18. Sethian, J.A.: Level Set Methods and Fast Marching Methods—Evolving Interfaces in Computational Geometry, Fluid Mechanics, Computer Vision, and Materials Science, 2nd edn. Cambridge University Press, Cambridge (1999)
19. Sussman, M.: A parallelized, adaptive algorithm for multiphase flows in general geometries. Comput. Struct. 83(6-7), 435–444 (2005)
20. Sussman, M., Smereka, P., Osher, S.: A level set approach for computing solutions to incompressible two-phase flow. J. Comput. Phys. 114(1), 146–159 (1994)
21. Tryggvason, G., Bunner, B., Esmaeeli, A., Juric, D., Al-Rawahi, N., Tauber, W., Han, J., Nas, S., Jan, Y.J.: A front-tracking method for the computations of multiphase flow. J. Comput. Phys. 169(2), 708–759 (2001)
22. Wang, K., Chang, A., Kale, L.V., Dantzig, J.A.: Parallelization of a level set method for simulating dendritic growth. J. Parallel Distrib. Comput. 66(11), 1379–1386 (2006)

Optimization of Aircraft Wake Alleviation Schemes through an Evolution Strategy

Philippe Chatelain[*], Mattia Gazzola, Stefan Kern[**], and Petros Koumoutsakos

Chair of Computational Science
ETH Zurich, CH-8092 Zurich, Switzerland
philippe.chatelain@uclouvain.be, mattia.gazzola@inf.ethz.ch,
kerns@ge.com, petros@inf.ethz.ch
http://www.cse-lab.ethz.ch

Abstract. We investigate schemes to accelerate the decay of aircraft trailing vortices. These structures are susceptible to several instabilities that lead to their eventual destruction. We employ an Evolution Strategy to design a lift distribution and a lift perturbation scheme that minimize the wake hazard as proposed in [6]. The performance of a scheme is measured as the reduction of the mean rolling moment that would be induced on a following aircraft; it is computed by means of a Direct Numerical Simulation using a parallel vortex particle code. We find a configuration and a perturbation scheme characterized by an intermediate wavelength $\lambda \sim 4.64$, necessary to trigger medium wavelength instabilities between tail and flap vortices and subsequently amplify long wavelength modes.

Keywords: Large Scale Simulations in CS&E, Parallel and Distributed Computing, Numerical Algorithms for CS&E.

1 Introduction

Aircraft trailing vortices are powerful flow structures inherent to the very production of lift along the wing. These structures live long after an aircraft has flown by and constitute a potential hazard to any following aircraft. As a consequence, they require the enforcement of strict separation distances in particular at take-off and landing. This phenomenon is the limiting constraint on airport traffic, not without environmental consequences: longer traffic patterns lead to more noise and air pollution in particular.

The design of schemes to accelerate the decay of trailing vortices have been the topics of several theoretical [7,5], experimental [6] and numerical investigations [1,16].

The design and optimization of a wake alleviation scheme is a complex engineering problem. Wake decay simulations require solving the full three-dimensional Navier-Stokes well into the non-linear regime of vortex instabilities.

---

[*] Present address: Institute of Mechanics, Materials and Civil Engineering, Université catholique de Louvain, B-1348 Louvain-la-Neuve, Belgium.
[**] Present address: GE Global Research - Europe, D-85748 Garching bei München, Germany.

J.M.L.M. Palma et al. (Eds.): VECPAR 2010, LNCS 6449, pp. 210–221, 2011.

These results then allow the measurement of the hazard level. This is a highly non-linear and multi-modal optimization problem, which does not lend itself to gradient-based methods.

In this work, we couple a derandomized Evolution Strategy with Covariance-Matrix Adaptation (CMA-ES) to a fast parallel Navier-Stokes solver in order to design and optimize a wake alleviation scheme. We base our work on an approach proposed in [6]. We note that integrated optimization approaches have been used before, albeit at a smaller scale [14].

This paper is organized as follows. Section 2 presents the problem of wake alleviation and its statement as an optimization problem. Section 3 presents the optimization and numerical tools of our study. We present and discuss our results in Section 4 to finally conclude in Section 5.

2 Optimization of Wake Alleviation

2.1 Alleviation Scheme

We investigate the scheme proposed in [6]. This scheme relies on the periodic deflection of wing control surfaces (flaps) in order to perturb the near wake of the aircraft where there are several pairs of trailing vortices. The periodic control surface motions redistribute some lift between the inboard and outboard sections of the wing. This redistribution conserves the total lift –although not necessarily the pitching moment of the wing–, the circulation, and a zero rolling moment. The effect is a periodic oscillation of the positions of the tip and inboard flap vortices. This forced an accelerated reconnection of the tip vortices, at a rate which can be about twice as high as the regular Crow instability[6].

We will use the same perturbation amplitude as in [6] and redistribute $\Delta C_L/C_L = 6\%$ of the total wing lift.

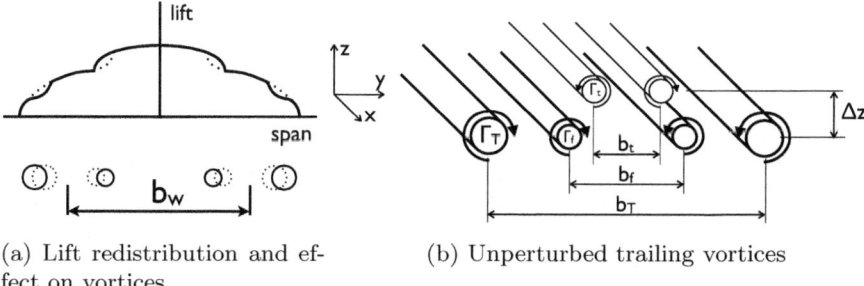

(a) Lift redistribution and effect on vortices

(b) Unperturbed trailing vortices

Fig. 1. Wing and wake configuration

2.2 Optimization of the Lift Distribution and Perturbation

In this section we describe the cost function, the parameterization of the problem and the search space.

Cost function. In the context of our optimization procedure, we will approximate the hazard posed to a following aircraft (with a wing span b_{follow}) by the maximum rolling moment averaged in the streamwise direction. We define the induced rolling moment as

$$C_{\text{roll}}(x, y, z, t) = \int_{y-1/2b_{\text{follow}}}^{y+1/2b_{\text{follow}}} (y' - y)u_z(x, y, z, t)dy' \tag{1}$$

and its streamwise average as

$$\langle C_{\text{roll}} \rangle_x (y, z, t) = \frac{1}{L_x} \int_0^{L_x} C_{\text{roll}}(x, y, z, t)dx . \tag{2}$$

We opt to define our cost function as the maximum average rolling moment

$$f_{\text{obj}} = \max_{y,z\in[-\infty,+\infty]} \langle C_{\text{roll}} \rangle_x (y, z, \tau_{\text{obj}}) \tag{3}$$

taken at a fixed dimensionless time $\tau_{\text{obj}} = 5$ and for a $b_{\text{follow}} = 1/2b_w$. This descent time value corresponds to a downstream distance of $\sim 4nm$, which matches the ICAO Standard Separation Distance between large jumbo jets. This mandatory separation grows to $6nm$ if the following aircraft is a light aircraft, justifying wake destruction within this time and space interval.

Parameterization and search space. We study the time evolution of the trailing vortices under the approximation of a streamwise periodic flow. The wake configuration is sketched in Fig. 1b. We account for the wing lift distribution through the geometry of the flap and tip vortices; they have, respectively, the circulations Γ_f and Γ_T and the spans b_f and b_T. The wing circulation and equivalent span can then be written as

$$\Gamma_w = \Gamma_T + \Gamma_f \tag{4}$$

$$b_w = \frac{\Gamma_T}{\Gamma_w}b_T + \frac{\Gamma_f}{\Gamma_w}b_f \tag{5}$$

The negative lift of the horizontal tail plane (HTP) is manifested by a third vortex pair with circulation Γ_t and span b_t. Because this pair is generated downstream of the wing, we assume it to be positioned Δz_t above the wing and flap vortices. These vortices are assumed to be Gaussian with core sizes σ_T, σ_f and σ_t.

The dimensionality of our search space will be sensibly smaller than the number of parameters outlined above as we choose to constrain several engineering characteristics of the problem. The total wing lift, proportional to $\Gamma_w b_w$, and the root wing circulation Γ_w have to be preserved. The lift redistribution is kept at $\Delta C_L/C_L = 6\%$ of the total wing lift. The HTP keeps the same negative lift and the vortex core sizes do not change. The resulting search space then counts 4 parameters

$\alpha = 2\pi/\lambda$ is the wavenumber of the perturbation;
$\beta = b_t/b_w$ is the span of the HTP vortices;
$\gamma = \Gamma_f/\Gamma_w$ is the circulation ratio of the flap vortices;
$\delta = (b_T - b_f)/(2b_w)$ is the separation between the tip and flap vortices.

The remaining parameters are kept constant and listed in the Table 1a. We bound the configuration parameters in order to avoid unfeasible or physically irrelevant configurations. The bounds are summarized in Table 1b.

Table 1. Parameters

(a) Constants

Parameter	Value
Γ_w/ν	2500
$\Gamma_t b_t$	$-0.0836\,\Gamma_w b_w$
σ_T	$0.05\,b_w$
σ_f	$0.05\,b_w$
σ_t	$0.025\,b_w$

(b) Ranges

Parameter	Minimun	Maximum
α	0.5	5.0
β	0.2	0.5
γ	0.1	0.5
δ	0.25	0.5

3 Methodology

3.1 Vortex Particle Method

We consider a three dimensional incompressible flow and the Navier-Stokes equations in its velocity ($\mathbf{u}$)-vorticity ($\boldsymbol{\omega} = \nabla \times \mathbf{u}$) form:

$$\frac{D\boldsymbol{\omega}}{Dt} = (\boldsymbol{\omega} \cdot \nabla)\,\mathbf{u} + \nu\nabla^2\boldsymbol{\omega} \tag{6}$$

$$\nabla \cdot \mathbf{u} = 0 \tag{7}$$

where $\frac{D}{Dt} = \frac{\partial}{\partial t} + \mathbf{u} \cdot \nabla$ denotes the Lagrangian derivative and ν is the kinematic viscosity. Vortex methods discretize the vorticity field with particles, characterized by a position $\mathbf{x}_p$, a volume V_p and a strength $\boldsymbol{\alpha}_p = \int_{V_p} \boldsymbol{\omega}d\mathbf{x}$. Particles are convected by the flow field and their strength is modified to account for vortex stretching and diffusion.

Using the definition of vorticity and the incompressibility constraint the velocity field is computed by solving the Poisson equation

$$\nabla^2\mathbf{u} = -\nabla \times \boldsymbol{\omega} . \tag{8}$$

This equation will be solved on a grid by means of a Fourier solver that allows for mixed periodic (x) and unbounded directions (y and z). We use remeshing[3,10,15] in order to remedy the loss of accuracy due to Lagrangian distortion. Remeshing consists in the periodic regularization onto a grid of the particle set via high order interpolation In the present work, remeshing is performed at the end of each time step and uses the third order accurate M_4' interpolation formula of [11]. The grid/mesh allows for additional advances: differential operators (such as those for stretching and diffusion) are evaluated on the mesh using fourth order finite differences and the Poisson equation (Eq. 8) is solved on the grid. The results of these calculations on the grid are then interpolated back onto the particles. We refer to [1,2] for details on the parallel implementation and the periodic-unbounded Poisson solver.

The vortex particle method is particularly well-suited for our flow configuration. It exploits the compact support of vorticity: particles are only needed where vorticity is non-zero. Likewise, the grid of the unbounded-periodic Poisson solver tracks the support of vorticity and grows or shrinks accordingly in the transverse directions. Finally, the method exhibits accuracy, robustness and relaxed stability properties for advection[4].

3.2 Evolution Strategy

We use a state-of-the-art Evolution Strategy with Covariance Matrix Adaptation (CMA-ES)[8]. CMA-ES belongs to the class of Evolutionary Algorithms comprising methods that are inspired by the principles of natural evolution to solve optimization and learning problems. It is operating with real valued parameters and adapts a Gaussian sampling distribution from the information acquired in the course of the optimization.

The gradient of the cost function in the search space is not readily available in the present investigation: an adjoint approach would be impractical and Finite Differences involve a stepsize selection procedure. The need for robustness and the likelihood of local minima in the cost function therefore close the case for CMA-ES. This requirement of robustness and the dimensionality of the problem (4) impose the population size of the Evolution Strategy, i.e. the number of function evaluations needed at every iteration of CMA-ES. We set it to 10 based on the investigations in [9].

Finally, we note that the search space is bounded through the constraints of Table 1b. These boundaries are enforced by biasing the sampling distributions, i.e. through a rejection algorithm.

3.3 Coupling and Computation

Every evaluation of the cost function is carried out by our parallel vortex particle code[1] and can involve running on hundreds of processors for several hours. Our approach consisted in dissociating this evaluation process from the optimization code. The latter is not computationally intensive and can easily run on a personal workstation; the former requires access to parallel architectures counting several hundred cores, typically in a supercomputing center, enabling the fast evaluation of several candidates of a population in parallel. This allows us to use an existing CMA-ES matlab implementation[1] and to only implement the evaluation of the cost function.

---

[1] Available at www.cse-lab.ethz.ch

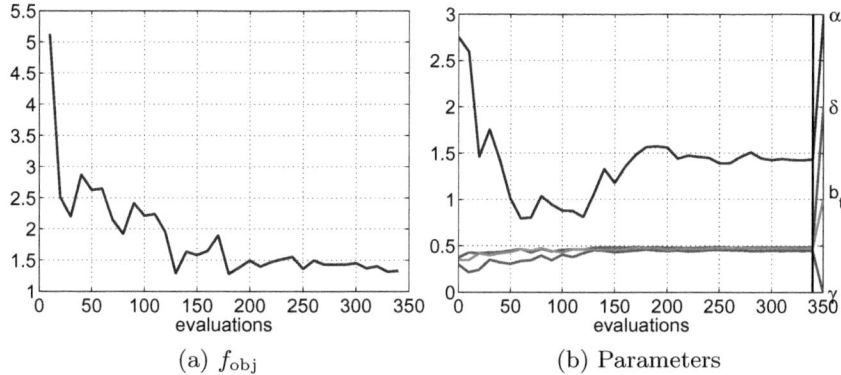

(a) f_{obj} (b) Parameters

Fig. 2. CMA-ES optimization history: evolution of the best cost function achieved in current generation and parameters

This matlab function determines the computational problem size from well-resolvedness considerations and then chooses a supercomputer partition size that keeps the wallclock duration of a simulation approximately constant (here between 4 and 12 hours). It generates the control files and scripts necessary to submit the parallel job on the super-computer queue, copies them and submits the job remotely. Several jobs –10 in this study–are submitted at the same time as they correspond to the function evaluations inside an iteration of CMA-ES. Their statuses are monitored and upon completion, their results are copied back in order to post-process them and return a scalar $f_{\mathrm{obj}}(\mathbf{x})$.

4   Results

4.1   Optimization

The history of the optimization is shown in Fig. 2. We see that CMA-ES went through 34 iterations, or 340 evaluations. The evaluations resulted in simulations running on Cray XT5 partitions ranging from 64 to 256 cores for run times between 6 and 12 wallclock hours. This represents a total of $270,000$ CPU hours.

The optimization was initialized in the center of our parameter intervals and converged (see Fig. 2) to a point which reduces the wake hazard by a factor of 4 with respect to the initial guess.

4.2   Optimum Parameter Set

The best candidate found over the course of the optimization is the case 174; it is described by the parameters $(\alpha, \beta, \gamma, \delta) = (1.3544, 0.48186, 0.47542, 0.48261)$. After encountering this point, the Evolution Strategy searches its neighborhood and eventually converges to this point (see Fig. 2b).

This candidate is characterized by a wavelength $\lambda = 4.64\,b_w$ sensibly smaller than the wavelength of the Crow instability for the equivalent wing vortex pair $\lambda_{\text{Crow}} \sim 8\,b_w$. Fig. 3 shows the evolution of the flow. The early phase is characterized by the fast growth of medium wavelength instabilities between the tail and flap vortices (see Fig. 3b and Fig. 3c). The reconnections generate dipoles similar to Ω-loops[12] which perturb the tip vortices and reconnect with them (Fig. 3d).

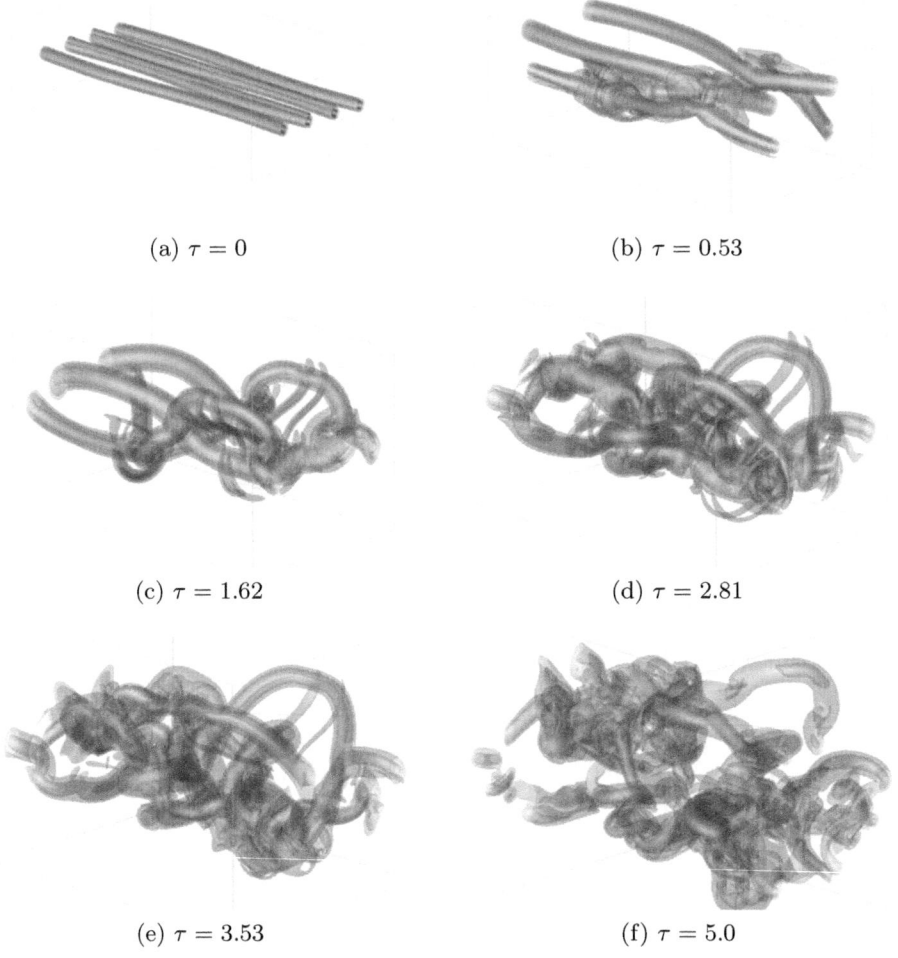

(a) $\tau = 0$ (b) $\tau = 0.53$

(c) $\tau = 1.62$ (d) $\tau = 2.81$

(e) $\tau = 3.53$ (f) $\tau = 5.0$

Fig. 3. Optimum parameter set (case 174): isosurfaces of vorticity norm $\|\boldsymbol{\omega}\| = 0.01$, 0.02, 0.04, $0.08\,\Gamma_w/(\pi\sigma_T^2)$

The outstanding features of the optimum become more apparent in a comparison with another less performant candidate. Fig. 4 shows the development of the best candidate of the first iteration (case 2), described by $(\alpha, \beta, \gamma, \delta) =$ $(2.1606, 0.37654, 0.13392, 0.42296)$ and thus a wavelength $\lambda = 2.91$. This case shows that even though its vortex dynamics produce a fast growing medium-wavelength instability between the flap and tail vortices, they do not perturb the tip vortices appreciably. The flow generates large dipoles (Fig. 4b and 4c) which get twirled around the tip vortices (Fig. 4d). This leads to fairly large secondary structures (Fig. 4e) but keeps the tip vortices relatively straight and unaffected.

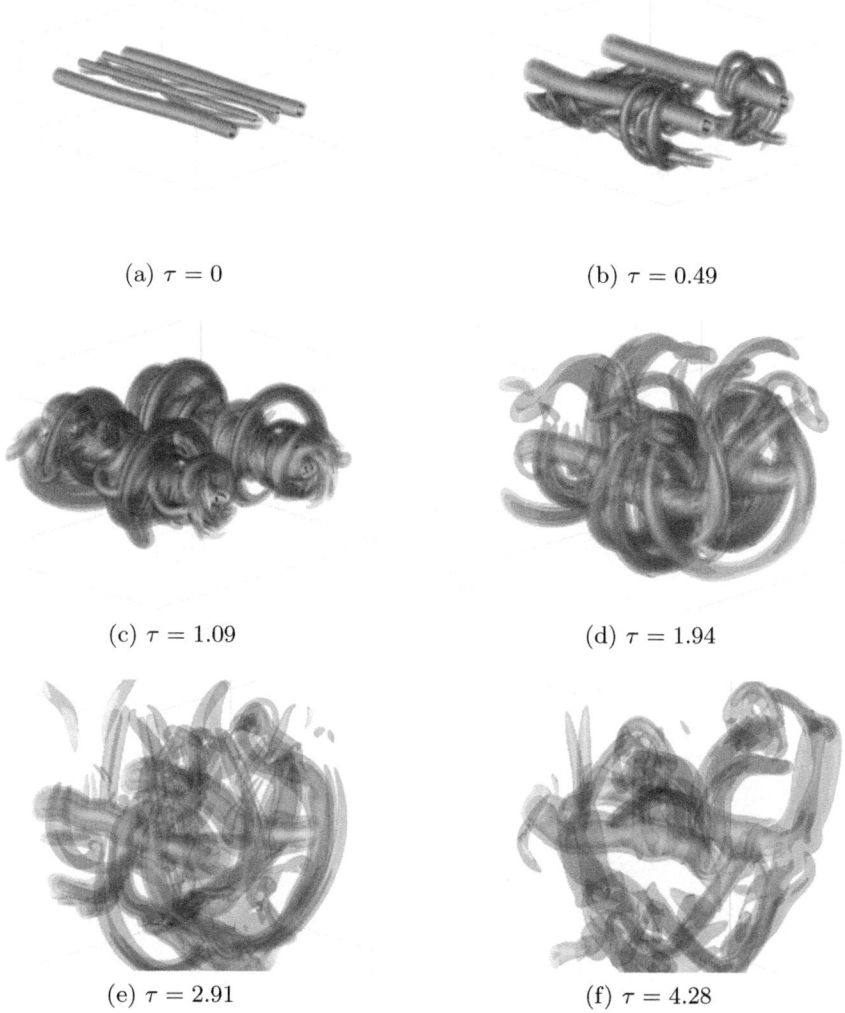

(a) $\tau = 0$ (b) $\tau = 0.49$

(c) $\tau = 1.09$ (d) $\tau = 1.94$

(e) $\tau = 2.91$ (f) $\tau = 4.28$

Fig. 4. Best candidate of the first generation (case 2): isosurfaces of vorticity norm $\|\boldsymbol{\omega}\| = 0.01,\ 0.02,\ 0.04,\ 0.08\ \Gamma_w/(\pi\sigma_T^2)$

In the optimum case, the transverse structures are smaller but more importantly, the tip vortices are displaced vertically over a half wavelength (Fig. 3c to 3f). In fact, this segmentation of the tip vortices is even apparent in the contours of the average rolling moment, shown in Fig. 5. The cores are distinguishable at two levels (Fig. 5d) thus causing the average moment to be roughly halved along the axes of these cores.

This effect appears to contribute substantially to the overall dissipation of the wake. And even more so if we consider the rolling moments of case 2 where there is no vertical spreading of the cores or halving of the average moment(Fig. 6).

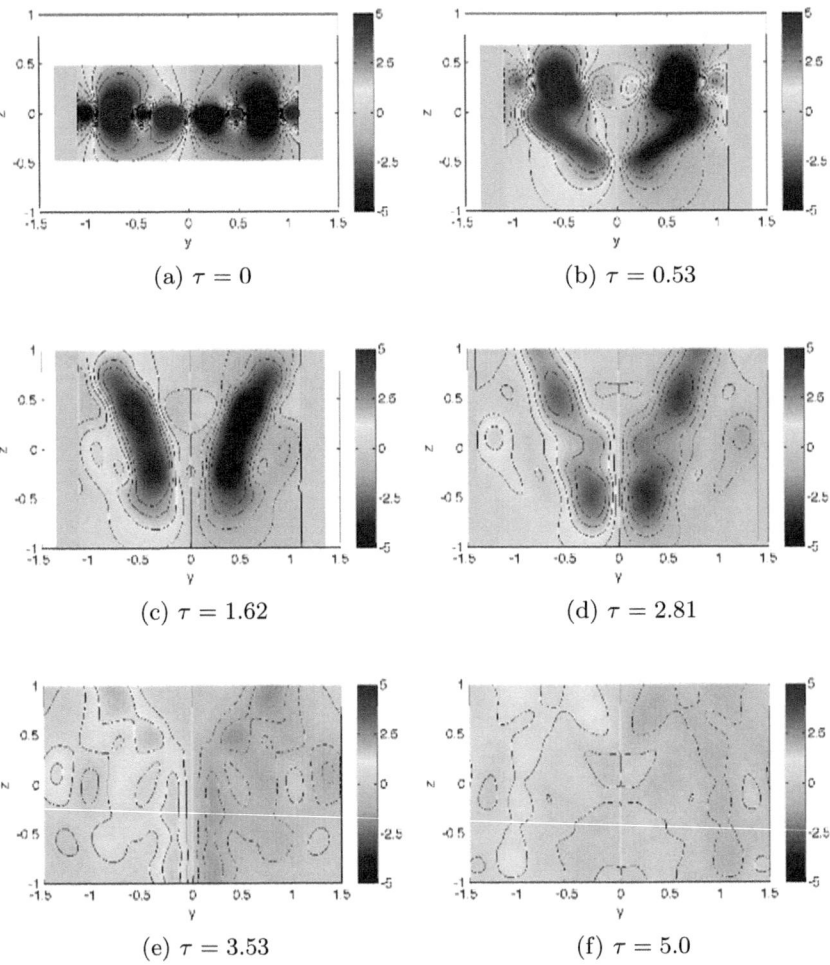

Fig. 5. Optimum parameter set (case 174): streamwise-averaged rolling moment $\langle C_{\text{roll}} \rangle_x$

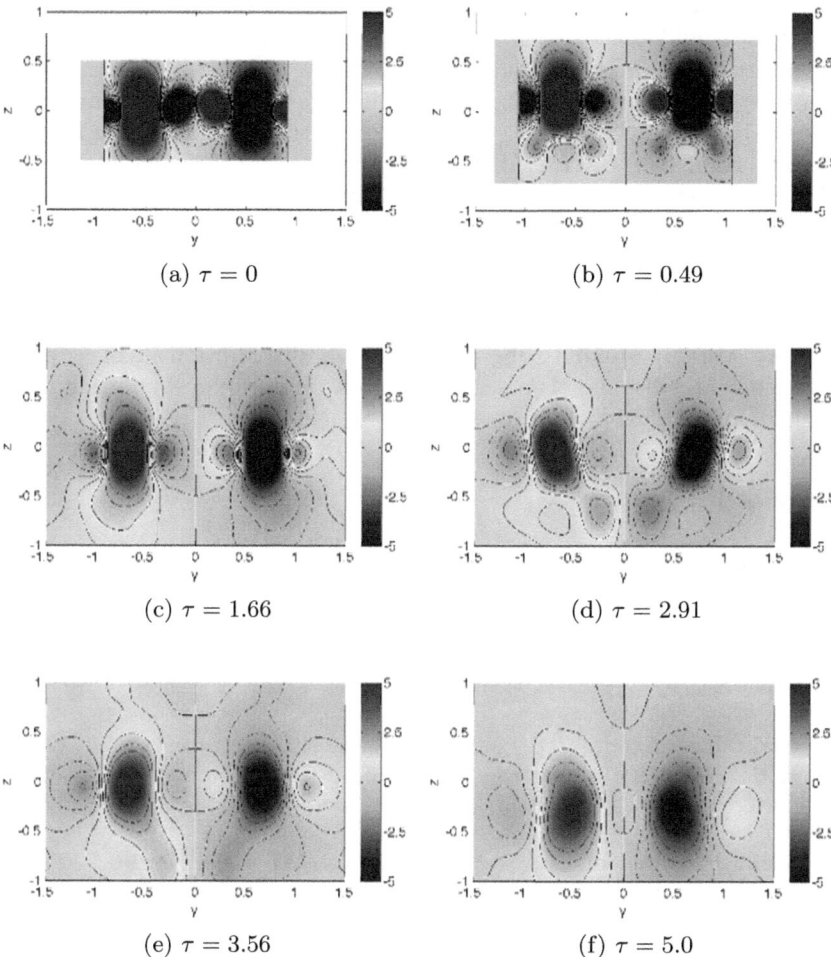

Fig. 6. Best candidate of the first generation (case 2): streamwise-averaged rolling moment $\langle C_{\mathrm{roll}} \rangle_x$

5   Conclusions

We have coupled a derandomized Evolution Strategy and an efficient parallel Navier-Stokes solver in order to optimize a wake alleviation scheme. The optimization relied on the parameterization of the wake configuration and the use of a wake hazard measurement for the cost function. Convergence of the ES required hundreds of function evaluations which were computed remotely on a supercomputing cluster.

An optimum was found at an intermediate wavelength $\lambda = 4.64 b_w$. For typical approach speeds, this corresponds to an actuation frequency which is in the sub-Hertz range $f \sim 0.2 - 0.4$Hz. The perturbation triggers fast-growing medium wavelength instabilities and vortex reconnections. The resulting flow disrupts the tip vortices and smears their induced rolling moment.

The present results were obtained from Direct Numerical Simulations at a moderate Reynolds number of 2500. While it may be argued that this mimics a uniform turbulent viscosity (see [13]), this constitutes a very crude RANS and future simulations will be carried out with an actual LES model.

Other future work areas include the addition of noise in the initial conditions in order to favor robust alleviation schemes over the course of the optimization. In addition, the cost function based on a fixed time measurement will be abandoned in favor of a time window average of the wake hazard. Finally, we plan to account for the spatial development of the flow and track the actuation effects more realistically. We will simulate the perturbed lift distribution itself, capture its effect in the near wake and then start a streamwise periodic simulation from the established vortex wake field.

Acknowledgments

The computational resources were provided by the Swiss Supercomputing Center (CSCS).

References

1. Chatelain, P., Curioni, A., Bergdorf, M., Rossinelli, D., Andreoni, W., Koumoutsakos, P.: Billion vortex particle direct numerical simulations of aircraft wakes. Computer Methods in Applied Mechanics and Engineering 197(13), 1296–1304 (2008)
2. Chatelain, P., Koumoutsakos, P.: A Fourier-based elliptic solver for vortical flows with periodic and unbounded directions. Journal of Computational Physics 229(7), 2425–2431 (2010)
3. Cottet, G.-H.: Artificial viscosity models for vortex and particle methods. J. Comput. Phys. 127(2), 299–308 (1996)
4. Cottet, G.-H., Koumoutsakos, P.: Vortex Methods, Theory and Practice. Cambridge University Press, Cambridge (2000)
5. Crouch, J.D.: Instability and transient growth for two trailing-vortex pairs. Journal of Fluid Mechanics 350, 311–330 (1997)
6. Crouch, J.D., Miller, G.D., Spalart, P.R.: Active-control system for breakup of airplane trailing vortices. AIAA Journal 39(12), 2374–2381 (2001)
7. Crow, S.C.: Stability theory for a pair of trailing vortices. AIAA Journal 8(12), 2172–2179 (1970)
8. Hansen, N., Muller, S.D., Koumoutsakos, P.: Reducing the time complexity of the derandomized evolution strategy with covariance matrix adaptation (CMA-ES). Evolutionary Computation 11(1), 1–18 (2003)
9. Kern, S.: Bioinspired optimization algorithms for the design of anguilliform swimmers. PhD thesis, ETH Zurich (2007)
10. Koumoutsakos, P.: Inviscid axisymmetrization of an elliptical vortex. J. Comput. Phys. 138(2), 821–857 (1997)

11. Monaghan, J.J.: Extrapolating b splines for interpolation. Journal of Computational Physics 60(2), 253–262 (1985)
12. Ortega, J.M., Bristol, R.L., Savas, Ö.: Experimental study of the instability of unequal-strength counter-rotating vortex pairs. Journal of Fluid Mechanics 474, 35–84 (2003)
13. Owen, P.R.: The decay of a turbulent trailing vortex. Aeronautical Quarterly 21, 69–78 (1970)
14. Stumpf, E.: Study of four-vortex aircraft wakes and layout of corresponding aircraft configurations. J. Aircraft 42(3), 722–730 (2005)
15. Winckelmans, G.: Vortex methods. In: Stein, E., De Borst, R., Hughes, T.J.R. (eds.) Encyclopedia of Computational Mechanics, vol. 3, John Wiley and Sons, Chichester (2004)
16. Winckelmans, G., Cocle, R., Dufresne, L., Capart, R.: Vortex methods and their application to trailing wake vortex simulations. C. R. Phys. 6(4-5), 467–486 (2005)

On-Line Multi-Threaded Processing of Web User-Clicks on Multi-Core Processors

Carolina Bonacic[1], Carlos Garcia[1], Mauricio Marin[2],
Manuel Prieto[1], and Francisco Tirado[1]

[1] Depto. Arquitectura de Computadores y Automatica
Universidad Complutense de Madrid
cbonacic@fis.ucm.es, {garsanca,mpmatias,ptirado}@dacya.ucm.es
[2] Yahoo! Research Latin America
Universidad de Santiago de Chile
mmarin@yahoo-inc.com

Abstract. Real time search — a setting in which Web search engines are able to include among their query results documents published on the Web in the very recent past — is a clear evidence that many of the off-line computations performed so far on conventional search engines need to be moved to the on-line arena. This is a demanding case for parallel computing since it is necessary to cope efficiently with thousands of concurrent read and write operations per unit time, all requiring latency times within a fraction of a second. To our knowledge, computations related to capturing user preferences through their clicks on the query result webpages and include this feature in the document ranking process are currently performed in an off-line manner. This is effected by pre-processing very large logs containing millions of queries submitted by actual users in a time scale of days, weeks or even months. The outcome is score data for the set of documents indexed by the search engine which were selected by users in the past. This paper studies the efficiency of this process in the on-line setting by evaluating a set of strategies for concurrent read/write operations executed on a multi-threaded multi-core architecture. The benefit of efficient on-line processing of user clicks is making it feasible to include user preference in document ranking also in a real-time fashion.

1 Introduction

Conventional Web Search Engines track user clicks performed on the URLs listed on the webpages containing search results to improve the quality of the document ranking process. User clicks are monitored along time to detect document popularity trends so that ranking can be updated accordingly to refine the results of subsequent queries; previously high-ranked pages that are not attracting user clicks are eventually demoted, while previously low-ranked pages that capture the interest of visitors are rewarded with a rank boost.

However, most of the optimizations to the ranking process that are based on click rates are still performed in an off-line manner. This means that the effects

J.M.L.M. Palma et al. (Eds.): VECPAR 2010, LNCS 6449, pp. 222–235, 2011.

of previous clicks on the present ranking process only become visible at regular intervals of the order of hours or even days.

In essence, the problem consists on efficiently ranking the URLs clicked by previous users that submitted similar queries to the search engine. For this purpose, clicks themselves are indexed and now concurrency conflicts appear among the continuous stream of click updates over the index and the required operations needed to process time-consuming tasks such as determination of similar queries and related clicks on documents.

By "similar" we mean queries correlated in some probabilistic way that considers clicked URLs and respective query terms. The calculation of the probabilities query-to-query, query-to-URL, and URL-to-URL can be very demanding in execution time and memory requirements with the additional challenge that must be performed on-the-fly for each user query.

In this paper we present algorithms for dealing with the problem of on-line indexing and querying a continuous stream of queries generated by users of a large Web search engine. We concentrate on what happens on a multicore-based click-ranking node dealing with this work-load and in particular we focus on how to organize the associated concurrent read/write operations submitted from the different threads running on such a click-ranking node. Our main contribution is a comparative study of different strategies that trade-off parallelism granularity and data locality.

The remaining of the paper is structured as follows. In Section 2 we provide some background and discuss related work. In Section 3 we describe different strategies explored by this research. In Section 4 we analyze our experimental results and in Section 5 we conclude summarizing our findings.

2   Background and Problem Setting

Web Search Engines use the inverted file data structure to index the text collection and speed up query processing. An inverted file is composed of a vocabulary table and a set of posting lists. The vocabulary table contains the set of relevant terms found in the collection. Each of these terms is associated with a posting list which contains the document identifiers where the term appears in the collection along with additional data used for ranking purposes. To solve a query, it is necessary to get the set of documents *ids* associated with the query terms and then perform a ranking of these documents so as to select the top K documents as the query answer. On conventional search engines, the posting list are update off-line and consequently query operations are exclusively read-only requests upon the inverted file.

A number of papers have been published reporting experiments and proposals for efficient parallel query processing upon inverted files which are distributed on a set of P **nodes** [1,2,3,16,15,19,17]. The two dominant approaches to distributing an inverted file are (a) the **document partitioning** strategy (also called local indexing), in which the documents are evenly distributed onto the set of available nodes and an inverted index is constructed in each processor using the respective subset of documents, and (b) the **term partitioning** strategy (called global indexing), in which a single inverted file is built from the whole text collection to then evenly distribute the terms and their respective posting lists onto the processors. Document partitioning is usually employed since it has better scalability. These strategies have been devised for distributed memory systems in which nodes have a share nothing architecture. Note, however, that their main principles can also be re-used on a multi-core setting in which nodes consists of several cores interconnected through a shared memory hierarchy.

Our context differs from a conventional set-up in several key aspects.

- First, our clicks engine has an inverted file that indexes URLs clicked by users in previous actual queries submitted to the Web search engine. The vocabulary table of that inverted file is formed by the query terms – or queries themselves treated as single units – and the associated postings lists contain references to the URLs clicked by users together with other data used for ranking. Furthermore, as shown in Figure 1 we need double inverted indexing so that from terms we can reach clicked URLs and vice-versa. A key operation is to start from the query term to reach a set of clicked URL, then these URLs are used to get a new set of terms from the second inverted file and from these terms get more URLs. The resulting sets of terms and URLs are then operated each other to generate a list of ranked URLs.

- Second, by "query" we mean a number of operations performed on the inverted file that tracks the relevance of the click made by users and this means that we need to process both read and write requests upon the inverted file. As mentioned above, in conventional search engines queries are usually read-only requests since the update of the posting list is performed off-line. However, for real-time indexing of click-through data, it is mandatory to process on-line write requests upon the inverted file to keep posting list up-to-date all the time. As new queries arrive to each click-ranking node it is necessary to detect if the clicked URLs are already being indexed. If so, the clicks count of the respective URLs must increased and the item promoted to the front of the posting lists associated with the query terms. We use a transposition heuristic on the posting list to promote highly clicked URLs to the front of the posting list. This is used as a low cost indication of how recently the URLs have been clicked which is useful for ranking purposes.

- Third, our engine prototype is explicitly designed to exploit the available thread level parallelism available in current multicore processors. A straightforward approach to transparently take advantage of such architectures is to rely on virtualization technology and use as many single-CPU virtual nodes per processor as cores are available. Unfortunately, this involve additional overheads that become an overkill in the extremely demanding arena of search engines. Note, however, that explicit parallelism complicated click-ranking node design and implementation, especially in this setting with concurrent read and white request upon the inverted file.

Our focus in this paper is to explore different alternatives to implement such a parallel click-ranking node. Intuitively, the simplest approach consists in

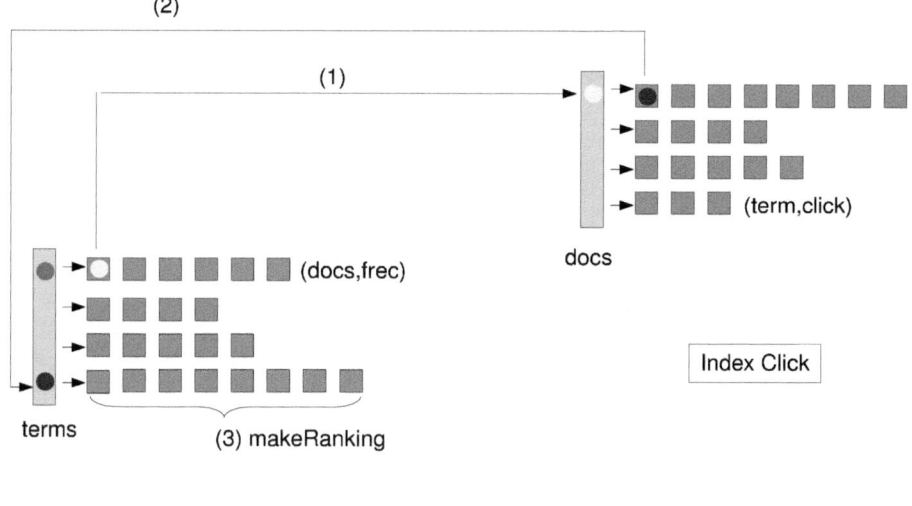

Fig. 1. Double inverted file organization. The first index (left) is a standard search engine index in which *doc* is a clicked URL and *frec* is the total number of times users have clicked the URL for queries containing the same term. The second index (right) enables the mapping from clicked URLs to query terms that caused the clicks on the URLs, where *click* indicates the average position in the result webpage of the respective URL. The sequence given by the labels (1), (2) and (3) indicates that from a given term (1) it is possible to reach a new term (2), which in turn leads to a new set of documents (3) to be included in the ranking process. For each posting list item in the first index, this sequence is repeated for each posting list item of the second index. Therefore, queries expand the set of active terms during a period of time and new query arrivals cause the modification of the posting lists of both indexes which potentially causes read/write conflicts.

exploiting thread level parallelism at the query level, i.e. assigning an independent thread per incoming query. This approach could perform well from a parallel implementation point of view as long as there are always enough simultaneous queries to keep all cores busy. In fact, this is the interesting case since performance only becomes critical when the engine operates under heavy query traffic. When traffic is sufficiently low, it does not really matter that a given strategy is less efficient than another provided that the response time of individual operations is below an upper limit. However, even assuming peak traffic, if we need to update ranking information online based on clicks made by users, read writer synchronization may jeopardize parallel performance and it is unclear if exploiting query-level parallelism will be enough.

As an alternative we need to evaluate if the processing of a single query itself can be organized to exploit intra-query parallelism. Intuitively, this is also

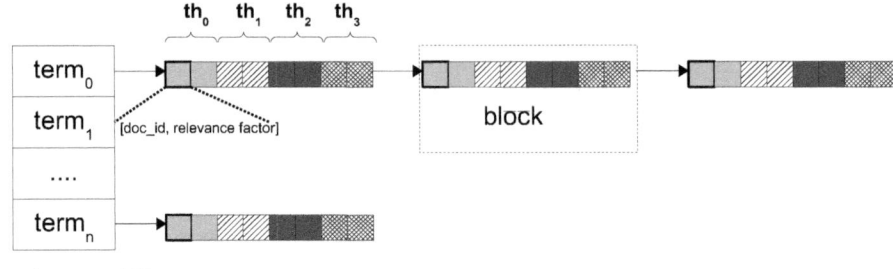

Inverted file

Fig. 2. Inverted file organization where each posting list is stored in a number of blocks and where each block is logically divided in chunks that are assigned to threads. Each chunk is composed of a number of posting list items and each item is a pair (doc_id, relevance factor). In this example, the first block of *term_0* is processed in parallel by threads *th0*, *th1*, *th2* and *th3*.

possible since the posting list of the inverted file are large enough and data parallelism can be exploited when traversing such lists. This is the idea illustrated graphically in Figure 2. Items on the posting lists (URL references) are usually kept ordered by a relevance factor that accounts for the frequency of clicks made by users to them, but for efficiency reasons, instead of keeping a fully ordered list, we group items into different blocks and kept the list sorted by the relevance factor just across blocks, i.e. the set of items kept in block i are all of higher relevance than the values associated with the set of items kept in block $i + 1$. With this organization, the idea is that posting list processing can be perform on a block basis and within each block, we can exploit parallelism at item level distributing each block into chunks with are assigned to the available threads.

In summary, in this paper we have tried to answer to the following complementary questions:

1. Is query-level parallelism enough to achieve satisfactory parallel performance under heavy query traffic?
2. If query-level parallelism is not enough, is it efficient to exploit parallelism at the item level?
3. How to implement the concurrency control mechanism required by the read write synchronization inherent to the online update?

3   Strategies for Read-Write Synchronization

As mentioned above, we have studied a number of strategies for implementing real-time parallel indexing of click-though data which (1) exploit parallelism at different levels (either query level or item level or both) and (2) implement different concurrency control policies to satisfy read write synchronization issues. All of them use either locks or barriers as synchronization mechanism and their main characteristics are summarized in Table 1.

The most restrictive strategies guarantee that the scheduling of regular searches on the inverted file, which we denoted as read transactions, and ranking update operations, which we denote as write transactions, are serializable, i.e. they maintain the illusion of serial execution. The other strategies does not enforced serialization, which potentially will allow for better performance. Nevertheless, the proposed strategies make use of data locality in different ways and this also have a strong influence on performance.

The "**Bulk Processing (BP)**" strategy does not exploit parallelism at the query level, i.e. either regular searches on the inverted file or updates of the posting lists, are processes serially without any kind of concurrency. Instead, it exploits parallelism a much finer level. As described above, the posting lists are statically divided into blocks. When serving read transactions, these blocks are logically partitioned into smaller chunks, which are processed in parallel by the available threads within the click-ranking node. A master thread performs a short sequential phase to merge the results of those threads and finally it gets the same outcome as a conventional single CPU algorithm. The block size is a key parameter in this approach. Intuitively, the larger the block size, the coarser the parallelism, but smaller blocks tend to improve data locality and some trade-off should be found.

The "**Concurrent-Reads (CR)**" does exploit query level parallelism but just for regular searches. However, before a write operation could take place, the CR strategy waits for all of the current reads being solved to end. In this way, read transactions exploit the available (intuitive) parallelism between independent queries but as in BP, write serialization is also guaranteed by isolating the execution of write transactions. In fact, write transactions are handle the same way in both approaches.

The "**Term-Level-Parallelism (TLP)**" strategy allows concurrency of both read and write operations as long as they involve different terms of the vocabulary and they are not correlated. Concurrent transactions are assigned on demand to the available threads of the click-ranking node and ideally, a different lock protects the posting list of each vocabulary term to control the concurrency. To enforce serializability, a thread does not proceed with a transaction till it acquires all its associated locks at once. In practice, since locks are an expensive resource, this approach uses hashing to map several terms onto the same lock. We have explored two alternative variations. The first one always allow concurrent reads (*TLP1*), whereas the second, which is much easier to implement, does not overlap read transactions that have some terms in common (*TLP2*).

Finally, we have also explored a couple of strategies which relax serializability requirements. The first one, which is denoted as "**Relaxed-Term-Level-Parallelism (RTLP)**", is similar to the TLP approach but without forcing threads to acquire all the locks of a given transaction at once and removing the atomic commitment of transactions. Our second approach, which we have denoted as "**Relaxed-Block-Level-Parallelism (RBLP)**", uses a similar strategy but controls concurrency at the block level (a hashing function maps several blocks of the posting lists to the same lock) to reduce potential imbalances caused by the Zipf Law. Obviously, the degree of concurrency of both RTLP and RBLP is much higher than the other strategies but they may introduce a non-serializable scheduling. Nevertheless, in click-ranking it usually does not matter much if some inconsistencies appear.

Table 1. Different strategies for implementing a multicore-based click-ranking node supporting online updates of the inverted index. The first column indicates if the strategy introduces non-serializable scheduling of read and write transactions, the second column indicates the source of parallelism and the third column indicates the inner mechanism used to control concurrency and enable online updates.

	Serializable	Parallelism	Implementation
BP	Yes	Fine Grain	Barrier based
CR	Yes	Concurrent Searches	Barrier based
TLP1	Yes	Concurrent Searches and Updates	Term Lock based
TLP2	Yes	Concurrent Searches and Updates	Term Lock based
RTLP	No	Concurrent Searches and Updates	Term Lock based
RBLP	No	Concurrent Searches and Updates	Block Lock based

4   Experiments

The computing platform used in our simulations is a dedicated cluster node equipped with two *Intel's Quad-Xeon* processors, whose main characteristics are summarized in Table 2. The search node prototype have been developed in C++, using Linux *POSIX Threads* as explicit threading API.

Our evaluation has focused on studying the performance under heavy query traffic since the ultimate goal is to deploy index nodes able to efficiently cope with drastic peaks in traffic. In particular, we have evaluated two different scenarios that emulated extreme cases:

1. **Workload 1** – Heavy traffic but limited concurrency –. In the most adverse scenario we have evaluated, there is a high probability that subsequent queries (in time) actually become quite similar from a semantic point of view. For instance, this emulates traffic when suddenly, many people becomes interested on the same topic and the engine receives a large stream of similar queries. In this case ranking updates and regular accesses to the same posting lists (i.e. read and write transactions) occur almost simultaneously very often, which limits the available parallelism of some of the strategies.
2. **Workload 2** – Heavy traffic, high concurrency –. This workload emulates a more benign scenario in which the chances of simultaneous read and write transactions to the same posting lists is much lower, which can be considered as an average case. Since subsequent queries are less correlated, there is a much coarser parallelism and it is expected that relaxed schemes can be benefited from it.

Table 2. Main features of the target computing platform

	Intel Quad-Xeon (2.66 GHz)	
Processor	L1 Cache (per core)	4x32KB+4x32KB (inst.+data) 8-way assoc. 64 byte/line
	L2 Unified Cache	2x4MB 16-way assoc. 64-byte/line
Memory	16 GBytes (4x4GB) 667 MHz FB-DIMM memory 1333 MHz system bus	
Operating System	GNU Debian System Linux kernel 2.6.22-SMP for 64 bits	
Intel C/C++ v.10.1 Compiler Switches	-fast Parallelization with POSIX Threads: -lpthreads	

Simulations have been performed with a text collection from an United Kingdom Web database sample. Query traces have been build using a real query-log containing user searches, which in turn, have been used to build synthetic click-through traces using different user behavior models. Obviously, users do not click on links at random, but make an informed choice that depends on many factors, but for the focus of this paper, we believe that this random model is enough to obtain useful insights on the comparative performance of the strategies tested in this work. Nevertheless, we are aware of our simplifications and it is clear that precise modeling of click-through behavior or real click-logs will be necessary to refine our experiments.

In our experiments we have assumed than users click on average just on one of the most promising top-k links presented in the search results page, but we have also evaluated more demanding workloads in which for each search query users clicks on more than one link and hence more ranking updates are necessary for a single search request.

There are many parameters that have a noticeable influence on the performance of our click-ranking node but for the sake of clarity we only summarize here our major findings and report results using optimal parameters.

One of the most important parameters is the block size used within the posting lists. Intuitively, the larger the block size, the coarser the parallelism for strategies like BP, but smaller blocks tend to improve data locality so a trade-off is in place. Empirically, we have found that blocks of 128 elements provide the best performance rates across all strategies and this is the block size used in all our experiments.

Another important parameter is the number of locks used in strategies such as *TLP* and *RTLP*. As locks resources are limited and their administration is expensive in terms on running time, the strategies TLP1, TLP2 and RTLP use hashing to reduce the number of actual locks created by the program. To assess the impact of hash collisions in this application setting, we tried with different number of locks and fortunately, we found that even with a moderate number of identifiers, namely an array of locks (hash table) with no less than 4099 entries, the performance degradation over having unlimited locks becomes negligible (less than 2%).

Performance Results

Figure 3 shows the scalability of the different strategies using the workload 1. Surprisingly, BP is the strategy that scale the best in our platform and reaches an impressive speedup of 7.75x for eight threads, in spite of only exploiting what we have denoted above as item level parallelism. This is an important finding since intuitively, the other approaches are able to exploit coarser parallelism. However, click-ranking is a demanding application since read-write synchronization is very frequent and in practice, even if we are assuming high query traffic, query level parallelism is not enough to keep many concurrent threads busy. Figure 4 further demonstrate this issue. If we assume that users click on average on to four link for each search, more ranking updates will be needed and less query level parallelism will be actually available.

Figures 5 and 6 focused on 8-thread experiments and show the actual query throughput *and* index-update throughput achieved with workloads 1 and 2 respectively. Under a less demanding workload (Figure 6), query level parallelism is higher, but even in that case, BP is able to outperform the other strategies since read-write synchronization overheads are still large enough.

Finally, Figure 7 shows running times of individual query and index-update operations. However, these results has to be read with precaution since they measure the time elapsed between the instant in which the query/index-update operation starts execution and the instant it is completely processed. Precision of measures can be compromised given the tiny values of running times. However, they show a general trend from which we can explain the differences in overall query throughput observed in the above experiments. First, the curves in Figure 7 [top] show that average running time per operation tends to be very similar each other across strategies. They are smaller for 5 clicks indicating that index-update operations are faster than query operations and the trace executed in that case is populated by more of these faster operations. Except for the BP strategy, all other strategies basically assign one thread to process sequentially the query/index-update operation. One would expect BP to outperform all others in this case because it uses all threads to process each query/index-update operation. However, BP has the burden of barrier synchronizing the threads in order to start with the next operation, and the results show that this cost is significant. On the other hand, the points in Figure 7 [bottom] clearly show that all other strategies tend to consume a significant amount of time for some operations which is an indication that they suffer, from time to time, from long delays due to lock contention among the active threads. The BP strategy does not suffer from this problem because it simply processes one operation at a time and,

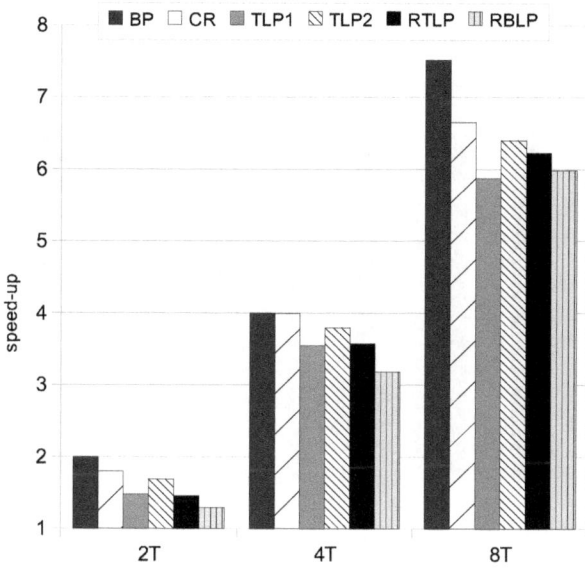

Fig. 3. Scalability of the different approaches for workload 1 and assuming users just click on one link per search on average. Results for 2, 4 and 8 threads (T).

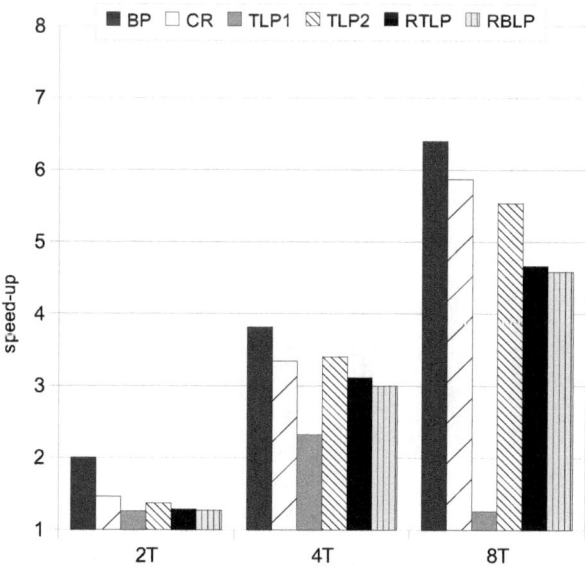

Fig. 4. Scalability of the different approaches for workload 1 and assuming a more demanding experiments in which users click on four links per search on average. Results for 2, 4 and 8 threads (T).

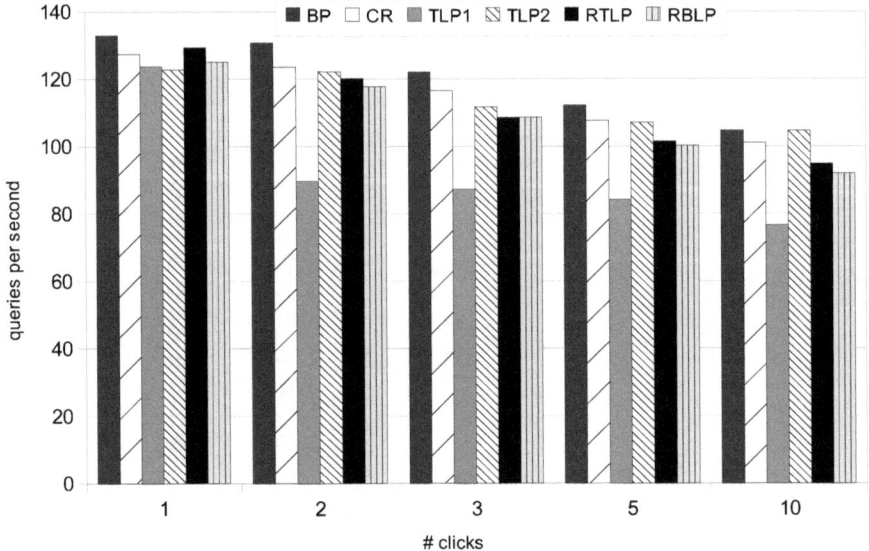

Fig. 5. Query/index-update throughput achieved with workload 1 under different click-through behavior from users. Results for 8 threads.

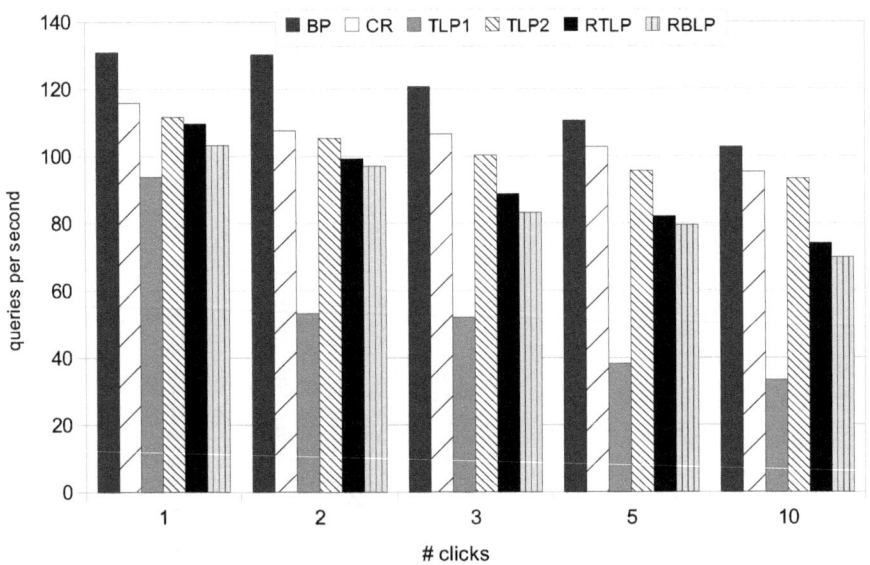

Fig. 6. Query/index-update achieved with workload 2 under different click-through behavior from users. Results for 8 threads.

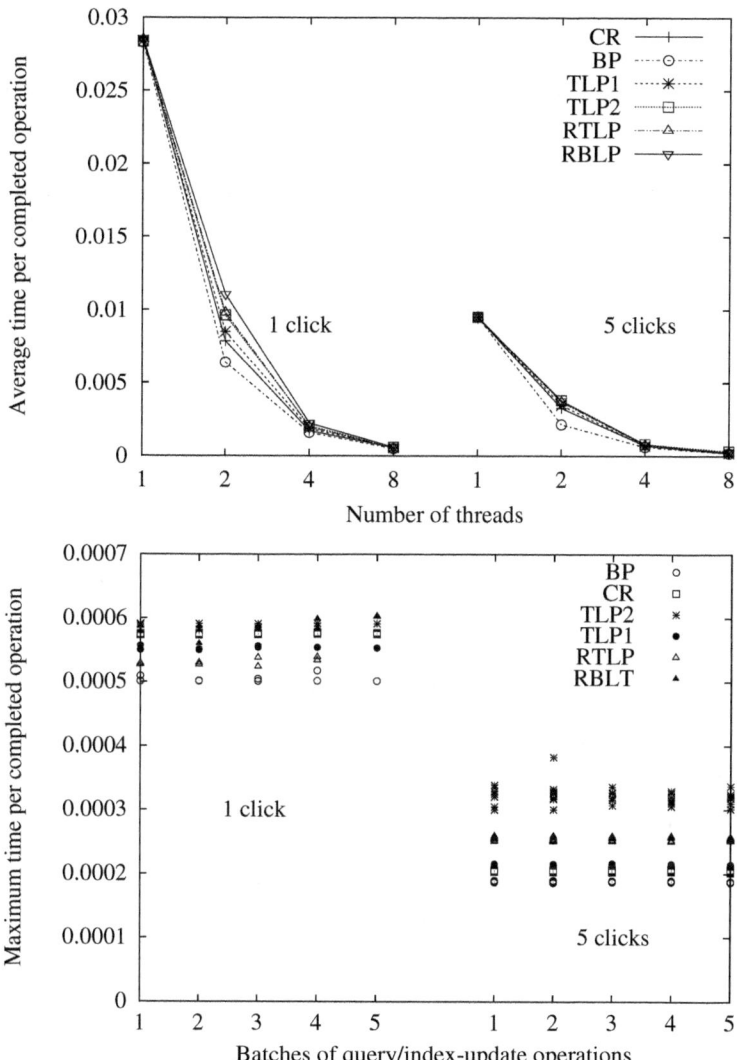

Fig. 7. Individual query times. The figure at the top shows results for average running time per query/update-index operation for 1, 2, 4 and 8 threads. The figure at the bottom shows the maximum running time observed every 1,000 query/update-index operations for 8 threads.

while it uses locks for implementing oblivious barrier synchronization, the threads do not have to compete each other to acquire locks at the end of each operation. This explains the better performance of BP with respect to the relevant performance metric for our case, namely query/index-update throughput.

5   Conclusions

We have presented an experimental study that compares different concurrency control strategies devised to process user clicks in an on-line manner. The results show that the BP approach based on exploiting the full parallelism available from the cores for processing one single query or index update at a time, is the best alternative. This strategy outperforms the more intuitive approach found out as current practice in multi-threaded search engine nodes. Namely, the strategy in which each active query or index update is handled by an independent concurrent thread that is mapped to one of the available cores. We tested different variants of the intuitive approach. Each one representing differing degrees of compromise between fully concurrent operation and strict serialization of read/write transactions. The results show that even for the very relaxed strategies, the BP approach (which is serializable) performs better.

References

1. Arusu, A., Cho, J., Garcia-Molina, H., Paepcke, A., Raghavan, S.: Searching the web. ACM Trans. 1(1), 2–43 (2001)
2. Badue, C., Baeza-Yates, R., Ribeiro, B., Ziviani, N.: Distributed query processing using partitioned inverted files. In: SPIRE, pp. 10–20. IEEE-CS, Los Alamitos (2001)
3. Barroso, A., Dean, J., Olzle, U.H.: Web search for a planet: The google cluster architecture. IEEE Micro. 23(2), 22–28 (2002)
4. Baeza-Yates, R., Gionis, A., Junqueira, F., Murdock, V., Plachouras, V., Silvestri, F.: Design trade-offs for search engine caching. ACM TWEB 2(4), 1–28 (2008)
5. Baeza, R., Ribeiro, B.: Modern Information Retrieval. Addison-Wesley, Reading (1999)
6. Ding, S., He, J., Yan, H., Suel, T.: Using Graphics Processors for High Performance IR Query Processing. In: WWW, pp. 421–430 (2009)
7. Dragicevic, K., Bauer, D.: A survey of concurrent priority queue algorithms. In: IPDPS, pp. 1–6 (2008)
8. Fagni, T., Perego, R., Silvestri, F., Orlando, S.: Boosting the performance of Web search engines: Caching and prefetching query results by exploiting historical usage data. ACM TOIS 24(1), 51–78 (2006)
9. Gan, Q., Suel, T.: Improved Techniques for Result Caching in Web Search Engines. In: WWW, pp. 431–440 (2009)
10. Jeong, B.S., Omiecinski, E.: Inverted file partitioning schemes in multiple disk systems. IEEE Trans. Parallel and Distributed Systems 16(2), 142–153 (1995)
11. Lempel, R., Moran, S.: Predictive caching and prefetching of query results in search engines. In: WWW, pp. 19–28 (2003)
12. Long, X., Suel, T.: Three-level caching for efficient query processing in large Web search engines. In: 14th WWW, pp. 257–266 (2005)
13. MacFarlane, A., McCann, J., Robertson, S.: Parallel search using partitioned inverted files. In: SPIRE 2002, pp. 209–220. IEEE CS, Los Alamitos (2002)
14. Markatos, E.: On caching search engine query results. Computer Communications 24(7), 137–143 (2000)
15. Marin, M., Gil-Costa, V.: High-performance distributed inverted files. In: Proc. CIKM, pp. 935–938 (2007)

16. Marin, M., Bonacic, C., Gil-Costa, V., Gomez-Pantoja, C.: A Search Engine Accepting On-Line Updates. In: Kermarrec, A.-M., Bougé, L., Priol, T. (eds.) Euro-Par 2007. LNCS, vol. 4641, pp. 348–357. Springer, Heidelberg (2007)
17. Zobel, J., Moffat, A.: Inverted Files for Text Search Engines. ACM Computing Surveys 38(2) (2006)
18. Marin, M., Paredes, R., Bonacic, C.: High-Performance Priority Queues for Parallel Crawlers. In: 10th ACM International Workshop on Web Information and Data Management (WIDM 2008), California, US, October 30 (2008)
19. Moffat, W., Webber, J., Zobel, J., Baeza-Yates, R.: A pipelined architecture for distributed text query evaluation. Information Retrieval (August 2007)
20. Persin, M., Zobel, J., Sacks-Davis, R.: Filtered document retrieval with frequency-sorted indexes. Journal of the American Society for Information Science 47(10), 749–764 (1996)
21. Ribeiro-Neto, B., Barbosa, R.: Query performance for tightly coupled distributed digital libraries. In: ACM Conference on Digital Libraries, pp. 182–190 (1998)
22. Xi, W., Sornil, O., Luo, M., Fox, E.A.: Hybrid partition inverted files: Experimental validation. In: Agosti, M., Thanos, C. (eds.) ECDL 2002. LNCS, vol. 2458, pp. 422–431. Springer, Heidelberg (2002)

Performance Evaluation of Improved Web Search Algorithms*

Esteban Feuerstein[1], Veronica Gil-Costa[2],
Michel Mizrahi[1], and Mauricio Marin[3]

[1] Departamento de Computación, Universidad de Buenos Aires, Argentina
{efeuerst,mmizrahi}@dc.uba.ar
[2] Universidad Nacional de San Luis, Argentina and CONICET, Argentina
gvcosta@unsl.edu.ar
[3] Yahoo! Research Latin America and Universidad de Santiago, Chile
mmarin@yahoo-inc.com

Abstract. In this paper we propose an evaluation method for parallel algorithms that can be used independently of the used parallel programming library and architecture. We propose to predict the execution costs using a simple but efficient framework that consists in modeling the strategies via a BSP architecture, and estimating the real costs using as input real query traces over real or stochastically generated data. In particular we apply this method on a 2D inverted file index used to resolve web search queries. We present results for OR queries, for which we compare different ranking and caching strategies, and show how our framework works. In addition, we present and evaluate intelligent ranking and caching algorithms for AND queries.

1 Introduction

Inverted files [1] are the most widely used index data structures to cope efficiently with high traffic of queries upon huge text collections. An inverted file consists of a vocabulary table which contains the set of relevant terms found in the text collection, and a set of posting lists that contain the document identifiers where each term appears in the collection along with additional data used for ranking purposes. To answer a query, in a Web search engine context, a broker machine sends the query to a set of search processors in order to get the set of documents associated with the query terms. These processors perform a ranking of these documents in order to select the top-K documents.

Many different methods for distributing the inverted file onto P processors or computers and their respective query processing strategies have been proposed in the literature [1]. The different ways of doing this splitting are mainly variations of two basic dual approaches: document partition (a.k.a local index) and term partition (a.k.a global index). Variants of these two basic schemes have been proposed in [5] which focus on optimizing for particular situations.

In a previous paper [2] we introduced a novel distributed architecture for indexing, named the 2D index, that consists in arranging a set of processors in

---

* Partially supported by Universidad de Buenos Aires' UBACYT project X436.

J.M.L.M. Palma et al. (Eds.): VECPAR 2010, LNCS 6449, pp. 236–250, 2011.

a two-dimensional array, applying term-partitioning at row level and document-partitioning at column level. We showed in that paper that, for AND queries, choosing the adequate number of rows (R) and columns (C) given the available number of processors it is possible to obtain significant improvements in the performance against the basic architectures of term and document partitioning with the same number of processors.

In this paper we propose to use the bulk-synchronous model of parallel computing (BSP) [8] as the cornerstone of a simple but efficient framework in which we blend empiric analysis and theoretical tools to predict the cost of real system executions, which we apply to variations of the 2D index. In that framework, we combine the usage of real input logs, average-cost analysis of certain operations and stochastic generated values to compute realistic estimations of the cost of the real system.

Our framework has two important features. The first is precisely the ability to scale the system architecture obtaining reliable results without the need to acquire additional resources, permitting for example to predict what happens when there are thousands of processors with billions of documents. This is an important issue as deploying and running algorithms on large clusters to determine their efficiency and performance is extremely expensive, and prohibitive in many cases, as it is sometimes impossible to access the required resources. Our methodology enables us to validate the feasibility and convenience of an architecture or algorithm more easily than constructing a real system or building complex simulations involving the use of queuing theory, real-time managing techniques and, not less involved, the real implementation of all the operations.

The second important characteristic is the possibility of applying this framework to any kind of system, meaning synchronous or asynchronous, with distributed or centralized ranking. Besides, in this way we can ensure a fair comparison between algorithms independently of the hardware used to run the codes and particular implementation details. The result is that our framework based on the BSP model may be seen as a unifying setting to fairly compare different algorithms and implementations.

Apart from BSP, several general-purpose parallel models have been presented in the literature, like LogP or QSM [8]. As it has been proved, the different models can be reciprocally simulated efficiently, and we base our evaluation framework on BSP, which has a well developed literature, libraries and other resources that provide an efficient way of evaluating algorithms for a wide variety of problems. We insist on the fact that we are not choosing one model over the others as a real platform over which the algorithms would actually run, but choosing one of them as the "computational model" that will be used to compare the behavior of different strategies or configurations.

The rest of this paper is organized as follows: Section 2 presents generalities about distributed search architectures and some improvements proposed to speedup query response time. Some of these improvements were introduced in previous papers, while Clairvoyant Distributed Ranking and the re-using intersections cache are original contributions of this paper. Section 3 presents our methodology using the BSP model. Section 4 shows the features of the query log and the parameters used in the experiments, as well as the results of the experiments we conducted. Finally, Section 5 presents our conclusions.

2   Search Architecture

Queries are introduced to the system via a receptionist machine named the *broker*. The broker is in charge of routing the queries to the clusters *search processors* and receiving the respective answers. It decides to which processor a given query will be routed, normally by using a load balancing heuristic. The parallel processing of queries consists in a series of operations that will be executed in different processors, the results of which will be combined to get the final answer to the query. These are the primitive operations of broadcast or communication, list intersection, list merging, list ranking, etc. Each of these operations has a cost depending on its actual parameters. The combination of these costs for all the queries in an execution environment conforms the total cost of a sequence of queries.

Several features may be added to the basic setting just described to further speed up the search process or provide fault tolerance, as for example the already mentioned partitioning schemes, replication, particular policies for routing the queries among the processors, different ranking algorithms, and various types of caches. In the following we describe some of these standard features as well as some original ones that we introduce in this paper, and will be tested using our evaluation framework.

Partitioning and Replications. We already mentioned in the introduction the exploitation of parallelism over the text database that is achieved by different variations or evolutions of the two "extreme" settings of document and term partitioning, in particular the 2D scheme of [2]. Fault-tolerance is generally supported by means of replication. To this end, and assuming a P-processor parallel search cluster, D copies for each of the P processors are introduced. This can be seen as adding an extra dimension to whichever configuration was initially chosen. Then the system can be seen as a $P \times D$ matrix in which each processor in the D dimension maintains an identical copy of the respective partition of the index. Upon receipt of a query by the broker node in the cluster, the broker sends it to all the processors that must evaluate it, and for each of them one replica is selected. The processors hit by the query work on their part of the solution to then merge all results to produce the final answer.

The basic model of concurrency in here is the distribution of the query search among the distributed knowledge of the P principal nodes and also internally to each of the D replicas, which contain the same data (global knowledge). The question arises on how to utilize the available resources to best execute search algorithms in the $P \times D$ arrangement of computers. A naive approach would be to attempt a text data partitioning throughout the system and hope that the multiplicity of computational resources can provide a speedup by virtue of simultaneous computations. Another approach would be to share in a more subtle way the resources of the D replicas. We explore these cases in this paper.

Caching. A typical way of reducing the number of processors involved in each query processing is by using caching in several ways. First, the broker machine may have a results cache (RCache) of previous queries, to avoid repeatedly computing the most frequent ones. The state of the art strategy for the results cache is the SDC strategy [6], which keeps a static section to store the answer for queries that are frequent along large periods of time and a dynamic part with the objective of capturing queries that become very popular during short periods of time. Additionally, we can reduce the number of processors participating on each query by sending the query only to a selected group of them as in [7] or [2]. The latter works propose to keep a compact location cache (LCache) at the broker machine.

Processors may maintain also caches of posting lists to reduce secondary memory accesses. Examples of these can be found in [6,4,3]. Alternatively there is an intersection cache [4] which keeps in each search node the intersection of the posting lists associated with pairs of terms frequently occurring in queries. The intersection operation is useful to detect the documents that contain all of the query terms, which is a typical requirement for the top-K results in major search engines. Figure 1 shows the five caches defining a hierarchy which goes from the most specific pre-computed data for queries (results cache) to the most generic data required to solve queries (posting list cache). When a new query arrives to the system, if the broker finds the new query in its RCache, the query is answered with no further computation. If that is not the case, the broker searches the query in its LCache to reduce the number of processors that will be involved in its processing. Finally, the broker selects a processor to send the query to. After that, the query is processed according to the partitioned index approach and the ranking scheme.

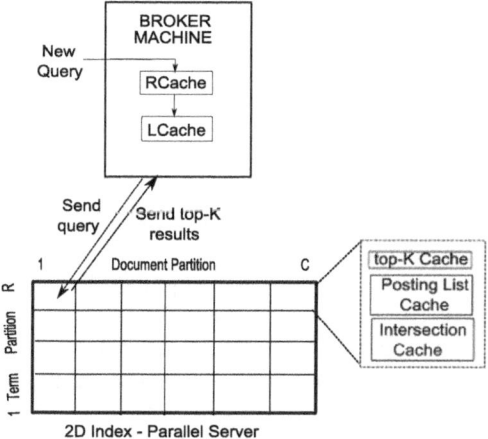

Fig. 1. Query processing using a 2D index and a five level cache hierarchy

Distributed vs. Centralized Ranking. There are basically two types of rankings under distributed architectures: distributed or centralized ranking. For clarity we explain the ranking schemes without considering caching at any level. In the distributed ranking scheme, the broker sends the query to a *manager-merger* processor. This processor sends the query to each other processor involved in the processing of the query (i.e. each processor owning terms of the query if the term partition scheme is used, or all processors when we use the document partition scheme). Then, these processors fetch the posting lists for each term and compute their union. In the last phase, they rank the local results and send the top-K documents identifier to the manager-merger processor which merges the partial results in order to build the global top-K results that are sent to the broker.

Using the centralized ranking strategy, there is one processor (named the manager-ranker processor) that globally ranks the pieces of posting lists that are sent to it by the processors involved in the processing of the query. In general it is enough that each involved processor sends a small list to the manager-ranker processor to allow him compute a barrier S_{MAX} that document frequencies for a certain term must surpass to qualify for the final result list. When posting lists are kept sorted by frequency, this allows to directly skip whole list-tails consisting of documents with frequencies smaller than the barrier.

To map these basic strategies to a 2D architecture we need first to describe the main computational path that each query will follow. At column-level a union (resp. intersection) is computed in a sequential way by sending the shortest list from the processor that owns it to the processor owning the second shortest list, who computes the union (resp. intersection) with its term (or terms) and sends the result to the following one and so on, until the last processor is reached. When using the distributed ranking scheme we perform a ranking over local data before sending partial results to the next processor. This allows reducing the communication cost. This mechanism requires that some processor, with the information of the lengths of the posting lists at that column, prepares the route the query will follow. Then the basic ranking strategies become:

Distributed Ranking for 2D index:

1. The broker machine sends the query to the manager-merger processor using a hash function over the query terms.
2. The manager-merger sends the query to a random processor at each column (which we call the manager-ranker).
3. The manager-ranker of each column prepares the route for the query and sends it to the first processor of that route.
4. Each processor that receives the query fetches the postings lists of its term(s), computes their union or intersection, makes the ranking and sends the result to the following processor in the routing list.
5. The last processor of the routing list sends the results to the manager-ranker of the column, which performs the final distributed ranking (in order to obtain the top-K postings of the column).
6. The manager-merger receives the results of all the columns and merges them to obtain the global top-K results.

Centralized Ranking for 2D index:

1. The broker machine sends the query to a manager-ranker processor.
2. The manager-ranker sends the query to a random processor at each column (which we call the column manager).
3. The column manager of each column prepares the routing list for the query and sends it to the first processor of that route.
4. Each processor that receives the query fetches the postings lists of its term(s), makes the union or intersection with the partial list it receives from previous processors from the same column, and sends the result to the following processor in the list.
5. The last processor sends the results to the column manager, who sends the results to the manager-ranker.
6. The manager-ranker receives the results of all the columns and ranks them.

Clairvoyant Distributed Ranking for AND queries. The main advantage of the centralized ranking approach with respect to the distributed one arises from the possibility of excluding documents that would not do it to the top-K results, by means of initially considering short prefixes of all the lists and then updating the global barrier as needed. The price to be paid for that gain is the extra communication and overhead due to the extra need of interaction among the processors. This suggests that it would be very valuable to know a-priory how many elements to consider of each list. In that case we could send exactly the required postings and do the minimum work. Considering that each successive intersection that is computed will further reduce the length of the resulting lists, just limiting the number of elements of the *first* list that must be sent would be useful to reduce the total computation and communication time. We can exploit the information of the query logs to *estimate* the number of elements of the shortest list that should be sent at the beginning of the computation to ensure that the appropriate number of results will be found. Of course we cannot be sure we will effectively obtain that number, but we have two possible solutions to that. First, we could choose the number of postings to send so as the probability of having enough elements exceeds a certain threshold, admitting the (rare) possibility of having less results than needed. A second idea would be to choose that number and, in case of not having enough elements in the intersection, behave as in the centralized case, i.e. asking for extra results to the first processor and repeating the process. We implemented the first of these ideas and compared it with the Distributed and Centralized strategies above.

Clairvoyant Distributed Ranking:

1. The manager-merger of the query sends it to a random processor at each column, that will be the manager-ranker.
2. The manager-ranker of each column prepares the route for the query and sends it to the first processor of that route.
3. Each processor that receives the query fetches a certain number of postings of its term(s) according to the estimation, intersects them with the partial list it receives and sends the result to the following processor in the list.

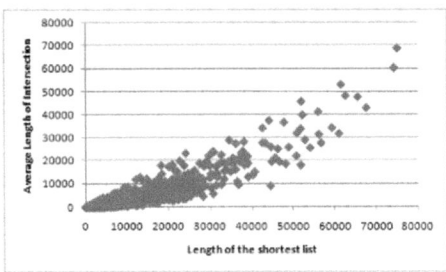

Fig. 2. Average length of the intersection vs length of the shortest list, two-term queries

4. The last processor sends the results to the manager-ranker of the column, which makes the ranking and sends the results to the manager-merger.
5. The manager-merger receives the results of all the columns and merges them.

We now explain how we did to compute the parameters of Clairvoyant Ranking, i.e. the number of postings of the shortest list that need to be sent to ensure that with high probability the top-K elements will be found. We used the same data sets and query logs described in Section 4 to compute the average length of the intersection for queries with different number of terms, and analyzed the results as a function of the length of the shortest posting list involved in the intersection. Figure 2 shows the resulting values for 2-term queries. We then computed min-square approximations, and another linear function that is a lower-bound to the average length for more than 90% of the queries. The inverse of the linear lower bound acts as an upper bound to the average length of the resulting list, so fetching that number of postings from the shortest list would, on average, be enough for more than 90% of the queries. We decided to fetch twice that number to cope with those intersections whose results are smaller than average. The values needed to have top-K results in at least 90% of the queries are those depicted in table 1. Those were the values used in the experiments described in section 4. For queries with 4 or more terms, we used the same values as for 4 term intersections.

Re-using Intersections. Although as we mentioned before replication is normally introduced to improve fault tolerance and throughput (due to the added processing capacity), the D replicas of a processor may contain intersections between pairs of terms that could be useful to solve new queries. It is then pertinent

Table 1. # of postings of the shortest list sent to ensure 90% of the queries have enough results

# of terms	# of postings sent
2	$(K/C + 1000) * 5$
3	$(K/C + 1000) * 22$
4	$(K/C + 400) * 50$

to try to exploit this advantage by first making a sort of tour along the replicas to "see" if there is something useful to solve a current query in the column. That way, the caches of the replicas behave in some way as a common shared resource. In the following we propose a strategy to achieve this goal.

Each processor will be responsible of a set of terms, given by a hashing function H. This function H must be such that it distributes terms among processors trying to cluster together terms frequently co-occurring in queries. H is extended from terms to queries, in some reasonable way (for example, defining $H(q) = H(t)$, where t is the term of q with the longest individual posting list among those forming q). Each processor p will have a cache of posting lists, divided in two parts: a static part where it will hold the most useful posting lists (according to the logs), among the terms t such that $H(t) = p$, and a dynamic part, both for terms t such that $H(t) = p$ and for terms t such that $H(t) <> p$. Besides, each processor will maintain a dynamic cache of intersections (pair intersections), which will contain only queries or partial queries q such that $H(q) = p$. Upon computing complex intersections, p will eventually cache pair-intersections in it's own cache and also may cache pair-intersections in other processor's intersection cache. Finally, each processor will have a results-cache, where it will hold results of previous queries assigned to it. The following algorithm shows the behavior of a processor upon receiving a query.

- Given a query $q = t_1, \ldots, t_r$ (abusing notation we denote by q also the set of terms $\{t_1, \ldots, t_r\}$)
- q is assigned to $p = H(q)$
- If p has q in its results cache then p updates the validity of q in cache and answers
- else
  If p has q in it's own intersection cache then p updates the validity of q in cache and answers
  - else
    - Let T_Y be the set of terms of q that appear already in an intersection with another term of q in the intersection cache of p
    - If $T_Y = q$ (i.e. p has all the pairs needed to compute the answer in its own cache)
      - p computes the answer, updates the validity of all the used pairs, caches q in the results cache and answers
    - else
      - Let $T_N = q - T_Y$
      - Let $T_N^p = T_N \cap H^{-1}(p)$ (i.e. the set of terms that are not cached and must be looked for in p)
      - Let $\overline{T_N^p} = T_N \cap \overline{H^{-1}(p)}$ (i.e. the set of terms that are not cached and must be looked for in other processors)
      - p asks processors in $H(\overline{T_N^p})$ for intersections or posting lists useful for computing q
      - Each processor answers, and blocks the contents of the cache for a superstep
      - Let T be the set if terms in $\overline{T_N^p}$ such that NO useful intersections are kept in the caches.
      - p couples together all the terms in T_N^p (putting together pairs with maximum frequency or maximum frequency × length of the posting lists), computes and caches their intersections.
      - If $|T_N^p|$ is odd, the uncoupled term is coupled with one term in T, p computes their intersection and caches the result where it corresponds according to H.
      - p couples together the rest of the terms in T (putting together pairs with maximum frequency or with maximum frequency × length of the posting lists), computes the intersection of each pair and caches each result where it corresponds according to H.
      - p computes the query using the information already in the caches of the other processors and its own cache
      - p communicates to the other processor which information it has used, the other processors update the validity of the information
      - p caches q in the results-cache and answers

We must also say what a processor does in case another processor makes a request: If I have a pair or a partial result useful for the query: (a) Answer to the asking processor, (b) Block it during one superstep, (c) When the processor tells me that it has used the provided pair or result, update the cache accordingly.

Finally, we must say what's the meaning of "caches the result where it corresponds according to H." When a processor computes a pair that does not belong to him accordingly, it will cache it in the owner's cache. This means that upon a request of caching, the other processor p' must: (a) Receive the list, (b) Make place in the cache according to the caching policy, and (c) Update the cache with the new list.

3   A Cost Estimation Methodology

Our cost estimation framework is based on the bulk-synchronous model of parallel computing (BSP) [8], where computation is organized as a sequence of supersteps, in each of which the processors may perform computations on local data and/or send messages to other processors. The messages are available for processing at their destinations by the next superstep, and each superstep is ended with the barrier synchronization of the processors.

The total running cost of a BSP program is the cumulative sum of the costs of its supersteps, and the cost of each superstep is the sum of three components: computation, communication and barrier synchronization. Computation cost is given by the maximum computation cost of a processor in the superstep, and will be denoted by a quantity w. This will include also disk accesses, whose unit cost will represented by a constant δ. Besides, we note that just for participating in the processing of a query, a processor incurs a certain overhead β, which must also be taken in consideration when accounting for the actual cost of the query. Communication cost is given by the maximum number of word-length messages sent or received by a processor during the superstep (denoted by h) times the cost of communicating one word (γ). Hence this will be denoted by the product $h * \gamma$. Finally, we will use L for the synchronization cost. Parameters γ, L, β and δ take into account the characteristics of the particular computer architecture that is emulated.

The impact of uneven load balance in the cost is taken into account by considering maximum cost among all the processors for each of the costs above. Moreover, we can compute the load work and communication efficiency using this model as follows: for any performance measure X, in each superstep we compute its value for each processor, and define BSP efficiency for X as the ratio average(X)/maximum$(X) \leq 1$, over the P processors. This gives an idea of how well we distribute the workload among processors.

We end this section with the list of the primitive components that together conform the costs (computation and communication) that are charged to each query:

- $t_i(x, y)$: Expected time employed by a processor to compute the intersection of two lists of lengths x and y respectively.
- $t_m(x)$: Expected time employed by a processor to merge a set of lists of total length x.

- $t_r(x)$: Expected time employed by a processor to rank a list of length x.
- $I(x, y)$: Expected length of the intersection of two lists of length x and y.
- γ: time employed to transmit a unit of information from one processor to another.
- δ: cost of accessing a unit of information in disk.
- β: overhead due to the participation of a processor in a query.
- L: time for the barrier synchronization at the end of each superstep.

4   Experimental Setting

We did our experiments to estimate the cost of our algorithm using P=512 processors with different combinations of $P = C \times R$. The number of rows ranged from 1 (local index) to 512(global index). Queries where selected from a query log with 36,389,567 queries submitted to the AOL Search service between March 1 and May 31, 2006. We preprocessed the query log following the rules applied in [3] by removing stopwords and completely removing any query consisting only of stopwords. We also erased duplicated terms and assumed that two queries are identical if they contain the same words no matter the order. The resulting query trace has 16,900,873 queries, where 6,614,176 are unique queries and the vocabulary consists of 1,069,700 distinct query terms. 75% of these queries occur only once, and 12% of them occur twice. The index was built using a 1.5TB sample of the UK's web obtained from Yahoo! searches.

We used caching in several ways as we mentioned before. For the experiments we set the overall cache sizes so as to maintain information of at most 10% of the processed queries. In all the cases the broker machine has a cache of results of previous queries (RCache), and the processors maintain caches of posting lists to reduce secondary memory accesses. In the experiments reported in Section 4 we considered an additional level of caching introduced in Section 2. For the algorithms using replication, to improve the throughput and fault tolerance (see for example [6]), we set the number of replications $D = 4$, so in these cases we have $P \times D$ processors available to compute queries.

We defined particular costs for the different primitive functions and constants, based on benchmarking runs we did on the same collection. The values are expressed relative to a base-line in terms of ranking time defined as $t_r(x) = x$. Merge operations require in average $t_m(x) = x/6$. An intersection between two lists of lengths x, y can be done in time proportional to $\min(x \log y, x + y)$. However, the constants discovered in the benchmark make us count $t_i(x, y) = \min(x \log y, x + y)/6$. For I, the random variable that models the expected length of the intersection of lists, we used a power-law distribution, with parameters varying according to the number of lists being intersected. The actual values of these parameters were determined via sampling from real queries. We also established that the communication cost for transmitting a query is, on average, 1/4 of the time required to transmit a K-word list. For $P = 512$, the communication time for one unit under a heavily loaded network was $\gamma \approx 0.453$, and the overhead constant $\beta \approx 1.5$. The cost for the synchronization was $L \approx 8$.

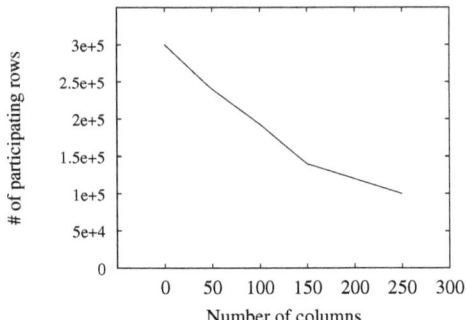

Fig. 3. Number of participating rows as a function of the number of columns

We give as example a succinct explanation of the costs charged to each processor in the Distributed Ranking case for AND queries, the other cases are analogous. For simplicity we omit the parameters of random variable I.

Distributed ranking costs:

1. Send the query to a processor at each column. Computation: β; Communication: $K/4 \times \gamma$
2. Send the query to the first processor in the route. Computation: β; Communication: $K/4 \times \gamma$
3. Fetch and intersect.(At each participating processor) Computation: $\delta \times K$ for disk access $+ t_i(I) + \beta$; Communication: $I \times \gamma$.
4. Rank. Computation: $t_r + \beta$; Communication: $I \times \gamma$
5. Merge. Computation: $t_m + \beta$; Communication: $K \times \gamma$

OR Queries with iterations. In this experiment we considered Centralized and Distributed ranking for OR queries, with an additional restriction imposed in order to balance communication, secondary memory recovery and computation costs among processors: we limit the number of operations of each type a processor may execute in each superstep. Operations not performed in one superstep due to that limitation, are delayed until the next one. Thus, the processing of a query is separated in successive *iterations*. The estimation of the costs is performed in each superstep, charging always the maximum cost for each operation. We used the RCache at the broker and top-K cache plus posting list cache at search processors' side.

As the number of columns augments, the probability that the query terms are located in fewer rows augments (see Figure 3), and therefore the communication costs tends to decrease, while the computation overhead increases because each query is processed by more columns. There must naturally exist a trade-off between the two extremes of the 2D index. Figure 4[Left] shows the cost for the centralized and distributed ranking algorithms when processing OR queries with an asynchronous system for different $P = R \times C$ combinations. Results show that when the index is working as a term partition (matrix size of 512×1) both ranking strategies tend to perform well, but a better configuration is 16×32. A document partition index has a higher cost due to the overhead of dealing with every query in all processors. The centralized ranking has a lower cost in each extreme of the matrix because it requires less iterations to finish the processing of a query when using Persin filters.

Figure 4[Right] shows the average efficiency measured as $\sum(w_{p_i}/Max_{wp})/P$ in each superstep, where w_{p_i} is the computation performed by processor p_i, and Max_{wp} is the maximum computation performed in that superstep. Results show that using a large number of rows improves both ranking's efficiency, but when working with 512 columns and one row per column, the distributed ranking scheme reports a better efficiency than the centralized one.

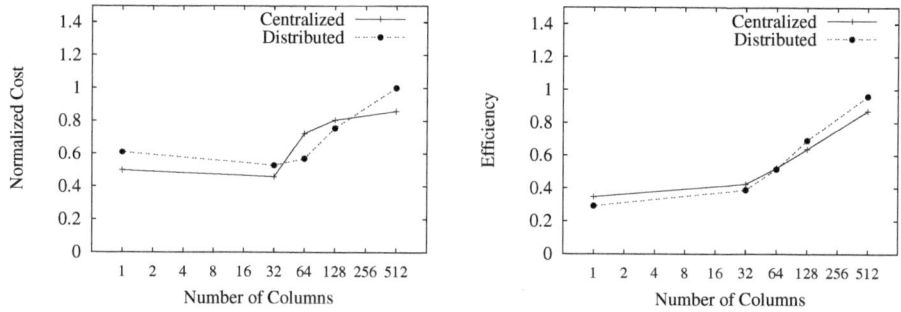

Fig. 4. OR queries [Left] Normalized costs. [Right] Computation efficiency.

OR Queries with complete lists. Figure 5 shows results for OR queries when we accessed to the whole posting lists of each term to perform the ranking operation, i.e. without the iterations defined in the previous experiment. The same cache settings were used. As we may see, in this case the term partition index with centralized ranking presents high costs due to the imbalance it has both in communication (because it has to transmit complete posting lists from the search processors to the manager) and in secondary memory access (when retrieving the posting lists for each term). On the other extreme, the document partition index is more balanced in both costs, but its overhead is clearly higher. The Distributed ranking scheme presents more balanced communication with all configurations (ranking is performed in each processor before sending results). But the term partitioned index presents some imbalance in secondary memory retrieval as ranking is performed over larger posting lists. Again, we see the trade-off communication/overhead, with the optimal configuration realized by a 128×4 arrangement of the processors.

Different ranking strategies for AND Queries. Figure 6 shows the performance of Distributed, Centralized and Clairvoyant strategies with different configurations. As we may see, the best configurations are those in which the 256 processors are arranged in 32 or 64 columns, depending on the ranking strategy, the optimal combination being the clairvoyant strategy with 64 columns. Again, as the number of columns augments, the probability that the query terms are

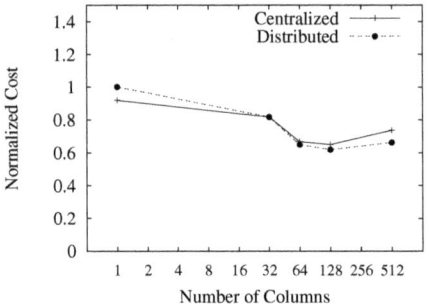

Fig. 5. Normalized costs for OR queries with complete lists

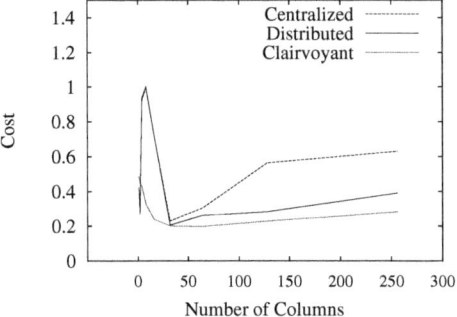

Fig. 6. Normalized costs vs. number of columns, for different ranking strategies

located in fewer rows augments (see Figure 3), and therefore the communication costs tends to decrease, while the computation overhead increases because each query is processed by more columns. There is naturally a trade-off between the two extremes. Both the distributed and centralized approaches suffer from the inconvenient that each time a new group of postings must be considered for the intersection, they must be compared with *all* the previously fetched terms, with a cost that is almost quadratic in the required number of iterations. The clairvoyant strategy is free of that problem.

Intelligent Caching in a Replicated Local Index. In this section we report the results of an experiment in which we compare the effects of different caching protocols for AND queries in a $1 \times P \times D$ setting (i.e. a local index with replication, $D = 4, 8$). The base-line was the architecture having all the caching levels considered before, except the intersection cache (No Intersection). We then considered adding the intersection cache, but without any particular strategy for routing the queries and their corresponding caching decisions (Intersection). Finally, we considered the usage of intelligent distribution of the traffic among the rows, using the hashing function H as described in Section 4.2.1 (Intersection-h). Fig. 7(a) shows the throughput obtained by all three intersection algorithms. This graphics shows the improvement (almost 30% with K=128 and more

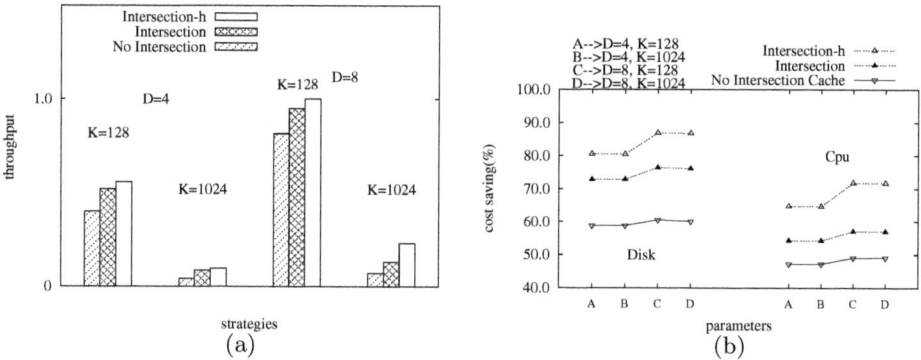

Fig. 7. Different caching strategies, AND (a) Throughput (b) Disk/CPU savings

notorious when with K=1024 is 60%) obtained when using a more intelligent algorithm to reassign the intersections operations. Fig. 7(b) shows the disk and CPU cost saving relative to an algorithm that uses no caching at all. As it may be seen, all three algorithms obtain visible differences in both aspects with respect to that base line. However, the use of intersection caches helps improve that in an extra 20%. The most significant savings are obtained for bigger values of D. In those cases, the number of disk accesses needed by the best algorithm is about 1/4 of the number needed without caching of intersections.

5   Conclusions

We have presented a methodology for predicting the costs of different algorithms and architectures for distributed indexes of different sizes. We provided results for some interesting experiments regarding a novel two-dimensional architecture and some new ranking and caching strategies. The results obtained allow us to be optimistic on the further development and applications of the cost framework, as for example putting together the different features we have tested separately, like replication, caching, etc. The cost framework can be used for analyzing and comparing our work with other authors when evaluating the scalability of alternative strategies.

References

1. Baeza-Yates, R., Ribeiro-Neto, B.: Modern Information Retrieval. Addison-Wesley, Reading (1999)
2. Feuerstein, E., Marin, M., Mizrahi, M., Gil-Costa, V., Baeza-Yates, R.: Two-dimensional distributed inverted files. In: Karlgren, J., Tarhio, J., Hyyrö, H. (eds.) SPIRE 2009. LNCS, vol. 5721, pp. 206–213. Springer, Heidelberg (2009)
3. Gan, Q., Suel, T.: Improved techniques for result caching in web search engines. In: WWW 2009, pp. 431–440. ACM Press, New York (2009)

4. Long, X., Suel, T.: Three-level caching for efficient query processing in large web search engines. In: WWW 2005, pp. 257–266 (2005)
5. Marin, M., Gil Costa, V.: High-performance distributed inverted files. In: CIKM 2007, pp. 935–938. ACM Press, New York (2007)
6. Marin, M., Gil Costa, V., Gomez-Pantoja, C.: New caching techniques for web search engines. In: ACM HPDC (2010)
7. Puppin, D., Silvestri, F., Perego, R., Baeza-Yates, R.: Tuning the capacity of search engines: Load-driven routing and incremental caching to reduce and balance the load. ACM Transactions on Information Systems (TOIS) 28, 2 (2010)
8. Ramachandran, V., Grayson, B., Dahlin, M.: Emulations between QSM, BSP and LogP: a framework for general-purpose parallel algorithm design. J. Parallel Distrib. Comput. 63, 1175–1192 (2003)

Text Classification on a Grid Environment

Valeriana G. Roncero, Myrian C.A. Costa, and Nelson F.F. Ebecken

COPPE / Federal University of Rio de Janeiro - UFRJ
Cidade Universitária - Centro de Tecnologia - Bloco I, Sala I-248
Ilha do Fundão – P.O. Box 68516 Rio de Janeiro, RJ - CEP 21941-972, Brazil
{valery,myrian}@nacad.ufrj.br, nelson@ntt.ufrj.br

Abstract. The enormous amount of information stored in unstructured texts cannot simply be used for further processing by computers, which typically handle text as simple sequences of character strings. Text mining is the process of extracting interesting information and knowledge from unstructured text. One key difficulty with text classification learning algorithms is that they require many hand-labeled documents to learn accurately. In the text mining pattern discovery phase, the text classification step aims at automatically attribute one or more predefined classes to text documents. In this research, we propose to use an algorithm for learning from labeled and unlabeled documents based on the combination of Expectation-Maximization (EM) and a naïve Bayes classifier on a grid environment, this combination is based on a mixture of multinomials, which is commonly used in text classification. Naïve Bayes is a probabilistic approach to inductive learning. It estimates the a posteriori probability that a document belongs to a class given the observed feature values of the documents, assuming independence of the features. The class with the maximum a posteriori probability is assigned to the document. Expectation-Maximization (EM) is a class of iterative algorithms for maximum likelihood or maximum a posteriori estimation in problems with unlabeled data. The grid environment is a geographically distributed computation infrastructure composed of a set of heterogeneous resources. The semi-supervised learning classifier in the grid is available as a grid service, expanding the functionality of Aîuri Portal, which is a framework for a cooperative academic environment for education and research. Text classification mining methods are time-consuming by using the grid infrastructure can bring significant benefits in learning and the classification process.

Keywords: grid computing, text classification, expectation-maximization and naïve bayes.

1 Introduction

Text mining is a relatively new practice derived from Information Retrieval (IR) [1, 2] and Natural Language Processing (NLP), Baeza-Yates *et al* [3]. The strict definition of text mining includes only the methods capable of discovering new information that is

J.M.L.M. Palma et al. (Eds.): VECPAR 2010, LNCS 6449, pp. 251–262, 2011.

not obvious or easy to find out in a document collection, i.e., reports, historical documents, e-mails, spreadsheets, papers and others. Text mining executes several processes, each one consisting of multiple phases, which transform or organize an amount of documents in a systematized structure. These phases enable the use of processed documents later, in an efficient and intelligent manner. The processes that compose the text mining can be visualized in fig. 1 that is summarized version of the figure model from Han *et al* [4] on page 6.

Text classification has become one of the most important techniques in text mining. The task is to automatically classify documents into predefined classes based on their content. Many algorithms have been developed to deal with automatic text classification. One of the common methods is the Naïve Bayes, Mitchell [5]. Although the naïve Bayes works well in many studies [6, 7, 8], it requires a large number of labeled training documents for learning accurately. In the real world task, it is very hard to obtain the large labeled documents, which are mostly produced by humans. Nigam *et al.* [9] apply the Expectation-Maximization (EM) algorithm to improve the accuracy of learned text classifiers by augmenting a small number of labeled training documents with a large pool of unlabeled documents. The EM algorithm uses both labeled and unlabeled documents for learning. Their experimental results show that using the EM algorithm with unlabeled documents can reduce classification error when there is a small number of training data.

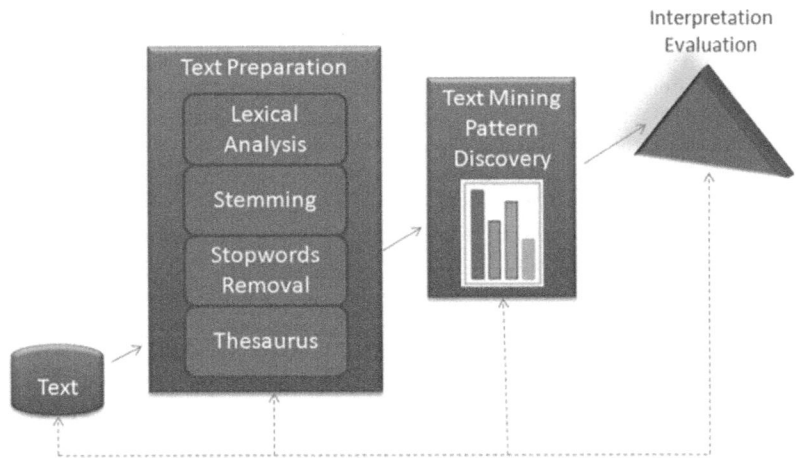

Fig. 1. Shows a summary of the text mining phases

Unfortunately, the EM algorithm is too slow when it performs on very large document collections. In order to reduce the time spent, we propose to use the grid infrastructure to improve the computational time in learning and classifying process. The text classification task uses an algorithm based on the combination of EM algorithm and the Naïve Bayes classifier, Dempster *et al* [10]. This can bring significant benefits and expands the functionality of Aîuri Portal, which uses this semi-supervised learning algorithm as a grid service, which will be explained in section 3.2.

Implementation of text mining techniques in distributed environment allows us to access different geographically distributed data collections and perform text mining tasks in distributed way.

This paper is organized as follows. In section 2, we present an overview of text classification task with the classification algorithms. In section 3, we briefly present an overview of grid computing. Section 4 describes the distributed implementation of naïve Bayes classifier via the EM algorithm on a grid and we briefly conclude on Section 5.

2 Text Classification

Text categorization or classification aims to automatically assign categories or classes to unseen text documents [11, 12], some classification techniques are naïve Bayes classifier [5], k-nearest neighbor, Yang [13], and support vector machines, Joachims [14]. The Naïve Bayes algorithm requires a large number of labeled training documents, but to obtain training labels is expensive, while large quantities of unlabeled documents are readily available. The combination of EM algorithm and a Naïve Bayes classifier can make use of unlabeled documents to training. This new algorithm first trains a classifier using the available labeled documents, and probabilistically labels the unlabeled documents. It then trains a new classifier using the labels for all the documents, and iterates to convergence, fig 2.

In this section, we briefly review the naïve Bayes classifier and the EM algorithm that is used for making use of unlabeled data.

2.1 Naïve Bayes Classifier

Naïve Bayes Classifier is a probabilistic learning algorithm that derives from Bayesian decision theory describing by Mitchell [12], which by default assumes observations are independent. It is easy to build a Naïve Bayes Classifier when you have a large number features. Researchers have shown that Naïve Bayes Classifier is competitive with other learning algorithms in many cases and in some cases it outperforms the other methods [8]. Learning in Naïve Bayes Classifier involves estimation of the parameters for a classifier, using the labeled document only. The classifier then uses the estimated parameters to classify unobserved documents.

First we will introduce some notation to describe text. Let D be a set of text documents $D = \{d_1, d_2, d_{|D|}\}$, and c_k be a possible class from a set of predefined classes $C = \{c_1, c_2, c_{|C|}\}$. The probability of a document d being in class c, $P(c_k \mid D)$, is computed as

$$P(c_k \mid D) = P(c_k) \times P(D \mid c_k) \tag{1}$$

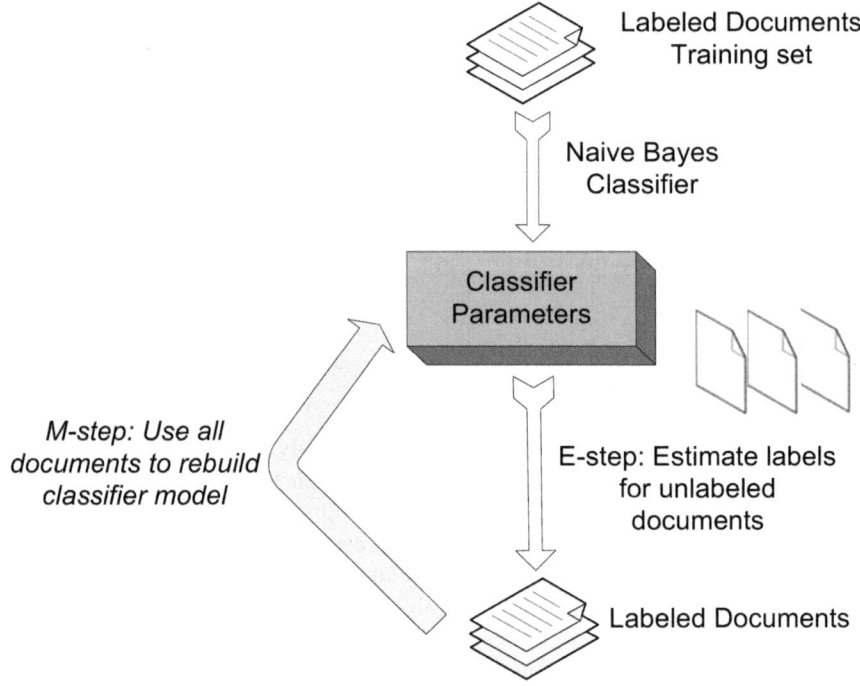

Fig. 2. Expectation-Maximization algorithm with Naïve Bayes classifier

Class probability $P(c_k)$ can be estimated from training data. However, direct estimation of $P(c_k/D)$ is impossible in most cases because of the sparseness of training data.

By assuming the conditional independence of the elements of a vector, $P(D/c_k)$ is decomposed as follows,

$$P(D \mid c_k) = \prod_{j=1}^{k} P(d_j \mid c_k),$$ (2)

where d_j is the jth element of a set of text documents D. Then Equation (1) becomes

$$P(c_k \mid D) = P(c_k) \times \frac{\prod_{j=1}^{k} P(d_j \mid c_k)}{P(D)}.$$ (3)

With this equation, we can calculate $P(c_k/D)$ and classify D into the class with the highest $P(c_k/D)$.

Note that the naïve Bayes classifier assumes the conditional independence of features. This assumption however does not hold in most cases. For example, word occurrence is a commonly used feature for text classification. However, obvious strong dependencies exist among word occurrences. Despite this apparent violation of the assumption, the naïve Bayes classifier exhibits good performance for various natural language processing tasks.

2.2 Expectation-Maximization Algorithm

One disadvantage of the Naïve Bayes Classifier is that it requires a large set of the labeled training documents for learning accurately. The cost of labeling documents is expensive, while unlabeled documents are commonly available. By applying the EM algorithm, we can use the unlabeled documents to augment the available labeled documents in the training process. Figure 3 shows the procedure of modified EM algorithm.

Input: Training Documents
Output: Classification Model

---

1. Train the classifier using only labeled data.
2. Classify unlabeled documents, assigning probabilistic-weight class labels to them.
3. Update the parameters of the model. Each probabilistically labeled document is counted as its probability instead of one.
4. Go back to (2) until convergence.

Fig. 3. Modified EM algorithm

The EM algorithm is a type of iterative algorithm for maximum likelihood or maximum a posteriori estimation in problems with incomplete data [10, 15, 16]. This algorithm can be applied to minimally supervised learning, in which the missing values correspond to missing labels of the documents, McLachlan *et al* [17]. In our task, the class labels of the unlabeled documents are considered as the missing values.

The EM algorithm consists of the E-step in which the expected values of the missing sufficient statistics given the observed data and the current parameter estimates are computed, and the M-step in which the expected values of the sufficient statistics computed in the E-step are used to compute complete data maximum likelihood estimates of the parameters [10].

The EM algorithm starts using the Naïve Bayes Classifier to initialize the parameters feature probabilities and class priors using the labeled documents. The E-step and M-step are iterated until the change in class labels for the unlabeled documents is below some threshold (i.e. the algorithm converges [16] to a local maximum). The E-step almost dominates the execution time on each epoch, since it estimates the class labels for all the training documents [9].

3   Grid Environment

A grid is a geographically distributed computation infrastructure composed of a set of heterogeneous machines, often with separate policies for security and resource use, Qi *et al* [18], that users can access via a single interface. Grids therefore, provide a common resource-access technology and operational services across widely distributed virtual organizations composed of institutions or individuals that share resources.

Today grids can be used as effective infrastructures for distributed high-performance computing and data processing, Foster *et al* [19].

3.1   NACAD Grid Environment

The NACAD Grid uses Globus Toolkit 4 (GT4) [20] as a grid middleware, which is a widely used middleware in scientific and data-intensive grid applications, and is becoming standard for implementing grid systems. The toolkit addresses security, information discovery, resource and data management, communication, fault detection, portability issues and is based on grid services. Grid services is a technology based on the concepts and technologies of grids and web services and can be defined as a web service that delivers a set of interfaces that follows specific conventions, fig. 4. This technology was originated from the necessity to integrate services through virtual, heterogeneous and dynamic organizations, composed of distinct resources, whether within the same organization or by resource sharing.

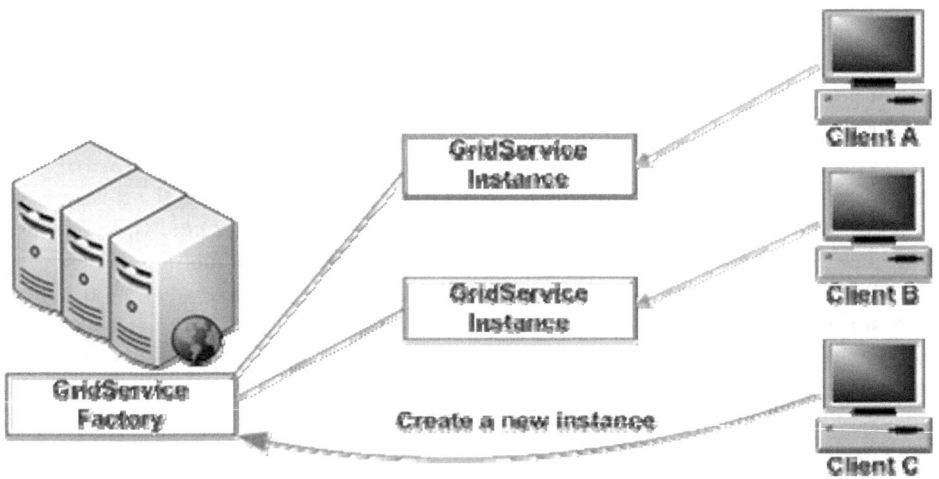

Fig. 4. Grid services

3.2   Grid Services

A grid service is a web service that conforms to a set of conventions (interfaces and behaviors) that defines how a client interacts with a grid service. A web service converts an application into a web-application, which is published, found, and used through the web facilitating the communication between applications.

In this work we propose to implement the classifier model as a grid service in the Aîuri Portal [21], which is a framework for a cooperative academic environment for education and research. The components are grid Text Mining services and in the future Data Mining components will be added. The classifier service will be included in the Aîuri Portal showing that many distinct algorithms can be easily added and accessed through this Portal, fig 5.

Fig. 5. The classifier service as a component service in the Aîuri Portal

4   Naïve Bayes Classifier via the EM Algorithm on a Grid Environment

The enormous amount of information stored in huge document databases in unstructured format or semi-structured format cannot simply be used for further processing by computers, which typically handle text as sequences of character strings. Text mining provides some methods, like classification, able to extract interesting information and knowledge from unstructured text. One key difficulty with text classification learning algorithms is that they require many hand-labeled documents to learn accurately. Using the Naïve Bayes Classifier via the EM algorithm we can use the unlabeled documents to increase the available labeled documents in the training process. Implementation of text mining techniques in distributed environment allows us to access different data collections that are geographically distributed and perform text mining tasks in distributed way. Figure 6 shows the distributed EM algorithm for text classification on a grid environment.

Input: Training Documents
Output: Classification Model

---

1. The Portal receives from the user the parameters input file, the labeled documents and the unlabeled documents.
2. The Portal sends a request to the service container to create the grid service instance, which is composed with the Naïve Bayes Classifier and the EM application.
3. The classifier grid service builds the initial global classifier from only the labeled documents using the grid nodes.
4. Each grid node receives a pre-defined set of training documents from the Portal.
5. Iterate until the classifier parameters are still improved:
    5.1. E-step: Each grid node estimates the class of each document by using the current global classifier.
    5.2. M-step: Each grid node re-estimates its own local classifier given the estimated class of each document.
6. Sum up the local classifier to obtain the new global classifier and return them to all grid nodes.

Fig. 6. The distributed EM algorithm for text classification on a grid environment

5   Results

In this study we used the Reuters text collection to evaluate the proposed approaches. Term stemming and stopwords removal techniques are used in the prior stage of text preprocessing.

Reuters-21578 collection, Distribution 1.0 [22], is a financial corpus with news articles averaging 200 words each. In this corpus there are about 12000 classified stories into 118 possible categories, this collection has articles with none or at least one category. For our study, we use The Modified Apte Split to generate a single-topic split, in which articles with a single topic and with the text tag not empty were assigned for training, testing and unlabeled sets. For the training and unlabeled sets, we used articles with attributes: lewissplit = "train" and topics = "yes" of the Reuters tag; and for testing set used articles with attributes lewissplit = "test" and topics = "yes". Considering only articles with this single-topic split and the categories which still have at least one train and one test article, we have 8 of the 10 most frequent categories. In this single-topic split we used 4936 articles for training, 2189 articles for unlabeled set and 549 articles for testing. Table 1 presents the distribution of articles per most frequent categories, the number of training and testing articles and the total quantity of articles.

Table 1. Number of training, unlabeled and testing articles from Reuters' most frequent categories

Categories	# Train	# Unlabeled	# Test	# Total
grain	37	10	4	51
trade	226	75	25	326
money-fx	185	87	21	293
crude	228	121	25	374
interest	171	81	19	271
ship	97	36	11	144
acq	1436	696	160	2292
earn	2556	1083	284	3923
Total	4936	2189	549	7674

5.1 Performance Criteria

The simulation results were evaluated using Accuracy, Recall, Precision and F measures, which were computed in the unlabeled and testing sets for each category. The information used was:

TP – is the number of correct predictions of a positive article.
TN – is the number of correct predictions of a negative article.
FP – is the number of incorrect predictions of a positive article.
FN – is the number of incorrect predictions of a negative article.

Accuracy is the percentage of correct classifications obtained (4).

$$Accuracy = \frac{TP + FP}{TP + TN + FP + FN} \tag{4}$$

Recall is the percentage of total documents for the given topic that are correctly classified (5).

$$Re\,call = \frac{TP}{TP + FN} \tag{5}$$

Precision is the percentage of predicted documents for the given topic that are correctly classified (6).

$$Precision = \frac{TP}{TP + TN} \tag{6}$$

F measures is the weighted harmonic mean of precision and recall (7).

$$F = \frac{2 * Precision * Recall}{Precision + Recall} \tag{7}$$

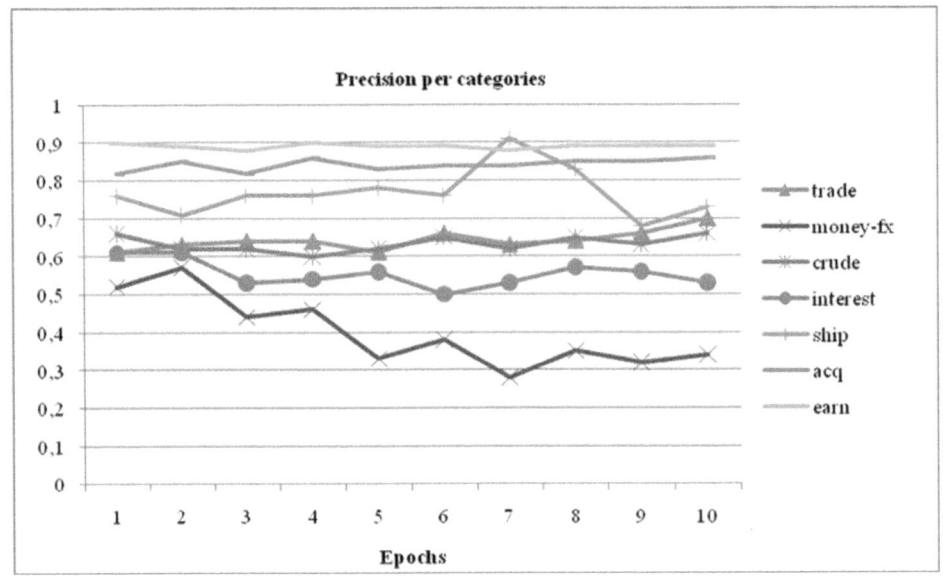

Fig. 7. The categories precision measure evolution across 10 epochs

A confusion matrix [23] contains information about actual and predicted classifications done by a classification system. Performance of such systems is commonly evaluated using the data in the matrix. Table 2 shows the confusion matrix for the eight most frequent categories classifier.

Table 2. Confusion Matrix of the eight categories

	grain	trade	money-fx	crude	interest	ship	acq	earn
grain	0	0	2	0	0	2	0	0
trade	0	24	1	0	0	0	0	0
money-fx	0	1	16	0	3	0	0	1
crude	0	2	1	20	0	0	1	1
interest	0	1	10	0	8	0	0	0
ship	0	0	0	0	0	11	0	0
acq	1	4	5	1	0	1	124	24
earn	8	2	11	9	4	1	18	231

6 Conclusion

In this study, we propose to use a combination [9] of Expectation-Maximization (EM) [10] and a Naïve Bayes classifier on a grid environment. The semi-supervised algorithm implemented as a grid service, is an extension of the Aîuri Academic Portal.

Using this portal on grid, we can explore distributed services across grid environments, categorize a large collection of textual documents, extract consistent knowledge, visualize the results and set the parameters to allow better understanding of the behavior of different knowledge discovery tasks.

References

1. Salton, G., McGill, M.J.: Introduction to Modern Information Retrieval. McGraw-Hill Book Company, New York (1983)
2. Baeza-Yates, R., Ribeiro-Neto, B.: Modern Information Retrieval. ACM Press Books, New York (1999)
3. Kao, A., Poteet, S.R.: Natural Language Processing and Text Mining. Springer, Heidelberg (2007)
4. Han, J., Kamber, M.: Data Mining: Concepts and Techniques. Morgan Kaufmann, San Francisco (2001)
5. Mitchell, T.M.: Bayesian Learning. In: Machine Learning, ch. 6, pp. 154–200. McGraw-Hill, New York (1997)
6. Joachimes, T.: A probabilistic analysis of the Rocchio algorithm with TFIDF for text categorization. In: Proc. of the 14th Int. Conf. on Machine Learning, pp. 143–151 (1997)
7. Lewis, D., Ringuette, M.: A comparison of two learning algorithms for text categorization. In: 3rd Annual Symposium on Document Analysis and Information Retrieval, pp. 81–93 (1994)
8. McCallum, A., Nigam, K.: A comparison of events models for Naive Bayes text classification. In: AAAI 1998 Workshop on Learning of Text Categorization, pp. 41–48. AAAI Press, Menlo Park (1998)
9. Ningam, K., McCallum, A., Thrun, S., Mitchell, T.: Text classification from labeled and unlabeled documents using EM. Machine Learning 39(2/3), 103–134 (2000)
10. Dempster, A.P., Laird, N.M., Rubin, D.B.: Maximum likelihood from incomplete data via the EM algorithm. Journal of the Royal Statistic Society, Series B 39(1), 1–38 (1977)
11. Yang, Y., Pedersen, J.O.: A comparative study on feature selection in text categorization. In: Proc. of the 14th Int. Conf. on Machine Learning, pp. 412–420. Morgan Kaufmann Publishers, San Francisco (1997)
12. Mitchell, T.M.: Machine Learning. McGraw-Hill, New York (1997)
13. Yang, Y.: An evaluation of statistical approaches to text categorization. Journal of Information Retrieval 1, 67–88 (1999)
14. Joachims, T.: Text categorization with Support Vector Machines: learning with many relevant features. In: Nédellec, C., Rouveirol, C. (eds.) ECML 1998. LNCS, vol. 1398, pp. 137–142. Springer, Heidelberg (1998)
15. Duda, R., Hart, P., Stork, D.: Pattern Classification. Wiley Interscience, Hoboken (2001)
16. Bilmes, J.A.: A gentle tutorial of the EM algorithm and its application to parameter estimation for gaussian mixture and hidden markov models. Technical Report TR-97-021. In: International Computer Science Institute, University of California, Berkeley (1998)
17. McLachlan, G.J., Krishnan, T.: The EM algorithm and extensions. John Wiley & Sons, New York (1997)
18. Qi, L., Jin, H., Foster, I., Gawor, J.: HAND: Highly Available Dynamic Deployment Infrastructure for Globus Toolkit 4. In: Proc. of the 15th Euromicro Int. Conf. on Parallel, Distributed and Network-Based Processing, pp. 155–164 (2007)

19. Foster, I., Kesselman, C., Tuecke, S.: The Anatomy of the Grid: Enabling Scalable Virtual Organizations. International Journal of Supercomputer Applications 15(3) (2001)
20. The Globus Toolkit, http://www.globus.org/toolkit/
21. Serpa, A.A., Roncero, V.G., Costa, M.C.A., Ebecken, N.F.F.: Text Mining Grid Services for Multiple Environments. In: Palma, J.M.L.M., Amestoy, P.R., Daydé, M., Mattoso, M., Lopes, J.C. (eds.) VECPAR 2008. LNCS, vol. 5336, pp. 576–587. Springer, Heidelberg (2008)
22. Lewis, D.D.: Reuters-21578 - Distribution 1.0 (2004),
    http://www.daviddlewis.com/resources/testcollections/reuters21578/
23. Kohavi, R., Provost, F.: Glossary of Terms. In: Editorial for the Special Issue on Applications of Machine Learning and the Knowledge Discovery Process, Machine Leanring, vol. 30(2/3) (1998)

On the Vectorization of Engineering Codes Using Multimedia Instructions

Manoel Cunha[1], Alvaro Coutinho[2], and J.C.F. Telles[2]

[1] Federal University of Paraná, Brazil
manoel@ufpr.br
[2] Federal University of Rio de Janeiro, Brazil
alvaro@nacad.ufrj.br

Abstract. After years dominating high performance computing, expensive vector computers were gradually replaced by more affordable solutions and the use of vectorization techniques once applied to many scientific codes also faded. This paper addresses the vectorization of engineering codes using Streaming SIMD Extensions (SSE) also known as multimedia instructions. This particular kind of vectorization differs from the old vectorization techniques in the sense that it relies on hardware features and instruction sets only present on modern processors. Evolving from Intel MMX, SSE is the fifth generation of an established technology highly suited to handle computing intensive tasks like encryption/decryption, audio and video compression, also including digital signal processing but not so well explored for scientific computing, specially among engineering programmers. To demonstrate the benefits of vector/SIMD computing and its implementation on existing scalar algorithms, the authors present this technique applied to an engineering program for the solution of two dimensional elastostatic problems with the boundary element method. Taking an application from a well-know reference on the area, the paper focus on the programming techniques and addresses common tasks used in many other codes, like Gauss integration and equations system assembling. Thus, the vectorization guidelines provided here may also be extended to solve many other types of problems, using other numerical methods as well as other multimedia instruction set extensions. Numerical experiments show the effectiveness of the proposed approach.

1 Introduction

Over decades, since the invention of the first computers, hardware and software resources have been created or modified to follow the increasing complexity of engineering and scientific problems. During many years, vector computers dominated high performance computing but its technologies have been recently superseded by more affordable architectures. In the current world of off-the-shelf-built clusters and multi-core blades, only two major manufacturers, NEC and Cray, still offer expensive solutions. Nowadays, vector computers share less than 1% of the high performance computer market.[1]

---

[1] www.top500.org

J.M.L.M. Palma et al. (Eds.): VECPAR 2010, LNCS 6449, pp. 263–270, 2011.

Despite the fading use of such computers the benefits of vector computing were not forgotten by hardware engineers who brought the old technology into the new processors present on today's clusters, servers, desktops and notebooks. The first step in this direction came in the late 90's with instructions sets such as Intel MMX, AMD 3D-Now and Apple Altivec. Originally developed for multimedia applications, these Single-Instruction-Multiple-Data (SIMD) instructions quickly evolved into powerful extensions highly suited to applications processing data streams such as encryption/decryption, audio and video compression, digital image and signal processing, among others. Today, Streaming SIMD Extensions (SSE) represent an established technology with significant impact in processor performance.

Unfortunately, developers of engineering codes still do not take full advantage of the resources offered by modern processors and compilers, such as auto-vectorization or explicit language instructions. This fact has called the attention of the present authors that address this work to engineering programmers willing to speed their applications with the use of multimedia instructions. Here, this vectorization technique is applied to a well known boundary element application to solve two-dimensional elastostatic problems, although the concepts can also be implemented to other kind of problems and alternative numerical methods.

The present text is organized as follows: the next section presents an outline of the boundary element theory and the following section describes the selected application. Section 4 introduces the Streaming SIMD Extensions while Section 5 details the SSE implementation of the code. In Section 6 a performance analysis is presented. The paper ends with a summary of the main conclusions.

2   Outline of the Boundary Element Theory

The boundary element method (BEM) [1] is a technique for the numerical solution of partial differential equations with initial and boundary conditions.

Using a weighted residual formulation, Green's third identity, Betty's reciprocal theorem or some other procedure, an equivalent integral equation can be obtained and converted to a form that involves only surface integrals performed over the boundary. The bounding surface is then divided into elements and the original integrals over the boundary are simply the sum of the integrations over each element, resulting in a reduced dense and non-symmetric system of linear algebraic equations.

The discretization process involves selecting nodes on the boundary, where unknown values are considered. Interpolation functions relate such nodes to the approximated displacements and tractions distributions on the respective boundary elements. The simplest case positions a node in the center of each element and defines an interpolation function that is constant over the entire element. For linear 2-D elements, nodes are placed at, or near, the end of the elements and the interpolation function is a linear combination of the two nodal values. High-order elements, quadratic or cubic, can be used to better represent curved boundaries using three and four nodes, respectively.

Once the boundary solution has been obtained, interior point results can be computed directly from the basic integral equation in a post-processing routine.

3   The Application

The program reviewed here is a well-known code presented by Telles [1] for the solution of two dimensional elastostatic problems using linear boundary elements.

The INPUT routine reads the program data, MATRX routine generates the equation system, while the OUTPT routine prints the boundary solution, computes and prints boundary stresses and internal displacements and stresses.

The solver is usually the most time consuming routine in BEM programs. High-performance solvers are available from standard libraries and LAPACK's solver SGESV [2] is used here.

However, the generation of the equations system as well as the computing of internal points together can take the most part of the processing time [10] and demand special care. Those two procedures are usually implemented by the researchers and greatly limit the speedup if not properly optimized. Hence, the vectorization programming techniques are here applied to the generation of the system of equations and the evaluation of internal point results, since they can be implemented following the same techniques.

4   The Streaming SIMD Extensions

Computers have been originally classified by Flynn's taxonomy [7] according to instructions and data streams as Single-Instruction-Single-Data (SISD), Single-Instruction-Multiple-Data (SIMD), Multiple-Instruction-Single-Data (MISD) and Multiple-Instruction-Multiple-Data (MIMD).

As the name suggests, the SIMD model applies to systems where a single instruction processes a vector data set, instead of scalar operands.

One of the first SIMD implementations was the MMX technology introduced with Pentium computers intended to enhance the performance of multimedia applications. The 57 MMX instructions can process simultaneously 64-bit data sets of integer type.

With the release of the Pentium III processor an instruction set called Streaming SIMD Extensions (SSE) was added to Intel 32-bit architecture. SIMD extension comprises 70 instructions, 12 for integer operations and 50 for single-precision floating-point operations. The remaining 8 instructions are for cache control and data prefetch.

While MMX operates on integers and SSE are essentially single-precision floating-point instructions, the SSE2 set introduced with Pentium 4 added support for double-precision (64-bits) floating-point operations and extended MMX integer instructions from 64 to 128 bits. The 144 instructions implemented by SSE2 were followed by 13 SSE3 instructions, that also support complex floating-points operations.

4.1 Auto-vectorization Compilers

The first and quickest way to implement Streaming SIMD Extensions is auto-vectorization, a feature present on most recent compilers such as Intel Fortran and C++.

The SSE technology is now also incorporated by Fortran and C/C++ compilers, which offer compiler options to generate vectorized SSE code automatically. Once SSE compiler options are set, the compiler will search the code for vectorization opportunities, automatically replacing scalar operations by vector instructions whenever possible.

The implementation of auto-vectorization techniques in the application here considered is object of a previous work [11] and is not discussed here.

4.2 Compiler Intrinsics

Auto-vectorization compiler options are available in Fortran as well as in C/C++ compilers. While Fortran users must rely on the compiler ability to generate SSE executables, the C/C++ language alternatively allows to insert explicit vector functions in the code, giving programmers more control to the vectorization process.

C/C++ compiler intrinsics provide the user with new data types and a set of vector functions. Thus, one benefit of SSE intrinsics is the use of C/C++ syntax of function calls and variables instead of assembly instructions and hardware registers. Intrinsics are expanded inline to eliminate call overhead.

Vector addition, subtraction, multiplication and division can be performed using the intrinsic functions _mm_add_ps, _mm_sub_ps, _mm_mul_ps and _mm_div_ps, respectively. These functions perform one operation on two sets of four floating-point single-precision values, simultaneously, as illustrated in Fig. 1.

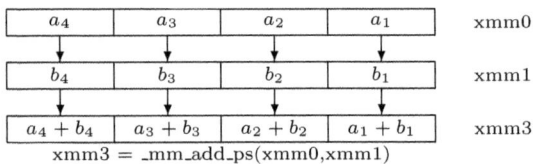

$$xmm3 = \_mm\_add\_ps(xmm0,xmm1)$$

Fig. 1. SSE Packed Addition

SSE does not include instructions to evaluate sines, cosines, logarithms and other trigonometric functions. To bypass this limitation, vector implementations make calls to the Short Vector Math Library (SVML), an Intel library intended for use by the Intel compiler vectorizer.

In this section only a few instructions are presented. However, SSE provides a large set of vector operations. For a full description of all SSE instruction set the reader is referred to Bik [3,4]. A more extensive C/C++ vector implementation of the original Fortran code with SSE compiler intrinsics will be addressed in the next section.

5   An SSE Implementation

In the application under study, an equation system is generated in routine MATRX with its influence coefficients computed by subroutine FUNC. This routine evaluates all the non-singular element integrals using Gauss integration. For elements with the singularity at one of its extremities the required integrals are computed analytically. In the first case, a set of small matrix operations are initially computed, as follows:

$$
\begin{bmatrix} UL_{11} & UL_{12} \\ UL_{21} & UL_{22} \end{bmatrix} = -C1 \left[\begin{bmatrix} C2\ logR & 0 \\ 0 & C2\ logR \end{bmatrix} - \begin{bmatrix} DR_{11} & DR_{12} \\ DR_{21} & DR_{22} \end{bmatrix} \right]
$$

Those 2x2 matrices can be converted into vectors of size 4 and matrix operations can be performed with vector instructions. Thus, a very simple approach is to use SSE to evaluate those matrices leaving some intermediate operations to be executed with scalar instructions.

In the original algorithm, those matrices are computed from 2 to 6 times, accordingly to the number of Gauss integration points defined by an empiric formula. Alternatively, a fully vector implementation of the matrix computation above can be achieved by using 4 Gauss integration points and evaluating all four values of each coefficient at once, including the intermediate values.

A possible SSE implementation of the vector computation discussed is presented in Listing 1.

Listing 1. An SSE implementation

```
xmm0 = _mm_set_ps1(DXY[0]);   //                        DXY1
xmm1 = _mm_set_ps1(DXY[1]);   //                        DXY2
xmm2 = _mm_load_ps(CTEv4);    //          .5 * (XI + 1)
xmm0 = _mm_mul_ps(xmm0,xmm2); //          .5 * (XI + 1) * DXY1
xmm1 = _mm_mul_ps(xmm1,xmm2); //          .5 * (XI + 1) * DXY2
xmm3 = _mm_set_ps1(X[II]-XS); //                              X[II] - XS
xmm4 = _mm_set_ps1(Y[II]-YS); //                              Y[II] - YS
xmm0 = _mm_add_ps(xmm0,xmm3); // XMXY = .5 * (XI + 1) * DXY1 + X[II] - XS
xmm1 = _mm_add_ps(xmm1,xmm4); // YMYI = .5 * (XI + 1) * DXY2 + Y[II] - YS
xmm2 = _mm_mul_ps(xmm0,xmm0); // XMXI^2
xmm3 = _mm_mul_ps(xmm1,xmm1); //                  YMYI^2
xmm2 = _mm_add_ps(xmm2,xmm3); // XMXI^2 + YMYI^2
xmm2 = _mm_sqrt_ps(xmm2);     // R = sqrt(XMXI^2 + YMYI^2)
xmm0 = _mm_div_ps(xmm0,xmm2); // DR1 = XMXI / R
xmm1 = _mm_div_ps(xmm1,xmm2); // DR2 = YMYI / R
xmm6 = _mm_set_ps1(BN[0]);    //               BN1
xmm7 = _mm_set_ps1(BN[1]);    //                        BN2
xmm3 = _mm_mul_ps(xmm0,xmm6); //          DR1 * BN1
xmm4 = _mm_mul_ps(xmm1,xmm7); //                   DR2 * BN2
xmm5 = _mm_mul_ps(xmm0,xmm1); //       UL12 = DR1 * DR2
xmm3 = _mm_add_ps(xmm3,xmm4); // DRDN = DR1 * BN1 + DR2 * BN2
xmm6 = _mm_mul_ps(xmm6,xmm1); //                              DR2 * BN1
xmm7 = _mm_mul_ps(xmm7,xmm0); //                              DR1 * BN2
_mm_store_ps(ul12v4,xmm5);
xmm0 = _mm_mul_ps(xmm0,xmm0); //          DR1 * DR1
xmm1 = _mm_mul_ps(xmm1,xmm1); //          DR2 * DR2
xmm5 = _mm_add_ps(xmm5,xmm5); //       2 * DR1 * DR2
xmm7 = _mm_sub_ps(xmm7,xmm6); //                        DR1 * BN2 - DR2 * BN1
xmm4 = vmlsLn4(xmm2);         //                  log R
xmm5 = _mm_mul_ps(xmm5,xmm3); //       2 * DR1 * DR2 * DRDN
_mm_store_ps(ul11v4,xmm0);
_mm_store_ps(ul22v4,xmm1);
```

For each integration point i, **UL** is used to compute another matrix, **G**, as follows:

$$
\begin{bmatrix} G_{11} & G_{12} & G_{13} & G_{14} \\ G_{21} & G_{22} & G_{23} & G_{24} \end{bmatrix} = \begin{bmatrix} G_{11} & G_{12} & G_{23} & G_{24} \\ G_{21} & G_{22} & G_{23} & G_{24} \end{bmatrix} + \left[\begin{bmatrix} UL_{11}^{i} & UL_{12}^{i} \\ UL_{21}^{i} & UL_{22}^{i} \end{bmatrix} * B_1^i \begin{bmatrix} UL_{11}^{i} & UL_{12}^{i} \\ UL_{21}^{i} & UL_{22}^{i} \end{bmatrix} * B_2^i \right] * W^i
$$

Each one of the 2x4 matrices above can be splitted into two 2x2 matrices, as indicated below:

$$\begin{bmatrix} G_{11} & G_{12} \\ G_{21} & G_{22} \end{bmatrix} = \begin{bmatrix} G_{11} & G_{12} \\ G_{21} & G_{22} \end{bmatrix} + \begin{bmatrix} UL_{11}^i & UL_{12}^i \\ UL_{21}^i & UL_{22}^i \end{bmatrix} * B_1^i * W^i$$

Since all values of **UL** are stored in vectors, it is quite simple to perform the multiplications of each value by the respective four values stored in B_1 and W.

However, there is no SSE instruction to perform the sum of the elements of a vector needed in the computation of **G**.

Using the SSE unpack and move instructions, the values stored on four vectors can be reordered to obtain the same effect of a matrix transposition, although here the operations are performed on vectors.

Listing 2 shows how SSE unpack and move instructions can be used in the computation of one half of matrix **G**.

Listing 2. SSE matrix transposition

```
// computing G1
xmm7 = _mm_set_ps1(C);
xmm6 = _mm_set_ps1(C1);
xmm6 = _mm_mul_ps(xmm6,xmm7);
xmm5 = _mm_load_ps(B1Wv4);
xmm5 = _mm_mul_ps(xmm5,xmm6);
xmm0 = _mm_load_ps(ul11v4);
xmm1 = _mm_load_ps(ul22v4);
xmm2 = _mm_load_ps(ul21v4);
xmm3 = _mm_load_ps(ul12v4);
xmm0 = _mm_mul_ps(xmm0,xmm5);
xmm1 = _mm_mul_ps(xmm1,xmm5);
xmm2 = _mm_mul_ps(xmm2,xmm5);
xmm3 = _mm_mul_ps(xmm3,xmm5);
xmm4 = _mm_unpackhi_ps(xmm0,xmm1);
xmm5 = _mm_unpackhi_ps(xmm2,xmm3);
xmm6 = _mm_unpacklo_ps(xmm0,xmm1);
xmm7 = _mm_unpacklo_ps(xmm2,xmm3);
xmm0 = _mm_movelh_ps(xmm6,xmm7);
xmm1 = _mm_movehl_ps(xmm7,xmm6);
xmm2 = _mm_movelh_ps(xmm4,xmm5);
xmm3 = _mm_movehl_ps(xmm5,xmm4);
xmm0 = _mm_add_ps(xmm0,xmm2);
xmm1 = _mm_add_ps(xmm1,xmm3);
xmm0 = _mm_add_ps(xmm0,xmm1);
_mm_store_ps(v4G1,xmm0);
```

Well-known optimization techniques, usually applied to scalar codes, can also be used in the implementation of vector algorithms in order to replace long latency instructions and to reduce data dependence. High performance techniques are presented by the authors in previous works [8,9] and will not be addressed here.

6   Results Summary

The implementation evaluated here runs on a SGI Altix XE 1200 cluster with 11 nodes, comprising 8 Quad-Core Xeon 2.66GHz (X5355) processors and 8 GB memory per node. The operating system is Linux SLES 10.1, ProPack 5 SP3 and Intel C/C++ 10 compiler is used.

The case study corresponds to a square plate under biaxial load, as found in [1], with nodes distributed along the boundary.

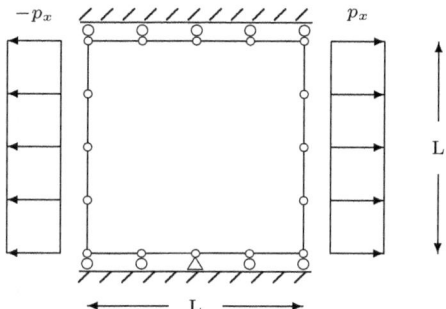

Fig. 2. A square plate under biaxial load

To evaluate the performance of the techniques presented here, a 10.000 boundary nodes study case was used. This particular discretization yields a 20.000 square dense nonsymmetric system of linear equations.

In its original version the program spent 45.822 seconds to run while the executable version generated with autovectorization compiling options took 32.918 seconds to complete the same task. A fully vectorized implementation, using vector intrinsic functions solved the same problems within a runtime of 12.88 seconds.

7   Conclusions

This paper introduces the Streaming SIMD Extensions (SSE), also known as multimedia instructions, and its application to engineering codes. The SSE instruction set enhanced the Intel IA-32 architecture with instructions that handle a set of floating-point values stored in vectors, simultaneously, instead of scalar variables. These vector operations can enhance the performance of modern processors significantly.

In the first part of the work [11] auto-vectorization techniques were presented. Here, explicit vector/SIMD instructions or compiler intrinsics are addressed in some detail and its use is demonstrated in a numerical application to solve two-dimensional elastostatic problems. The implementation proposed illustrates the basic concepts underlying SSE and provides guidelines to generate vector executables with C/C++ compiler intrinsics. The techniques presented are applied to a boundary element code but other methods can equally be addressed with the same techniques.

The results show a reduction in the runtime of 30% using auto-vectorization techniques while the implementation with SSE intrinsics yields a reduction of over 70% when compared to the original code.

SIMD extensions are currently found in most current processors, hence the knowledge of SIMD programming appears to be a decisive factor in the future of high performance computing [5,6]. The implementation of boundary element codes on the IBM Cell Broadband Engine processor using SIMD instructions has been presented in a previous work [12]. Intel next microarchitecture will include Advance Vector Extensions (AVX), supporting 256-bit wide vector registers.

Thus, the procedures introduced here arise as an additional and important optimization tool for numerical applications on today's and future processor architectures.

References

1. Brebbia, C.A., Telles, J.C.F., Wrobel, L.C.: Boundary elements techniques: theory and applications in engineering. Springer, Berlin (1984)
2. Dongarra, J., et al.: LAPACK users guide, 3rd edn. SIAM, Philadelphia (1999)
3. Bik, A.J.C.: Software vectorization handbook. Intelpress, Hillsboro (2004)
4. Gerber, R., Bik, A.J.C., Smith, K.B., Tian, X.: The software optimization cookbook, 2nd edn. Intel Press, Hillsboro (2006)
5. Patterson, D.A., Hennessy, J.: Computer organization and design: the hardware/software interface, 3rd edn. Elsevier-Morgan Kaufmann (2007) (revised)
6. Hennessy, J.L., Patterson, D.: Computer arquitecture: a quantitative approach, 4th edn. Elsevier-Morgan Kaufmann (2007)
7. Flynn, M.J.: Very high-speed computing systems. Proc. IEEE 54(12), 1901–1909 (1966)
8. Cunha, M.T.F., Telles, J.C.F., Coutinho, A.L.G.A.: On the parallelization of boundary element codes using standard and portable libraries. Engineering Analysis with Boundary Elements 28(7), 893–902 (2004), doi:10.1016/j.enganabound.2004.02.002
9. Cunha, M.T.F., Telles, J.C.F., Coutinho, A.L.G.A.: A portable implementation of a boundary element elastostatic code for Shared and Distributed Memory Systems. Advances in Engineering Software 37(7), 893–902 (2004), doi:10.1016/j.advengsoft.2004.05.007
10. Cunha, M.T.F., Telles, J.C.F., Coutinho, Á.L.G.A.: Parallel boundary elements: A portable 3-D elastostatic implementation for shared memory systems. In: Daydé, M., Dongarra, J., Hernández, V., Palma, J.M.L.M. (eds.) VECPAR 2004. LNCS, vol. 3402, pp. 514–526. Springer, Heidelberg (2005)
11. Cunha, M.T.F., Telles, J.C.F., Ribeiro, F.L.B.: Streaming SIMD extensions applied to boundary element codes. Advances in Engineering Software (2008), doi: 10.1016/j.advengsoft, 01.003
12. Cunha, M.T.F., Telles, J.C.F., Coutinho, A.L.G.A.: On the implementation of boundary element engineering codes on the Cell Broadband Engine. In: Palma, J.M.L.M., Amestoy, P.R., Daydé, M., Mattoso, M., Lopes, J.C. (eds.) VECPAR 2008. LNCS, vol. 5336. Springer, Heidelberg (2008)

Numerical Library Reuse in Parallel and Distributed Platforms

Nahid Emad, Olivier Delannoy, and Makarem Dandouna

PRiSM Laboratory, University of Versailles
45, avenue des États-unis, 78035 Versailles cedex, France

Abstract. In the context of parallel and distributed computation, the currently existing numerical libraries do not allow code reuse. Besides, they are not able to exploit the multi-level parallelism offered by many numerical methods. A few linear algebra numerical libraries make use of object oriented approach allowing modularity and extensibility. Nevertheless, those which offer modularity together with sequential and parallel code reuse are almost non-existent. We analyze the lacks in existing libraries and propose a design model based on a component approach and the strict separation between computation operations, data definition and communication control of applications. We present then an implementation of this design using YML scientific workflow environment jointly with the object oriented LAKe (Linear Algebra Kernel) library. Some numerical experiments on GRID5000 platform validate our approach and show its efficiency.

Keywords: large scale distributed systems, numerical library, code reusability, design model.

1 Introduction

To solve linear algebra problems on large scale distributed systems an application can rely on existing libraries such as LAPACK[1] which provides a set of routines that can be used to create solvers. Parallel solvers for distributed memory architectures can be built on top of the services provided by sequential libraries. The approach consists in building the parallel version by using distributed versions of the basic operations used in the library. For example in LAPACK the parallelization is done by the parallelization of BLAS. Nevertheless, these libraries allow neither data type abstraction nor code reuse between the parallel and sequential versions of the applications. That means the subroutines of the solvers are not able to adapt their behaviors depending on the data types. Those subroutines must be defined once for use in sequential and once again in parallel. The component approach used in libraries such as PETSc (Portable, Extensible Toolkit for Scientific Computation)[2] or Trilinos [6] increased drastically the modularity, interoperability and reusability of high level components within the libraries as well as in the user applications. It increases the ease of use, code reuse, and maintainability of libraries. Nevertheless, it doesn't allow scalable sequential/parallel

J.M.L.M. Palma et al. (Eds.): VECPAR 2010, LNCS 6449, pp. 271–278, 2011.

code reusability. The Linear Algebra Kernel (LAKe)[9] is an object oriented library in C++ which makes use of MPI. It introduces code reuse between the sequential and the parallel versions of an application. However LAKe doesn't allow to use concurrently the parallel and sequential versions of a code inside the same application. This feature is required to build hybrid numerical methods in the context of distributed computing. These methods are defined by a set of collaborating classical iterative methods called co-methods. Each co-method aims at decreasing the number of iterations required by the method to compute its results. An extended version of LAKe proposed in [4] allows it to support the hybrid methods. Nevertheless, as in PETSc, the scalability of the reusability offered by this extension is limited.

In this paper, we propose a design model for numerical libraries allowing their reuse on parallel and distributed systems. Our approach is based on three levels of abstraction concerning computation aspect, data definition and communication control of an application. The simultaneous reusability between the sequential and the parallel codes is possible thanks to this abstraction. We show that our design can be mapped on some scientific workflow environments. We present then the implementation of our approach using YML scientific workflow environment (http://yml.prism.uvsq.fr/) jointly with LAKe library. We will see that the approach makes possible to exploit the hybrid methods in the context of large scale distributed systems. Finally, we give the results of some experiments in order to validate our solution.

2   Linear Algebra Libraries

2.1   Imperative Numerical Libaries

In order to implement numerical solvers, one can use libraries such as LAPACK [1] and ARPACK[7] written in FORTRAN using a traditional imperative programming style. They consist in a set of routines which provides the individual steps of the iterative methods. Parallel solvers for distributed memory architectures can be built on top of the services provided by the aforementioned libraries. This approach has been used to build libraries such as ScaLAPACK[3] and P_ARPACK[8]. The parallel solvers exploit intra co-method parallelism. Nevertheless, these libraries allow neither data type abstraction nor code reuse between the parallel and sequential versions of the applications. That means the subroutines of the solvers are not able to adapt their behaviors depending on the data types. Consequently, those subroutines must be defined once for use in sequential and once again in parallel and then the application code is different if using the parallel or sequential library.

2.2   Object Oriented Numerical Libraries

The object oriented approach used in libraries such as PETSc [2] or Trilinos [6] enforced drastically the modularity, interoperability and reusability of high level components within the libraries as well as in the user applications. Using PETSc or Trilinos, the application specifies the building blocks of the solver. However the solver is provided by the library which provides parallel and sequential solvers and allows to make use of one and/or the other in the same application. However these parallel and sequential solvers are still limited by their scalability in the context of hybrid methods.

LAKe is an object oriented library written in C++. It defines a framework to implement iterative solvers. The design approach of this library is based on a strict separation between the computation part, which is composed by numerical algorithms and services, and the data management and communication part of the application code. The latter are used to represent both sequential and parallel data type used by LAKe computation part. Using the object oriented approach and template based generic programming provided by C++, LAKe allows the computation part to be common to both sequential and parallel versions of the application. The computation part of the library is identical in the case of a sequential data set or distributed data set. The parallel version of LAKe makes use of the message passing interface (MPI) standard version 1 [5] for communication between the various involved computation processes. However the use of MPI is completely transparent to the user. He/she can switch from a sequential solver to a parallel one by changing the type of the matrix representing the data. LAKe achieves code reuse between sequential and parallel versions thanks to a strict separation between the computation and the data/comunication management aspects of applications.

The intra-method communication of the parallel version of LAKe increases the time performance when handling huge matrices. However LAKe is not suitable to implement hybrid methods. For hybrid methods we need two levels of parallelism. Using MPI means we need local communication at the co-method level and global communication between the co-methods composing the hybrid method. LAKe provides no access to the MPI communicator to client applications nor to the computation part of the library.

An extension of LAKe has been proposed to support hybrid methods. It makes possible the use of sequential and parallel co-method processes concurrently within the same application. This extension described in [4] discusses a solution that matches the architecture design of LAKe. The user of the library must explicitly define the number of processors allocated to each process representing a co-method. However, its use is not easy due to the configuration of the communicators (with MPI standard version 1). This limits the scalability of the solution proposed to only a few number of co-method processes. Experiments have been done up to three concurrent co-methods.

3 A Reusable Numerical Library Design Model

Most of previously mentioned libraries suffer from many problems. Imperative numerical libraries lack portability, modularity, interoperability. Despite modularity and reusability of their high level components, object oriented libraries such as LAKe do not allow the simultaneous reusability of components between the sequential and the parallel versions of an application. Extended LAKe,

PETSc and Trilinos allow this kind of reusability but it is not scalable. We notice that all these libraries lack an additional level of abstraction which is necessary to achieve such a kind of reusability.

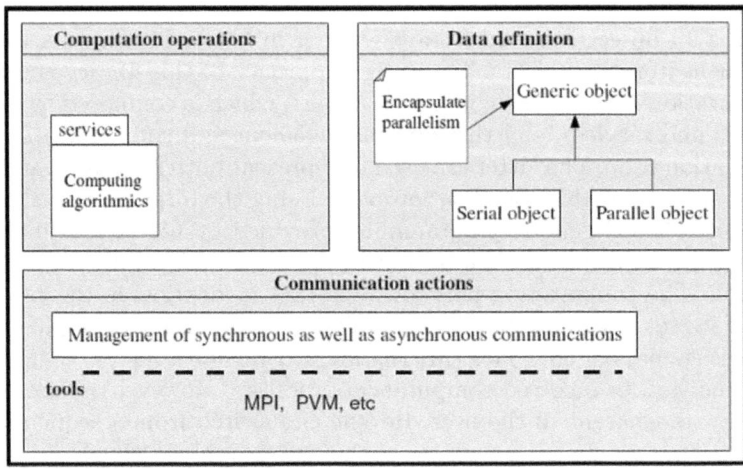

Fig. 1. Reusable numerical libraries design

To remedy to these problems, we propose a library design model based on three levels of abstraction. That means, a model which separates strictly the computation aspect, the data definition and the communication actions of applications (see figure 1). The data definition includes data types abstraction. The computation aspect represents all computation components. These two components communicate through the communication actions. Our main goal is to achieve the simultaneous reusability between sequential and parallel components, so in data definition part we encapsulate the parallelism in a common generic object which has the same interface in parallel and in serial. Then, parallel objects can be used polymorphically. Components of the computation part will be clients of these objects. We want to allow the code to be the same between the sequential and parallel versions of an application. Thereby every function is implemented once and used either in sequential or in parallel. Additionally, the maintainability of the library implemented according to this model would be simplified using this approach.

3.1   Library Integration in Scientific Workflow Environment

A scientific workflow environment describes an application along three aspects: a set of services, the control flow and the data flow of the application. Based on these informations, it orchestrates the execution of the application. A service is a public interface associated to an implementation. The public interface describes how a service interacts with clients. Each service defines a set of input and output parameters also known as communication channels. A service can be stateless or not depending on the underlying middleware capabilities in that respect. The control flow consists in describing the order of execution of the services involved in the application. It does not contain the computation code, only the order of computation. It is a coarse grained description of the application where computation is handled by services as defined above. The data flow consists in describing the exchange of data between the services. Some workflow environment mixes the data flow and control flow together. Data migration from data repositories are managed transparently by the workflow environment. To provide a solution independently from the underlying middleware, that services have to be supposed stateless.

We target a scientific workflow environment which model is defined by three main layers: a layer to interact with end users. A second layer which includes workflow manager and an integrator of services such as databases and computation codes. Finally, there is a layer to interact with middleware. In the environments, based on this model, the user can make use of large scale distributed architectures transparently and independently from the deployed middleware. The computation and data components in the model are represented by some services. The communication between these services would be done by the maddeware. Note that the strict separation between computation aspect, data definition and communication actions required by our library design model matches easily with environments realized according to aforementioned model.

YML is a scientific workflow environment based on the above model [10]. It permits to represent computation and data definitions of our library design model by the corresponding YML components. Besides, it confides the communication actions of our model to the middeware. The activities graph which defines a solver will then be described by YML workflow langage. As a consequence, the solvers are independent from the communication mechanisms used by the middleware (MPI or others). In order to achieve our objective of simultaneous serial and parallel code reusability, we integrate the computation and data components of LAKe library in YML. This solution allows to exploit the multi level parallelism of hybrid methods on parallel and distributed systems.

4   Experiments

For our experiments, we selected two matrices from the MatrixMarket collection. The used matrices are summarized in the table 1. *NNZ* corresponds to the number of non zero elements of the matrix, we also added the Froebenius norm which impacts on the convergence criterion. In order to validate our approach, we evaluated YML/LAKe on the Grid'5000 platform. Grid'5000 is a French national testbed dedicated to large scale distributed system experiments. We make use of computing resources of the Grid Explorer cluster of Grid'5000. We demonstrate the validity of our approach (a) by presenting the feasibility, (b) by decreasing the number of iterations needed to converge when the number of co-methods increases and (c) by showing the scalability of the solution in regards of the

Table 1. List of matrices used for experiments

Matrix name	Size	NNZ	Froebenius norm
pde490000	490000	2447200	10^{+3}
pde1000000	1000000	4996000	10^{+3}

matrix sizes and of the number of co-methods used to solve a large eigenproblem with MERAM (multiply explicitly restarted Arnoldi method).

MERAM is a hybrid method composed of several instances of the same iterative method ERAM. In other words, this method is based on a set of p instances of ERAM. The latter is an iterative method which computes a few eigenvalues and eigenvectors of a large sparse non-Hermitian matrix. The instances of ERAM work on the same problem but they are initialized with different subspace size $m_i, i \in [1, k]$. MERAM defines a set of parameters, the most significant ones are the matrix A, n the size of this matrix, r the number of desired eigenelements, $m_1, .., m_p$ the subspace sizes of the co-methods ERAM composing MERAM also noted MERAM($m_1, ..., m_p$) and *tol* denoting the tolerance expected for the results. An horizontal line denotes the tolerance or the error allowed on the results. The vertical axis represents the estimated error of the solution obtained at each iterations. The horizontal axis represents the number of iterations. In figure 2, we present two executions of MERAM for the matrix pde490000. The two executions differ in the number of involved co-methods. In the first execution, MERAM(10,30,50) requires 98 iterations to converge while MERAM(10, 20, 30, 50) requires 91 iterations. In other words, the increase in the number of co-methods decreases the iteration count of the hybrid method. We notice that by making use of YML/LAKe we are able to overcome the limitation in the number of co-methods composing a hybrid method. One of the motivation of YML/LAKe is the scalability issue in regards of the number of co-methods composing an hybrid one and the size of the problems to be solved. Figure 3 illustrates the progresses made in that regards. Using YML/LAKe we have been

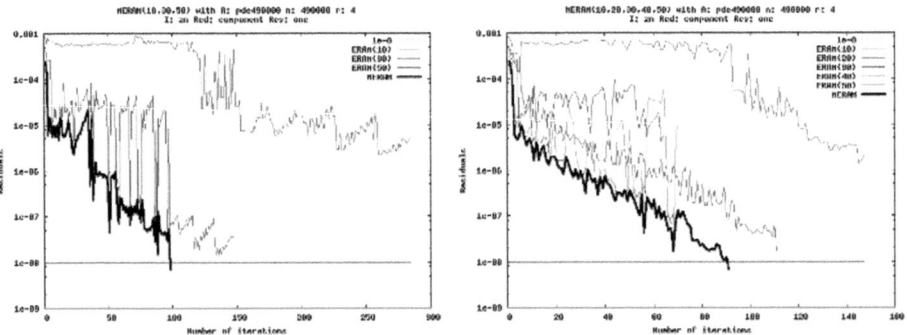

Fig. 2. Convergence of MERAM for matrix 490000 with different number of co-methods

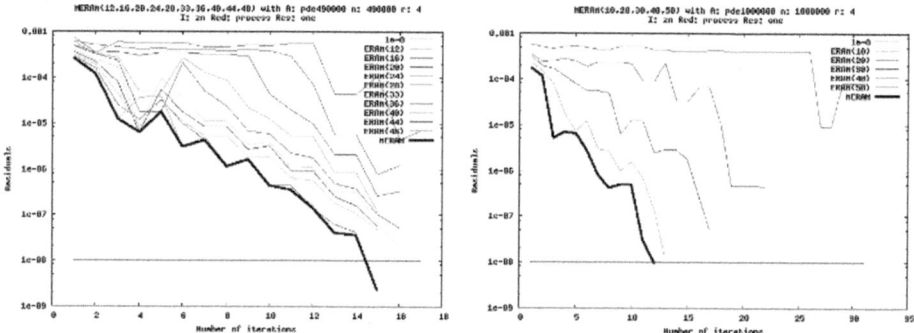

Fig. 3. Scalability of the solution: number of co-methods/size of A

able to solve eigenproblems with one million-order matrices. Our approach is based onto the fragmentation in blocks of the matrix of the problem and its distribution during the projection step of the iterative method. The second scalability issue relates to the number of co-methods used to solve an eigenproblem. Extended LAKe allows to test the hybrid methods composing by only a small number of co-methods (up to 3). Using YML/LAKe we have been able to test effortlessly with ten co-methods and it is possible to increase this number.

5 Conclusion

Hybrid methods and some of linear algebra applications are well adapted to parallel and distributed systems as well as large scale distributed memory architectures such as GRID and peer to peer systems. Such methods require several levels of parallelism in the same application organized in a tree. Existing numerical libraries are not able to exploit all these levels of parallelism. Their use on distributed systems is still difficult and complex. Their design doesn't allow the simultaneous reusability between the sequential and the parallel versions of an application. Moreover, they do not manage effectively communications in these complex environments; they combine communication with the definition of data and computations. We have presented a model to design reusable numerical libraries for parallel and distributed systems. Our approach is based on three levels of abstraction consisting of the computation, the data definition and the communication actions of applications. We involve in our solution the use of scientific workflow environments. These environments provide tools to orchestrate the execution on a set of distributed services and they allow the use of middleware transparently to end users. To solve simultaneous serial and parallel reuse, we proposed to map our library design model on such a scientific workflow environment. We realized this solution by the integration of LAKe library in YML framework.

We validated our approach with experiments using YML/LAKe. Future works will include the application of our approach to some existing libraries such as

PETSc or certain components of Trilinos. This is possible by including some interfaces in the library source code and its integration on a scientific workflow environment such as YML. According to this approach, we can obtain a set of numerical libraries which can cooperate together for the resolution of a problem. Applications developers can use these libraries more easily in the context of large scale distributed systems through the workflow environment.

Acknowledgment

Experiments presented in this paper were carried out using the Grid'5000 experimental testbed, an initiative from the French Ministry of Research through the ACI GRID incentive action, INRIA, CNRS and RENATER and other contributing partners (see `https://www.grid5000.fr/`).

References

1. Anderson, E., Bai, Z., Bischof, C., Blackford, S., Demmel, J., Dongarra, J., Du Croz, J., Greenbaum, A., Hammarling, S., McKenney, A., Sorensen, D.: LAPACK Users' Guide, 3rd edn. Society for Industrial and Applied Mathematics, Philadelphia (1999)
2. Balay, S., Gropp, W.D., McInnes, L.C., Smith, B.F.: Efficient management of parallelism in object oriented numerical software libraries. In: Arge, E., Bruaset, A.M., Langtangen, H.P. (eds.) Modern Software Tools in Scientific Computing, pp. 163–202. Birkhäuser Press, Beijing (1997)
3. Dongarra, J.J., Hammarling, S., Petitet, A.: Case studies on the development of ScaLAPACK and the NAG numerical PVM library, pp. 236–248 (1997)
4. Emad, N., Sedrakian, A.: Toward the reusability for iterative linear algebra software in distributed environment. Parallel Comput. 32(3), 251–266 (2006)
5. Message Passing Interface Forum. MPI: A message-passing interface standard. Technical Report UT-CS-94-230 (1994)
6. Heroux, M.A., Bartlett, R.A., Howle, V.E., Hoekstra, R.J., Hu, J.J., Kolda, T.G., Lehoucq, R.B., Long, K.R., Pawlowski, R.P., Phipps, E.T., Salinger, A.G., Thornquist, H.K., Tuminaro, R.S., Willenbring, J.M., Williams, A., Stanley, K.S.: An overview of the trilinos project. ACM Trans. Math. Softw. 31(3), 397–423 (2005)
7. Lehoucq, R., Sorensen, D., Yang, C.: Arpack users' guide: Solution of large scale eigenvalue problems with implicitly restarted arnoldi methods (1997)
8. Maschhoff, K.J., Sorensen, D.C.: P_ARPACK: An efficient portable large scale eigenvalue package for distributed memory parallel architectures. In: Madsen, K., Olesen, D., Waśniewski, J., Dongarra, J. (eds.) PARA 1996. LNCS, vol. 1184, pp. 478–486. Springer, Heidelberg (1996)
9. Noulard, E., Emad, N.: A key for reusable parallel linear algebra software. Parallel Computing Journal 27(10), 1299–1319 (2001)
10. Emad, N., Delannoy, O., Petiton, S.: Workflow global computing with YML. In: The 7th IEEE/ACM International Conference on Grid Computing, Barcelona, Spain, September 28th-29th (2006)

Improving Memory Affinity of Geophysics Applications on NUMA Platforms Using Minas

Christiane Pousa Ribeiro[1], Márcio Castro[1],
Jean-François Méhaut[1], and Alexandre Carissimi[2]

[1] University of Grenoble
LIG Laboratory - INRIA
Grenoble, France
{pousa,bastosca,Jean-Francois.Mehaut}@imag.fr
[2] Universidade Federal do Rio Grande do Sul
Porto Alegre, Brazil
asc@inf.ufrgs.br

Abstract. On numerical scientific High Performance Computing (HPC), Non-Uniform Memory Access (NUMA) platforms are now commonplace. On such platforms, the memory affinity management remains an important concern in order to overcome the memory wall problem. Prior solutions have presented some drawbacks such as machine dependency and a limited set of memory policies. This paper introduces Minas, a framework which provides either explicit or automatic memory affinity management with architecture abstraction for ccNUMAs. We evaluate our solution on two ccNUMA platforms using two geophysics parallel applications. The results show some performance improvements in comparison with other solutions available for Linux.

1 Introduction

The increasing number of cores per processor and the efforts to overcome the hardware limitations of classical Symmetric Multiprocessors (SMP) parallel systems remain a problem. Due to this, Non-Uniform Memory Access (NUMA) platforms are becoming very common computing resources for numerical scientific High Performance Computing (HPC). A NUMA platform is a large scale multi-processed system in which the processing elements are served by a shared memory that is physically distributed into several memory banks interconnected by a network. Thus, memory access costs may vary depending on the distance between cpus and memory banks. The effects of this asymmetry can be reduced by optimizing memory affinity [1,2].

Memory affinity is assured when a compromise between threads and data is achieved by reducing either the number of remote accesses (latency optimization) or the memory contention (bandwidth optimization). In the past, researches have led to many different solutions on user and kernel space. However, such solutions present some drawbacks, such as: platform dependency (developers must have prior knowledge of the target architecture), they do not address different memory

J.M.L.M. Palma et al. (Eds.): VECPAR 2010, LNCS 6449, pp. 279–292, 2011.

accesses and they do not include optimizations for numerical scientific data (i.e., array data structures) [1,2,3].

To overcome these issues, our research have led to a new solution named Minas: *an efficient and portable framework for managing memory affinity* on cache-coherent NUMA (ccNUMA) platforms. Minas enables explicit and automatic control mechanisms for numerical scientific HPC applications. Beyond the architecture abstraction, this framework also provides several memory policies allowing better memory access control. In this paper, we evaluate its portability and efficiency by performing experiments with two Geophysics applications on two ccNUMA platforms. The results are compared with Linux solutions for ccNUMAs (*first-touch*, *numactl* and *libnuma*).

This paper is organized as follows: first, we discuss the related work (Section 2). After presenting the Minas design, its characteristics and implementation details (Section 3), we will show its performance evaluation (Section 4). We will then give a brief conclusion and present our future work (Section 5).

2   Related Work

In order to guarantee memory affinity and thus achieve better performance, developers usually spend significant time optimizing data allocation and placement on applications and ccNUMA platforms. As a consequence, research groups have studied different ways to simplify memory affinity management on such platforms using Linux [2]. Two approaches have been proposed for the Linux operating system, the explicit approach (libraries, interfaces and tools) and the automatic approach (memory policies in user or kernel spaces) [3,4,5,6].

On the Linux operating system, the explicit approach is a basic support to manage memory affinity on ccNUMAs which is composed of three parts: kernel/system calls, a library (*libnuma*) and a tool (*numactl*). The kernel part defines three system calls (*mbind()*, *set_mempolicy()* and *get_mempolicy()*) that allow the programmer to set a memory policy (bind, interleave, preferred or default) for a memory range. A memory policy is responsible for placing memory pages on physical memory banks of the machine. The use of such system calls is a complex task, since developers must deal with pointers, memory pages, sets of bytes and bit masks. The second part of this support is a library named *libnuma*, which is a wrapper layer over the kernel system calls. The limited set of memory policies provided by *libnuma* is the same as the one provided by the system calls. The last part, the *numactl* tool, allows the user to set a memory policy for an application without changing the source code. However, the chosen policy is applied over all application data (it is not possible to either express different access patterns or change the policy during the execution [3]). Additionally, providing a list of nodes (memory banks and cpus/cores), that are platform-dependent parameters, is mandatory when using this tool.

The automatic approach is based on the use of memory policies and it is the simplest way to deal with memory affinity, since developers do not have to take into consideration the memory management. In this approach, the operating system is responsible for optimizing all memory allocation and placement. *First-touch* is the default policy in the Linux operating system to manage memory affinity on ccNUMAs. This policy places data on the node that first accesses it [2]. To assure memory affinity using this policy, it is necessary to either execute a parallel initialization of all shared application data allocated by the master thread or allocate its data on each thread. However, this strategy will only present performance gains if it is applied on applications that have a regular data access pattern and if threads are not frequently scheduled to different cores/cpus. In case of irregular applications (threads do not always access the same data), *first-touch* will result in a high number of remote accesses.

Currently, there are some proposals concerning new memory policies for Linux. For instance, in [4,5,6], the authors have designed and have implemented the *on-next-touch* memory policy. This policy allows more local accesses, since each time a thread touches a data, the data migrates when needed. Its performance evaluation has shown good performance gains only for applications that have a single level of parallelism and large amount of data (e.g., matrices which size are higher than 8K x 8K). In case of multiple levels of parallelism (nested parallelism), each thread may create other threads. When these threads share a significant amount of data, several data migration can be performed, since each thread may be in a different machine node. These data migrations have presented an important overhead and they usually have lowered the application performance gains. Moreover, for small amount of data, *on-next-touch* policy have also not presented a good performance since the overhead with migrations is more expensive than the cost of remote accesses.

3   Minas

Minas [7] is an efficient and portable framework that allows developers to manage memory affinity in an explicit or automatic way on large scale ccNUMA platforms. In this work, efficiency means fine control of memory accesses for each application variable and similar performance on different ccNUMA platforms. As portability, we mean architecture and compiler abstraction and none or minimal source code modifications.

This framework is composed of three main modules: Minas-MAi, Minas-MApp and numarch. Minas-MAi, which is a high level interface, is responsible for implementing the explicit NUMA-aware application tuning mechanism whereas the Minas-MApp preprocessor implements an automatic mechanism. The last module, numarch, is responsible for extracting several information about the target platform. This module can be used by the developer to consult some important information about the architecture and it is also used by Minas-MAi and Minas-MApp mechanisms.

Minas differs from other memory affinity solutions [2,3] in at least four aspects. First of all, Minas offers code portability. Since numarch provides architecture abstraction, the developer do not have to specify nodes that will be used by Minas to place data. Secondly, Minas is a flexible framework since it supports two different mechanisms to control memory affinity (explicit and automatic tuning). Thirdly, Minas is designed for array oriented applications, since this data structure usually represents the most important variables in kernels/computations. Finally, Minas implements several memory policies to deal with both regular applications (threads always access the same data set) and irregular applications (threads access different data during the computations).

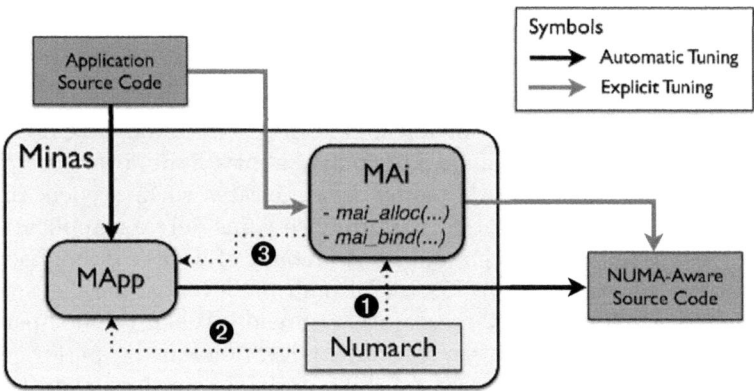

Fig. 1. Overview of Minas

Figure 1 shows a schema of Minas mechanisms to assure memory affinity. The original application source code can be modified by either using the explicit mechanism (gray arrows) or the automatic one (black arrows). The decision between automatic and explicit mechanisms depends on the developer's knowledge about the target application. One possible approach is to first use the automatic tuning mechanism and to verify whether the performance improvements are considered sufficient or not. If the gains are not sufficient, developers can then explicitly modify (manual tuning) the application source code using Minas-MAi.

Depending on the mechanism, numarch is used to retrieve different information. In explicit mechanism, Minas-MAi retrieves from numarch the number of nodes and cpus/cores as well as theirs physical identifiers in order to apply memory policies (dashed arrow 1). Differently, in the automatic mechanism, Minas-MApp gets from numarch the machine's NUMA factor, interconnection bandwidth, cache subsystem information and the amount of free memory of each node. These information are then used by the heuristic function to determine a suitable memory policy (dashed arrow 2). The chosen memory policy will be applied by using Minas-MAi memory policy functions (dashed arrow 3).

The current version of Minas is implemented in C. Minas has been tested on different ccNUMA architectures (Intel, AMD and SGI) with Linux as operating system. Minas supports C/C++ and Fortran and the following compilers: Intel, GNU and Portland.

3.1   MAi: Memory Affinity interface

MAi (Memory Affinity interface) is an API (Application Programming Interface) that provides a simple way of controling memory affinity [8]. It simplifies memory affinity management issues, since it provides simple and high level functions that can be called in the application source code to deal with data allocation and placement. All MAi functions are array-oriented, since MAi was designed for numerical scientific HPC applications.

The most important group of functions on MAi is the memory policies group, since it is responsible for assuring memory affinity. The interface implements eight memory policies that have as their memory affinity unit an array. The memory policies of MAi can be divided in three groups: bind, cyclic and random. Bind memory policies optimize latency, by placing data and threads as close as possible. Both, random and cyclic groups optimize bandwidth of ccNUMA platforms, since they minimize interconnect and memory contention.

Bind group has two memory policies, *bind_block* and *bind_all*. In *bind_block* memory policy, data is divided into blocks depending on the number of threads that will be used and on their placement within the machine. In *bind_all* memory policy, data is placed in one or a set of restrict nodes. Cyclic group is composed by *cyclic*, *skew_mapp* and *prime_mapp* memory policies. In *cyclic*, data is placed according to a linear round-robin distribution, using one memory page per round. In the *skew_mapp* memory policy, a page i is allocated on the node $(i + \lfloor i/M \rfloor + 1) \bmod M$, where M is the number of memory banks. The *prime_mapp* policy uses a two-phase strategy. In the first phase, the policy places data using *cyclic* policy on (P) virtual memory banks, where P is a prime greater or equal to M (real number of memory banks). In the second phase, the memory pages previously placed on virtual memory banks are reordered and placed on real memory banks also using the *cyclic* policy. In *random* policy, memory pages are placed randomly on CC-NUMA nodes, using a random uniform distribution.

The data distribution over the machines nodes can be performed using the entire array or an array tile (blocks distribution). A tile is a sub array which size can be specified by the user or automatically chosen by MAi. Such memory policies allows developers to express different memory access operations, such as write-only, read-only or read/write.

MAi also allows the developer to change the memory policy applied to an array during the application execution, allowing to express different patterns. Finally, any incorrect memory placement can be optimized through the use of MAi memory migration functions. The unit used for migration can be a set memory pages (automatically defined by MAi) or a set of rows/columns of an array (specified by the user).

3.2   MApp: Memory Affinity preprocessor

MApp (Memory Affinity preprocessor) is a preprocessor that provides a transparent control of memory affinity for numerical scientific HPC applications over ccNUMA platforms. MApp performs optimizations in the application source code considering the application variables and platform characteristics at compile time. Its main characteristics are its simplicity of use (automatic NUMA-aware tuning, no manual modifications) and its platform/compiler independence.

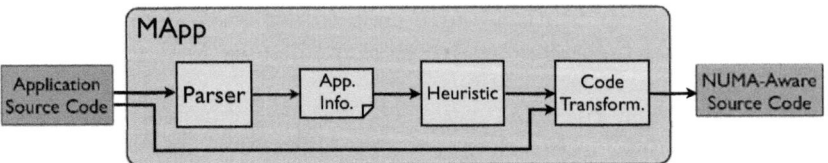

Fig. 2. Overview of MApp code transformation process

The code transformation process is divided into four steps (Figure 2). Firstly, it scans the application source code to obtain information about variables (App Info.). During the scanning process, MApp searches for shared static arrays that are considered large by Minas (eligible arrays). An eligible array is considered large if its size is equal or greater than the size of the highest level cache of the platform. Secondly, it fetches the platform characteristics, retrieving information from the numarch module (NUMA factor, nodes, cpus/cores, interconnection network and memory subsystem). During the third step, it chooses a suitable memory policy for each array. Finally, the code transformation is performed by including Minas-MAi specific functions for allocation and data placement.

The most important step of MApp automatic tuning process is the strategy used to decide which memory policy will be applied for each array. Based on empirical data from our previous works and experiments [8,9,10], we have designed an heuristic responsible for deciding which memory policy would be the most effective considering the underlying ccNUMA characteristics. On platforms with a high number of interconnections between nodes and small NUMA factor (ratio between remote latency and local latency to access data), the heuristic will apply cyclic memory policies. On the contrary, on platforms with low number of interconnections and high NUMA factor, the heuristic will opt for *bind_block* memory policies. Figure 3 shows a simple example of a code transformation generated by MApp. This example is a parallel code (C with openMP) that performs some operations in four arrays. However, as we can observe, MApp only applied memory policies for three of them (eligible arrays). Small variables such as integers *i,j* and *xux* will probably fit in cache so MApp will not interfere on compiler decisions (allocation and placement of variables). In this example, we suppose that the target ccNUMA platform has a small NUMA factor (remote latency is low) and a bandwidth problem for interconnection among nodes. Thus, on such a platform, optimizing memory accesses considering bandwidth instead of latency is important. Due to this, MApp has decided to spread memory pages of *vel*, *vxy* and *tem* with *cyclic* memory policy in order to optimize bandwidth.

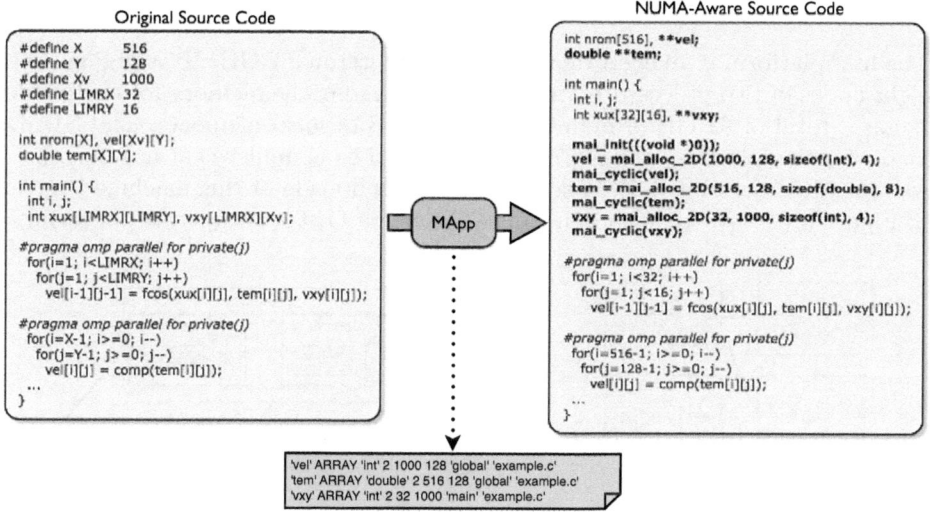

Fig. 3. Example of MApp source code transformation

3.3   Numarch: NUMA Architecture Module

The numarch module has an important role for Minas, since it retrieves the machine information that are necessary to place data on memory banks. This module extracts information about the interconnection network (number of links and bandwidth), memory access costs (NUMA factor and latency) and architecture characteristics (number of nodes, cpus/cores and cache subsystem). To retrieve such information, numarch parses the /sys/devices/ file system of the operating system. The retrieved information is stored in temporary files on the /tmp/ of the operating system. Using such information Minas-MApp places data among the machine nodes reducing latency costs (less remote accesses) and optimizing bandwidth (interconnect contention and memory contention).

This module can also be used as a library, since it provides some high level functions that can be called on the application source code to get some information of the target NUMA machine. The library is composed by a set of functions to retrieve information such as number of nodes, cache size, total of free memory on each node, number of cores per processor, the node of a core/cpu and cores and cpus on a node. Such information can be used by the developer to better understand the machine topology and characteristics.

4   Performance Evaluation

In this section, we present the performance evaluation of Minas and compare its results with other three memory affinity solutions for Linux based platforms. We first describe the two ccNUMA platforms used in our experiments. Then, we describe the two numerical scientific applications (ICTM [10] and Ondes 3D [9]) and their main characteristics. Finally, we present the results and their analysis.

4.1 Cache-Coherent NUMA Platforms

The first platform is an eight dual core AMD Opteron 2.2 GHz. It is organized in eight nodes of two processors with 2 MB of shared cache memory for each node. It has a total of 32 GB of main memory (4 GB of local memory). The NUMA factor for this platform varies from **1.2** to **1.5**. The compiler that has been used was the GCC (version 4.3). A schematic representation of this machine is given in Figure 4 (a). We have chosen to use the name **Opteron** for this platform.

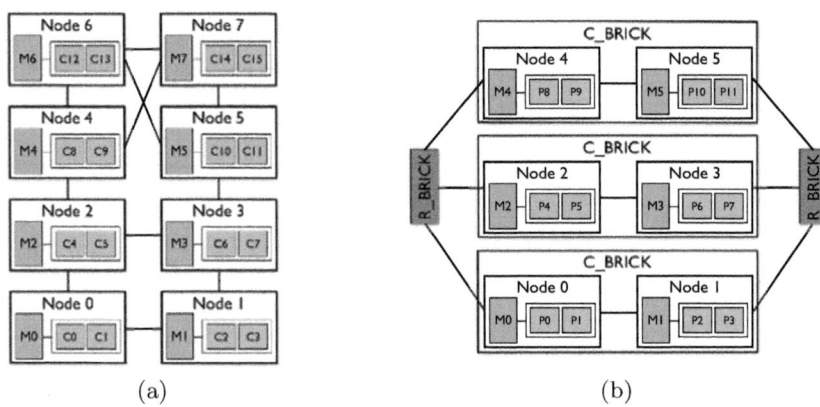

(a)   (b)

Fig. 4. NUMA platforms: (a) Opteron (b) SGI

The second ccNUMA platform is a SGI Altix 350 with twelve Itanium 2 processors of 1.5 GHz and 4 MB of shared cache memory each. It is organized in six nodes of two processors with a total of 24 GB of main memory (4 GB of local memory). The NUMA factor for this platform varies from **1.2** to **1.3**. The compiler that has been used was the ICC (version 9.0). A schematic representation of this machine is given in Figure 4 (b). We have chosen to use the name **SGI** to make reference to this platform. The operating system that has been used for both platforms is Linux 64-bits version with support for NUMA architecture.

4.2 Numerical Scientific Parallel Applications

In this section, we present applications Interval Categorizer Tessellation Model (ICTM) [10] and Simulation of Seismic Wave Propagation (Ondes 3D)[9]. Such applications represent important memory-bound numerical scientific problems. The applications have been implemented in C with OpenMP.

ICTM: Interval Categorizer Tessellation Model. ICTM is a multi-layered tessellation model for the categorization of geographic regions considering several different characteristics (relief, vegetation, climate, etc.). The number of characteristics that should be studied determines the number of layers of the model. In each layer, a different analysis of the region is performed. The input data is extracted from satellite images, in which the information is given in certain points referenced by their latitude and longitude coordinates. The geographic region is represented by a initial 2-D matrix of the total area into sufficiently small rectangular subareas. In order to categorize the regions of each layer, ICTM executes sequential phases. Each phase accesses specific matrices that have previously been computed and generates a new 2-D matrix as a result of the computation. Depending on the phase, the access pattern to other matrices can either be regular or irregular. Since the categorization of extremely large regions has a high computational cost, a parallel solution for ccNUMA platforms has been proposed in [10]. In this paper, we have carried out experiments using 6700x6700 matrices (2 Gbytes of data) and a radius of size 40 (number of neighbors to be analysed by status matrix phase). As shown in Figure 5 (a), the algorithm basically uses nested loops with short and long distance memory accesses (Figure 5 (b)) during the computation phases.

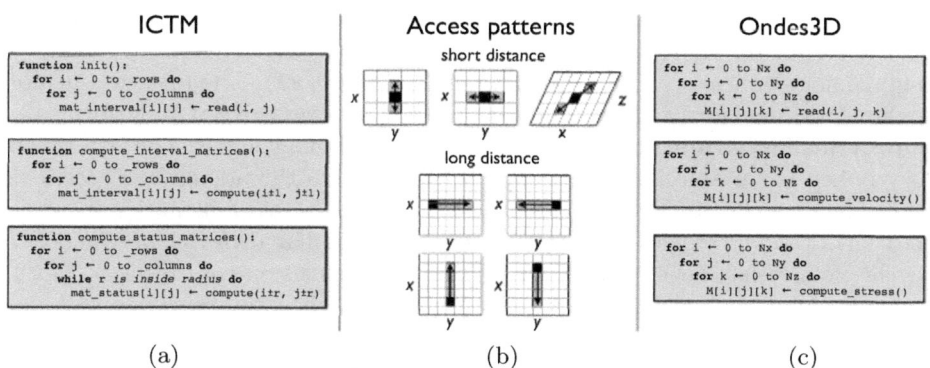

Fig. 5. Access patterns: ICTM and Ondes 3D

Ondes 3D: Simulation of Seismic Wave Propagation. Ondes 3D is an application that simulates seismic wave propagation in three dimensional geological media based on finite-difference discretization. It has been developed by the French Geological Survey (BRGM - www.brgm.fr) and it is mainly used for strong motion analysis and seismic risk assessment. The particularity of this simulation is to consider a finite computing domain even though the physical domain is unbounded. Therefore, the user must define special numerical boundary conditions in order to absorb the outgoing energy. Ondes 3D has three main steps: data allocation, data initialization and propagation calculus (composed by two calculus loops). During the first two steps, the three dimensional arrays are dynamically allocated and initialized (400x400x400, approximately 4.6 Gbytes of memory). During the last step, the two calculus loops compute velocity and stress of the seismic wave propagation. In all three steps, the three dimensional arrays are accessed in a regular way (same data access pattern) [9]. Figure 5 (c) presents a schema of the application with its three steps. On contrary to ICTM, Ondes 3D has only short distance memory accesses, as presented in Figure 5 (b).

4.3   Experimental Results

In this section we present results that have been obtained for each application
and platform. We have carried out series of experiments using Minas and three
Linux solutions (*first-touch* policy, *numactl* and *libnuma*).

The results have been obtained through the average of several executions
varying the number of threads from 2 to the maximum number of cpus/cores
of each platform. Our results are organized by application (ICTM and Ondes
3D). For each application, we have divided the results into two groups according
to the memory affinity management (automatic: First-Touch and Minas-MApp;
explicit: Minas-MAi, *numactl* and *libnuma*).

Regarding the explicit memory affinity solutions, we have changed applications source codes (Minas-MAi and *libnuma*) or their executions parameters (*numactl*). In order to use Minas-MAi and *libnuma*, the developer must add specific
data management functions. The results with Minas-MAi have been obtained
by applying the most suited memory policy for each array of the application.

Depending on the application and platform, we have chosen one of the following memory policies (*cyclic*, *prime_mapp* and *bind_block*). The first two memory
policies are ideal for irregular applications (ICTM) over ccNUMA platforms that
have a small NUMA factor, since they spread data among nodes. The latter memory policy is suitable for regular applications where threads always access the
same data set (Ondes 3D). Since *libnuma* has a limited set of memory policies, we
have used two strategies. The interleave policy (similar behavior of Minas-MAi
cyclic policy) has been applied for ICTM whereas the *numa_tonode_memory()*
function has been used for Ondes 3D. The last explicit solution, *numactl*, does
not require source code modifications. However, we had to change the execution
command line of all applications to specify which memory policy should have
been applied as well as the nodes and cpus lists.

Figure 6 shows the speedups for ICTM on Opteron and SGI platforms with
the automatic (Figure 6 (a) and (b)) and the explicit (Figure 6 (c) and (d))
memory affinity solutions. As it can be observed, Minas has outperformed all
other memory affinity solutions on both platforms.

Considering the automatic solutions applied to ICTM, Minas-MApp has presented satisfactory results on both platforms (Figure 6 (a) and (b)). Minas-MApp
heuristic has chosen *cyclic* memory policy to control data allocation and placement on both platforms. The chosen policy has resulted in better performance
gains than *first_touch* (on average, 10% Opteron and 8% on SGI). After a careful
analysis of these results and application characteristics, we have concluded that
first_touch policy has generated more remote accesses.

The explicit solutions have presented different behaviors depending on the
platform (Figure 6 (c) and (d)). On Opteron, the Minas-MAi *cyclic* memory
policy has presented the best results. However, there is not a significant difference between Minas-MAi and other explicit solutions (*libnuma* and *numactl*). It
can be explained by the fact that *libnuma* and *numactl* also offer a similar policy,
named interleave. It seems that the slight performance gains of Minas-MAi are

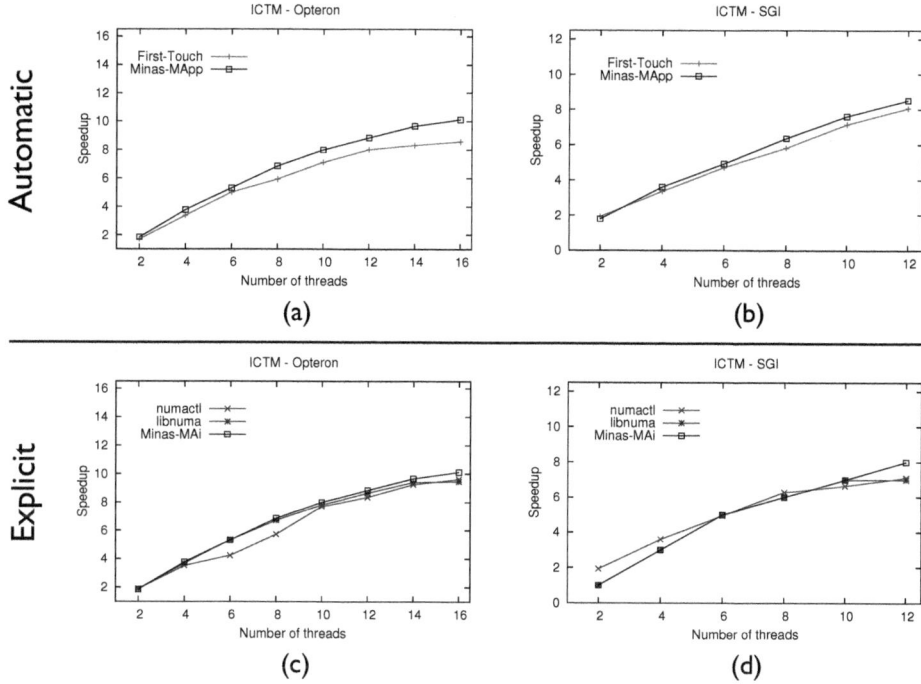

Fig. 6. Performance of ICTM on Opteron and SGI platforms

due to the array optimizations (specialized allocation functions and false sharing reduction). On SGI, Minas-MAi has also presented a better performance thanks to the array optimization included in allocation functions and memory policies. In the case of Minas-MAi, different cyclic memory policies (*cyclic* and *prime_mapp*) have presented equivalent performance gains. The network interconnection characteristics (short distance between memory banks) and the small NUMA factor of the platform can explain this insignificant difference. In this figure, we can also observe that Minas-MAi was the most scalable solution on both platforms in comparison to *libnuma* and *numactl*.

In Figure 7, we show the speedups for Ondes 3D application on Opteron and SGI platforms with the automatic (Figure 7 (a) and (b)) and the explicit (Figure 7 (c) and (d)) memory affinity solutions. On both platforms, Ondes 3D application with Minas has presented better performance gains than the other solutions for memory affinity control.

The results obtained with automatic solutions in Ondes 3D have shown that *first_touch* and Minas-MApp had similar performance gains. The Minas-MApp heuristic has chosen *cyclic* as the best policy according to the platform characteristics. However, as discussed before, the best policy for this application on such platforms is Minas-MAi *bind_block*. Since, *first_touch* and *bind_block* have similar behavior, their results are expected to be equivalent or superior to the Minas-MApp choice.

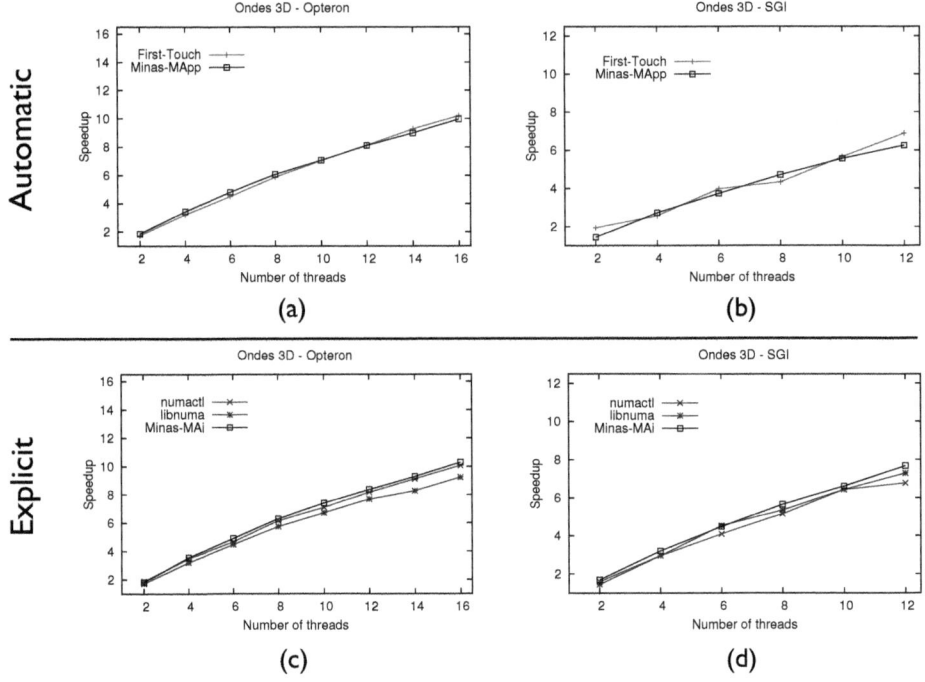

Fig. 7. Performance of Ondes 3D on Opteron and SGI platforms

Finally, the results with explicit solutions in Ondes 3D (Figure 7 (c) and (d)) have shown that *libnuma* and *numactl* have had a worse performance than Minas-MAi. Since this application has a regular memory access, it is important to keep both thread and their data as close as possible. In order to do so, data should be divided among NUMA nodes and threads should be fixed on cores/cpus of such nodes. This strategy can be achieved by either Minas-MAi or *libnuma*. However, *libnuma* demands considerable codification efforts, since developers must implement all data distribution algorithm and thread scheduling. Additionally, the same solution may not work on platforms with different architecture characteristics. In contrast with *libnuma*, Minas-MAi provides a specific policy for this purpose which is called *bind_block*. This policy automatically fixes threads and distributes data among the NUMA nodes (architecture abstraction). Thus, no source changes are needed when the same solution is applied on different platforms. *Numactl* is the less flexible of all explicit solutions and it does not provide such data distribution strategy (in this case we have used the interleave policy).

In Table 1, we present the minimum and maximum performance losses of Minas automatic tuning mechanism (Minas-MApp) in comparison with Minas explicit tuning mechanism (Minas-MAi) for each application and platform. We can notice that in some cases, Minas-MApp had an insignificant impact in terms of performance in relation with Minas-MAi (ICTM on both platforms and Ondes 3D on Opteron). However, according to our experiments, the performance loss

Table 1. Impact of Minas automatic tuning (Minas-MApp) mechanism

	ICTM	Ondes 3D
Opteron	$[0\%; 0\%]$	$[0\%; 3\%]$
SGI	$[0\%; 0\%]$	$[10\%; 13\%]$

may be important (up to 13%). Considering all the experiments and results, we can conclude that Minas-MApp can be a viable solution when developers do not choose to explicitly modify the application source code.

5   Conclusion and Future Work

In this paper we have focused our work on Minas, a memory affinity management framework to deal with memory placement on ccNUMA platforms for numerical scientific HPC applications. We have carried out some experiments over two ccNUMAs to evaluate the efficiency of Minas when used to guarantee memory affinity of two Geophysics applications. Such experiments have shown that Minas has improved the overall performance of applications in comparison with other solutions available on Linux. We have observed that the automatic mechanism of Minas (Minas-MApp) have presented improvements when compared with the Linux native *first_touch* policy. Considering the explicit mechanisms, Minas-MAi has shown better results than other explicit solutions (*numactl* and *libnuma*).

Our future work on Minas includes providing dynamic memory policies, providing a NUMA aware allocator for dynamic data structures [7] as tcmalloc [11], support of memory policies created by developers on Minas-MApp as well as a support for other runtime systems (e.g., Charm++ [12] and TBB [13]).

Acknowledgment

This research was supported by the French ANR under grant NUMASIS ANR-05-CIGC and CAPES (Brazil) under grant 4874-06-4.

References

1. Verghese, B., Devine, S., Gupta, A., Rosenblum, M.: Operating system support for improving data locality on CC-NUMA compute servers. In: ASPLOS-VII: Proceedings of the 7th International Conference on Architectural Support for Programming Languages and Operating Systems, pp. 279–289 (1996)
2. Antony, J., Janes, P.P., Rendell, A.P.: Exploring thread and memory placement on NUMA architectures: Solaris and linux, ultraSPARC/FirePlane and opteron/HyperTransport. In: Robert, Y., Parashar, M., Badrinath, R., Prasanna, V.K. (eds.) HiPC 2006. LNCS, vol. 4297, pp. 338–352. Springer, Heidelberg (2006)
3. Kleen, A.: A NUMA API for Linux, Tech. Rep. Novell-4621437 (2005), http://whitepapers.zdnet.co.uk/01000000651,260150330p,00.htm

4. Löf, H., Holmgren, S.: Affinity-on-next-touch: Increasing the Performance of an Industrial PDE Solver on a cc-NUMA System. In: ICS 2005: Proceedings of the 19th Annual International Conference on Supercomputing, pp. 387–392. ACM, New York (2005), http://portal.acm.org/citation.cfm?id=1088149.1088201

5. Terboven, C., Mey, D.A., Schmidl, D., Jin, H., Reichstein, T.: Data and Thread Affinity in OpenMP Programs. In: MAW 2008: Proceedings of the 2008 workshop on Memory access on future processors, pp. 377–384. ACM, New York (2008), http://dx.doi.org/10.1145/1366219.1366222

6. Goglin, B., Furmento, N.: Enabling High-Performance Memory Migration for Multithreaded Applications on Linux. In: IEEE (ed.) MTAAP 2009: Workshop on Multithreaded Architectures and Applications, held in conjunction with IPDPS 2009, Rome, Italie (2009), http://hal.inria.fr/inria-00358172/en/

7. Ribeiro, C.P., Méhaut, J.-F.: Minas Project - Memory affInity maNAgement System (2009), http://pousa.christiane.googlepages.com/Minas

8. Ribeiro, C.P., Castro, M., Fernandes, L.G., Carissimi, A., Méhaut, J.-F.: Memory Affinity for Hierarchical Shared Memory Multiprocessors. In: 21st International Symposium on Computer Architecture and High Performance Computing - SBAC-PAD, IEEE, São Paulo (2009)

9. Dupros, F., Pousa, C., Carissimi, A., Méhaut, J.-F.: Parallel Simulations of Seismic Wave Propagation on NUMA Architectures. In: ParCo 2009: International Conference on Parallel Computing, Lyon, France (2009)

10. Castro, M., Fernandes, L.G., Ribeiro, C.P., Méhaut, J.-F., de Aguiar, M.S.: NUMA-ICTM: A Parallel Version of ICTM Exploiting Memory Placement Strategies for NUMA Machines. In: PDSEC 2009: Parallel and Distributed Processing Symposium, International, pp. 1–8 (2009)

11. Google, Google-perftools: Fast, multi-threaded malloc() and nifty performance analysis tools (2009), http://code.google.com/p/google-perftools/

12. Gürsoy, A., Kale, L.V.: Performance and modularity benefits of message-driven execution. J. Parallel Distrib. Comput. 64(4), 461–480 (2004)

13. Intel, Intel Threading Building Blocks (2010), http://www.threadingbuildingblocks.org/

HPC Environment Management: New Challenges in the Petaflop Era

Jonas Dias and Albino Aveleda

Federal University of Rio de Janeiro
Cidade Universitária, Centro de Tecnologia, I-248,
Caixa Postal 68516, Ilha do Fundão, Rio de Janeiro, Brazil
Telephone and Fax: +552125628080
{jonas,bino}@nacad.ufrj.br

Abstract. High Performance Computing (HPC) is becoming much more popular nowadays. Currently, the biggest supercomputers in the world have hundreds of thousands of processors and consequently may have more software and hardware failures. HPC centers managers also have to deal with multiple clusters from different vendors with their particular architectures. However, since there are not enough HPC experts to manage all the new supercomputers, it is expected that non-experts will be managing those large clusters. In this paper we study the new challenges to manage HPC environments containing different clusters with different sizes and architectures. We review available tools and present LEMMing [1], an easy-to-use open source tool developed to support high performance computing centers. LEMMing integrates machine resources and the available management and monitoring tools on a single point of management.

Keywords: parallel and distributed computing.

1 Introduction

High performance computing (HPC) systems are typically found on universities and research centers. Broadly used to process large-scale scientific experiments and complex simulations, they are also very popular among enterprises. About 62% of the 500 most powerful supercomputers in the world are in the industry sector [2]. Recently, the PetaFlop barrier has been broken [3]. In other words, it is the capacity to sustain more than 10^{15} floating point operations per second. That was achieved using supercomputers with more than 100,000 processing cores. They are on the top of the list of the 500 biggest supercomputers in the world. Since 1993, the TOP500.org portal collects information about the 500 world's most powerful systems. It publishes a ranking twice a year with the performance of many supercomputers measured by the LINPACK benchmark [4]. Observing the TOP500 statistics, it is possible to follow the evolution and trends of the HPC technology.

Figure 1 shows the historical growth of the number of processors per system to obtain more powerful computers. Today, the vast majority of the TOP500 supercomputers have more than two thousand processors. Nowadays, the development of new

J.M.L.M. Palma et al. (Eds.): VECPAR 2010, LNCS 6449, pp. 293–305, 2011.
© Springer-Verlag Berlin Heidelberg 2011

technologies produces better hardware. However, the greater the number of nodes in a supercomputer, the greater is the chance of a failure in any of these nodes. Thus, management processes need to be faster and easier to handle failures quickly.

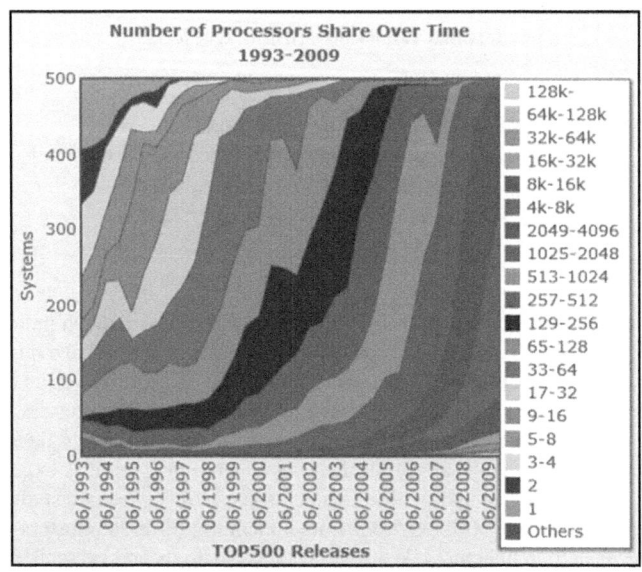

Fig. 1. Growth of the number of processors on supercomputers. (Source: TOP500.org historical charts)

Most cluster management and monitoring tools can handle huge supercomputers. However, they usually do not present very organized information and they are not user friendly. For example, most tools present the listing of the compute nodes of a cluster sequentially. On small clusters, it is not a problem, since it is easy to check several nodes listed on a screen. But on huge supercomputers, to find a single node on a sequential list with hundreds or even thousands of nodes is not practical. To solve this problem, the nodes can be easily organized on a hierarchy, but no tool has ever presented such solution. The user interface usability may not be so important for an expert manager, but with the spread of HPC, many supercomputers are managed by non-expert administrators, which may need an easy-to-use interface.

To manage a supercomputer includes ensuring the full access to its functionalities and resources, handling defects and problems as fast as possible. A queue system may be offered to distribute the access between users equally or due to some internal policy. For example, on an engineering company, the engineer group may have greater priority over the development group. It is also interesting to have monitoring tools to show the status of the machines over time and store critical scenarios that may be repeated. HPC centers usually have a complex and non-uniform infrastructure with different types of supercomputers. They use a variety of tools, including their own, to manage their HPC environment. It may be very hard to keep control of many resources using multiple tools. Each tool has its own user interface, web address and features. Often the system administrator may have to deal with more than two different tools to accomplish a single management process. For example, the network monitoring software alerts that a cluster is reporting an error regarding some of its nodes being down. The manager may need to access the specific monitoring cluster tool and, after detecting where the problem is, access the cluster server through a secure shell to solve the problem through command lines.

There are few software for analysis and administration of HPC centers in the market. The available solutions are mostly proprietary and usually do not integrate with software from other vendors. Therefore, a center with heterogeneous machinery may have to deal with multiple software systems. Some products also lacks in usability, especially when it is necessary to manage a huge number of nodes.

We believe that the development of an open source tool for HPC environments with improved usability and integration is needed. Thus, we propose LEMMing, an open source tool that let you administrate a HPC environment with multiple clusters and servers through a single Rich Internet Application (RIA). LEMMing is meant to be a single point of management. It supports deployment, management and monitoring of clusters and servers focused, mainly, on high performance computing, taking advantage of other available tools. It also lets the manager to customize the interface with the preferred resources, widgets and external tools.

The remainder of this paper is organized as follows. Section 2 provides an overall comparison between different types of management and monitoring tools for supercomputers. Section 3 describes the LEMMing project and section 4 concludes and discusses future work.

2 Available Tools

LEMMing takes advantage of other available tools to deploy, manage and monitor supercomputers. Thus, it is necessary to discuss these tools, but is not possible to give many details and discuss all of them. We will discuss here the tools we believe are the most important, according to our experience. Proprietary software is not referenced directly in accordance to trademark law restrictions.

2.1 Deployment Tools

There are many possibilities to setup a cluster but there are few tools that offer an easy way to do it. The most popular ones are OSCAR (Open Source Cluster Application Resource) [5], Rocks [6] and xCAT (Extreme Cloud Administration Toolkit) [7]. OSCAR runs a graphical user interface (GUI) on the cluster head node that configures and deploys the cluster in several steps. Alternatively, Rocks installation involves the Linux operating system and a customizable set of other packages. OSCAR and Rocks install useful tools for management and monitoring. Rocks has a group of tools and commands that simplifies management tasks like: components listing, node addition and removal and change in properties like IP address, hostname, gateway and DNS server. OSCAR supports these operations on the GUI or through scripts. xCAT provides support for the deployment and administration of high scale systems and also has plenty of useful scripts. The focus of xCat is not, actually, the ease of installation and configuration but robustness and scalability. It does not mean that OSCAR and Rocks are not robust and do not scale. It just means that xCAT can be harder to install and OSCAR and Rocks may need some adjustments after installation for improved results. Table 1 summarizes a comparison among OSCAR, Rocks and xCAT involving important characteristics for cluster deployment.

Table 1. Comparison of different cluster deployment tools

	Cluster Installation	*Node Adding*	*MPI*	*Queuing System*	*Monitoring Tool*
OSCAR	GUI	GUI + Network listening	Yes	Yes	Yes
Rocks	GUI	UI + Network listening	Yes	Yes	Yes
xCAT	Command Line	Command Line + Manual Adding	No	No	No

Both OSCAR and Rocks provide a GUI to deploy the cluster environment while xCAT is completely installed through command line and requires the manual editing of SQLite tables to set up cluster configurations and available resources. OSCAR installation GUI includes a node-adding step. Rocks provides an ASCII user interface to handle node adding. OSCAR and Rocks have the option to listen the network to automatically add new booted nodes. On xCAT the nodes must be manual added editing a set of database tables using command line. OSCAR and Rocks can install MPI libraries [8,9], a workload and resource-management tool (shortened as Queuing System) such as Sun Grid Engine [10] or TORQUE [11]. They both can also install a monitoring tool, usually Ganglia [12]. On xCAT, these tools must be installed and integrated separately.

2.2 Monitoring Tools

Network monitoring tools, such as Nagios [13] and Cacti [14], are very useful for HPC environment management. Nagios monitors the status of the environment hosts and possibly some of their services. If anything stops working, Nagios sends alerts to the administrators. Cacti shows statistical data on graphics about CPU load, network bandwidth and others metrics. Together, they provide a global view of the available resources and services on the whole network. Nagios and Cacti are prepared to be extended and customized through plug-ins with specialized functionalities. This level of customization is very interesting since different HPC centers have specific necessities, like the monitoring of its critical services or some specific cluster.

Another very popular and useful tool is Ganglia [12]. Ganglia shows many monitoring metrics about the cluster nodes on a web page. It plots graphics of the metrics over time. Ganglia uses XML to report the available metrics for each node of the cluster or grid, so it is easy to integrate Ganglia with other applications. Table 2 gives a comparison among Cacti, Ganglia, and Nagios monitoring tools.

All the described monitoring tools are web based, but none is a rich internet application (RIA). Only Nagios send alerts if something goes wrong on the network. However, Cacti has plug-ins that also do it. Nagios and Cacti can be extended by plug-ins. Ganglia only lets the user to add extra metrics to monitor the cluster. Ganglia monitoring focus differs from Nagios and Cacti, since it is designed to monitor a single cluster or grid, while Nagios and Cacti are prepared to manage a whole network. However, a supercomputer or a grid can also be considered a private LAN. Anyway, this focus difference implies that Ganglia provides more detailed information about the cluster or grid nodes while Nagios and Cacti may show an overview of the network status.

Table 2. Comparison of the mentioned monitoring tools

	Web Based	RIA	Send Alert	Plugins	Monitoring focus
Cacti	Yes	No	No	Yes	Network
Ganglia	Yes	No	No	No	Cluster/Grid
Nagios	Yes	No	Yes	Yes	Network

2.3 Proprietary Solutions

Proprietary cluster solutions normally use a set of open source systems and tools (Linux, Ganglia, compilers and MPI Libraries). Many vendors use OSCAR, Rocks or xCAT to handle its base cluster installation. In addition, they tune the cluster configuration and install their proprietary environments to administrate the cluster. These tools are designed for specific machines and do not deal with other types of hardware. Actually, different machine models from the same vendor commonly come with distinct software to handle administration tasks. Sometimes there is a separate tool for each part of the system, for example: a tool to handle the storage system, other to handle the compute nodes and yet others to handle users and job queues.

Some proprietary software provides integration offering the management of varied equipment through a single graphical user interface. The difficulties found on this software are the usability and incompatibility with varied hardware. Since they are all proprietary, there is usually no collaboration. These tools typically do not have a web interface. As an illustration, one of these integrated management software understands each node of a cluster as an isolated resource. It then lists the nodes as items in the same level of a tree on a side panel. For example, there is a tree item called MyCluster and all the nodes are appended under this item as leaves. There is no problem if the cluster has few nodes, but on a large-scale system, it is impractical to find a specific node in the middle of a huge list of names.

Many of the discussed software are not prepared to handle the new challenges to manage large-scale supercomputers. These challenges are related to the increasing number of processors on clusters, the heterogeneous environments, the particular tools an environment needs to use and the different type of administrators, with varied level of knowledge, managing these clusters. HPC centers with huge clusters need a tool that presents the cluster resources as a whole but presents it organized and customized by the user. It must be prepared to show thousand of nodes smartly, highlighting problems and warnings and providing an easy mapping to the hardware position on the equipment room. The software needs to be easy to use, to configure and to customize, also prepared to integrate with other tools easily. It should make easier for a non-expert manager to handle the main operations of a HPC center environment. Integration, flexibility and great usability are the most important features to deal with these new challenges.

The need to have a free tool that integrates many resources and tools providing an easy-to-use web interface for easy HPC environment administration lead us to design LEMMing. LEMMING uses the open source collaboration model to integrate HPC management and monitoring software stack with improved usability.

3   The LEMMing Project

The LEMMing project is inspired on the Zimbra Collaboration Suite[15]. Zimbra is a groupware [16] centered on e-mail and contact exchanging. It combines a set of open source tools, configuring them together. Zimbra offers an easy way to deploy and manage an e-mail server and web client with improved usability. Taking advantages of AJAX (Asynchronous JavaScript And XML) [17] features, it offers a rich web interface that behaves like a desktop application.

We decided to implement LEMMing instead of extending another tool because we need an application with great usability enhanced by a RIA. No cluster management and monitoring open source tool is developed as a RIA. Extending another tool would also make LEMMing dependent on it. For example: neither OSCAR nor Rocks installs Nagios as a monitoring tool; if LEMMing was a Nagios plug-in, it would compel the administrators to install Nagios to use LEMMing on a OSCAR or Rocks cluster system. However, administrators may not want to monitor the cluster using Nagios. They may prefer Ganglia. OSCAR and Rocks install Ganglia to monitor the cluster and LEMMing, actually, depends on Ganglia.

LEMMing is an acronym of Linux Enterprise Management and Monitoring but also makes reference to a species of small rodents that live near the Arctic region. The lemming population increases drastically on a period of the year and the little animals start searching for a new habitat. They normally die on this process and the population is reduced to few specimens. Since the size of clusters is also increasing quickly, raising the probability of failure, we found the analogy interesting. The hardware technology evolves fast and computers are more robust. Nevertheless, considering a hypothetical failure probability of one component in a thousand, a cluster with thousands of nodes has many failures. If these defects are not noticed and corrected quickly, the cluster machine may be just like the lemming population – reduced to few nodes.

The main purpose of LEMMing is to consolidate multiple clusters management tools on a single web interface of easy access and control. To achieve this objective, LEMMing is flexible and provides great usability. LEMMing structure is also prepared to incorporate new features and components easily using a plug-in approach. LEMMing is designed as a web application to keep the remote access aspect of clusters for management tasks. To attend the usability requirement, we developed a rich

Internet application (RIA) using the AJAX model. Among many toolkits and frameworks, after some research, we decided to use the ZK [18] framework. ZK was chosen since it provides an easy development model based on interface components and lets you produce AJAX content very easily without programming Javascript. It means that both server and client side of the web application can be written in Java.

Since the AJAX model, alone, does not enhance usability, it was also necessary to study and discuss a group of user interface components that fit the application necessities but that are also familiar to users. Thus, we decided to keep LEMMing interface very simple and similar to a navigational file manager desktop application (such as Gnome Nautilus, Mac OS Finder or Windows Explorer). Most computer users are familiar to file manager applications since it is an elementary application when using an operating system. The web interface is better discussed on section 3.2.

The web interface is the main component of LEMMing, but, in order to improve flexibility, the part of LEMMing responsible by the operations inside the cluster is detached from the web interface. The interface may, then, communicate with the managed clusters accessing the LEMMing cluster component through Web Services. The web services are better described on section 3.1. This approach makes it easier to manage multiple supercomputers on a single interface, balancing the workload between the LEMMing web server and the cluster head node server, since LEMMing web application just delegates to the component inside the cluster what they should do inside the cluster. Together, the web and the cluster component of LEMMing provide the following features:

- Freeware
- Web Service based
- AJAX interface design
- Integration of other tools
- Single point of management
- Support for many cluster topologies organization
- Integrated with workload management
- Parallel shell tools
- Customizable Dashboard

Being freeware is an important feature since it improves collaboration and let the advanced managers of a cluster to improve LEMMing code according to some specific necessity. LEMMing is also easy to integrate with other tools such as Ganglia, Cacti and Nagios. A single point of management means that LEMMing concentrates the management of many components of many supercomputers and servers on a single interface. It also concentrates other tools that may be associated to a specific resource (the Ganglia page of a cluster, for example, or a local particular tool) or may represent the whole status of the HPC environment (the Nagios or Cacti web pages). LEMMing also reduces complexity of cluster management, providing a quick highlight of the status of available resources.

LEMMing is designed to be flexible, i.e., to handle multiple clusters and its components with different architectures and installed systems. It is not feasible to build a huge tool that can support all the available cluster and server systems solution out of the box. Thus, we decided to split LEMMing in two modules. The first module

(LEMM-WS) is inevitably associated with a cluster system and is designed to implement a given application programming interface (API) using the cluster system available tools. The second module (LEMM-GATE) is the web application that uses the API to manage the machines. Thus, to support a new system, only the first module may need to be adapted. The cluster system remains transparent to the second module that only has to deal with the known API.

We used web services technology to implement the idealized architecture. The module within the cluster exposes a set of web services that work as an API to the web module. The administrator access the web application (LEMM-GATE) through a browser and the web module access the web services (LEMM-WS) to get cluster information and to execute operations on it. This architecture lets the LEMM-GATE server to be accessible through an external network while the clusters may remain accessible only through internal network. This approach prevents the cluster servers from external attack.

3.1   LEMMing Web Services (LEMM-WS)

The idea of LEMMing is to take advantage of existent and available tools, before obtaining data manually. Thus, the web services deployed inside the cluster system explores the reports from Ganglia, from the queue system, like Torque [11] or SGE (Sun Grid Engine) [10] and available proprietary systems. The available web services take advantages of macro commands and scripts available to execute operations on the cluster. They can also connect to database systems or read configuration files. The LEMMing web services module is very dependable of the installed cluster system, since these operations usually differ between cluster architectures. Since the cluster systems are naturally different, there is no way to build a single web service capable to control all types of clusters. To amortize this issue, LEMM-WS contains a set of libraries to support traditional interfacing mechanisms such as shell command execution, database access, XML parsing, and file access. These packages were designed to reduce the effort while developing LEMM-WS for a new cluster system. Currently, LEMM-WS fully supports Rocks clusters, but we are already deploying it on other cluster systems.

LEMM-WS works as an API to handle the cluster systems. This API is divided into two web services: the Informant and the Operator. The informant web service implements the API methods regarding the cluster configuration and status. The operator web service implements the methods that can change the configuration or status of the cluster. For example, the informant web service provides the set of available nodes of the cluster, their load, processors and memory capacity. It also provides information regarding the jobs submitted to the cluster and the users of the cluster. The operator web service provides methods such as node renaming, rebooting, adding and removal. The implementation of these methods may vary depending on the cluster system, but it is not very difficult since LEMM-WS provides libraries with traditional integrations approaches.

An important aspect of LEMMing is the node position on the cluster. This aspect influences how the web interface displays the cluster structure through an organized manner. Thus, LEMMing uses the concept of dimensionality on clusters. The dimensionality let the system specify the position of the nodes using spatial coordinates.

LEMMing also suggests that the position of the nodes on the cluster should be related the hostname of the nodes. For example, a 1-dimensional cluster is usually a small cluster with a set of PCs that may be referred as node-00, node-01, node-02 and so on. Their position can be easily mapped using a single dimension. On a 2-dimensional cluster, the machines may be organized on racks. The nodes of the cluster may then be referred as node-0-0, node-0-1, node-1-0 and so on. The first number indicated the number of the rack where the machine is, and the second number indicates the position of the machine inside that rack. The 3-dimensional and 4-dimensional are analogous and may be used on cluster systems that divide the machines in chassis and blades. The informant web service may then, provide the coordinates of a node in the cluster. For example, if the web service method returns (7,5,3,1), it means that the node is the machine number one of the blade number three of the chassis number five of the rack number seven of the cluster. This approach makes it easier to organize the cluster on the graphical interface, especially when there are many nodes involved. Moreover, it makes it easier to keep the names of the nodes more consistent, if the cluster is expanded or shrunk.

LEMM-WS must be deployed inside the cluster system in order to access its configuration, status and resources locally. Thus, it needs a web service container available on the cluster head node. LEMM-WS can be easily deployed on a web server like Apache Tomcat [19].

3.2 LEMMing Web Application (LEMM-GATE)

The web application module is responsible to access the web services provided by LEMM-WS, showing data obtained from the informant web service and work as the graphical interface to access the operations provided by the operator web service. LEMM-GATE is a rich Internet application, i.e., it is a web page but behaves like a desktop application. LEMM-GATE user interface was designed to be familiar to the user. Thus, it is similar to a navigational file manager desktop application. On the left side, there is a cluster overview with a navigable tree where the leaves represent the nodes of the cluster. On a file manager, the leaves are usually the file system folders. The left side panel containing the cluster tree can be resized or hidden to maximize the right side panel area, if the user needs more space to visualize the panel.

LEMM-GATE is prepared to organize the cluster tree on the left panel according to the dimensionality of the cluster, which means that the compute nodes are organized on a hierarchy that divides the whole set into racks, chassis and blades. Resources reporting errors or warnings are highlighted to alert the administrators to examine them. The organization of the cluster nodes as a hierarchical tree is designed to enhance the navigation over the cluster structure. On machines with a huge number of nodes, it is especially necessary because the cluster is displayed in parts, making it easier to inspect. This organization is also very important on a cluster expansion, since it does not require renaming the old nodes. However, if the nodes are named sequentially, but the rack is not full, when new nodes are added on the empty spaces, there may be a name inconsistence.

On the right side of LEMM-GATE interface there is a detailed view of the item selected on the tree. On a file manager, when a folder is selected on the tree, the right side panel shows the content of that folder. On LEMMing, when a tree item is selected, the right side panel shows the configuration of that item, the status and additional information that may be customized. The user can associate tabs to the selected tree item. A tab may display a whole web page from another tool (for example, the Nagios page), or it may display some selected widgets. LEMMing offers some widgets by default, like the cluster or node load monitoring widget, the CPU usage share widget and the cluster queue viewer widget. It is easy to download new widgets and also easy to create them according to local needs. A widget is a self-contained file specified in the ZUL language [18], which is an easy XML-like language to specify web components based interfaces. The ZUL language lets the specification of scripts that can be written in Java to perform more advanced operations, involving, for example, the integration of other tools. The ZUL file just needs to be added to the widget folder of LEMMing to become available to use.

The user may customize a different set of tabs for each different type of item of the tree. For example, items representing a whole HPC environment may have tabs with Nagios and Cacti; items representing a cluster may have tabs with Ganglia and widgets monitoring cluster load and queue; and compute node items may also display monitoring widgets and the Ganglia page for that specific node. Other customizations are possible adding content to items representing racks, chassis and blades. Since it is possible to integrate different tools inside LEMMing, the interface complexity may grow. However, since the interface is customizable and each administrator may have its own LEMMing account, an administrator can build its own workspace on LEMMing on a complexity level that is adequate to his skills. All customizations made by each user are recorded on XML files. So, when a user logs into LEMMing, it finds its own customized workspace.

LEMM-GATE is also the interface to access the operations of the cluster provided by the operator web service. The management operations can be executed accessing the context menus of the tree items. The action calls the operator web service on LEMM-WS to execute the operation. For example, to reboot a node, the user right clicks it on the item and selects the 'Reboot' option on the context menu. This action calls the web service and executes the reboot method passing the selected node as an argument. A video demonstration of LEMMing can be found at the project page [1]. Figure 3 shows two sample screens of LEMMing to give an idea of its interface.

The web services module being developed detached from the AJAX interface allows the building of some specific deployment scenarios. As said, LEMM-GATE and LEMM-WS are independent modules and can be deployed on different servers. However, they can also be put in the same server on the same web container. On a simple environment with only one cluster, for example, it is reasonable to deploy both LEMMing modules on the cluster head node. However, on big HPC environments, the web application should be deployed on a web server and the web services inside each cluster. Therefore the single web application will access all the LEMM-WS modules deployed on every cluster and manage the clusters through them. The scalability of LEMMing has not been measured yet. However, LEMMing has already been used to manage three different clusters simultaneously and it worked nicely. Although we still cannot affirm that LEMMing scales well, we believe that the load balancing provided by the LEMM-GATE being deployed separately from LEMM-WS brings good performance to the system.

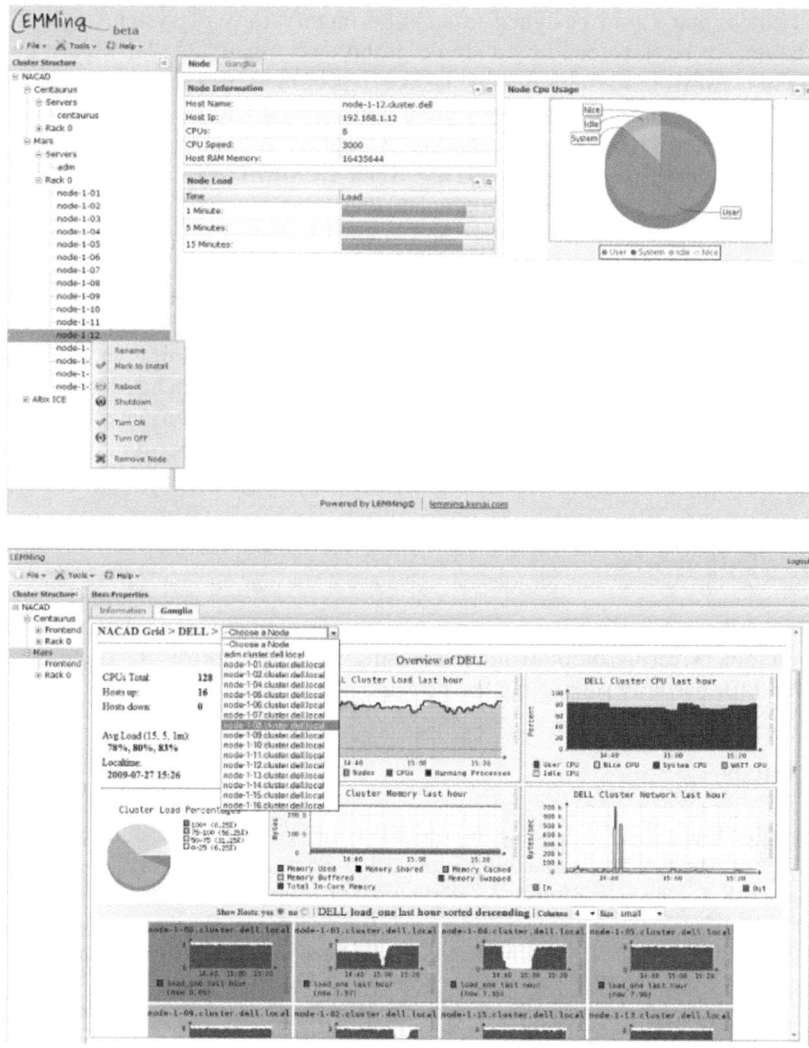

Fig. 3. LEMMing web interface

4 Conclusion

The new challenges to manage HPC environments with high scale supercomputers require new solutions to support multiple and heterogeneous resources on a single point of management with improved usability. We are proposing LEMMing, a tool to integrate the cluster management and monitoring software stack and operate it through a rich Internet application. LEMMing does not replace existing tools or any proprietary tool, it just combine them on a smart interface to make it easier to manage a huge supercomputer.

We believe that a tool designed to attend a huge variety of systems and architectures should not be developed by a single group or company. Collaboration is essential to take LEMMing to other platforms. We expect that, being open source, it grows up with the community and support a great number of environments. LEMMing uses great available tools (Rocks, Torque/SGE, Ganglia) through a flexible architecture and a smart and easy-to-use web interface. Since LEMM-GATE, the web interface, is detached from LEMM-WS, the component that integrates with the cluster, another possible scenario is where the vendors build their LEMM-WS module for their cluster systems and provide it to their clients.

We are now working on LEMMing to support different cluster systems. We are also adding new features like an integrated IPMI [20] support, users administration and queuing system management. For future work, we are planning an experiment to measure the scalability of LEMMing. We believe LEMMing attends most HPC management scenarios, even with multiple supercomputers with thousands of processors. More information and a video demonstrating LEMMing can be found on the LEMMing project web page at http://lemm.sf.net/ [1].

Acknowledgments

The authors thank the High Performance Computing Center (NACAD-COPPE/UFRJ) and Professor Alvaro Coutinho for supporting LEMMing Project. They also thank Dell Brazil for partially supporting the project.

References

[1] Dias, J., Aveleda, A.: LEMMing Project (2010), http://lemm.sf.net/
[2] TOP500, TOP500 Supercomputing Sites (2010), http://www.top500.org/
[3] Barker, K.J., Davis, K., Hoisie, A., Kerbyson, D.J., Lang, M., Pakin, S., Sancho, J.C.: Entering the petaflop era: the architecture and performance of Roadrunner. In: Proceedings of the 2008 ACM/IEEE Conference on Supercomputing, Austin, Texas, pp. 1–11 (2008)
[4] Petitet, A., Whaley, R., Dongarra, J., Cleary, A.: HPL - A Portable Implementation of the High-Performance Linpack Benchmark for Distributed-Memory Computers Dispon?vel em (2010), http://www.netlib.org/benchmark/hpl/
[5] des Ligneris, B., Scott, S., Naughton, T., Gorsuch,N.: Open Source Cluster Application Resources (OSCAR): design, implementation and interest for the [computer] scientific community. In: Proceedings of the 17th Annual International Symposium on High Performance Computing Systems and Applications and the OSCAR Symposium, Sherbrooke, Quebec, Canada, May 11-14, p. 241 (2003)
[6] Papadopoulos, P.M., Katz, M.J., Bruno, G.: NPACI Rocks: Tools and Techniques for Easily Deploying Manageable Linux Clusters. In: IEEE International Conference on Cluster Computing, Los Alamitos, CA, USA, p. 258 (2001)
[7] xCAT.: Extreme Cloud Administration Toolkit (2010),
http://xcat.sourceforge.net/
Dispon?vel em: http://xcat.sourceforge.net/

[8] Gropp, W.: MPICH2: A New Start for MPI Implementations. In: Recent Advances in Parallel Virtual Machine and Message Passing Interface, pp. 37–42 (2002)

[9] Gabriel, E., Fagg, G.E., Bosilca, G., Angskun, T., Dongarra, J.J., Squyres, J.M., Sahay, V., Kambadur, P., Barrett, B., et al.: Open MPI: Goals, Concept, and Design of a Next Generation MPI Implementation. In: Recent Advances in Parallel Virtual Machine and Message Passing Interface, pp. 353–377 (2004)

[10] Gentzsch, W.: Sun Grid Engine: Towards Creating a Compute Power Grid. In: Ccgrid 2001: Proceedings Of The 1St International Symposium On Cluster Computing And The Grid, p. 35 (2001)

[11] Staples, G.: TORQUE resource manager. In: Proceedings of the 2006 ACM/IEEE Conference on Supercomputing, Tampa, Florida, p. 8 (2006)

[12] Massie, M.L., Chun, B.N., Culler, D.E.: The ganglia distributed monitoring system: design, implementation, and experience. Parallel Computing 30(7), 817–840 (2004)

[13] Barth, W.: Nagios: System and Network Monitoring. No Starch Press (2008)

[14] Shivakumar, P., Jouppi, N.P., Shivakumar, P.: CACTI 3.0: An Integrated Cache Timing, Power, and Area Model (2001)

[15] Gagné, M.: Zimbra collaboration suite, Version 4.5. Linux J. 157, 14 (2007)

[16] Ellis, C.A., Gibbs, S.J., Rein, G.: Groupware: some issues and experiences. ACM Commun. 34(1), 39–58 (1991)

[17] Paulson, L.D.: Building Rich Web Applications with Ajax. Computer 38(10), 14–17 (2005)

[18] Chen, H., Cheng, R.: ZK Step-By-Step: Ajax without JavaScript Framework. Apress (2007)

[19] Tomcat, Apache Tomcat (2010), `http://tomcat.apache.org/`. Dispon?vel em: `http://tomcat.apache.org/`

[20] Intel Corporation, Dell, Hewlett-Packard, NEC Corporation (2010), IPMI - Intelligent Platform Management Interface, `http://www.intel.com/design/servers/ipmi/`. Dispon?vel em: `http://www.intel.com/design/servers/ipmi/`

Evaluation of Message Passing Communication Patterns in Finite Element Solution of Coupled Problems

Renato N. Elias[1], Jose J. Camata[2], Albino Aveleda[2], and Alvaro L.G.A. Coutinho[2]

[1] IM, Multidisciplinary Institute,
Federal Rural University of Rio de Janeiro,
60 Capitão Chaves St., Nova Iguaçu, RJ 26221-010
renato@gmail.com
[2] NACAD, High Performance Computing Center,
Federal University of Rio de Janeiro,
P.O. Box 68506, Rio de Janeiro, RJ 21945-970
{camata,bino,alvaro}@nacad.ufrj.br

Abstract. This work presents a performance evaluation of single node and subdomain communication schemes available in EdgeCFD, an implicit edge-based coupled fluid flow and transport code for solving large scale problems in modern clusters. A natural convection flow problem is considered to assess performance metrics. Tests, focused in single node multi-core performance, show that past Intel Xeon processors dramatically suffer when large workloads are imposed to a single node. However, the problem seems to be mitigated in the newest Intel Xeon processor. We also observe that MPI non-blocking point-to-point interface sub-domain communications, although more difficult to implement, are more effective than collective interface sub-domain communications.

Keywords: Parallel Computing, Message Passing, Communication Patterns, Coupled Problems, Edge-Based.

1 Introduction

In 2008 the petascale barrier has been broken. According to Kogge [5] such systems can carry out real computations 1,000 times more challenging than those computable by early terascale systems. The size of such systems raises particular challenges, including performance on each node, scalable programming models, performance and correctness debugging, and improving fault tolerance and recovery. On the applications side, Gropp [6] stresses the fact that when discussing such systems researchers often overlook the increasing complexity of individual nodes, processors and the underlying network. Particular applications may benefit from the sheer power of such systems, but the majority of them have to be re-examined. Again, according to Gropp [6], researchers are creating new tools to develop, debug, and tune applications, as well as creating new programming models and languages that could enhance scalability by reducing communication overhead. The Computational Fluid Dynamics (CFD) community is aware of these new developments [7].

J.M.L.M. Palma et al. (Eds.): VECPAR 2010, LNCS 6449, pp. 306–313, 2011.

In Brazil there is a growing need to understand complex processes in the oil and gas industry. Particularly, understanding these processes is therefore critical to effective exploration for oil and gas in the recently discovered pre-salt fields in ultra-deep waters offshore in Brazil. Several of such complex processes can be recast in the general framework of fluid-structure interaction and coupled fluid flow and transport problems, involving multiple spatial and temporal scales. This paper presents a parallel performance evaluation of computation and communication models implemented in EdgeCFD, an implicit edge-based coupled fluid flow and transport solver for large-scale problems in modern clusters. EdgeCFD currently supports stabilized and multiscale finite element formulations and has been used in problems ranging from Newtonian and non-Newtonian fluid flows, free-surface flow simulations with fluid-structure interaction, gravity currents and turbulence (details available in [2] and references therein). Of particular interest in the present work is EdgeCFD's performance in the current multi-core processors, particularly process placement within processors and the impact of several subdomain communication models. The target machines are modern clusters with the latest processor and network technologies, paving the way towards sustained petascale performance. Following [4] and [10], where strategies for massive parallelism computations in unstructured grids are discussed, EdgeCFD adopts peer-to-peer non-blocking communication among processes.

The remainder of this paper is as follows. Next section details the benchmark software and communication models currently supported. The natural convection problem used to access parallel performance metrics is given in Section 3. The paper ends with a summary of our main conclusions.

2 EdgeCFD: The Benchmark Software

EdgeCFD was chosen to evaluate performance in several aspects, such as: parallel models, system architecture and processors. EdgeCFD was developed to exploit parallel architectures in four different ways, which broaden the range of machines that can be efficiently used. Three of them rely on message passing interface (MPI) implementations while the fourth one is based on threaded parallelism for shared memory systems or system components such as many-cores processors with shared memory at cache levels.

For the message passing implementations, the three variants differ in how data are split and message exchanges are scheduled among processors. The simplest case is based on collective communication calls. In this strategy, nodes are divided in two groups: parallel interface and internal nodes. Interface nodes, all hollow and solid vertices in Figure 1a, are known by all processors. On the other hand, nodes owned exclusively by a processor are internal nodes. Computation of matrix-vector products and residuals are performed in just one, blocking, step where the parallel interface values are combined using MPI_ALLREDUCE operations. This collective communication model was further extended to remove the need of excessive data exchange as well as redundant information storage. In this case, globalization operations are also performed in one step but, now, using MPI_ALLGATHER calls. The more complex message

passing parallel model employs peer-to-peer (p2p) message exchanges among processors and takes advantage of communication and computation overlapping. Following
[4], the point-to-point (p2p) communication strategy is based on a master-slave relationship between processors. This relationship is established by creating a hierarchy
based on host partition numbers. Thus, the processor P_i is slave of P_j if P_i and P_j are
neighbors and $i < j$. Note that a processor can be slave and master at the same time,
depending only on the number that identifies it in relation to its neighbors. Figure 1a
illustrates a two dimensional mesh which is decomposed into four partitions. The hollow vertices denote the nodes, or degrees of freedom of the system of equations, that
will be sent to the receiver (a master processor). On the other hand, the solid vertices
represent the nodes or degrees of freedom that will be received from donors (a slave
processor). Figure 1b shows the communication graph corresponding to this mesh. In
this case, P_1 is slave of P_2 and P_3. P_2 is slave of P_3 and P_4 but it is master of P_1. P_3 is
slave of P_4 and master of P_1 and P_2. Finally, P_4 is master of P_2 and P_3.

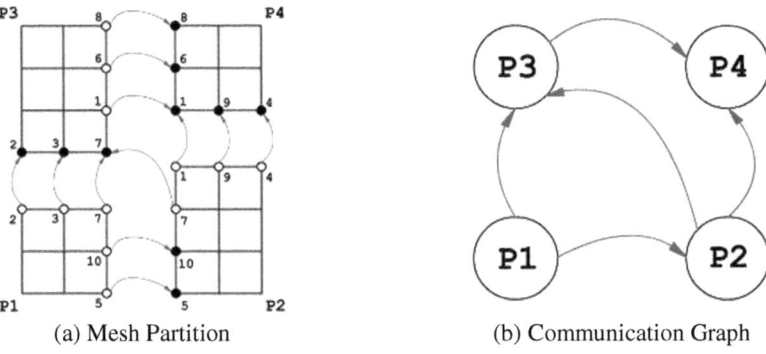

(a) Mesh Partition   (b) Communication Graph

Fig. 1. Master-Slave relationship

The information exchange among neighboring processors is implemented in two
stages: in the first stage, slave processes send their information to be operated by
masters (where the interface contributions are accumulated) and, in the second stage,
the solution values are copied back from masters to slaves. In addition, EdgeCFD uses
non-blocking send and receive MPI primitives, which allow communication and
computation overlapping.

3 Performance Tests

The three dimensional Rayleigh–Benard convection problem is used to investigate code
performance in situations ranging from small to large scale simulations in different
architectures and system configurations. The problem consists in a fluid, initially at rest,
contained in a 3D rectangular domain with aspect ratio 4:1:1 (Length:Depth:Height) and
subjected to a unity temperature gradient [3]. For a 4:1:1 container aspect ratio, with
no-slip boundary conditions at walls, the flow is three dimensional and must gives rise
to four convection cells. The fluid properties are set to result in Rayleigh and Prandt

numbers of 30,000 and 0.71 respectively. For the performance tests two meshes with different discretizations are used. The coarser mesh (MSH1) is formed by 178,605 tetrahedra elements and 39,688 nodes while the finer one (MSH2) is made by splitting the domain in 501×125×125 divisions, which gives rise to 39,140,625 tetrahedral elements. In both cases the solution is evolved towards steady-state using EdgeCFD's block sequential implicit time-marching scheme. In this scheme the Navier-Stokes block is solved by the Inexact-Newton method and the temperature block by simple multicorrection iterations. The inner iterative driver for both Navier-Stokes and temperature transport is an edge-based preconditioned GMRES method. A nodal block-diagonal preconditioner is used for the Navier-Stokes equations while a simple diagonal preconditioning is employed for the temperature equation. GMRES tolerance for the temperature is fixed at 10^{-3} while the maximum tolerances for the inexact Newton method is set to 0.1. For both, flow and transport, the number of Krylov vectors is fixed in 25. We consider that steady state is achieved when the relative velocity increment differs by less than 10^{-5}.

Tests are carried out in three different Intel Xeon-based HPC systems. All systems are equipped with quad core CPUs.

- *SGI Altix ICE cluster* with 32 compute nodes. Each node has eight 2.66 GHz cores (in two Intel Xeon Processor Quad-Cores, Clovertown - X5355). L2 cache size 8MB on-die for Quad-Core; 4MB per core pair; shared by the two cores. Memory Blade: 16GB. All nodes are interconnected using InfiniBand technology in a Hypercube topology.
- *DELL cluster PowerEdge M1000e* with 16 compute nodes (M600). Each node has eight 3.00 GHz (in two Intel Xeon Processor Quad-Cores, Harpertown - E5450). L2 cache size 12MB on-die for Quad-Core; 6MB per core pair; shared by two cores. Memory Blade: 16GB. All nodes are interconnected using InfiniBand technology in a full-CLOS topology.
- *Intel Nehalem server* with eight 2.8 GHz (in two Intel Xeon Processor Quad-Cores, Nehalem – X5560). L3 cache size 8 MB shared by all cores. Memory: 12 GB.

All systems have similar configurations in terms of number of sockets per node and cores per socket. Tests were performed using Intel Fortran compiler using the same compilation options in all cases. The cache sharing scheme in older Intel Xeon processors, although behaving as Quad-core are, in fact, two Dual-core processors put together. Mesh entities are ordered to improve data locality as described in [1].

In the performance tests, initially, we evaluate single node performance and multicore performance according to the parallel models described in Section 2. Tests are primarily conducted in the coarser mesh (MSH1), for runs using up to 8 intra-node cores. In other words, no network connection was employed in order to reveal intrinsic CPU aspects, such as: cache size, memory sharing, load balance, process placement, etc. Figure 2 shows the speedup for EdgeCFD, running the parallel schemes presented in section 2, using 2 different systems, SGI Altix ICE 8200 (Figure 2a) and the Nehalem server (Figure 2b). The systems are chosen to show the performance evolution between 2 Xeon family processors, X5355 and X5560 respectively.

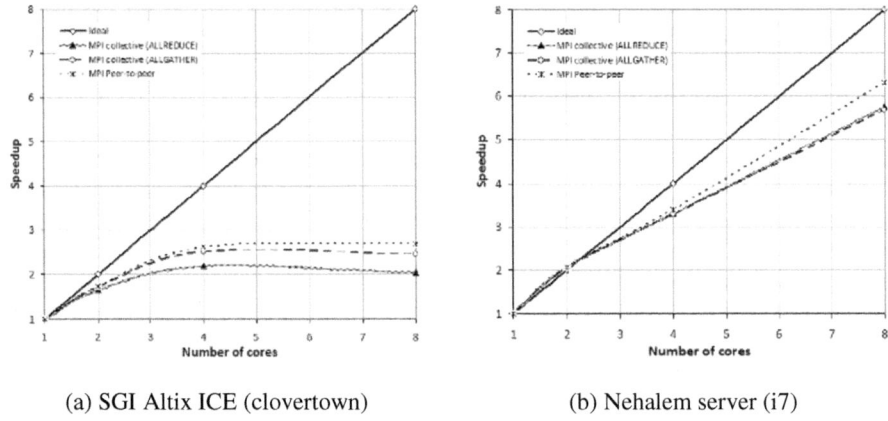

(a) SGI Altix ICE (clovertown)   (b) Nehalem server (i7)

Fig. 2. Speedup for two Xeon systems running up to 8 intra-node cores

As can be seen from Figure 2, the overall performance is much better in the newer Xeon processor (Nehalem or i7) than in previous version (Clovertown/Harpertown). Regarding the parallel scheme, the peer-to-peer model resulted in the best perform-ance, as expected, reaching a particularly good speedup in the Nehalem processor. However, for the Clovertown CPU, the poor performance led us to investigate this issue from other aspects, such as: cache size and sharing, node architecture, etc. In both architectures peer-to-peer message passing is clearly superior. Earlier experi-ments [11] on a HP ProLiant DL145 G3 cluster with 912 cores powered with Opteron 2218 dual core processors and Gigabit network have shown that p2p was also faster than collective communication when using more than one computational node.

Figure 3a show a raw comparison among the benchmark systems when running se-rial cases (using one core). It clearly shows the performance increase for the Intel Xeon processor between releases from 2006 (Clovertown) until 2009, when Nehalem processor was launched. The wall time reduction, in our tests, reached 37% for proc-essors with 3 years of difference. The performance gains are even more pronounced when we analyze the multi-core case running in a single node (Figure 3b). Other in-teresting result is the parallel performance shown by Nehalem which is faster than the Cluster Dell and SGI Altix-ICE systems, where older Intel Xeon CPUs are present. It seems that these results are mainly influenced by the cache memory system that Nehalem processor has.

Motivated by the results presented so far, tests were also conducted considering different combinations of cores per nodes. Figure 4a shows the elapsed wall time spent to solve the Rayleigh-Benard problem in parallel (message passing with peer-to-peer scheme), using 8 cores for several arranges of cores × nodes. Tests were per-formed in Cluster Dell but similar results were also obtained in the SGI Altix ICE 8200.

From Figure 4a, we note that diminishing the number of cores per node, the per-formance increases substantially, which points out that older Xeon processors suffer when all cores are simultaneously busy. It is important to remember that the main

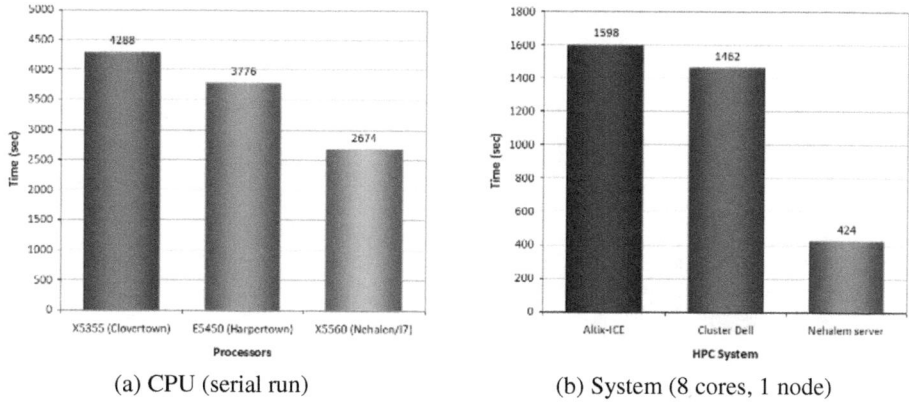

(a) CPU (serial run)   (b) System (8 cores, 1 node)

Fig. 3. System comparisons for CPU performance in serial runs (a) and peer-to-peer MPI performance using 8 cores in 1 node (b)

EdgeCFD's kernels (matrix-vector product, stiffness matrix build up and assembly of elements residual) strongly relies in indirect memory addressing operations and are, thus, influenced by how mesh entities are accessed and used during these operations. In EdgeCFD, mesh entities are reordered to makes efficient use of cache memory as explained in details in [1]. However, due to the complexity of the main loops of the software, cache misses are expected even for reordered meshes. To better understand the meshing ordering effect we have also run this problem considering two mesh configurations: original nodal ordering and nodes reordered using Reverse Cuthill Mckee (RCM). In the latter case, edges and elements were ordered in ascending order of edge (element) nodes. All tests were made in a single node. For the first case, it was necessary 24:26 (mm:ss) to solve the Rayleigh-Benard problem on MSH1, while for the reordered, the wall time decreased to 17:57 (mm:ss). The parallel profiling information was obtained using TAU (Tuning Analysis Utilities [9]) and, for the case using one core per node, where all communication is made through InfiniBand network, the time spent in MPI_WaitAll calls was around 3.2% of the total wall time. For the case using all cores available in one node, where all communication is made using memory bus, the largest MPI cost, due to MPI_WaitAll routine, was around 2.4%. This may be an indication of the MPI inability to provide efficient communication in non-homogeneous systems (here memory bus/InfiniBand) as described in [8].

Figure 4b presents the speedup curve obtained with p2p communication pattern and using the best combination of cores per nodes in the Cluster Dell system. Comparing with Figure 2a, which presents the same metric for the SGI Altix ICE 8200, but using only one node, we can conclude that using a large number of cores per node dramatically reduces performance. Note that using one core per node, 12.5% of the theoretical processing power, an ideal speed-up is reached. This also supports the previous argument, because in this case, communication is homogeneous and uses only InfiniBand network.

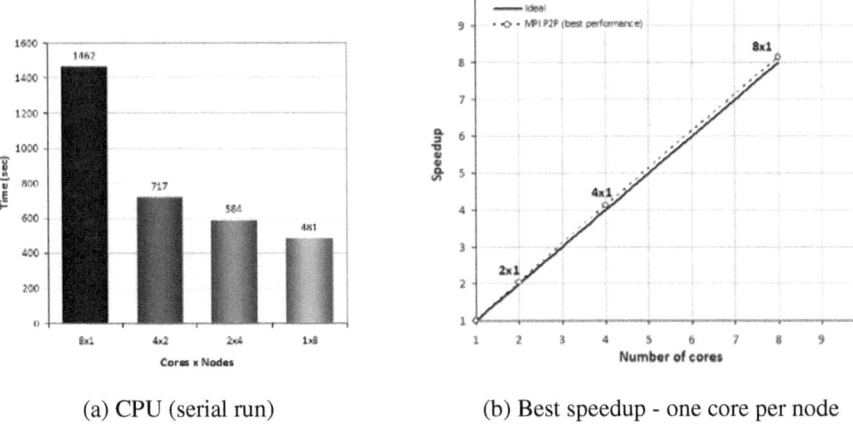

(a) CPU (serial run)   (b) Best speedup - one core per node

Fig. 4. Performance impact according to cores x nodes distribution on Cluster Dell

In order to illustrate the impact of the performance issues discussed in previous sections, tests are also conducted for the same problem in the finer mesh (MSH2) described in section 3. For this test, 64 cores are employed to solve 31,140,625 flow equations per nonlinear iteration per time step and 7,843,248 transport equations for each multi-correction iteration per time step. The number of time steps considered was enough to make the initialization process negligible. Two runs with different combinations of cores × node are used. In the first run, all cores of 8 nodes are used in order to exhaust nodes resources. In the second case, only 4 cores are used from 16 nodes, 50% of each node resources. In the first case, it is necessary 01:49:13 (hh:mm:ss) to solve 10 time steps while in the second case, considering only 50% of each node capacity, the walltime substantially decreased to 00:57:05. All runs are performed in the Cluster Dell, which uses Intel Xeon E5450.

4   Concluding Remarks

This work presented several performance tests for different versions of the Intel Xeon processor family running EdgeCFD, a stabilized finite element software for solving incompressible flows coupled (or not) to advection-diffusion transport problems. Tests focused in single node multi-core performance showing that past Intel Xeon processors dramatically suffers when large workloads are imposed to a single node. However, the problem seems to be mitigated in the newest Intel Xeon processor (codename Nehalem or commercial name Core I7) by the inclusion of a third level (L3) of shared cache memory. Other important change made by Intel, in its newest Xeon processor, was the construction of a fast linking channel among processors called Quick Path Interconnect, QPI for short. As a consequence, performance dramatically decreases when systems built with older Intel Xeon processors are subjected to large workloads. In the other hand, excellent performances may be reached when placement policies, such as using a smaller number of cores per node, are adopted as shown in Figure 4.

We also investigated message-passing parallelism. We observed that peer-to-peer two-stage information exchange among neighboring processors, using non-blocking send and receive MPI primitives, present the best performance. Experiments also demonstrate the difficulty of MPI to handle heterogeneous communication. Moreover, by setting a suitable MPI process distribution, execution time can be reduced by more than one half as we observe in the large run with the finer mesh on the Cluster Dell. A possible direction to tackle such difficulties is to set-up an architecture aware mesh partition, but this remains to be explored in EdgeCFD.

Acknowledgements. This work is partially supported by CNPq, Petrobras, ANP, Dell and Intel.

References

1. Coutinho, A.L.G.A., Martins, M.A.D., Sydenstricker, R.M., Elias, R.N.: Performance comparison of data-reordering algorithms for sparse matrix–vector multiplication in edge-based unstructured grid computations. International Journal for Numerical Methods in Engineering 66, 431–460 (2006)
2. Lins, E.F., Elias, R.N., Guerra, G.M., Rochinha, F.A., Coutinho, A.L.G.A.: Edge-based finite element implementation of the residual-based variational multiscale method. International Journal for Numerical Methods in Fluids 61(1), 1–22 (2009)
3. Kessler, R.: Nonlinear transition in three-dimensional convection. Journal of Fluid Mechanics 174, 357–379
4. Sahni, O., Zhou, M., Shepard, M.S., Jansen, K.E.: Scalable Implicit Finite Element Solver for Massively Parallel Processing with Demonstration to 160K Cores. In: Proceedings of the Supercomputing Conference, Portland, OR, USA (2009)
5. Kogge, P.: The Challenges of Petascale Architectures. IEEE Computing in Science Engineering, 10–16 (November-December 2009)
6. Gropp, W.D.: Software for Petascale Computing System. IEEE Computing in Science Engineering, 17–21 (November-December 2009)
7. Biswas, R.: Proceedings of the 21st International Conference on Parallel Computational Fluid Dynamics, Moffet Field, CA (May 2009)
8. Berti, G., Traff, J.L.: What MPI Could (and Cannot) Do for Mesh-Partitioning on Non-homogeneous Networks. In: Mohr, B., Träff, J.L., Worringen, J., Dongarra, J. (eds.) PVM/MPI 2006. LNCS, vol. 4192, Springer, Heidelberg (2006)
9. Shende, S., Malony, A.D.: The TAU Parallel Performance System. International Journal of High Performance Computing Applications 20(2), 287–311 (2006)
10. Houzeaux, G., Vázquez, M., Aubry, R., Cela, J.M.: A Massively Parallel Fractional Step Solver for Incompressible Flows. Journal of Computational Physics 228, 6316–6332 (2009)
11. Elias, R.N., Camata, J.J., Paraizo, P., Coutinho, A.L.G.A.: Communication and Performance Evaluation of Edge-Based Coupled Fluid Flow and Transport Computations, EC-COMAS, Coupled Problems, Barcelona (2009)

Applying Process Migration on a BSP-Based LU Decomposition Application

Rodrigo da Rosa Righi[1], Laércio Lima Pilla[1], Alexandre Carissimi[1],
Philippe Olivier Alexandre Navaux[1], and Hans-Ulrich Heiss[2]

[1] Institute of Informatics, Federal University of Rio Grande do Sul, Brazil
[2] Kommunikations- und Betriebssysteme, Technical University Berlin, Germany

Abstract. Process migration is an useful mechanism to offer load balancing. In this context, we developed a model called MigBSP that controls processes rescheduling on BSP applications. MigBSP is especially pertinent to obtain performance on this type of applications, since they are composed by supersteps which always wait for the slowest process. In this paper, we focus on the BSP-based modeling of the widely used LU Decomposition algorithm as well as its execution with MigBSP. The use of multiple metrics to decide migrations and adaptations on rescheduling frequency turn possible gains up to 19% over our cluster-of-clusters architecture. Finally, our final idea is to show the possibility to get performance in LU effortlessly by using novel migration algorithms.

1 Introduction

A possibility to increase performance on dynamic and heterogeneous environments comprises the processes' relocation. Generally, process migration is implemented within the application. This organization results in a close coupling between the application and the algorithms' data structures, which makes this approach non-extensible. Even more, some initiatives use explicit calls in the application code [2]. A different approach for migration happens at middleware level, linking the balancer tool with the programming library directly [16]. Commonly, this approach does not require changes in the application code nor previous knowledge about the system.

A typical scheduling middleware applies mechanisms to allocate the processes with longer computing times to faster machines. Nevertheless, this approach is not the best one for irregular applications and dynamic distributed environments, because a good process-resource assignment performed in the beginning of the application may not remain efficient with time [1,13,15]. At this moment, it is not possible to recognize either the amount of computation of each process nor the communication patterns among them. Besides fluctuations in the processes' computation and communication actions, the processors' load may vary and networks can become congested while the application is running. Therefore, an alternative is to perform process rescheduling through their migration to new resources, offering application runtime load balancing [7,11,17].

J.M.L.M. Palma et al. (Eds.): VECPAR 2010, LNCS 6449, pp. 314–326, 2011.

In this context, we designed a process rescheduling model called **MigBSP** that works over **BSP** (Bulk Synchronous Parallel) applications [9,20,22]. It explores the automatic and transparent load (processes) balancing at middleware level. To make decisions about load balancing, the model considers data about the infrastructure, the processes' behavior as well as migration costs. MigBSP was organized to work with BSP applications, once they are based on synchronous phases (supersteps). Thus, the main idea of the model is to reduce the duration of each superstep to decrease the application time. MigBSP contributions are twofold: (i) combination of multiple metrics to select the candidates for migration and; (ii) minimization of the model's overhead with adaptation that act over the rescheduling frequency.

The BSP model is mainly used for the development of scientific applications such as data mining, sorting and fluid dynamics [4]. Particularly, this paper presents the modeling of a parallel BSP-based version of the widely used **LU Decomposition** method [3]. In addition, it describes the execution of this application when linked to MigBSP over a cluster-of-clusters environment. The LU decomposition splits a matrix A in the product of a lower triangular matrix L and an upper triangular matrix U. LU is employed to turn the calculation of linear equations easier, since the solution of a triangular set of equations is trivial. Besides its usage, we choose LU because some initiatives impose changes in the code and/or extra executions when offering load balancing for this application[2,12,19]. Thus, the paper's final idea is to show the possibility for getting performance in LU application effortlessly by using novel migration algorithms.

2   MigBSP: Process Rescheduling Model

MigBSP manages load balancing issues, where the load is represented by BSP processes. It aims to reduce the time of each superstep of the application. Its key idea is to migrate processes which have a long computation time, perform several communication actions with other processes whose belong to a same site (*e.g.*, a cluster) and present low migration costs. Figure 1 (a) illustrates a superstep k of a BSP application in which the processes are not balanced among the resources. Figure 1 (b) shows the expected result with the rescheduling of the processes after superstep k, which will influence the execution of the next supersteps, (including $k+1$, $k+2$ and so on).

MigBSP's architecture is heterogeneous and composed by clusters and supercomputers. This architecture is assembled with the abstractions of Sets (different sites) and Set Managers. Set Managers are responsible for scheduling, capturing data from a specific Set and exchanging it among other managers. MigBSP can be seen as a scheduling middleware. There is no need for changes in the application code. All data necessary for its functioning may be captured directly in both communication and barrier functions as well as in other sources like the operating system. We described the first ideas of MigBSP in [8]. However, such work presents an evaluation with a synthetic application with a reduced number of supersteps (up to 400) and processes (up to 10).

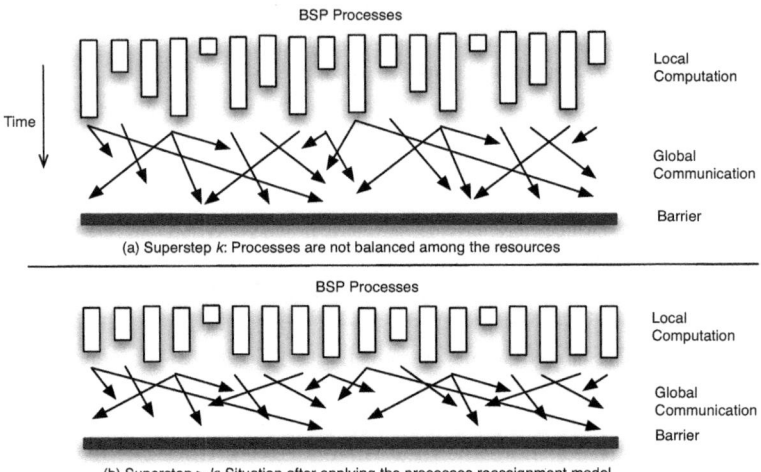

Fig. 1. BSP Supersteps in two different situations

The decision for process remapping is taken at the end of a superstep, after the barrier. At this moment, we can analyze data from all BSP processes. Aiming to generate as less intrusiveness in application as possible, we applied two adaptations that control the value of α - the adaptive period between rescheduling calls. The adaptations' ideas are: (i) to postpone the rescheduling call if the system is stable (processes are balanced) or to turn it more frequent, otherwise; (ii) to delay this call if a pattern without migrations in ω calls is observed. A variable D is used to indicate a percentage of how far the slowest and the fastest processes may be from the average. Our second adaptation works on increasing D. The higher its value, the greater the odds to increase α.

We employed a decision function called **Potential of Migration** (PM) to select the candidates for migration. Each process i computes q functions $PM(i,j)$, where q is the number of Sets and j means a specific Set. The main idea consists in performing a subset of the processes-resources tests at the rescheduling moment. $PM(i,j)$ is found using the Computation, Communication and Memory metrics (see Equation 1). Computation metric - $Comp(i,j)$ - considers a Computation Pattern $P_{comp}(i)$ that measures the regularity of a process i at its computation phase. This value is close to 1 if the process performs a similar number of instructions at each superstep and close to 0 otherwise. This metric also performs a computation time prediction based on data between rescheduling calls. In the same way, Communication metric - $Comm(i,j)$ - computes the Communication Pattern $P_{comm}(i,j)$ between processes and Sets. Furthermore, it uses a communication time prediction considering data between the rebalancing calls. Memory metric - $Mem(i,j)$ - considers process memory, transferring rate between the process and the manager of target Set, as well as migration costs.

$$PM(i,j) = Comp(i,j) + Comm(i,j) - Mem(i,j) \qquad (1)$$

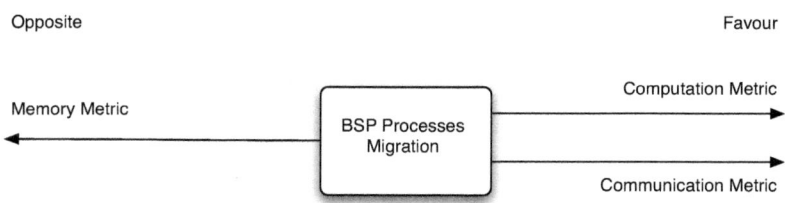

Fig. 2. Operation of the metrics to evaluate the Potential of Migration (PM) of a process: (i) Computation and Communication metrics act in favor of migration; (ii) Memory metric works in the opposite direction as a migration cost

Figure 2 depicts the operation of the considered metrics on process migration. Firstly, the BSP processes calculate $PM(i,j)$ locally. At each rescheduling call, each process passes its highest $PM(i,j)$ to its Set Manager which exchanges the PM of its processes among other managers. We used a heuristic to choose the candidates which is based on a decreasing ordered list of PMs. The processes with PM higher than $MAX(PM).x$ are candidates, where x is a percentage. The $PM(i,j)$ of a candidate process i is associated to a Set j. Therefore, the manager of Set j will select the most suitable processor to receive this process. Before a migration, its viability is verified by computing two times: t_l and t_d. t_l means the local execution of process i, while t_d encompasses its prediction of execution on the destination processor and includes the migration costs. For each candidate is chosen a new resource (if $t_l > t_d$) or its migration is canceled.

3   LU Decomposition Application

Consider a system of linear equations $A.x = b$, where A is a given $n \times n$ non singular matrix, b a given vector of length n, and x the unknown solution vector of length n. One method for solving this system is by using the LU Decomposition technique. This technique comprises the decomposition of the matrix A into a lower triangular matrix L and an upper triangular matrix U such that $A = LU$. A $n \times n$ matrix L is called unit lower triangular if $l_{i,i} = 1$ for all $i, 0 \leq i < n$, and $l_{i,j} = 0$ for all i, j where $0 \leq i < j < n$. An $n \times n$ matrix U is called upper triangular if $u_{i,j} = 0$ for all i, j with $0 \leq j < i < n$.

On input, A contains the original matrix A^0, whereas on output it contains the values of L below the diagonal and the values of U above and on the diagonal such that $LU = A^0$. Figure 3 illustrates the organization of LU computation. The values of L and U computed so far and the computed sub-matrix A^k may be stored in the same memory space of A^0. Algorithm 1 presents a sequential algorithm for producing L and U in stages. Stage k first computes the elements $u_{k,j}$, $j \geq k$, of row k of U and the elements $l_{i,k}$, $i > k$, of column k of L. Then, it computes A^{k+1} in preparation for the next stage. Algorithm 2 presents the functioning of the previous algorithm using just the elements from matrix A. Figure 3 also presents the data that is necessary to compute $a_{i,j}$ in the last

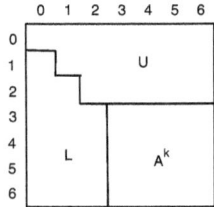

Fig. 3. L and U matrices decomposition using the same memory space of the original matrix A^0

Algorithm 1. Algorithm for LU Decomposition

1: **for** k=0 to n-1 **do**
2:    **for** j=k to n-1 **do**
3:       $u_{k,j} = a_{k,j}^k$
4:    **end for**
5:    **for** i=k+1 to n-1 **do**
6:       $l_{i,k}^k = \frac{a_{i,k}^k}{u_{k,k}}$
7:    **end for**
8:    **for** i=k+1 to n-1 **do**
9:       **for** j-k+1 to n-1 **do**
10:          $a_{i,j}^{k+1} = a_{i,j}^k - l_{i,k} \cdot u_{k,j}$
11:       **end for**
12:    **end for**
13: **end for**

Algorithm 2. Algorithm for LU Decomposition using the same matrix A

1: **for** k=0 to n-1 **do**
2:    **for** i=k+1 to n-1 **do**
3:       $a_{i,k} = \frac{a_{i,k}}{a_{k,k}}$
4:    **end for**
5:    **for** i=k+1 to n-1 **do**
6:       **for** j-k+1 to n-1 **do**
7:          $a_{i,j} = a_{i,j} - a_{i,k} \cdot a_{k,j}$
8:       **end for**
9:    **end for**
10: **end for**

---

statement of the Algorithm 2. Besides its own value, $a_{i,j}$ is updated using a value from the same line and another from the same column.

4   BSP-Based LU Application Modeling

This section explains how we modeled the LU sequential application on a BSP-based parallel one. Firstly, the bulk of the computational work in stage k of the sequential algorithm is the modification of the matrix elements $a_{i,j}$ with $i, j \geq k + 1$. Aiming to prevent the communication of large amounts of data, the update of $a_{i,j} = a_{i,j} + a_{i,k}.a_{k,j}$ must be performed by the process whose contains $a_{i,j}$. This implies that only elements of column k and row k of A need to be communicated in stage k in order to compute the new sub-matrix A^k.

An important observation is that the modification of the elements in row $A(i, k + 1 : n - 1)$ uses only one value of column k of A, namely $a_{i,k}$. The provided notation $A(i, k + 1 : n - 1)$ denotes the cells of line i varying from column $k + 1$ to $n - 1$. If we distribute each matrix row over a limit set of N processes, then the communication of an element from column k can be restricted to a multicast to $N - 1$ processes. Similarly, the modification of the elements in column $A(k + 1 : n - 1, j)$ uses only one value from row k of A, namely $a_{k,j}$. If we distributed each matrix column over a limit set of M processes, then the communication of an element of row k can be restricted to a multicast to $M - 1$ processes.

Considering the statements above, we are using a Cartesian scheme for the distribution of matrices. The square cyclic distribution is used as particularly suitable for matrix computations such as LU decomposition [3]. For them, it is natural to organize the processes by two-dimensional identifiers $P(s, t)$ with $0 \leq s < M$ and $0 \leq t < N$, where the number of processes $p = M.N$. Figure 4 depicts a 6×6 matrix mapped to 6 processes, where $M = 2$ and $N = 3$. Assuming that M and N are factors of n, each process will store nc (number of cells) cells in memory (see Equation 2).

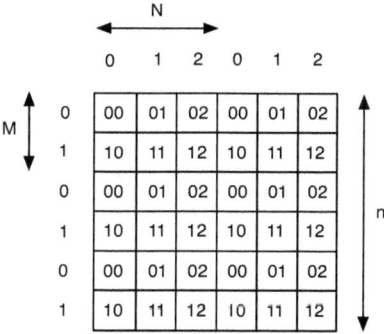

Fig. 4. Cartesian distribution of a 6×6 ($n \times n$) matrix over 2×3 ($M \times N$) processors. The label "st" in the cell denotes its owner, process $P(s, t)$.

$$nc = \frac{n}{M}.\frac{n}{N} \qquad (2)$$

A parallel algorithm uses data parallelism for computations and the need-to-know principle to design the communication phase of each superstep. Following the concepts of BSP, all communication performed during a superstep will be completed when finishing it and the data will be available at the beginning of the next superstep [4]. Concerning this, we modeled our algorithm using three kinds of supersteps. They are explained in Table 1. The element $a_{k,k}$ is passed to the process that computes $a_{i,k}$ in the first kind of superstep.

Table 1. Modeling three types of supersteps for LU computation

Type of superstep	Steps and explanation
First	Step 1.1 : $k = 0$
	Step 1.2 - Pass the element $a_{k,k}$ to cells which will compute $a_{i,k}$ ($k + 1 \leq i < n$)
Second	Step 2.1 : Computation of $a_{i,k}$ ($k + 1 \leq i < n$) by cells which own them
	Step 2.2 : For each i ($k + 1 \leq i < n$), pass the element $a_{i,k}$ to other $a_{i,j}$ elements in the same line ($k + 1 \leq j < n$)
	Step 2.3 : For each j ($k + 1 \leq j < n$), pass the element $a_{k,j}$ to other $a_{i,j}$ elements in the same column ($k + 1 \leq i < n$)
Third	Step 3.1 : For each i and j ($k+1 \leq i,j < n$), calculate $a_{i,j}$ as $a_{i,j} + a_{i,k}.a_{k,j}$
	Step 3.2 : $k = k + 1$
	Step 3.3 : Pass the element $a_{k,k}$ to cells which will compute $a_{i,k}$ ($k + 1 \leq i < n$)

The computation of $a_{i,k}$ is expressed in the beginning of the second superstep. This superstep is also responsible for sending the elements $a_{i,k}$ and $a_{k,j}$ to $a_{i,j}$. First of all, we pass the element $a_{i,k}$, $k + 1 \leq i < n$, to the other $N - 1$ processes that execute on the respective row i. This kind of superstep also comprises the passing of $a_{k,j}$, $k + 1 \leq j < n$, to the other $M - 1$ processes which execute on the respective column j. The superstep 3 considers the computation of $a_{i,j}$, the increase of k (next stage of the algorithm) and the transmission of $a_{k,k}$ to $a_{i,k}$ elements ($k + 1 \leq i < n$). The BSP application will execute one superstep of type 1 and will follow with the interleaving of supersteps 2 and 3. Concerning this, a $n \times n$ matrix will trigger $2n + 1$ supersteps in our LU modeling.

5   Evaluation Methodology

We applied simulation in three scenarios: (i) Application execution simply; (ii) Application execution with MigBSP without applying migrations; (iii) Application with MigBSP allowing migrations. Scenario ii consists in performing all scheduling calculus and all decisions about which processes will really migrate, but it does not comprise any actual migrations. Scenario iii enables migrations and adds the migrations costs on those processes that migrate from one processor to another. The difference between scenarios ii and i represents exactly the overhead imposed by MigBSP. The analysis of scenarios i and iii will show the final gain or loss of performance when process migration is applied.

We are using the SimGrid Simulator [6] (MSG module), which makes possible application modeling and process migration. This simulator is deterministic, where a specific input always results in the same output. We assembled an infrastructure with five Sets, which is depicted in Figure 5. Each node has a single processor. These Sets are based on a real cluster-of-clusters infrastructure located at the Federal University of Rio Grande do Sul, Brazil. Clusters Labtec, Corisco and Frontal have their nodes linked by Fast Ethernet, while ICE and Aquario are clusters with a Gigabit connection. The migration costs are based on real executions with AMPI [13].

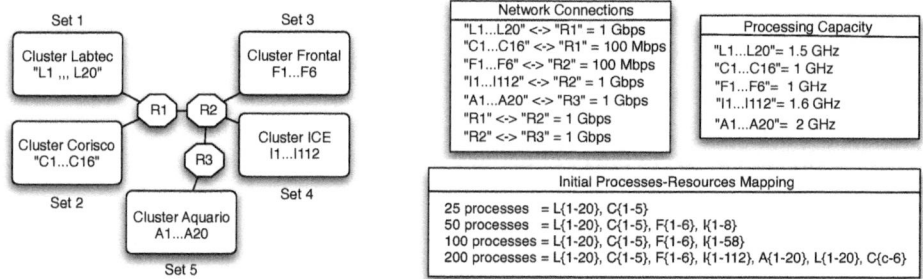

Fig. 5. Cluster-of-clusters environment with five Sets and the initial-processes mapping

Figure 5 presents the initial processes-resources mappings for each number of BSP processes. When the number of processes is greater than processors, the mapping begins again from the first Set. We modeled the Cartesian distribution $M \times N$ in the following manner: 5×5, 10×5, 10×10 and 20×10 for 25, 50, 100 and 200 processes, respectively. Moreover, we are applying simulation over square matrices with orders 500, 1000, 2000 and 5000. Lastly, the tests were executed using $\alpha = 4$, $\omega = 3$, $D = 0.5$ and $x = 80\%$.

6   Results Analysis

Table 2 presents the results when evaluating the 500×500, 1000×1000 and 2000×2000 matrices. The tests with the first matrix size show the worst results. Formerly, the higher the number of processes, the worse the performance, as we can observe in scenario i. The reasons for the observed times are the overheads related to communication and synchronization. Secondly, MigBSP indicated that all migration attempts were not viable due to low computing and communication loads when compared to migration costs. Considering this, both scenarios ii and iii have the same time results.

When testing a 1000×1000 matrix with 25 processes, the first rescheduling call does not cause migrations. After this call at superstep 4, the next one at superstep 11 informs the migration of 5 processes from cluster Corisco. They were all transferred to cluster Aquario, which has the highest computation power. MigBSP does not point migrations in the future calls. α always increases its value at each rescheduling call since the processes are balanced after the mentioned relocations. MigBSP obtained a gain of 12% of performance with 25 processes when comparing scenarios i and iii. With the same size of matrix and 50 processes, 6 processes from Frontal were migrated to Aquario at superstep 9. Although these migrations are profitable, they do not provide stability to the system and the processes remain unbalanced among the resources. Migrations are not viable in the next 3 calls at supersteps 15, 21 and 27. After that, MigBSP launches our second adaptation on rescheduling frequency in order to alleviate its impact and α begins to grow until the end of the application. The tests with 50 processes obtained gains of just 5% with process migration. This is explained by the fact that the computational load is decreased in this configuration when compared to the one with 25 processes. In addition, the bigger the number of the superstep, the smaller the computational load required by it. Therefore, the more advanced the execution, the lesser the gain with migrations. The tests with 100 and 200 processes do not present migrations owing to the forces that act in favor of migration are weaker than the Memory metric in all rescheduling calls.

Table 2. First results when executing LU linked to MigBSP (time in seconds)

Processes	500×500 matrix			1000×1000 matrix			2000×2000 matrix		
	Scen. i	Scen. ii	Scen. iii	Scen. i	Scen. ii	Scen. iii	Scen. i	Scen. ii	Scen. iii
25	1.68	2.42	2.42	11.65	13.13	10.24	90.11	91.26	76.20
50	2.59	3.54	3.34	10.10	11.18	9.63	60.23	61.98	54.18
100	6.70	7.81	7.65	15.22	16.21	16.21	48.79	50.25	46.87
200	13.23	14.89	14.89	28.21	30.46	30.46	74.14	76.97	76.97

The execution with a 2000×2000 matrix presents good results because the computational load is increased. We observed a gain of 15% with process relocation when testing 25 processes. All processes from cluster Corisco were migrated to Aquario in the first rescheduling call (at superstep 4). Thus, the application can take profit from this relocation in its beginning, when it demands more computations. The time for concluding the LU application is reduced when passing from 25 to 50 processes as we can see in scenario i. However, the use of MigBSP resulted in lower gains. Scenario i presented 60.23s while scenario iii achieved 56.18s (9% of profit). When considering 50 processes, 6 processes were transferred from cluster Frontal to Aquario at superstep 4. The next call occurs at superstep 9, where 16 processes from cluster Corisco were elected as migration candidates to Aquario. However, MigBSP indicated the migration of only 14 processes, since there were only 14 unoccupied processors in the target cluster. The execution of 100 processes presented the same behavior of the execution with 50 processes. Nevertheless, the performance gain was reduced to 4% with 100 processes given the reduction of the workload per process.

We observed that the higher the matrix order, the better the results with process migration. Considering this, the evaluation of a 5000×5000 matrix can be seen in the Figure 6. The simple movement of all processes from cluster Corisco to Aquario represented a gain of 19% when executing 25 processes. The tests with 50 processes obtained 852.31s and 723.64s for scenario i and iii, respectively.

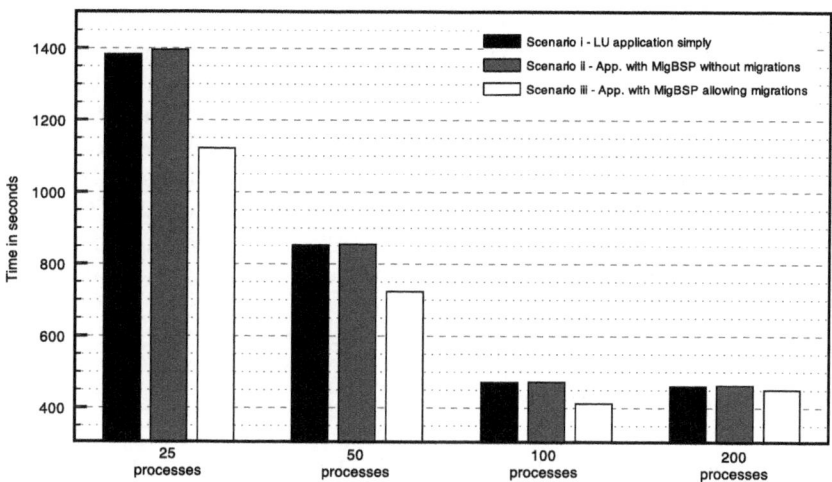

Fig. 6. Performance graph with our three scenarios for a 5000 × 5000 matrix

The same migration behavior found on the tests with the 2000 × 2000 matrix was achieved in Scenario *iii*. However, the increase of matrix order represented a gain of 15% (order 5000) instead of 10% (order 2000). This analysis helps us to verify our previous hypothesis about performance gains when enlarging the matrix. Finally, the tests with 200 processes indicated the migration of 6 processes (p195 up to p200) from cluster Corisco to Aquario at superstep 4. Thus, the nodes that belong to Corisco just execute one BSP process while the nodes from Aquario begin to treat 2 processes. The remaining rescheduling calls inform the processes from Labtec as those with the higher values of *PM*. However, their migrations are not considered profitable. The final execution with 200 processes achieved 460.85s and 450.33s for scenarios *i* and *iii*, respectively.

7 Related Work

Bhandarkar, Brunner and Kale presented a support for adaptive load balancing in MPI-based LU application [2]. Periodically, the MPI application transfers control to the load balancer using a special call MPI_Migrate(). Processes reorganization on LU application is proposed by Ennes et al [19]. Such work imposes an extra execution for getting parameters for a communication-graph construction.

Concerning the BSP scope, Jiang, Tong and Zhao presented resource load balancing based on agents [14]. Load balancing is launched when a new task is inserted and it is based on the load rank of the nodes. Scheduling service sends this new task to the current lightest node. Load value is calculated taking such information: CPU, memory resource, number of current tasks and number of network links. In addition, we can cite two works that present migration on BSP applications. The first one describes the PUBWCL library, which exploits the computing cycles of idle computers [5]. PUBWCL can migrate a process during its computation phase and after the barrier. All algorithms just use data about the nodes and consider the computation times from each process.

Other work on BSP context comprises the implementation of the PUB library [4]. The author explains that a load balancer decides when to launch the processes migration, but this issue is not addressed in [4]. He proposed both centralized and distributed strategies for load balancing. In the distributed approach, every node chooses c nodes randomly and asks for their load. One process is migrated if the minimum load found is smaller than the load of its current node. Both strategies take into consideration neither the processes communication nor the migration costs.

8   Concluding Remarks

Scheduling schemes for multi-programmed parallel systems can be viewed in two levels [10,18,21]. In the first level processors are allocated to a job. In the second level processes from a job are (re)scheduled using this pool of processors. MigBSP can be included in this last scheme, offering algorithms for load (BSP processes) rebalancing among the resources during application runtime. Our model can be seen as a scheduler middleware that does not insert changes in the application code nor needs knowledge about it or the system infrastructure. Especially, this paper presented MigBSP shortly as well as a modeling and an execution of a BSP-based LU application. The tests when linking it to the LU application enabled us to conclude encouraging results: gains of performance and a short overhead of MigBSP. Contrary to existing works[2,12,19], these results are obtained without modifying the application code and without extra executions to feed the load balancing model.

The short overhead of MigBSP is enabled mainly by using efficient adaptations and through the rapid calculus of the scheduling decisions. Firstly, PM (Potential of Migration) considers processes and Sets (different sites), not performing all processes-resources tests at the rescheduling moment. Meanwhile, our adaptations were crucial to enable MigBSP as a viable scheduler. Instead of performing the rescheduling call at each fixed interval, they manage a flexible interval between calls based on the behavior of the processes. Their concepts are: (i) to postpone the rescheduling call if the system is stable (processes are balanced) or to turn it more frequent, otherwise; (ii) to delay this call if a pattern without migrations in ω calls is observed.

For example, the low overhead of MigBSP may be expressed when executing 50 processes and a 2000×2000 matrix. In this context, it adds 3% of costs (MigBSP algorithms are enabled but no migrations are performed). Firstly, this feature is due to the simplicity of the PM evaluation, since it considers the hierarchy notion and employs heuristics. Secondly, MigBSP adaptations work to turn the model viable, especially when migrations cause performance gains but the system remains unbalanced. This occurred with a matrix of order 1000 and 50 processes. Besides this, we observed that the larger the matrix size, the bigger the gain with migrations. Thus, MigBSP obtained the best results with a

5000×5000 matrix. In this situation, we can observe gains larger then 15% when applying our migrations decisions on application execution. Gains of 19% and 15% were obtained when running 25 and 50 processes with migrations to the fastest cluster. Moreover, contrary to other situations, this matrix size enables migrations when using 200 processes due to its larger computing grain.

Acknowledgements

This work was funded by the following Brazilian Agencies: CNPq and CAPES.

References

1. Aggarwal, G., Motwani, R., Zhu, A.: The load rebalancing problem. In: SPAA 2003: Proceedings of the Fifteenth Annual ACM Symposium on Parallel Algorithms and Architectures, pp. 258–265. ACM Press, New York (2003)
2. Bhandarkar, M.A., Brunner, R., Kale, L.V.: Run-time support for adaptive load balancing. In: IPDPS 2000: Proceedings of the 15 IPDPS 2000 Workshops on Parallel and Distributed Processing, pp. 1152–1159. Springer, London (2000)
3. Bisseling, R.H.: Parallel Scientific Computation: A Structured Approach Using BSP and MPI. Oxford University Press, Oxford (2004)
4. Bonorden, O.: Load balancing in the bulk-synchronous-parallel setting using process migrations. In: 21th International Parallel and Distributed Processing Symposium (IPDPS 2007), pp. 1–9. IEEE, Los Alamitos (2007)
5. Bonorden, O., Gehweiler, J., auf der Heide, F.M.: Load balancing strategies in a web computing environment. In: Wyrzykowski, R., Dongarra, J., Meyer, N., Waśniewski, J. (eds.) PPAM 2005. LNCS, vol. 3911, pp. 839–846. Springer, Heidelberg (2006)
6. Casanova, H., Legrand, A., Quinson, M.: Simgrid: A generic framework for large-scale distributed experiments. In: Tenth International Conference on Computer Modeling and Simulation (uksim), pp. 126–131. IEEE Computer Society, Los Alamitos (2008)
7. Chen, L., Wang, C.-L., Lau, F.: Process reassignment with reduced migration cost in grid load rebalancing. In: IEEE International Symposium on Parallel and Distributed Processing, IPDPS 2008, pp. 1–13 (April 2008)
8. da Rosa Righi, R., Pilla, L., Carissimi, A., Navaux, P.O.A.: Controlling processes reassignment in bsp applications. In: 20th International Symposium on Computer Architecture and high Performance Computing (SBAC-PAD 2008), pp. 37–44. IEEE Computer Society, Los Alamitos (2008)
9. da Rosa Righi, R., Pilla, L.L., Carissimi, A., Navaux, P., Heiss, H.-U.: Migbsp: A novel migration model for bulk-synchronous parallel processes rescheduling. In: 10th IEEE International Conference on High Performance Computing and Communications, pp. 585–590 (2009)
10. Frachtenberg, E., Schwiegelshohn, U.: New Challenges of Parallel Job Scheduling. In: Frachtenberg, E., Schwiegelshohn, U. (eds.) JSSPP 2007. LNCS, vol. 4942, pp. 1–23. Springer, Heidelberg (2008)
11. Galindo, I., Almeida, F., Badía-Contelles, J.M.: Dynamic load balancing on dedicated heterogeneous systems. In: Lastovetsky, A., Kechadi, T., Dongarra, J. (eds.) EuroPVM/MPI 2008. LNCS, vol. 5205, pp. 64–74. Springer, Heidelberg (2008)

12. Gustavson, F.G.: High-performance linear algebra algorithms using new generalized data structures for matrices. IBM J. Res. Dev. 47(1), 31–55 (2003)
13. Huang, C., Zheng, G., Kalé, L., Kumar, S.: Performance evaluation of adaptive mpi. In: PPoPP 2006: Proceedings of the Eleventh ACM SIGPLAN Symposium on Principles and Practice of Parallel Programming, pp. 12–21. ACM Press, New York (2006)
14. Jiang, Y., Tong, W., Zhao, W.: Resource load balancing based on multi-agent in servicebsp model. In: Shi, Y., van Albada, G.D., Dongarra, J., Sloot, P.M.A. (eds.) ICCS 2007. LNCS, vol. 4489, pp. 42–49. Springer, Heidelberg (2007)
15. Low, M.Y.-H., Liu, W., Schmidt, B.: A parallel bsp algorithm for irregular dynamic programming. In: Xu, M., Zhan, Y.-W., Cao, J., Liu, Y. (eds.) APPT 2007. LNCS, vol. 4847, pp. 151–160. Springer, Heidelberg (2007)
16. Maassen, J., van Nieuwpoort, R.V., Kielmann, T., Verstoep, K., den Burger, M.: Middleware adaptation with the delphoi service. Concurrency and Computation: Practice & Experience (2006)
17. Pontelli, E., Le, H.V., Son, T.C.: An investigation in parallel execution of answer set programs on distributed memory platforms: Task sharing and dynamic scheduling. Comput. Lang. Syst. Struct. 36(2), 158–202 (2010)
18. Qin, X., Jiang, H., Manzanares, A., Ruan, X., Yin, S.: Communication-aware load balancing for parallel applications on clusters. IEEE Trans. Comput. 59(1), 42–52 (2010)
19. Silva, R.E., Pezzi, G., Maillard, N., Diverio, T.: Automatic data-flow graph generation of mpi programs. In: SBAC-PAD 2005: Proceedings of the 17th International Symposium on Computer Architecture on High Performance Computing, Washington, DC, USA, pp. 93–100. IEEE Computer Society, Los Alamitos (2005)
20. Valiant, L.G.: A bridging model for parallel computation. Commun. ACM 33(8), 103–111 (1990)
21. Wieczorek, M., Podlipnig, S., Prodan, R., Fahringer, T.: Bi-criteria scheduling of scientific workflows for the grid. ccgrid, 9–16 (2008)
22. Xhafa, F., Abraham, A.: Computational models and heuristic methods for grid scheduling problems. Future Gener. Comput. Syst. 26(4), 608–621 (2010)

A P2P Approach to Many Tasks Computing for Scientific Workflows*

Eduardo Ogasawara[1,2], Jonas Dias[1], Daniel Oliveira[1], Carla Rodrigues[1],
Carlos Pivotto[1], Rafael Antas[1], Vanessa Braganholo[3], Patrick Valduriez[4],
and Marta Mattoso[1]

[1] COPPE/Federal University of Rio de Janeiro, Rio de Janeiro, Brazil
Cidade Universitária, Centro de Tecnologia, H-319,
Caixa Postal 68511, Ilha do Fundão, Rio de Janeiro, Brazil
Tel.: +552125628672; Fax: +552125628676
[2] Federal Center of Technological Education, Rio de Janeiro, Brazil
[3] Fluminense Federal University
[4] INRIA & LIRMM, Montpellier, France
{ogasawara,jonasdias,danielc,carlarod,pivotto}@cos.ufrj.br,
{zuquim,marta}@cos.ufrj.br, vanessa@ic.uff.br,
Patrick.Valduriez@inria.fr

Abstract. Scientific Workflow Management Systems (SWfMS) are being used intensively to support large scale *in silico* experiments. In order to reduce execution time, current SWfMS have exploited workflow parallelization under the arising Many Tasks Computing (MTC) paradigm in homogeneous computing environments, such as multiprocessors, clusters and grids with centralized control. Although successful, this solution no longer applies to heterogeneous computing environments, such as hybrid clouds, which may combine users' own computing resources with multiple edge clouds. A promising approach to address this challenge is Peer-to-Peer (P2P) which relies on decentralized control to deal with scalability and dynamic behavior of resources. In this paper, we propose a new P2P approach to apply MTC in scientific workflows. Through the results of simulation experiments, we show that our approach is promising.

Keywords: Scientific experiments, scientific workflows, Scientific Workflow Management Systems (SWfMS), many tasks computing (MTC), peer-to-peer (P2P).

1 Introduction

The evolution of computer science in the last decade has enabled the exploration of a new type of scientific experiments based on computer simulations, known as *in silico* experiments [1]. In such experiments, scientists may use different programs to perform an activity. In these scenarios data produced by one activity needs to be passed as input to another activity, and conversion steps may need to be performed along the execution. This chain of programs that composes a scientific experiment is usually

---

* Work partially sponsored by CNPq and INRIA within the Sarava équipe associée and by CAPES.

J.M.L.M. Palma et al. (Eds.): VECPAR 2010, LNCS 6449, pp. 327–339, 2011.

represented as scientific workflows. A scientific workflow is an abstraction that allows the structured composition of programs as a sequence of activities aiming a desired result [1]. This sequence is supported by Scientific Workflow Management Systems (SWfMS), which are software packages that enable to define, execute, and monitor scientific workflows. SWfMS are responsible for coordinating the invocation (also called orchestration) of programs (specified within scientific workflows) to be executed either locally or in remote environments.

Due to the exploratory nature of the scientific method [2], an *in silico* experiment may require the exploration of a certain scientific workflow using different parameters or input data. This situation occurs, for example, in Monte Carlo simulations, parameter sweep, and data mining, where the same workflow is exhaustively executed until the exploration finishes. Recently, these experiments were grouped into a new computational paradigm called Many Tasks Computing (MTC) [3]. It consists on using various computing resources over short periods of time to accomplish many (dependent or independent) computational tasks.

SWfMS have successfully exploited workflow parallelization in homogeneous computing environments, such as multiprocessor or cluster systems. These environments rely on centralized control of resources which eases parallelization and exploit high-throughput, low-latency communication networks which bring performance in order to confirm or refute a hypothesis. However, there are now many different heterogeneous computing environments which could be exploited for executing scientific workflows, for instance, grids, desktop grids, volunteers computing projects (such as BOINC [4] and World Community Grid [5]), or hybrid clouds [6], which may combine users' own computing resources with multiple edge clouds. The main problem is that each of these computing environments requires different efforts, resources and scientists skills to be applied within a solution for workflow parallelization. However, a major requirement for scientists is to model activities or workflows to be parallelized in an implicit way, independent of the target environments. Another important requirement for scientists is provenance gathering [7], *i.e.*, the capability of reproducing the results of a scientific experiment. Nevertheless, provenance gathering in heterogeneous distributed environments is still an open problem.

Peer-to-Peer (P2P) is a promising approach to address the aforementioned challenges of applying MTC to scientific workflows. Unlike traditional distributed (client-server) computing, P2P relies extensively on decentralized control and the ability of any node (peer) to perform any task which makes it possible to deal with scalability and dynamic behavior of nodes (including churns, *i.e.*, frequent joins and leaves of the P2P network).

In previous work [8], the authors have already discussed how P2P techniques can be very useful to large-scale grids. More generally, we believe that, in heterogeneous computing environments, P2P is a powerful approach to support MTC of scientific workflows, with integrated provenance gathering.

In this paper, we explore a new P2P approach to support MTC through scientific workflows in heterogeneous computing environments. We propose SciMule, a P2P middleware that allow transparent parallelization of workflow tasks and dynamic scheduling of MTC. To evaluate our approach, we developed a SciMule simulation environment and performed several experiments under typical scientific workflow scenarios.

The paper is organized as follows. Section 2 introduces a comparative study based on qualitative aspects of the state of art P2P approaches suitable to deal with workflow activity distribution. Section 3 describes the SciMule conceptual architecture and presents our parallel workflow execution strategies. Section 4 presents experimental results. We conclude in Section 5.

2 Backgrounds on P2P Networks

P2P systems can be classified into two types: centralized and decentralized [9]. Centralized systems use one or more servers (also called super-peers) which are responsible for storing information about other peers and solving P2P network requests such as joining the network. Decentralized systems do not rely on specific peers for network control, because information is totally distributed between all peers. Furthermore, there are hybrid solutions which combine characteristics of the two aforementioned types.

The hierarchical P2P approach is a hybrid solution to decrease complexity and topology maintenance overhead of large-scale decentralized P2P systems. This approach creates several smaller network groups, where the peers still form a decentralized network. Some peers from a group on a given hierarchy level connect with peers from a higher hierarchy level, thus forming the hierarchical structure.

Canon [10] is a DHT-based hierarchical approach. In Canon, each peer creates links to other peers on its own hierarchical level generating a domain. The hierarchy is constructed by adding links from each peer in one domain to some set of peers in other domains, resulting in a higher level group. Messages are routed inside domains. If a query regarding some resource discovery cannot be resolved locally, then it is routed to a peer p which keeps connection with the next higher level. At Canon, churns at lowest levels do not affect the entire network and only internal neighborhood tables need be updated.

Garcés-Erice *et al.* [11] proposes a hierarchical approach using the super-peer model. They propose a network formed by two types of nodes: ordinary nodes and super-peers. Ordinary nodes are organized into groups and exchange messages only inside their group. Super peers are gateways between groups. They create links to others super-peers for inter-group message exchanging and keep information about ordinary nodes and their content.

Considering the super-peer model, hierarchical structures can be used in Grid Systems as shown in Mastroianni et al. [12], where each virtual organization in the Grid has a single super-peer which establishes connection with other super-peers to form a network at a higher level. Super peers control membership requests and resource discovery services on the Grid. A similar architecture is discussed in a previous work [13], where a CAN-based P2P network is built according to the super-peer model using a more sophisticated message routing. Super-peers keep information about local nodes and neighboring super-peers.

The main focus of this work is on executing scientific workflows in large-scale P2P networks and includes the network type specification. There are many aspects that should be considered to select a suitable approach to deal with distributed workflow activities execution. A single workflow activity can generate thousands of tasks that would be distributed over the P2P network. Thus, the *load balancing* factor is very important to keep the network stable without overloading a set of peers. *Scalability* is also important since it is a large-scale network that deals with thousands of nodes. P2P is supposed to be fault-tolerant but churn events are predicted. So, the *churn risk*, i.e., how impactful is a churn event on the worst case scenario in the network, is also very important on any P2P system. *Maintenance cost* of the P2P topology may also be an important aspect since it affects the scheduling process, since the node submitting a workflow activity should know where it can distribute tasks. Table 1 summarizes an analysis considering the types of P2P networks we previously discussed and the following factors: (i) Load Balancing; (ii) Scalability; (iii) Churn Risks and; (iv) Maintenance Cost.

Table 1. Analysis of the P2P networks approaches

Factor \ Network type	Centralized	Decentralized	Hierarchical
Load Balancing	None	High	Moderate
Scalability	Low	Moderate	High
Churn Risks	Total Failure	Ponctual Failures	Domain Failures
Maintenance Cost	Low	High	Moderate

Centralized P2P networks have no load balancing since the super-peers centralize resource discovery and search services. The decentralized approach fully distributes control over the network, thus providing high load balancing. The hierarchical approach centralizes part of the services on some special (inter-group) peers, thus providing a moderate load balancing. The centralized approach relies on the central node capacities, so it does not scale very well. The decentralized approach has a moderate scalability since it is hard to maintain a huge network with fully distributed control. The hierarchical approach scales better since it establishes part of the control on the inter-group peers.

Regarding the churn risks, in a centralized network, if a churn happens on the central nodes, the whole network fails. In a decentralized network, the churn represents just a punctual failure, since the nodes are independent. On a hierarchical P2P network, if a churn happens on an inter-group peer, the whole group fails. The maintenance cost of a centralized network is low, since only the information on the central nodes has to be updated. In the decentralized approach, though, the information is distributed and updates usually involve flooding algorithms. On the hierarchical approach, the inter-group peer keeps some information about its groups and it is the gateway between its group and the others. Flooding may happens only inside groups.

Considering these aspects, we believe hierarchical P2P networks may be the most adequate solution for distributing scientific workflows activities, especially for large-scale networks that demand great performance.

3 Design of SciMule

SciMule is a middleware designed to distribute, control and monitor the execution of activities of a scientific workflow in a P2P environment. We consider P2P environment as a distributed and heterogeneous computing environment where activities, data and parameters are distributed over the network to promote workflow/activity parallelization. These activities can be programs or even independent scientific workflows. SciMule was designed considering a hierarchical structured approach. We choose the hierarchical approach since it establishes a good tradeoff between the centralized and decentralized approaches, which generally scales well for large scale P2P networks, while tolerating churn effects. SciMule has a three-layer architecture: (i) a submission layer that dispatches activities to be executed in the P2P environment through a generic SWfMS, (ii) an execution layer that receives experiment packages (*i.e.* activities, parameters and data) that need to be executed, and (iii) the overlay layer that holds information about how peers are placed on the network and how they are related together. This type of information is important when a new node needs to be inserted on the overlay and also to keep the P2P network balanced.

During SciMule network lifetime, a peer may play several different roles. It may act as a client peer submitting new activities through the submission layer, or it may act as an executor peer, receiving tasks to be executed by the execution layer. A peer may also be elected as a gate peer. The gate peer is responsible for keeping and publishing the list of nearby peers and their subjects. A subject is an abstraction that represents a set of programs related to a certain domain of knowledge. The gate peer role is managed by SciMule through the overlay layer. Each role is strongly coupled to a specific layer, but they are all independent. Peers may act only as a client, others may act only as executors, but they usually act as both. An elected gate peer may also act as client and/or executor.

SciMule aims at isolating scientists from the complexity of distributing workflow activities (or entire workflows) using MTC paradigm over a P2P network. This is done by offering a transparent and explicit structure to distribute scientific workflows activities that demand high computation. In this way, SciMule is an adaptation of Hydra [14] for the P2P environment. Hydra is also a middleware which provides a set of components to be included on the workflow specification of any SWfMS to control parallelization of activities in clusters. While Hydra and SciMule share many conceptual behavior, their architecture design is completely different due to the intrinsic characteristics of P2P versus client server architecture. For example, in SciMule, a computer may distribute a set of tasks that compose an activity acting as a client peer, but it may latter run tasks that arrive from other peers acting as an executor peer. This behavior is not supported by Hydra.

3.1 SciMule Architectural Features

SciMule shares many important features with Hydra [14]. It was also designed to provide two different types of parallelization: parameter sweep parallelism and data parallelism. These two types of parallelism may normally represent a barrier for the scientists to control and register provenance, since they require a great effort and discipline to manage too much information when executed in an ad-hoc manner over any distributed environment. SciMule provides a systematic approach to support both types of parallelism with heterogeneous distributed provenance gathering. Some of these features are also available in Hydra, but SciMule aims a broader execution experience running asynchronously and distributed over the dynamic and heterogeneous P2P network. Meantime, SciMule architecture is still simple to deploy and able to be architecturally linked to any SWfMS with minimum effort. The entire architecture is described in the following sub-sections.

3.2 SciMule Conceptual Architecture

Figure 1 presents the SciMule architecture. As mentioned before, it is composed by three layers: submission, execution and overlay. It is important to observe that each peer has all the three layers. Numbers alongside the arrows of Figure 1 denote the execution sequence of the architecture components in the scope of each layer.

Submission Layer Components. SciMule submission layer components provide transparent ways to parallelize scientific workflows and to distribute activities through neighbor peers using MTC paradigm. It is divided in two basic parts: workflow components and MTC controller components. The workflow components represent generic modules that are included in the SWfMS to enable the interaction with SciMule. SciMule has two main components to be plugged into the SWfMS: *Client Setup* and *Client Dispatch*.

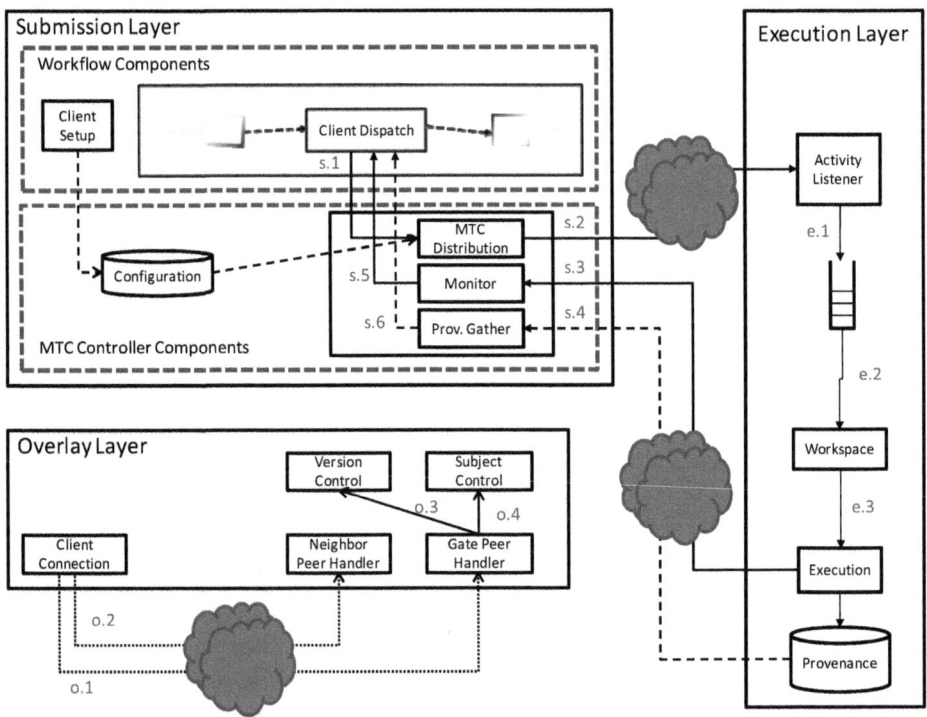

Fig. 1. SciMule Architecture

Similar to Hydra, the *Client Setup* is the component responsible for general configuration of the type of parallelism to be used during workflow execution and the P2P environment, which includes both data files and template files [14] to be transferred over the P2P network. All these configurations are used by SciMule MTC controller. The *Client Dispatch* distributes activities over the P2P network through the MTC controller during workflow executions. The execution on the P2P network is transparent while using this component. Once the distributed execution is finished, the *Client Dispatch* returns control to the SWfMS.

The MTC Controller Components represents the SciMule MTC engine that is responsible for distributing tasks, gathering distributed provenance data and handling churns during workflow execution. The MTC controller components are invoked by the Client Dispatch (event s.1 of Figure 1). An activity that needs to be distributed is divided into a set of tasks. A task is an experiment package which includes enough information for the execution, *i.e.*, workflow code, data and parameter for execution. Each task is scheduled considering the status of the peer neighbors (computing power, number of tasks being processed and the available bandwidth). The status information may be obtained through gossip [15]. After this point, each task contains the executor peer address. The peer address may be the local machine, which means that the task is going to be executed locally. All the optimization process is static for each distributed activity. This means that once a task is scheduled to be executed by peer α, it is executed by α, unless α suffers a churn (in this case the task may be redistributed to another peer).

The MTC distribution component distributes the tasks (event s.2 of Figure 1) and monitors all peers that are running one of the distributed tasks (event s.3 of Figure 1). Once a task finishes its remote execution, provenance data is collected from the remote peer (event s.4 of Figure 1). Like in Hydra [14], the provenance repository was modeled to link the prospective provenance and the retrospective distributed provenance [7] collected during the scientific experiment life cycle. Once all tasks are completed, the control is returned to the SWfMS (event s.5 of Figure 1), together with the collected provenance data (event s.6 of Figure 1).

Execution Layer. On the execution layer, there are three main components: the activity listener, the workspace handler and the task handler. The activity listener is the component responsible for listening to the connections with other peers in order to receive a task. Each task received is put into a queue (event e.1 of Figure 1) to be consumed by the task handler.

The workspace handler is responsible for setting up the executor peer environment to execute the task, *i.e.*, unpack the experiment package for execution (it creates the directory structure, sets parameters, etc). Once the task is ready (event e.2 of Figure 1), the task handler invokes the corresponding program using the parameters and data that were packaged in the task (event e.3 of Figure 1). When the execution is finished, the control returns to the MTC distribution.

Overlay Layer. In order to support MTC, SciMule overlay is defined to balance peers according to locality principles [15] and to subject. Each subject is a preset tool of a certain domain (*e.g.*, bioinformatics). The goal is that peers keep communication to other peers that enroll the same subject, which means that they may have the same set of programs. The principle of locality establishes that it is also important that peers cooperate with other peers that are close to them in locality. In this way, SciMule is a scalable P2P MTC network that is both locality and subject aware.

In the SciMule approach, most information about the overlay is stored on gate peers. Gate peers keep a list of nearby nodes, *i.e.*, nodes that once contacted them to join the P2P network. A gate peer has one or more subjects associated to it. Gate peers control versions of their related subjects, replicate them on nearby gate peers and notify known peers enrolled with the same subject about the new versions, so they can update their files. The rules for deploying new versions of subjects assume concepts of reputation, *i.e.*, only Gate peers with greater reputation may deploy new versions of subjects. Gate peers are elected according to some metrics: they are assumed to have low churn frequency and high reputation. To avoid data loss, gate peers also have a backup node that replicates nearby peers list. Data management and replica control on large-scale P2P systems are well discussed by Akbarinia *et al.* [16]. Leader election occurs for a promotion of a peer into to a gate peer to guarantee a certain rate between gate peers and peers in the network.

In SciMule, each gate peer keeps a list of ordinary nodes that defines a group, and each ordinary node belongs to only one group, which means that it is registered in a single gate peer. However, a peer may have neighbors from its own group and from two other adjacent groups. Thus, when churn happens, three groups of the network are affected, at most. Any node inside a group can submit a task from a workflow activity directly to its neighbors. This decentralized submission enhances the load balancing.

The amount of gate peers in the network directly affects the maximum connectivity of a node in the network. On a given network with n peer nodes and g gate peers distributed on the network, the gate peers keep a mean of n/g nearby nodes. Each node is registered on the nearest gate peer, from where it obtains the list of other nearby nodes. The node also contacts another nearby gate peer to get a broader listing. Thus, a node registered on a given gate peer may have neighbors registered on the same gate peer and also from the other two adjacent gate peers. Since each gate peer has n/g nodes on its list, the maximum connectivity of the node is $3n/g$.

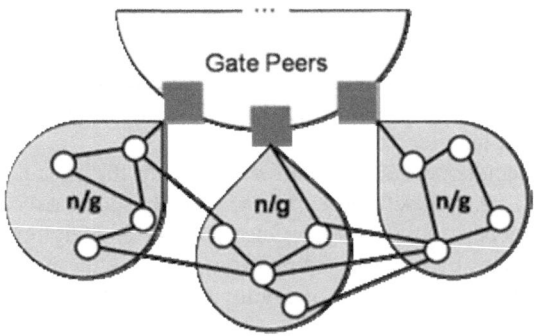

Fig. 2. P2P MTC configuration

Although each peer contacts some gate peers to join SciMule network, it does not necessarily establishes inter-peer connections with them, since other peers may be better suited to the selection criteria. SciMule limits the number of neighbors that an incoming peer can get. A novice peer only has about half the average of neighbors that older peers have. This limitation aims at avoiding free riding [17]. The number of inter-peer is also important to maintain the network balanced, which means that new incoming peers connect to peers with lower inter-peer connection cardinality. The cardinality information of a group is stored in the gate peer and is periodically refreshed. When a peer requests the nearby peer list, it comes ordered by cardinality. We are also studying the possibility to spread this information using gossip [15]. The goal is to balance the P2P network structure along time. Figure 2 presents an example of P2P MTC configuration.

4 SciMule Evaluation

We have built a simulation environment to evaluate the proposed architecture. We chose the simulation approach instead of a real life experiment since it is difficult and expensive to build a real P2P infrastructure with thousands of nodes just to test and evaluate our architecture [18].

SciMule Simulator was developed on top of PeerSim [19]. It was necessary to extend the PeerSim simulator and add some extra components to support SciMule architectural demands. The PeerSim node component was extended to store locality and subjects. It also got the ability to become a gate peer, which means that it can store a list of nearby nodes. We also modeled the link between nodes, the data package that transit on links, the workflow activity and its tasks. The data transferring system considers the bandwidth of both sender and receiver peers. Since a peer can perform multiple data transfers, the system uses a round robin approach to share the bandwidth. The workflow activity is modeled as an object that can be decomposed in a set of tasks. A task is an object that has a size and a processing cost. The size attribute is considered for data transfers while the cost is considered when the task is computed by one of the executor peers. Most parts of the simulator were developed following SciMule three-layer architecture.

The simulator allows the control of several variables, like the size of the network and characteristics of the workflow activities that are submitted during the simulation. For the sake of performance, the simulator uses PeerSim cycle-based approach. The variables were parameterized based on a real life experiment [14] assuming that a cycle corresponds to 30 seconds.

Our study has eight independent variables [20]: (i) the number of simulation cycles, (ii) the initial number of peers (n) in the network, (iii) average number (k) of neighbors, (iv) activities submission frequency (f) that obeys a Poisson distribution, (v) number of tasks (*tasks*) of an activity, (vi) task computational cost (*cost*) in processing units (p.u.), (vii) task data size (*size*) in kilobits and (viii) churn frequency (*churn*), also following a Poisson distribution. Five of these variables – k, *tasks*, *cost*, *size* and *churn* – are factors [20] which have from two to four treatments [20]. On this aspect, our simulation experiment is a five-factor, four-treatment study [20]. Table 2 summarizes our independent variables and factors with their respective treatments [14].

The combinations of all factors generated 384 instances of simulation. The dependent variables [20], *i.e.*, the values we assess, are the speed up and the time spent transferring data of each activity executed on the P2P network. However, our analysis focuses on speed up results, in order to evaluate the general performance of the network due to increase of number of peers involved on execution.

Table 2. Summary of the study variables

					Factors		
				Independent Variables			
cycles	*n*	*f*	*k*	*tasks*	*cost*	*size*	*churn*
14400	4096	0.01	32	128	4000	12000	0.00
			64	512	8000	24000	0.05
			128		16000	48000	0.10
			256		32000	96000	

Each one of the 384 instances was considered an independent simulation. All the simulations instances ran on a SGI Altix 8200 cluster using Hydra [14]. The simulation results were stored on a PostgreSQL database. We have made a statistical analysis of the data taking the average speed up from the completed executions of the distributed activities [21]. We have selected four representative activities for our experimental analysis. Two of them have Low task Cost (LCMS and LCBS) with different task sizes. And the other two have Small task Size (HCSS and MCSS) with different costs. They are described on Table 3. Since we are measuring speed up, the processing units (p.u.) do not need to be converted to a real life unit. On the simulations, the peers have a computational power that follows a gamma distribution with average of 80 p.u/cycle, with scale 30 and shape 2.

Table 3. Representative Activities for Performance Analysis

Activity Name	Task Cost (p.u.)	Task Size (MB)
LCMS - Low Cost Medium Size	4,000	6
LCBS – Low Cost Big Size	4,000	12
HCSS – High Cost Small Size	32,000	1.5
MCSS – Medium Cost Small Size	16,000	1.5

Figure 3shows the speed up curve of parameter sweep parallelization of the selected activities varying the churn events. It shows the scenarios without churn events, with a 5% and 10% frequency in a Poisson distribution. The first scenario is unrealistic, but it presents a baseline for measurements. In the first graphic, even without churn, activities with tasks of larger size do not scale very well. The peer that submits the activity is the responsible for transmitting all the data to the other selected execution peers. Transmitting several huge tasks may overload the submitter peer network bandwidth. Thus, the data set delivery takes more time, slowing down the overall execution process. It seems that involving fewer peers in the execution is more convenient, since, with the same data set, a peer may run different executions just by assigning a different set of parameters. Therefore, scheduling several tasks of an activity to a small set of peer saves data transmissions and speeds up the overall execution.

The scenarios with churn show a decrease in speed up, especially on the activities with larger size. The larger the number of nodes involved, the higher is the chance of a churn in the executor nodes group. When a churn happens, the task needs to be rescheduled and, possibly, the task data set needs to be retransferred to another peer, if the new peer does not have it. Since smaller tasks are easier to transport in the network, activities HCSS and MCSS suffer less impact from churn events because its tasks have only 1.5MB. LCMS and LCBS, in contrast, suffer more impact since their tasks have 6MB and 12MB, respectively.

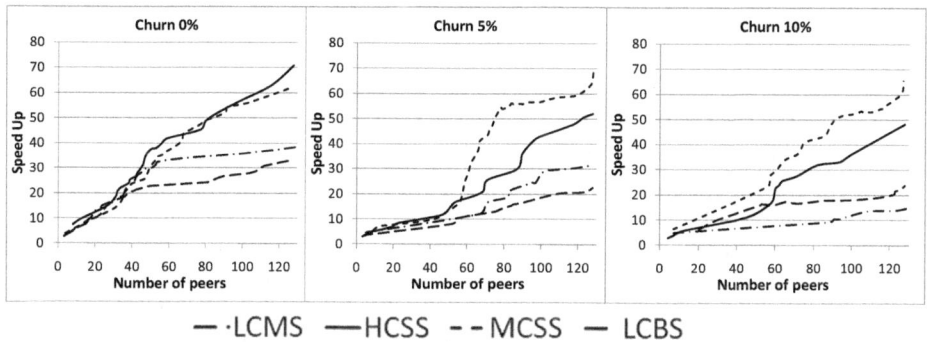

Fig. 3. Statistical results for parameter sweep for the selected activities

Figure 4 shows the speed up results for the same activities, but now considering the data fragmentation scenario. Different from the parameter sweep case, each task of the activity has a specific data set. Thus, involving fewer peers in the execution may not be the best strategy. In Figure 4, the speed up rate is positive in all scenarios. However, the impact of churn events is still clear. Compared to the parameter sweep cases, the data fragmentation is more sensitive to churn events. This is reasonable since, when a parameter sweep task is rescheduled, the chosen node has possibly the necessary data packages to process the new assigned task. On a data fragmentation task, though, the data set necessarily has to be retransmitted.

Just like in the parameter sweep case, activities with smaller tasks (HCSS and MCSS) also scale better with churn. The activity cost seems to have little influence in performance, since data transmission appears to be the major bottleneck. However, it is possible to assess that MCSS activity scales better than HCSS.

These initial results show that P2P networks are suitable environments to distribute workflow activities. However, the results also suggest that a lot of improvements need to be made in the scheduling and data transmission mechanisms. It is important to optimize the ideal number of peers involved in the processing of an activity, the choice of the more suitable node to receive a rescheduled task and a better data distribution to not overload the bandwidth of some peers. To minimize churn effects, it is possible to use replication of data sets on other available nodes. When a churn happens, the task can be quickly rescheduled to nodes that already have a replica.

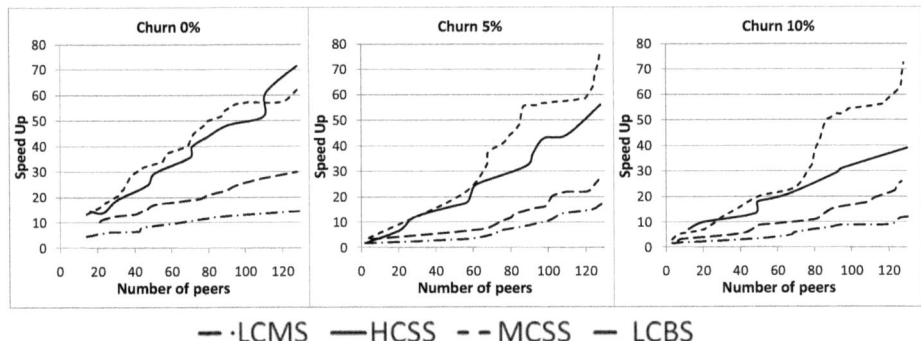

Fig. 4. Statistical results for data fragmentation for the selected activities

5 Conclusions

SciMule is a hierarchical P2P architecture that provides very good scalability for distribution of workflow activities in heterogeneous environments. We conducted a simulation to evaluate SciMule architecture and obtained positive results regarding the overall performance. From our initial results, we believe that hierarchical P2P is a promising approach to deal with MTC on heterogeneous environments such as hybrid clouds.

From the simulation results, we observed that different parallelization methods should have different approaches for execution. In the case of data fragmentation, it is preferable to have a large number of tasks of small size, than small numbers of tasks of larger size. This is promising since data fragmentation parallelism allows this kind of optimization. In the case of parameter sweep, it is preferable to restrict the number of nodes involved to avoid unnecessary data transfer and minimize the impact of churn events.

Finally, we observed that some issues must be addressed in future work, such as: improving the scheduling mechanism to choose a better number of peers to process an activity of a given size; and a better data discovery and distribution system to not overload the bandwidth of the submitter peers. The scheduler should consider data migration costs before scheduling a task to a node that do not have the data set. Many improvements are still possible to be explored using the SciMule simulator, but we have observed that our strategy is promising and may be an alternative to other heterogeneous distributed environments.

Acknowledgements

The authors are grateful to the High Performance Computing Center (NACAD-COPPE/UFRJ) where the experiments were performed.

References

[1] Deelman, E., Gannon, D., Shields, M., Taylor, I.: Workflows and e-Science: An overview of workflow system features and capabilities. Future Generation Computer Systems 25, 528–540 (2009)

[2] Jarrard, R.D.: Scientific Methods (2001),
    http://emotionalcompetency.com/sci/booktoc.html

[3] Raicu, I., Foster, I., Zhao, Y.: Many-task computing for grids and supercomputers. In: Workshop on Many-Task Computing on Grids and Supercomputers, pp. 1–11. Austin, Texas (2008)

[4] Anderson, D.: BOINC: a system for public-resource computing and storage. In: Proceedings. Fifth IEEE/ACM International Workshop on Grid Computing, Pittsburgh, USA, pp. 4–10 (2004)

[5] WCG, World Community Grid (2009), http://www.worldcommunitygrid.org

[6] Grid4All Consortium, Towards Hybrid Clouds - A Grid4All perspective on cloud computing, White paper (2009), http://www.grid4all.eu

[7] Freire, J., Koop, D., Santos, E., Silva, C.T.: Provenance for Computational Tasks: A Survey. Computing in Science and Engineering 10, 11–21 (2008)

[8] Pacitti, E., Valduriez, P., Mattoso, M.: Grid Data Management: Open Problems and New Issues. Journal of Grid Computing 5, 273–281 (2007)

[9] Lua, E.K., Crowcroft, J., Pias, M., Sharma, R., Lim, S.: A Survey and Comparison of Peer-to-Peer Overlay Network Schemes. IEEE Communications Surveys and Tutorials 7, 72–93 (2005)

[10] Ganesan, P., Gummadi, K., Garcia-Molina, H.: Canon in G Major: Designing DHTs with Hierarchical Structure. In: Proceedings of the 24th International Conference on Distributed Computing Systems (ICDCS 2004), Tokyo, Japan, pp. 263–272 (2004)

[11] Garcés-Erice, L., Biersack, E.W., Felber, P.A., Ross, K.W., Urvoy-Keller, G.: Hierarchical Peer-to-peer Systems. In: Proceedings of ACM/IFIP International Conference on Parallel and Distributed Computing, Klagenfurt, Austria, pp. 643–657 (2003)

[12] Mastroianni, C., Talia, D., Verta, O.: A super-peer model for building resource discovery services in grids: Design and simulation analysis. In: Sloot, P.M.A., Hoekstra, A.G., Priol, T., Reinefeld, A., Bubak, M. (eds.) EGC 2005. LNCS, vol. 3470, pp. 132–143. Springer, Heidelberg (2005)

[13] Martinez-Yelmo, I., Cuevas, R., Guerrero, C., Mauthe, A.: Routing Performance in a Hierarchical DHT-based Overlay Network. In: Euromicro Conference on Parallel, Distributed, and Network-Based Processing, pp. 508–515. IEEE Computer Society Press, Los Alamitos (2008)

[14] Ogasawara, E., Oliveira, D., Chirigati, F., Barbosa, C.E., Elias, R., Braganholo, V., Coutinho, A., Mattoso, M.: Exploring many task computing in scientific workflows. In: MTAGS 2009, pp. 1–10. ACM Press, Portland (2009)

[15] Dick, M.E., Pacitti, E., Kemme, B.: Flower-CDN: a hybrid P2P overlay for efficient query processing in CDN. In: Proceedings of the 12th International Conference on Extending Database Technology: Advances in Database Technology, pp. 427–438. ACM, Saint Petersburg (2009)

[16] Akbarinia, R., Pacitti, E., Valduriez, P.: Data currency in replicated DHTs. In: Proceedings of the 2007 ACM SIGMOD International Conference on Management of Data, pp. 211–222. ACM, Beijing (2007)

[17] Karakaya, M., Körpeoğlu, İ., Ulusoy, Ö.: A connection management protocol for promoting cooperation in Peer-to-Peer networks. Comput. Commun. 31, 240–256 (2008)

[18] Almeida, E.C.D., Sunyé, G., Traon, Y.L., Valduriez, P.: A Framework for Testing Peer-to-Peer Systems. In: Proceedings of the 2008 19th International Symposium on Software Reliability Engineering, Los Alamitos, CA, USA, pp. 167–176 (2008)

[19] Jelasity, M., Montresor, A., Jesi, G.P., Voulgaris, S.: The PeerSim simulator (2010), http://peersim.sourceforge.net

[20] Freedman, D., Pisani, R., Purves, R.: Statistics, 4th edn. W. W. Norton, New York (2007)

[21] Law, A.M.: Statistical analysis of simulation output data: the practical state of the art. In: Proceedings of the 39th Conference on Winter Simulation, pp. 77–83. IEEE Press, Washington D.C (2007)

Intelligent Service Trading and Brokering for Distributed Network Services in GridSolve

Aurélie Hurault[1] and Asim YarKhan[2]

[1] Institut de Recherche en Informatique de Toulouse - UMR 5505
University of Toulouse, France
hurault@enseeiht.fr
[2] Electrical Engineering and Computer Science
University of Tennessee, Knoxville, TN 37996
yarkhan@eecs.utk.edu

Abstract. One of the great benefits of computational grids is to provide access to a wide range of scientific software and a variety of different computational resources. It is then possible to choose from this large variety of available resources the one that solves a given problem, and even to combine these resources in order to obtain the best solution.

Grid service trading (searching for the best combination of software and execution platform according to the user requirements) is thus a crucial issue. Trading relies on the description of available services and computers, on the current state of the grid, and on the user requirements. Given the large amount of services that may be deployed over a Grid, this description cannot be reduced to a simple service name.

In this paper, a sophisticated service specification approach similar to algebraic data types is combined with a grid middleware. This leads to a transparent solution for users: they give a mathematical expression to be solved, and the appropriate grid services will be transparently located, composed and executed on their behalf.

1 Introduction

Grid computing and distributed computing projects have been very effective in exposing large collections of services and computational resources to users. But users still need to handle all the difficulties involved in finding and composing the appropriate resources to solve their problem. In Figure 1 we express the general problem of matching user service requests to the appropriate resources. On the right side of the figure, there is a set of services running on some computational resources, and on the left side there are end users that want to use these services to solve their problems. The goal is to find the services or combination of services that can solve the users problem accurately and efficiently, to execute these services and return the result to the user.

A corollary part of this problem is providing the user with the appropriate syntax to express their needs. This syntax has to be sufficiently precise to find relevant solutions, and it should also be easy for the user to express complex problems.

J.M.L.M. Palma et al. (Eds.): VECPAR 2010, LNCS 6449, pp. 340–351, 2011.
© Springer-Verlag Berlin Heidelberg 2011

Fig. 1. A simplified overview of the general problem of matching service requests to available resources

Our solution (see Figure 2) combines a service trader and Grid middleware that implements the emerging GridRPC API [SHM+02]. In this scenario, the end user asks the GridSolve [DLSY08, YSS+06] GridRPC middleware to solve some (possibly complex) problem request. The service trader uses its knowledge of the available services and matches the problem request with a service or a combination of services. GridSolve is then used to execute the services and the final result is returned to the user.

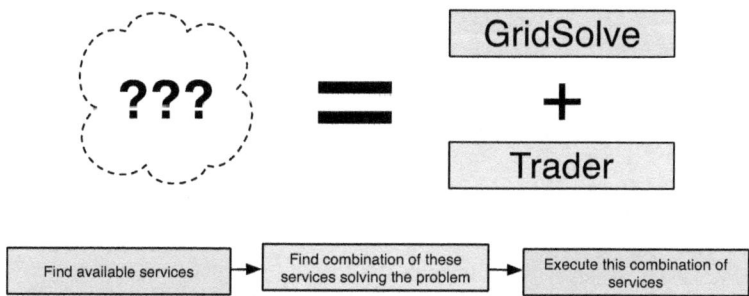

Fig. 2. Our solution takes complex user problems and using a service trader matches them to grid services exposed via GridSolve

2   Service Trading

A service trader acts on the behalf of a user to find a combination of the available computational services that will handle the users request. The details of the service trader used in this work are described in [HDP09]. For this research, improvements have been made in the way it searches for solutions to the users request. The computational complexity of the available services is used to select less computationally expensive solutions from the available services. This leads to an improvement in the execution time for the users request, since, in practical situations, we can prune the more computationally complex solutions.

2.1   Inputs for the Service Trader

The trader needs a description of the **application domain** using an algebraic specification. To this goal we use an order-sorted signature $(S, \leq, \Sigma)$ where:

S is a set of *sorts* (usually called *types* in programming languages, both terms will be used in this discussion);
$\leq$ is an order to express the subsorting (subtyping) link between sorts;
Σ is a set of symbols standing for constants and sorted (typed) functions.

For a complete definition see [GM92]. The operators used may be overloaded and some extra equations $\mathcal{E}$ are added to describe the properties of the operators (e.g., commutativity, associativity, neutral element). For example, a partial representation for linear algebra over scalars and matrices is: $S = \{Scalar, Matrix\}$

$$\Sigma = \{$$
$$0, I : Matrix$$
$$0, 1 : Scalar$$
$$+ : Matrix \times Matrix \rightarrow Matrix$$
$$+ : Scalar \times Scalar \rightarrow Scalar$$
$$* : Matrix \times Matrix \rightarrow Matrix$$
$$* : Scalar \times Matrix \rightarrow Matrix$$
$$* : Scalar \times Scalar \rightarrow Scalar$$
$$\}$$
$$\mathcal{E} = \{$$
$$Matrix \ x, Matrix \ y : x + y = y + x$$
$$Matrix \ x : 1 * x = x$$
$$Matrix \ x : 0 * x = 0$$
$$Matrix \ x : I * x = x$$
$$\}$$

The *Matrix* types can have lots of subtypes to describe and handle different matrix properties, for example, symmetric, dense, space, triangular and band.

The **computational services** are described as terms in an order-sorted signature. This leads to a really natural description, in particular in mathematical domains, since the notations are very similar. For example, the BLAS (Basic Linear Algebra Subroutines) [BDD⁺02] *saxpy* (addition) and *sgemm* (multiplication) functions can be described (in a simplified way) in the algebraic notation as:

$$Scalar\ \alpha, Matrix\ x,\ y : saxpy(\alpha, x, y) = \alpha * x + y$$
$$Scalar\ \alpha, \beta, Matrix\ x,\ y,\ z : sgemm(\alpha, x, y, \beta, z) = \alpha * x * y + \beta * z$$

A user **request** is also specified in this algebraic notation. For example a request to add three matrices would simply be expressed as:

$$Matrix\ a,\ b,\ c : a + b + c$$

The service trader is generic and the application domain can include anything that can be described using an algebraic specification. We have performed additional work with some optimization libraries [Hur06]. The main types in the optimization domain are functions and constraints, and the elements manipulated by these functions and constraints (e.g., `Real`, `Matrix`). The operators are minimization and maximization operators, the function description ($\rightarrow$), and the operators for constraints ($\leq$, $\&$, ...) and the operators for the manipulated elements ($*$, $+$, ...). Equations are used to express the constraints on the optimization.

The optimization libraries we have considered was the Matlab optimization toolbox[1] and the E04 package of the NAG[2] library.

2.2   The Trader Output

The trader generates a list of services and combination of services that satisfy the request. For example, given the linear algebra domain and *saxpy* and *sgemm* services described earlier, for the user request of $a + b + c$, the possible solutions satisfy the request include:

$$saxpy(1, a, saxpy(1, b, c))$$
$$saxpy(1, a, sgemm(1, I, b, 1, c))$$
$$sgemm(1, I, a, 1, saxpy(1, b, c)),$$

If the *saxpy* function is less computationally expensive than the *sgemm* function (which is true if for the BLAS functions), then the solution

$$saxpy(1, a, saxpy(1, b, c))$$

will be only solution returned by the trader, since the other possible solutions ($saxpy(1, a, sgemm(1, I, b, 1, c))$, $sgemm(1, I, a, 1, saxpy(1, b, c))$, ...) are more computationally expensive. To this end, the complexity of the services are input to the trader as well as the size of the matrices. Since before finishing the comparison we do not know what the parameters will be, we can't compute the exact cost. We use a medium size of the data to have an approximate cost and decide if it is interesting to do the comparison, or if we currently have better answers.

---

[1] http://www.mathworks.com/access/helpdesk/help/toolbox/optim/optim.shtml
[2] http://www.csc.fi/cschelp/sovellukset/math/nag/NAGdoc/fl/html/E04_fl19.html

We might lose some interesting responses, but by not doing some comparisons, we will win in the search time taken the trader.

For example, we are trying to find solutions to the request "$A * B$", where A and B are matrices. To evaluate if it is interesting to compare another solution with the *sgemm* service, we will approximate the cost of the *sgemm* service. To do that we use an estimated medium size for the matrices, since the trader does not know the sizes we will give when calling the *sgemm* service. This cost can be compared with the cost other known solutions.

The BLAS *saxpy* function has additional parameters not related to function signature (for example, the sizes of the matrices). These need to be available to do data transfer and execute the services. These parameters are considered after the analysis of the functional signature . The trader focuses on the functional aspect and considers the other parameters later. For example, considering the BLAS *strsm* (triangular solver) routine:

STRSM (SIDE, UPLO, TRANSA, DIAG, M, N, ALPHA, A, LDA, B, LDB)
M specifies the number of rows of the parameter B (matrix);
UPLO specifies if the matrix A is an upper or a lower triangular matrix;
SIDE specifies if the problem solved is $op(A) * X = \alpha * B$ or $X * op(A) = \alpha * B$.

To be able to find the value for those parameters, we need to complete the description of the services. For example we have to be able to specify that M is the number of rows of the parameter B, which is not possible with only one term for the description of the service. We add some possibilities in the description:

- When a procedure implements several functionalities, we introduce the "switch / case" functionality. The procedure will generate different services, each with one parameter set for a given value. In our example, one service with $SIDE$ set to 'left' and one service with $SIDE$ set to 'right' are generated.
- When some parameters depend on other, we describe this dependence, and when the value of the first parameter is known, the value of the second one is computed. In our example, when the exact type of A is known, $UPLO$ will be set to 'u' or 'l'.
- When some parameters specify the properties of other parameters or for default value, a term is assigned to a parameter, using only the constant of the domain and of the request. In our example, M will be assigned to $m(B)$ (number of rows of B). When B will be known, $m(B)$ will also be known and the value of M will be fixed.
- When there is several description are possible for the same parameter (for example when several matrices have the same size), there are all given with "||" (for "or") to separate them.

The full service description of *saxpy* is:

```
SAXPY( n,alpha,x,incx,y,incy ):
   y <- alpha * x + y
   n = m(x) * n(x) || m(y) * n(y);
   incx = default 1;
   incy = default 1;
```

Where $m(x)$ is the number of rows of the x matrices and $n(x)$ the number of columns. Some default values are given for $incx$ and $incy$.

2.3 Inside the Trader

As explained in [HDP09], the trader is based on equational unification and more particularly on the work of Gallier and Snyder [GS89]. It has a type system adapted to overloaded functions with subtyping, based on the $\lambda\&-$calculus defined by Castagna [CGL92]. The algorithm of Gallier and Snyder has been modified by adding an amount of energy to ensure that the computation will end. This amount is composed by the depth of combination and the number of equations that can be applied.

Since we are trying to compute an efficient combination of services, we first check the services that are less complex. To this end, we use the mathematical expression of the computational complexity of the services (as provided by GridSolve). The service complexity (e.g., $O(2n^2)$) uses the sizes of its parameters (e.g.,n) to estimate the execution time. We can thus compare the execution cost of the various service choices taking into account the size of the parameters. This has two major benefits:

- We find some efficient solutions;
- We prune branches on search tree, since we do not look at more expensive solutions.

3   Overview of GridSolve: A GridRPC Middleware

The purpose of GridSolve is to create the middleware necessary to provide a seamless bridge between computational scientists using desktop systems and the rich supply of services supported by the emerging Grid architecture. The goal is to make it easy for any general user to to take advantage of the many benefits (in terms of shared processing, storage, software, data resources, etc.) of using grids.

GridSolve is a client-agent-server system which provides a RPC interface (Remote Procedure Call) to software services that are deployed on GridSolve servers. GridSolve supports a variety of client side interfaces, namely C, Fortran, Matlab and IDL (Interactive Data Language). GridSolve attempts to make it easy to create, deploy and access software services over the network. More detailed information about GridSolve is available in other publications [DLSY08, YSS+06]. The general architecture of GridSolve is shown in Figure 3.

- The *Client* wishes to execute a remote procedure call from within an executable (C, Fortran) or scientific computing environment (Matlab or IDL).
- The *Server* provides software services that can execute functions on behalf of the clients. The server hardware can range in complexity from a uniprocessor to a MPP system and the functions executed by the server can be arbitrarily complex.
- The *Agent* acts as an intermediary between the clients and servers and maintains information about the available servers and services.

346 A. Hurault and A. YarKhan

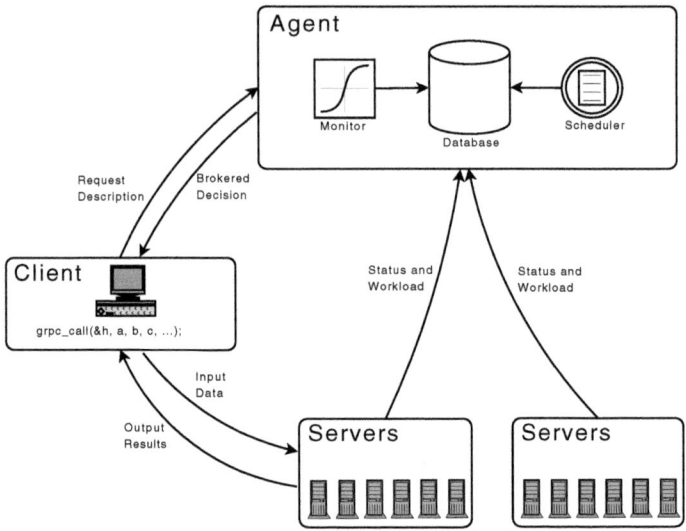

Fig. 3. Architectural overview of GridSolve

In practice, from the user's perspective the mechanisms employed by GridSolve make the remote procedure call fairly transparent. However, behind the scenes, a typical call to GridSolve involves several steps. The client sends a request for the desired service to the agent. The agent uses its knowledge of the available services and servers to return a list of appropriate servers to the client, ordered by the expected execution time. The client then selects a server from the list of servers and sends the service request. In the event of a failure, the client can resubmit the request to another server, providing a basic level of fault-tolerance. Finally, a server executes the function on the behalf of the client and returns the results.

4 Integration of Service Trading into GridSolve

As explained in the introduction, the goal is to facilitate a user job by making transparent calls to the grid, so that the user does not have to be knowledgeable about the available Grid resources or services.

To this aim, we proceed in four steps:

1. GridSolve provides information about available services.
2. The trader finds the combination of services that solves the user request.
3. The output of the trader is analyzed and the services are called.
4. The response is transferred back to the user.

4.1 Generating the Inputs of the Trader

GridSolve is responsible for providing information about the available services to the service trader. The service trader requires some additional semantic information that has to be added to the standard service descriptions provided by GridSolve.

This additional information is the mathematical expression of the functions computed by the service, and some information about the non-functional parameters. For example, for the BLAS *dgemm* and the *dsymm* functions, it will be:

```
SUBROUTINE dgemm
APPLICATION_DOMAIN="LinearAlgebra"
TRADER_DESCRIPTION="
c <-((alpha*((op transa a)*(op transb b)))+(beta*c)) ;
value m = (m c) || (m (op transa a)) ;
..."
```

```
SUBROUTINE dsymm
APPLICATION_DOMAIN="LinearAlgebra"
PARAMETERS_PROPERTIES = "a symmetric"
TRADER_DESCRIPTION="
c <- if (side='l') then ((alpha*(a*b))+(beta*c))
     if (side='r') then ((alpha*(b*a))+(beta*c)) ;
value m = (m c) ;
...
if ( a instanceof UpTriInvMatrix ) then ( uplo = 'u' );
..."
```

4.2   Discover the Combination of Services

Given a user request in the GridSolve client, GridSolve calls the service trader, which processes the request and returns a file containing a sequence of services calls that will satisfy the users request. For example, for the request $a + b + c$:

```
def, res2, copy c
def, res1, copy a
call, saxpy, m(b)*n(b), 1.0, copy b, 1, res2, 1
call, saxpy, m(res2)*n(res2), 1.0, res2, 1, res1, 1
```

The GridSolve client side system transparently transforms this sequence of request into a workflow DAG [LDSY08] and uses the GridSolve runtime to improve the performance and be able to execute different parts in parallel when possible.

4.3   Call the Services

To do the calls, GridSolve parses the output file. When it finds a "def", a new local variable is created and set to the second pointer. When it finds a "copy", a copy of the user data is created. And when it finds a "call", the GridRPC call is done with the parameters that follow. To do the call, some additional information is needed: type of data, sizes of data, pointer to the data. In the C and Matlab interfaces, those parameters are discovered in different ways:

- Information provided by the user in the C interface.
- Information found via Matlab data querying mechanisms in the Matlab interface.

In fact, a first analysis will create the temporary variables needed for the computation. Then a second analysis is done in order to make the GridRPC call. They will not be in sequence when parsing the file. The calls are executed using DAG interface of GridSolve [LDSY08] in order to improve the performance and be able to run different parts in parallel when possible.

4.4   The Service Trader C API

We have added two functions to the GridSolve C API for the service trader.

```
int gs_call_service_trader(char *req,... );
int gs_call_service_trader_stack(char * req, grpc_arg_stack *argsStack);
```

The first parameter is a string that expresses the request in a simple analytical or mathematical expression . The other arguments are pointers to the actual data, and information about the variable name and size. An example call using the service trader interface:

```
float *a = malloc (sizeof(float)*ma*na);
float *b = malloc (sizeof(float)*mb*nb);
...
gs_call_service_trader("(a+(b+a))","a",a,ma,na,...);
```

4.5   The Matlab Interface

The Matlab client interface is substantially simpler than the C interface since a variety of information about the variable names and sizes can be obtained by querying Matlab internal data representations. An example service trader call using the Matlab interface:

```
a=[1,2,3;4,5,6;7,8,9]
b=[10,20,30;40,50,60;70,80,90]
[output]=gs_call_service_trader("(a+(b+a))"),
output =
   12.  24.   36.
   48.  60.   72.
   84.  96.  108.
```

The output variable is integrated back into the Matlab workspace and can be used for later computation in Matlab.

5   Experiments

In these experiments we show the creation of a DAG from a users problem request, and the selection of the appropriate services using computational complexities to guide the selection.

For our small test experiment, the relevant Grid services are the previously described BLAS services (*saxpy*, *daxpy*, *sgemm* and *dgemm*) and a service implementing the Strassen-Winograd algorithm for matrix-multiplication (*sgemmb*). This variant has a lower computational complexity (approximately $O(n^{2.8})$) than standard matrix-multiplication algorithms ($O(n^3)$), but it only becomes efficient if the matrix size is sufficiently large. With the complexity information provided as an input to the trader, it will be able to choose the best algorithm as a function of the size of the matrices. The "exact" complexity information given to the trader is $7 * pow(n, log2(7)) + 3 * m * k - 6 * n * n$ for the Strassen-Winograd algorithm and $2.0 * m * n * k + 2.0 * m * k$ for the classic algorithm.

As an example, we want to compute $(((a * b) + c) + ((ba * bb) + c))$, where a, b and c are 3x3 matrices, ba is a 3x3000 matrix and bb is a 3000x3 matrix. Using the Matlab interface:

```
gs_call_service_trader("(((a*b)+c)+((ba*bb)+c)))")
```

The trader generates a sequence of 3 service calls that will solve this problem. GridSolve then creates a DAG based on the data dependencies between these calls and calls the services.

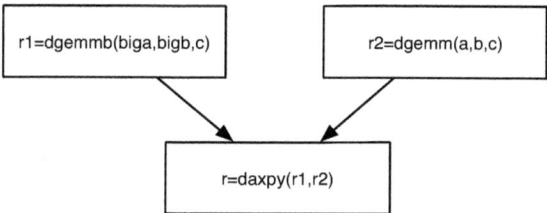

These events are totally transparent to the user, who simply needs to present the desired mathematical expression.

In this experiment, we observe that based on the size of the matrices, the service trader uses the complexity information to select the more efficient multiplication algorithm: the Strassen-Winograd matrix multiplication *dgemmb* for the bigger matrices and standard *dgemm* for the smallest ones.

6   Summary

In this paper we developed a combination of a service trader and a grid middleware system to enable a "novice" user to gain access a remote library, without knowing about grid computing or about the available library and services.

The key point for the end user is the transparency of all the details involved in this process. The fact that the services are evaluated at each call makes the solution more tolerant to a service crash. If a service disappear, the trader will find some other solution to the users problem with a different combination of services. For example, if the *dgemmb* service is not available, the service trader will be able to replace it by *dgemm*. If the *daxpy* is not available, the trader will be able to find a solution with *dgemm*. The user doesn't have to be aware of all the alternative services that may satisfy their request.

The execution of the users request is made more efficient by two factors; firstly, the service trader evaluates the computational complexity of the available services on the users specific data, and secondly, the GridSolve DAG execution system enables any parallelism in the execution of the services. There is a cost for the analysis done by the service trader in this current prototype, however we expect that this cost can be reduced in the future. Moreover, the major advantage provided by the work developed here is ease-of-use. The end user does not have to be knowledgeable in grid computing, mathematical libraries, algorithmic complexity, data dependency analysis, fault-tolerance, or any of the details that are transparently handled by this system.

References

[BDD⁺02]   Blackford, S., Demmel, J., Dongarra, J., Duff, I., Hammarling, S., Henry, G., Heroux, M., Kaufman, L., Lumsdaine, A., Petitet, A., Pozo, R., Remington, K., Whaley, C.: An Updated Set of Basic Linear Algebra Subprograms (BLAS). ACM Trans. Math. Softw. 28(2), 135–151 (2002)

[CGL92]   Castagna, G., Ghelli, G., Longo, G.: A calculus for overloaded functions with subtyping. In: Proceedings of the ACM Conference on Lisp and Functional Programming, vol. 5, pp. 182–192 (1992)

[DLSY08]   Dongarra, J., Li, Y., Seymour, K., YarKhan, A.: Users' Guide to GridSolve V0.19. Innovative Computing Laboratory. Technical Report, University of Tennessee, Knoxville, TN (June 2008)

[GM92]   Goguen, J., Meseguer, J.: Order-sorted algebra i: equational deduction for multiple inheritance, overloading, exceptions and partial operations. Theor. Comput. Sci. 105(2), 217–273 (1992)

[GS89]   Gallier, J., Snyder, W.: Complete Sets of Transformations for General E-Unification. Theor. Comput. Sci. 67(2-3), 203–260 (1989)

[HDP09]   Hurault, A., Daydé, M., Pantel, M.: Advanced Service Trading for Scientific Computing over the Grid. Journal of Supercomputing 49(1), 64–83 (2009)

[Hur06]   Hurault, A.: Courtage sémantique de services de calcul. PhD thesis, INPT, Toulouse (Décember 2006)

[LDSY08]   Li, Y., Dongarra, J., Seymour, K., YarKhan, A.: Request Sequencing: Enabling Workflow for Efficient Problem Solving in GridSolve. In: International Conference on Grid and Cooperative Computing (GCC 2008), pp. 449–458 (October 2008)

[SHM⁺02]   Seymour, K., Hakada, N., Matsuoka, S., Dongarra, J., Lee, C., Casanova, H.: Overview of GridRPC: A Remote Procedure Call API for Grid Computing. In: Parashar, M. (ed.) GRID 2002. LNCS, vol. 2536, pp. 274–278. Springer, Heidelberg (2002)

[YSS⁺06]   YarKhan, A., Seymour, K., Sagi, K., Shi, Z., Dongarra, J.: Recent Developments in GridSolve. International Journal of High Performance Computing Applications (IJHPCA) 20(1), 131–141 (2006)

Load Balancing in Dynamic Networks by Bounded Delays Asynchronous Diffusion

Jacques M. Bahi[1], Sylvain Contassot-Vivier[2,3], and Arnaud Giersch[1]

[1] LIFC, University of Franche-Comté, Belfort, France
jacques.bahi@univ-fcomte.fr, arnaud.giersch@univ-fcomte.fr
http://lifc.univ-fcomte.fr/equipe/display/1
[2] LORIA, University Henri Poincaré, Nancy, France
sylvain.contassotvivier@loria.fr
http://www.loria.fr/~contasss/homeE.html
[3] AlGorille INRIA Team, France

Abstract. Load balancing is a well known problem, which has been extensively addressed in parallel algorithmic. However, there subsist some contexts in which the existing algorithms cannot be used. One of these contexts is the case of dynamic networks where the links between the different elements are intermittent. We propose in this paper an efficient algorithm, based on asynchronous diffusion, to perform load balancing in such a context. A convergence theorem is proposed and proved. Finally, experimental results performed in the SimGrid environment confirm the efficiency of our algorithm.

Keywords: Load balancing, dynamic network, asynchronism.

1 Introduction

In the parallel computation domain, the load balancing is often a central issue to reach the optimal theoretical performances. As a consequence, that problem has been extensively studied since the beginning of parallelism in computer science [4,1,9,12]. It can be observed that the evolution of the load balancing algorithms has rather closely followed the one of the parallel architectures. Hence, balancing algorithms have muted from static and centralized distributions [10,7] to dynamic and/or decentralized ones [11,8].

Although those mutations allow for load balancing in numerous parallel contexts, there are always emergent architectures which require new breakthroughs in the balancing schemes in order to fully benefit from the endlessly increasing computational power. The parallel systems are more and more complex, often including heterogeneous computational units and interconnection networks. Moreover, the modularity of those systems sharply increases the number of possible parallel contexts. So, it becomes less and less interesting to design balancing algorithms specific to a given context. Thus, there is a strong demand for fully adaptive algorithms which are as generic as possible, that is to say, which can be used on any kind of parallel architecture without requiring major modifications.

J.M.L.M. Palma et al. (Eds.): VECPAR 2010, LNCS 6449, pp. 352–365, 2011.

Moreover, in addition to the complex architectures, the emergence of more and more dynamical systems has also been observed during the recent years. The dynamical aspect typically stands at the level of the communications as the links between the computing units are only intermittent.

The most suited strategies to such contexts are the neighborhood strategies based on diffusion algorithms [5]. Unfortunately, most of the current solutions are either synchronous [8] or assume a static network [3]. In the objective to respond to these two current issues, we propose in this paper a load balancing algorithm based on bounded delays asynchronous diffusion.

The following section presents the general computing model used to perform our theoretical study and design of our algorithm. In section 3, a detailed discussion on the balancing ratios to be used in our algorithm is given. Then, the algorithm is provided in section 4 as well as the proof of the balancing convergence in time in section 5.2. Finally, a quality evaluation of our algorithm, performed with the SimGrid environment, confirms the very good performances of our approach in section 6.

2   Model

As our balancing algorithm is iterative, its convergence must be proven in order to ensure that the load will be balanced in finite time till there are no modification of the state of the system during the balancing phase. When the system configuration dynamically evolves during the running of the algorithm being balanced, no load stabilization may be observable although our balancing algorithm will follow the evolution of the computational power repartition. Hence, as soon as the system configuration is stabilized, the load repartition will follow with a slight delay. This behavior of our algorithm is proved in a convergence theorem given below. Nevertheless, that theorem and its proof require some notations and a description of the temporal evolution of the system state.

2.1   Notations

For the sake of clarity, we distinguish two kinds of features: those of the platform and the elements related to the application.

Platform characteristics:

$P = \{1, \ldots, n\}$: the set of the n processors in the system.

$G(t) = (P, L(t))$: the undirected connection graph of the links between the n processors at time t.

$N_i(t)$: the set of processors directly connected to processor i at time t.

$d_j^i(t)$: the delay of j according to i at time t. By definition, it verifies $d_j^i(t) \leq t$.

B : the bound of the delays, i.e. $\forall i, j \in P \times P$, $\forall t \in \mathbb{N}$, $t - B < d_j^i(t) \leq t$.

Application related values:

$x_i(t)$: the load of processor i at time t.

$x_j^i(t) = x_j(d_j^i(t))$: the load of processor j at time $d_j^i(t)$. That information represents the evaluation at time t on processor i of the load on processor j.

$s_{ij}(t) = \alpha_{ij}(t)(x_i(t) - x_j^i(t))$: the amount of load sent by processor i to processor j at time t. Concerning $\alpha_{ij}(t)$, we have the following constraints: $\forall i, j \in P$, $\alpha_{ij}(t) \in [0, 1]$ and $\sum_{j=1}^{n} \alpha_{ij}(t) = 1$. Also, $s_{ij}(t) = 0$ if $j \notin N_i(t)$ or $x_i(t) \le x_j^i(t)$.

$r_{ij}(t)$: the amount of load received on processor j from processor i at time t.

$v_{ij}(t)$: the amount of load sent by processor i before time t and not yet received by processor j at time t.

2.2 General Load Balancing Scheme

First of all, we consider that we have an initial total load L such that

$$\sum_{i=1}^{n} x_i(0) = L \tag{1}$$

and that there is a conservation of the load in the sense that it is either on the processors or in transit in the interconnection network. In that context, we use the following decentralized scheme to balance the load in the system.

Algorithm 1. *At each time step t, each processor:*

1. *Compares its load to the loads of its connected neighbors*
2. *Determines the load quantities to send to its less loaded neighbors*
3. *Sends those amounts of load to the corresponding nodes*
4. *Potentially receives some load from its more loaded neighbors*

2.3 Dynamical Evolution of the System State

In the scope of that study, the main issue addressed is the temporal evolution of the interconnection network of the system. Contrary to classical parallel systems, we consider dynamic links between the different processing units. However, some constraints are necessary to ensure the diffusion of the loads through the system.

Concerning the network, we define the extended neighborhood of a processor i at time t as the set

$$\overline{N}_i(t) = \{j \mid \exists t' : t - B < t' \le t \text{ such that } j \in N_i(t')\}$$

This means that j has been connected at least one time to i during the time interval $\{t - B + 1, \ldots, t\}$.

Assumption 1. *There exists $B \in \mathbb{N}$ such that $\forall i, j \in P \times P$ and $t \ge 0$, $\max(0, t - B) \le d_j^i(t) \le t$ and the union of the communication graphs $\cup_{\tau=t}^{t-B+1} G(\tau)$ is a connected graph.*

This assumption, known as the jointly connected condition [8], implies that information can be exchanged between any couple of nodes i and j within any time interval of length B, and that the delay between two nodes cannot exceed B.

The example given in Fig. 1 shows the effect of that assumption. In the two consecutive times, the connection graphs $G(t)$ are not fully connected. However, their fusion yields a virtual graph which is actually connected.

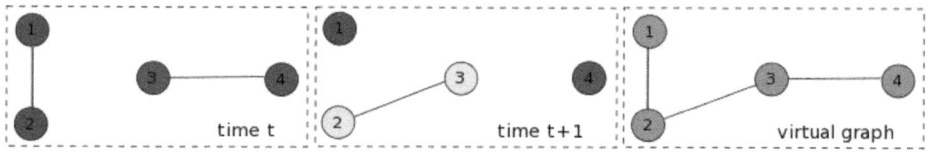

Fig. 1. Jointly connected graph

It is interesting to point out that the load balancing in such a context is similar to a kind of information percolation in an intermittent network [6].

Assumption 2. $\forall t \geq 0$, $\forall i \in P$ and $\forall j \in N_i(t)$, when $x_i(t) > x^i_j(t)$, there exists $\alpha_{ij}(t) > 0$ such that $\alpha_{ij}(t)(x_i(t) - x^i_j(t)) \leq s_{ij}(t)$.

Assumption 2 indicates that as soon as two nodes are connected, the more loaded sends a non negligible ratio of its load excess to the other node.

Assumption 3.

$$x_i(t) - \sum_{k \in N_i(t)} s_{ik}(t) \geq x^i_j(t) + s_{ij}(t), \quad \forall j \in N_i(t) \text{ s.t. } s_{ij}(t) > 0$$

Assumption 3 is essential to avoid the starvation and the ping-pong phenomena. It ensures that the remaining load on the sending processor will not become smaller than the loads on the receptors. A famine occurs when a node has no more workload. The ping-pong state is established when two nodes continually exchange load between each other without reaching an equilibrium.

These last two assumptions are similar to assumption 4.2 in [1].

So, according to these assumptions, we have $\forall i \in P$, $\forall t \in \mathbb{N}$, the following load evolution:

$$x_i(t+1) = x_i(t) - \sum_{j \in N_i(t)} s_{ij}(t) + \sum_{j \in \overline{N}_i(t)} r_{ji}(t) \tag{2}$$

where:

- $s_{ij}(t)$ is given by $\alpha_{ij}(t)(x_i(t) - x^i_j(t))$ with the constraints given above. The values $\alpha_{ij}(t) \in [0, 1]$ define the strategy of the load balancing algorithm. As already mentioned, they must be carefully chosen to ensure the convergence of the algorithm. Their determination is detailed in the following section,

– the last term corresponds to the total amount of load received by processor i from the processors in its extended neighborhood.

Although that equation describes the load evolution on the processors, it does not give any information on the loads in transit. For this, we have:

$$v_{ij}(t) = \sum_{s=0}^{t-1}(s_{ij}(s) - r_{ij}(s))$$

and $v_{ij}(0) = 0$. Moreover, the constraint of load conservation discussed above implies that:

$$\sum_{i=1}^{n}\left(x_i(t) + \sum_{j\in\overline{N}_i(t)} v_{ij}(t)\right) = L, \quad \forall t \geq 0 \tag{3}$$

Once again, it is worth noticing that although the convergence of our balancing is proved in the context of load conservation, our algorithm should provide also interesting results in a more general context of intermittent load evolutions. In fact, inside each time interval where the global load stays constant, our algorithm will tend to balance the current global amount of load among the processors. Thus, an overall gain in performance may be expected in numerous contexts of dynamic loads. However, it is obvious that the detailed behavior and applicability of our algorithm in such cases would require another complete study.

3   Choice of the Load Ratios

As seen above, the values of the $\alpha_{ij}(t)$ must be chosen such that the amount of load on every node converges to $\frac{L}{n}$.

Let's denote by j^*, the processor satisfying $x_{j^*}^i = \min_{k\in N_i(t)} x_k^i(t)$. It clearly appears that j^* depends on both time and processor i.

In order to correctly choose the $\alpha_{ij}(t)$, Assumptions 2 and 3 are used to deduce the constraints. Assumption 2 can be carried out by fixing an arbitrary constant $\beta \in [0, 1[$ and choosing:

$$\begin{cases} \sum_{k\neq j^*\in N_i(t)} \alpha_{ik}(t)(x_i(t) - x_k^i(t)) \leq \beta(x_i(t) - x_{j^*}^i(t)) \\ \alpha_{ij^*}(t) = \frac{1}{2}\left(1 - \frac{\sum_{k\neq j^*}\alpha_{ik}(t)(x_i(t)-x_k^i(t))}{x_i(t)-x_{j^*}^i(t)}\right) \end{cases} \tag{4}$$

And we deduce

$$\alpha_{ij^*}(t) \geq \frac{1-\beta}{2} = \alpha$$

Finally, it comes that $\forall i, j^*, t$ such that $j^* \in N_i(t)$ and $x_{j^*}^i(t) = \min_{k\in N_i(t)} x_k^i(t)$,

$$s_{ij^*}(t) = \alpha_{ij^*}(t)(x_i(t) - x_{j^*}^i(t)) \geq \alpha(x_i(t) - x_{j^*}^i(t)).$$

We can observe that the load sent cannot exceed the halve of the load difference between the sender and the receiver.

Furthermore, Assumption 3, avoiding ping-pong effects, implies to choose $\alpha_{ij}(t)$ such that $\forall t \geq 0$, $\forall i \in P$, and $j \neq j^* \in N_i(t)$ satisfying $x_i(t) > x_j^i(t)$,

$$0 \leq \alpha_{ij}(t) \leq \frac{1}{2} \left(1 - \frac{\sum_{k \neq j} \alpha_{ik}(t)(x_i(t) - x_k^i(t))}{x_i(t) - x_j^i(t)} \right) \tag{5}$$

4   Load Balancing Algorithm

The algorithmic scheme of our load balancing is given below.

Algorithm 2. *At each time step t, each processor:*

1. *Compares its load to the loads of its connected neighbors*
2. *Determines the $\alpha_{ij}(t)$ and deduces the $s_{ij}(t)$*
3. *Sends those amounts of load to the corresponding nodes*
4. *Receives some loads from more loaded nodes*

Although it does not appear directly, the heterogeneity of the processors can be taken into account in this algorithm, for example by introducing virtual processors of the same power (GCD of the actual powers) and distributing them among the actual processors according to their relative speeds. Finally, at each time t and on each node i, the load update is given by (2) and the global behavior of that algorithm is depicted by the following theorem.

Theorem 1. *Under Assumptions 1, 2 and 3, the asynchronous load balancing Algorithm 2 converges to* $x^* = \frac{1}{n} \sum_{i=1}^{n} x_i(0)$.

5   Proof of the Load Balancing Convergence

5.1   Technical Results

Let $m(t) = \min_i \min_{t-B < \tau \leq t} x_i(\tau)$. Note that $x_j^i(\tau) \geq m(t)$, $\forall i, j, t$. Lemma 1 and 2 below can be proven similarly to the lemma of pages 521 and 522 in [BT89].

From Assumption 1 we can conclude that the amount of load $v_{ij}(t)$ in the network before time t and not yet received consists in workloads sent in the time interval $\{t - B + 1, ..., t - 1\}$, so $v_{ij}(t) \leq \sum_{\tau=t-B+1}^{t-1} s_{ij}(\tau)$, $\forall i \in P, \forall j \in N_i(t)$.

Lemma 1. *The sequence $m(t)$ is monotone, non-decreasing and converges and $\forall i \in P, \forall s \geq 0$,*

$$x_i(t+s) \geq m(t) + \left(\frac{1}{n} \right)^s (x_i(t) - m(t))$$

Let $i \in P$, $t_0 \in \mathbb{N}$ and $t \geq t_0$, we say that the event $E_j(t)$ occurs if there exists $j \in \overline{N}_i(t)$ such that

$$x_j^i(t) < m(t_0) + \frac{\alpha}{2n^{t-t_0}} (x_i(t_0) - m(t_0)) \quad \text{and} \quad s_{ij}(t) \geq \alpha \left(x_i(t) - x_j^i(t) \right),$$

where α is deduced from (4).

Lemma 2. Let $t_1 \geq t_0$, if $E_j(t_1)$ occurs, then $E_j(\tau)$ doesn't occur for any $\tau \geq t_1 + 2B$.

Lemma 3. $\forall i \in P, \forall t_0 \in \mathbb{N}, \forall j \in \overline{N}_i(t)$,

$$t \geq t_0 + 3nB \quad \Rightarrow \quad x_j(t) \geq m(t_0) + \eta \left(\frac{1}{n}\right)^{t-t_0} (x_i(t_0) - m(t_0)).$$

where $\eta = \frac{\alpha}{2} \left(\frac{1}{n}\right)^B$.

Definition 1. We say that a node j is l-connected to a node i if it is logically connected to i by l communication graphs, i.e. if there exists a minimal sequence (without redundancy) $\{i_0(t_0), i_1(t_1), \ldots, i_l(t_l)\}$ such that $i = i_0(t_0)$, $i_{j-1} \in N_{i_j}(t_j) \ \forall j \in \{1, ..., l\}$, $i_l = j$ and $t_1 < t_2 < \cdots < t_l$.

Lemma 4. If node j is l-connected to node i then

$$\forall t \geq t_0 + 3nlB, \ x_j(t) \geq m(t_0) + \eta^l \left(\frac{1}{n}\right)^{(t-t_0)^l} (x_i(t_0) - m(t_0)).$$

5.2 Proof of Theorem 1

Consider a node i and a time t_0. Assumption 1 implies that node i is B-connected to any node $j \in P$ and Lemma 4 gives: $\forall t \in [t_0 + 3nMB, t_0 + 3nMB + B]$, $\forall j \in P$,

$$x_j(t_0 + 3nMB + B) \geq m(t_0) + \delta (x_i(t_0) - m(t_0)),$$

where $\delta > 0$. This inequality being true for all $i \in P$, it follows that

$$m(t_0 + 3nMB + B) \geq m(t_0) + \delta \left(\max_i x_i(t_0) - m(t_0)\right).$$

We show that

$$\lim_{t_0 \to \infty} \max_i x_i(t_0) - m(t_0) = 0,$$

otherwise we would have $\lim_{t_0 \to \infty} m(t_0) = +\infty$. As $\lim_{t \to \infty} m(t) = c$ and $m(t) \leq x_j(t) \leq \max_i x_i(t)$, we deduce that

$$\forall j \in P, \ \lim_{t \to \infty} x_j(t) = c,$$

which implies that

$$\lim_{t \to \infty} s_{ij}(t) = 0.$$

Thanks to Assumption 1, we deduce that

$$\lim_{t \to \infty} v_{ij}(t) = 0,$$

and thanks to (1) and (3), we deduce that

$$nc = \lim_{t \to \infty} x_i(t) = \sum_{i=1}^{n} x_i(0),$$

i.e.

$$c = \sum_{i=1}^{n} x_i(0)/n,$$

which leads to

$$\lim_{t \to \infty} x_i(t) = \frac{1}{n} \sum_{i=1}^{n} x_i(0) = \frac{L}{n},$$

proving Theorem 1.

6   Experimental Evaluation

In order to evaluate the efficiency of our load balancing algorithm, we have implemented it in the SimGrid environment [2]. This is a simulation-based complete framework for evaluating cluster, grid and P2P algorithms and heuristics. Among its numerous interests, let's point out realistic computations and communications models. So, the results presented here are fully representative of real results that should be obtained with a similar parallel architecture.

As mentioned in the introduction, our load balancing algorithm is quite generic. However, it should be more interesting in the context of parallel iterative algorithms in which a pool of tasks is repeatedly executed. In that context, we model the iterative process by associating to each task a number of iterations to be performed. Thus, a same task with a constant number of operations is repeatedly executed until its associated number of iterations becomes null. So, as the load balancing takes the form of the migration of those tasks from one node to one of its neighbors, a task may accomplish its iterations on different nodes.

Temporal dependencies between the tasks only occur in synchronous iterative algorithms and when there are some data dependencies between the tasks. Classically, we say that a task A depends on another task B if the computations of A require the knowledge of the data processed in B (typically at the previous iteration). In such a case, when two data-dependent tasks are migrated on different nodes, this implies a dependency between those two nodes. However, in the context of use of iterative algorithms, such dependencies very often already exist due to the domain decomposition induced by the parallel treatment of the problem. Indeed, the notion of neighbor between nodes is commonly related to those data dependencies.

In asynchronous iterative algorithms, there are no temporal dependency between the tasks, even if there are some data dependencies. This comes from the fact that each task performs its computations without waiting for the last version of its data dependencies but by using the version of those dependencies which are locally available at that time. In that way, asynchronous algorithms are much more flexible and provide better performances than synchronous ones in numerous parallel contexts.

Due to that last remark and to the fact that our balancing algorithm is also asynchronous by nature, the evaluations presented below take place in the case of asynchronous iterative algorithms. Moreover, we consider that the domain decomposition is regular and that all the tasks have the same amount of workload.

Before presenting the results, it is necessary to explain how the efficiency of the balancing is evaluated.

6.1   Efficiency Evaluation

In the scope of this study, we evaluate the efficiency of the balancing by comparing the performance gains obtained with our algorithm and with a near optimal scheduling. That theoretical optimal performance is deduced from the nodes speeds and the tasks workload. As the tasks are composed of a given number of iterations whose workload is always the same (in terms of flops), the problem is reduced to the choice of the correct node for each iteration of each task. Thus, the optimal makespan is deduced by using a list scheduling algorithm that successively places each iteration of each task on the node which is able to offer the soonest termination. It is important to notice that the makespan computed here has only a theoretical value to give us a hint on the ideal minimal makespan but it does not take into account any overhead due to the tasks scheduling and migrations.

6.2   Experimental Contexts

In the following, the efficiency of our algorithm is evaluated for three common topologies: a line, a ring and a complete graph. Although the line is a bit easier to manage from the algorithmic point of view, it is actually the worst case in terms of performances as the load diffusion will be the longest in that case.

The experiments have been conducted in the following conditions:

Cluster	
Size	10 and 50 machines
Powers	homogeneous or heterogeneous (ratio 10 between slowest and fastest)
Links	homogeneous
Initial distribution of the tasks	
or	All on a single node
	evenly distributed over the processors
Communications	
or	Always active (permanent links)
	Intermittent (while ensuring Assumption 1)
Tasks	
Number	10000
Data size	80 bytes per task
Iterations	randomly chosen in the range [100,500]
Flops	1600 per iteration

6.3   Results

Convergence of the load balancing scheme. Although the convergence of our load balancing scheme has been theoretically proved, it is interesting to get

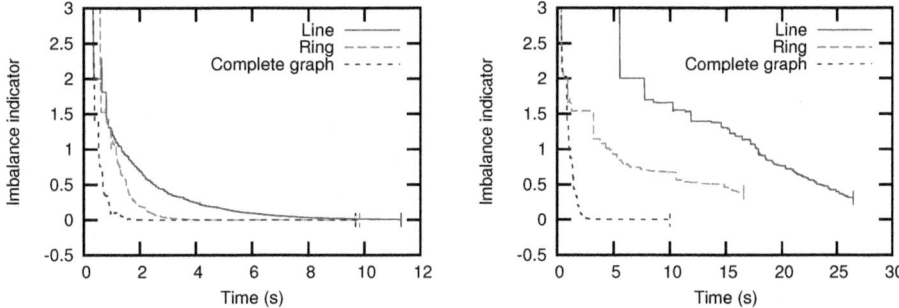

Fig. 2. Evolution of the load imbalance in the system with permanent links (left) and with intermittent links (right)

an experimental confirmation of this as well as an idea of its speed. In order to see the evolution of the balancing of the system during the algorithm execution, we define the imbalance indicator as the standard deviation at time t of the ratios between the actual load on each node at time t and the ideal load it should have. The ideal loads are deduced from the speeds of the nodes. So, the measure is equal to zero when the system is ideally balanced and the measure increases with the imbalance.

In Fig. 2 (left) are depicted the evolutions of the imbalance indicator for the three topologies considered, with a homogeneous cluster of 10 nodes and permanent links. In this context, the number of iterations is set to 500 for all the tasks. Moreover, since we want to observe the speed of the load diffusion in the system, the initial repartition must be as imbalanced as possible, this is why the entire load (all the tasks) is placed on a single node at the beginning. The small vertical tick at the end of each curve indicates the termination time of the program, once there is no more task to compute. It can be seen that the three curves tend to decrease toward zero, which confirms the convergence of the load balancing algorithm. Moreover, the curves do not decrease at the same rate but they follow the hierarchy of the diameters of the topologies, which is coherent with the respective speeds of data diffusion in those topologies. Finally, the decreases are not strict and some small fluctuations are observable on the curves, especially for the ring and complete graph topologies. Such fluctuations come from the decentralized and asynchronous nature of the load balancing scheme. Indeed, the decisions of load transfers are taken on each node according to its own vision of the system, deduced from information that may be partially outdated at the moment of the decision. This behavior is emphasized with more densely connected graphs because each node is directly linked to more others. In such contexts there is a greater probability that the load estimation sent by one node to the others change meanwhile they receive it and use it, implying potential oversized or undersized load transfers. This phenomenon is difficult to eradicate but it is strongly limited by the constraints applied on the ratios of load to be transferred. This is why those fluctuations stay small and do not prevent the global convergence of the load distribution.

As the most original feature of our load balancing scheme is to be asynchronous and then to support temporary interruptions of the links, it is interesting to see their impact on the evolution of the load imbalance during the execution of the program. Hence, the results obtained with the same cluster but with intermittent links are given in Fig. 2 (right). The general trend of the curves is similar to the previous context as they all decrease toward the load balance in the system. Also, as could be expected from the additional communication delays induced by the intermittent links, the balancing times and total execution times are longer than with permanent links. Finally, the hierarchy between the topologies is still respected according to their respective diffusion speeds although the differences between their convergence speeds are accentuated. In fact, in such a context of communication delays, the convergence toward the complete load balance may be longer than the total computation time of the tasks, resulting in a still perceptible imbalance indicator at the end of the execution, as can be seen for the line and ring topologies. Nonetheless, this does not mean in any way that the convergence is no longer ensured and if the execution time had been longer, the imbalance indicator would have continued to decrease.

This is depicted in Fig. 3 in which the number of iterations per task has been set to 3000 in place of 500. This does not modify the initial load balancing problem as the starting configuration is exactly the same (same system and same number of tasks gathered on the same node) but directly increases the total execution time. Finally, we can see in this figure that the convergence is systematically reached for sufficiently long computations, whatever the topology of the system is.

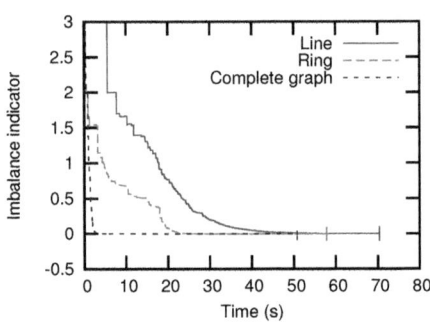

Fig. 3. Evolution of the load imbalance in the system with intermittent links and more iterations per task

Efficiency of the load balancing algorithm scheme. For the sake of clarity, we present our results in different tables for each topology used. In each cell of the tables, there are two percentages. The first one (on top) gives the relative overhead of our balancing relatively to the theoretically optimal one. As mentioned in Section 6.1, that reference time only includes the computation time of all the tasks but not any scheduling or task migration overhead. It is computed by using a best choice list algorithm at the level of the iterations inside the tasks. It is quite obvious that this time is not always attainable in practice but it gives a good reference for the evaluation of our load balancing algorithm. So, 10% indicates that our balancing makes the whole iterative process terminate in a time 10% greater than the optimal one. The second value, in italic, indicates the gain of our balancing relatively to the iterative process without any balancing.

Table 1 provides the results for the line topology. That topology is the most difficult case of load balancing as every node has at most two neighbors and this is the communication network with the largest diameter. Hence, that case will yield the longest load diffusions. It can be observed that the results are better for the smaller cluster. This could be expected as a load diffusion will always be longer in a larger system. Moreover, in large systems, each processor has less work to do and for the same initial amount of work, the makespan will be much smaller. This explains the higher ratios according to the optimal makespan. Also, the results are quite different according to the initial load distribution and it is interesting to see that our balancing does not imply any overhead in simple cases like the even distribution on homogeneous nodes. Finally, concerning the intermittent links, our balancing is farther from the optimal time, but this is normal for two reasons. The first one is that the load diffusion is more difficult and naturally longer in such contexts. The second one is that, as mentioned above, the optimal time is computed without taking into account the scheduling and migration costs, which are much more important with intermittent links. Moreover, the absolute performances of our algorithm stay very good in that context as large gains are still obtained relatively to the unbalanced version.

The results for the ring topology are presented in Table 2. As expected for a topology with a smaller diameter, the gains are better than with the linear topology in all the contexts with the small cluster. With the larger cluster, the results are similar or better, except for the intermittent links with the initial distribution of the work on only one node. This probably comes from the tuning of the parameters of our load balancing algorithm which may be optimized. However, the results stay globally satisfying.

In Table 3 are given the results for the complete graph topology. Here again, a good behavior can be observed for the small cluster whereas the algorithm gives rather deceptive results for the larger one. Those results tend to confirm that the local strategy of work distribution plays a major role. That strategy gives the rules of how a node distributes its overload to its less loaded nodes while respecting the constraints given in Section 3. So, it defines the β and α_{ij} values. For example, the use of a slightly different strategy taking into account the load average among the node and its less loaded neighbors to compute those parameters produces slightly better results in the context of the complete graph. In particular, there are no more loss of time in the already balanced cases as we obtain an overhead of only 4.67% and a gain of 0.92% in the case of an evenly distributed load on homogeneous processors with constant links and an overhead of 3.31% and a gain of 2.21% in the same context with intermittent links.

Finally, all those results point out the interest of our asynchronous load balancing algorithm in both contexts of constant and intermittent links. Also, they reveal that a single distribution strategy does not seem to be adapted to all the contexts of parallel systems. A deeper analysis of the behavior of our algorithm according to its inner parameters will be necessary to precisely identify the potential causes of inefficient results.

Table 1. Results obtained with a linear topology

	Initial tasks distribution	Homogeneous processors		Heterogeneous processors	
		10	50	10	50
Constant links	All tasks on one node	31.38	387.82	34.96	367.5
		86.86	*90.24*	*92.09*	*83.9*
	Even distribution	0.44	2.33	16.25	46.26
		0.97	*3.13*	*80.64*	*75.31*
Intermittent links	All tasks on one node	55.58	967.35	146.73	832.17
		84.44	*78.65*	*85.54*	*67.89*
	Even distribution	0.48	3.18	52.78	99.89
		0.93	*2.33*	*74.56*	*66.26*

Table 2. Results obtained with a ring topology

	Initial tasks distribution	Homogeneous processors		Heterogeneous processors	
		10	50	10	50
Constant links	All tasks on one node	11.55	292.48	23.43	370.14
		88.85	*92.15*	*92.76*	*83.8*
	Even distribution	0.26	2.08	2.78	44.39
		1.15	*3.37*	*82.89*	*75.63*
Intermittent links	All tasks on one node	23.75	1187.76	127.99	1116.72
		87.63	*74.24*	*86.64*	*58.09*
	Even distribution	0.54	3.45	34.94	80.62
		0.87	*2.07*	*77.53*	*69.51*

Table 3. Results obtained with a complete graph topology

	Initial tasks distribution	Homogeneous processors		Heterogeneous processors	
		10	50	10	50
Constant links	All tasks on one node	6.12	811.01	15.24	791.51
		89.39	*81.78*	*93.25*	*69.29*
	Even distribution	0.4	7.45	2.8	108.62
		1.01	*-1.72*	*82.89*	*64.79*
Intermittent links	All tasks on one node	28.11	4101.52	46.96	1085.86
		87.19	*15.97*	*91.39*	*59.15*
	Even distribution	0.31	6.74	7.93	331.93
		1.09	*-1.04*	*82.03*	*27.09*

7 Conclusion

An asynchronous decentralized load balancing algorithm has been presented. Its main advantages are to be usable on dynamic networks where the links are intermittent. Moreover, it is quite generic and can be applied to numerous computational algorithms.

The convergence of the balancing has been proved and experimentally confirmed in the context of load conservation. Also, it has been pointed out that in case of variable load, the algorithm will implicitly tend to balance the load during the time intervals in which the load stays constant and will thus globally follow the load variations.

Some simulations have been conducted within the SimGrid environment in the context of an asynchronous parallel iterative algorithm. Globally, the experiments confirm the interest of our algorithm, even in the most difficult context of data diffusion (linear topology). In most of the cases, our algorithm does not induce any significant overhead in already balanced contexts and provide sharp improvements in the other contexts.

However, as it has been pointed out by the simulations, there is still some room for a finer tuning of our algorithm, especially for the more densely connected topologies with a large number of elements. So, our next investigations will be focused on the optimization and auto-tuning of the inner parameters of our algorithm in order to provide the best efficiency in every context of use. Also, a deeper study of the context of variable load will be investigated as well as the management of integer loads and a theoretical evaluation of the maximal balancing time according to the bound of the communication delays and the load migration costs.

References

1. Bertsekas, D., Tsitsiklis, J.: Parallel and Distributed Computation. Prentice Hall, Englewood Cliffs (1999)
2. Casanova, H., Legrand, A., Quinson, M.: SimGrid: a Generic Framework for Large-Scale Distributed Experiments. In: 10th IEEE International Conference on Computer Modeling and Simulation (March 2008)
3. Cortes, A., Ripoll, A., Cedo, F., Senar, M.A., Luque, E.: An asynchronous and iterative load balancing algorithm for discrete load model. Journal of Parallel and Distributed Computing 62(12), 1729–1746 (2002)
4. Cybenko, G.: Dynamic load balancing for distributed memory processors. Journal of Parallel and Distributed Computing 7, 279–301 (1989)
5. Elsässer, R., Monien, B., Preis, R.: Diffusion schemes for load balancing on heterogeneous networks. Theory of Computing Systems 35, 305–320 (2002)
6. Fatès, N.: Directed percolation phenomena in asynchronous elementary cellular automata. In: El Yacoubi, S., Chopard, B., Bandini, S. (eds.) ACRI 2006. LNCS, vol. 4173, pp. 667–675. Springer, Heidelberg (2006)
7. Genaud, S., Giersch, A., Vivien, F.: Load-balancing scatter operations for grid computing. Parallel Computing 30(8), 923–946 (2004)
8. Bahi, J.M., Couturier, R., Vernier, F.: Synchronous distributed load balancing on dynamic networks. Journal of Parallel and Dist. Comp. 65(11), 1397–1405 (2005)
9. Kumar, V.: Introduction to Parallel Computing. Addison-Wesley Longman Publishing Co., Inc., Boston (2002)
10. Miguet, S., Robert, Y.: Elastic load-balancing for image processing algorithms. In: Zima, H.P. (ed.) ACPC 1991. LNCS, vol. 591, pp. 438–451. Springer, Heidelberg (1992)
11. Willebeek-Lemair, M.H.: Startegies for dynamic load balancing on highly parallel computers. IEEE Trans. on Parallel and Distributed Systems 4(9), 979–993 (1993)
12. Li, Y., Lan, Z.: A survey of load balancing in grid computing. In: Zhang, J., He, J.-H., Fu, Y. (eds.) CIS 2004. LNCS, vol. 3314, pp. 280–285. Springer, Heidelberg (2005)

A Computing Resource Discovery Mechanism over a P2P Tree Topology*

Damia Castellà, Hector Blanco, Francesc Giné, and Francesc Solsona

University of Lleida, C/ Jaume II, 69, Lleida, Spain
{dcastella,hectorblanco,sisco,francesc}@diei.udl.cat

Abstract. Peer-to-Peer (P2P) computing, the harnessing of idle compute cycles through Internet, offers new research challenges in the domain of distributed computing. In this paper, we propose an efficient computing resource discovery mechanism based on a balanced multi-way tree structure capable of supporting both exact and range queries, efficiently. Likewise, a rebalancing algorithm is proposed. By means of simulation, we evaluated our proposal in relation to other approaches of the literature. Our results reveal the good performance of our proposals.

Keywords: Parallel and distributed computing, P2P computing.

1 Introduction

P2P paradigm takes advantage of the under utilization of personal computers, integrating thousand or even millions of users into a platform based on the sharing of computational resources [1]. Some current research projects, such as CompuP2P [2], CHEDAR [3] or CoDiP2P [4], propose using the P2P paradigm for distributed computing. P2P computing is distinguished by a mutable amount of computational resources (CPU, Memory and Bandwidth) provided by each peer. Thus, the computational resource management becomes a research challenge.

The resource discovery mechanisms in P2P computing are classified in structured and unstructured ones [5]. The unstructured algorithms are characterized by the fact that they use only local information of their neighbors. The CHEDAR's searching mechanism [3] fits in this category. Although these systems adapt easily to frequent node joins and disjoins, they do not scale very well for very large networks. On the other hand, the searching mechanisms based on structured information are generally faster and have a predictable service search time. In this set, we can find the well known Chord algorithm [6], which is used by the CompuP2P platform [2]. Although the Chord protocol is very efficient for exact queries, this is not well suited for range queries since hashing destroys the ordering of data. Recent works, such as Squid [13] supports keyword searches, including wildcards, partial keywords and ranges queries, based on DHT. It uses a locality-preserving indexing scheme based on Space Filling Curves (SFC), where each data element is indexed and shared using a set of

---

* This work was supported by the MEyC-Spain under contract TIN2008-05913 and the CUR of DIUE of GENCAT and the European Social Fund.

J.M.L.M. Palma et al. (Eds.): VECPAR 2010, LNCS 6449, pp. 366–379, 2011.

keywords and mapped to a single point in its key space. Other works, such as [11, 7], propose network discovery services without the use of DHTs. Caron et al. [11] present a new architecture, Distributed Lexical Placement Table (DLP), based on a Prefix Tree which supports automatic completion of partial search string, range queries and multicriteria searches. Likewise, BATON [7] proposes a balanced tree structure overlay which supports exact and range queries, also without the use of DHTs. This emphasizes that adding a small number of links in addition to tree edges, they are able to obtain an excellent fault tolerance and a balanced congestion. Finally, Harren et al. [12] take advantage of DHTs to improve the scalability and adapts it to support complex queries in relational databases. Although these works are very optimized for resource discovery services, they are very restrictive in providing a discovery mechanism over a mutable environment. Note that, in a P2P computing environment, the shared resources change their disponibility over the time. So, it is necessary to check periodically the resources for better scheduling purposes.

Other works related to Grid environments, [14, 15] propose a distributed super-peer model for handling membership management and resource discovery service in large-scale Grids and exploit centralized/hierarchical information service provided by the Grid infrastructure of the local PO (Physical Organization). The related work describes that a Grid is viewed as a network interconnecting small-scale Grids, referred as PO's. For each PO, a subset of powerful nodes having high availability properties are used as super-peers. These nodes are responsible for the communications with the other POs and maintain metadata about all the nodes of the local PO. A structured P2P topology of super-peers implements the join and departure of Grid nodes and the resource discovery service. The super-peer model is similar to the manager's peer role used in our proposal, which controls a set of peers, named Areas, and stores information about discovery services. However, its organization is different because each peer of an area can be the manager of the immediate lower level area, whereas in a super-peer model, each PO is considered as a leaf node.

In this paper, we propose a new structured computing resource discovery mechanism, which provides exact and range query facilities and scalability features with a low algorithmic cost. Our approach is oriented to the CoDiP2P (P2P Distributed Computing) system developed by our group in previous works [4]. Following the CoDiP2P architecture, the proposed lookup mechanism follows an structured architecture. It is based on a balanced multi-way tree structure capable of supporting both exact and range queries efficiently. In addition, this paper proposes a rebalancing algorithm, which allows the tree to be maintained totally balanced and re-link any isolated area. Thus, CoDiP2P exploits efficiently the well known characteristics of a tree topology ($\theta(\log_{|Area|}(N_{tree}))$ lookup length, where N_{tree} is the total number of peers, and constant linkage cost) to manage the mutability of resources. We have analyzed the performance of our proposals by means of simulation in relation to the Chord algorithm by the case of exact queries, and the BATON algorithm by the case of range queries. In both cases, the obtained results reveal the competitiveness of our proposals.

The outline of this paper is as follows; Section 2 revises the CoDiP2P architecture. Section 3 presents the discovery mechanism used by CoDiP2P for exact and range queries. The rebalancing mechanism used for CoDiP2P system is described in Section 4. The efficiency measurements of our proposals are performed in Section 5. Finally, the main conclusions are explained in Section 6.

2   CoDiP2P Architecture

We present a review of the CoDiP2P architecture, explained in detail in [4]. In order to describe it, some previous concepts must be introduced:

- **Area** A_i is a logical space made up of a set of workers.
- **Manager** M_i manages an area and schedules tasks over the workers.
- **Worker** W_i is responsible for executing tasks scheduled by its manager.
- **Replicated managers** RM_i: Each area A_i maintains a set of Replicated Managers. Each RM_i maintains a copy of the same information kept by M_i. Thus, if M_i fails, then the oldest RM_i of the same area will replace it.

Fig. 1 shows the linked structure of peers in CoDiP2P based on a tree topology with an area size of 3 peers. Note that this type of structure allows a manager M_i of an area located at level i to be a worker W_j in an area located at level $i+1$ at the same time. In addition, this same node can be also a RM_i. The main functionalities of CoDiP2P system are *insertion and departure of peers*, *updating system information* and the *scheduling mechanism* to launch a parallel job from any peer in the system. These three algorithms are explained in detail in [4].

In order to understand better the *searching and rebalancing algorithms* explained later in this paper, some highlights of the *updating* and *departure algorithms* are needed.

2.1   Updating Algorithm

The main aim of the updating algorithm consists of maintaining the information about the computational resources available in the system. This is divided in two parts: the *manager_updating* and the *worker_updating.*

By means of the *manager_updating*, every manager M_i sends, every T seconds, a message to all the workers belonging to the same area A_i to notify that it is alive, together with the List of its Top Managers (LTM). The LTM is a list that contains the addresses of the managers located over the manager M_i in direction towards the root manager (M_1). For instance, the LTM of the manager M_8 shown in Fig. 1 would be M_4, M_2, M_1.

Whenever a $Peer_i$ receives a $message_{upd}$ from its manager M_i, it launches the *worker_updating* function. Each $Peer_i$, depending also on its role, sends to its manager M_i statistical information ($SttInf_j$) about its locally available computational resources or the available computational resources managed by such a peer (if $Peer_i$ is the manager of a lower area). Note that the information $SttInf_j$ is only sent when it differs from the previous one sent $SttInf_{j-1}$. The cost of the algorithm is $\theta(|Area|)$, where $|Area|$ is the number of peers in one area.

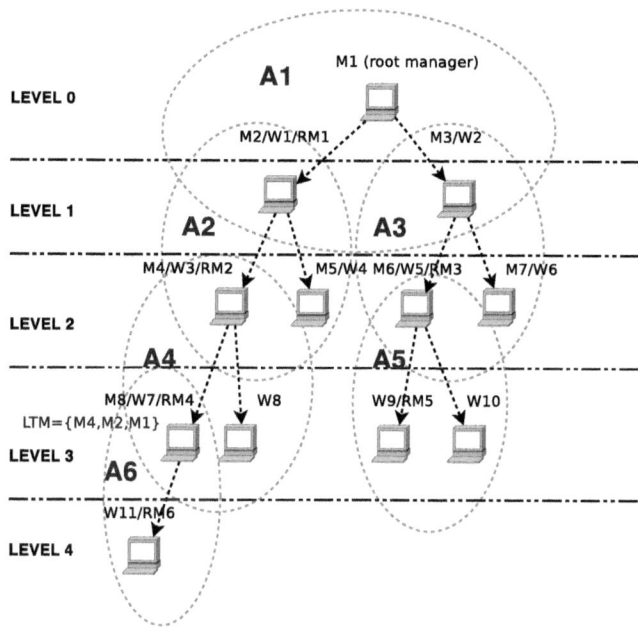

Fig. 1. Tree topology of peers in CoDiP2P

2.2   Departure of Peers

Whenever a peer leaves the tree, voluntarily or involuntarily, the manager of the disconnected peer detects the broken link in the updating algorithm, explained in Sec. 2.1. Whenever this happens, the manager checks if the peer is a worker node or a manager in the immediate lower level:

- In the worker case, there is no problem in restructuring the system because it is a final node, and no more work must be performed.
- In the manager case, the restructuring operation, described in Alg. 1, is applied by the replicated manager RM_i of the current area A_i. After T seconds, this detects that there is no answer from the faulting manager M_i and executes the mechanism.

Note that the algorithm selects the oldest replicated manager, by issues of peers reliability. Thus, the oldest peers, which the system considers more reliable, are in the upper levels of the tree, compared with the youngest peers, which are considered more irregular and are in the lower levels.

The cost of Alg. 1 is determined by the number of peers affected by the fall of the peer. The worst scenario happens when the replicated manager RM_i, selected to replace an output manager is also the manager of a lower level. The cost of the algorithm is $\theta(\log_{|Area|}(N_{tree}) - 1)$, where N_{tree} is the total number of peers in the CoDiP2P system and $|Area|$ is the size of the areas.

procedure RM_i.Manage_Departure_in_Tree()
Input: RM_i, A_i
begin
   if $|A_i - RM_i| \neq 0$ **then**
      $M_i := RM_i \in A_i$;
      RM_i notifies $\forall workers \in A_i$ that it will be manager;
      if RM_i *is also manager of a lower area* A_j **then**
         RM_i selects the oldest $RM_j \in A_j$;
         RM_i notifies RM_j that it will be manager of A_j;
         if RM_j *is manager of lower level* **then**
            RM_j.Manage_Departure_in_Tree();
         end
      end
   else
      RM_i becomes worker of upper area of A_i;
   end
end

Algorithm 1. Departure of peers in CoDiP2P.

3   Searching Algorithms

We have added a new resource discovery mechanism to the CoDiP2P service layer to provide two different kinds of searching, one based on exact queries and the other one based on range queries. Both algorithms look up the addresses of peers throughout the tree that have the desired CPU power available, although they can be used for looking up any kind of computing resources.

3.1   Exact Query Searching Algorithm

The searching mechanism is designed to take advantage of the topology and roles of peers. This algorithm is based on looking through the local database (DB) stored by each peer, which contains the computing resources characteristics of the peers located below it throughout the tree branch.

According to the Alg. 2, whenever a $Peer_i$ requests a CPU query, firstly it checks on its own DB if there is a peer with the required CPU power. In the case of a search failure, $Peer_i$ forwards the searching query to its manager M_i. If the search fails again, the next manager located on the branch continues the same search in a recursive way until it reaches the zero level (M_1). Finally, if the searching is successful then the CPU owners peer address is returned to the $Peer_i$. Note that the cost of this algorithm is $\theta(\log_{|Area|}(N_{tree}))$.

Note that one important problem related to the tree topology is the traffic congestion produced by the routing messages of the searching algorithm and the updating algorithm. Regarding to the searching traffic, only those searches which requested peers are located in another subtree of Level 1 arrives to the root peer. So, the updating algorithm can cause even more bottleneck in the root peer than the searching one. The searching congestion is measured in the Experimental Results Section (5.1). In addition, it is worth pointing out that

```
function Peer_i.Search_EQuery(CPU_query)
Result: Peer_Address
begin
    foreach register ∈ DB.CPU_Table do
        if register.CPU = CPU_query then
        |   Peer_Address := register.address;
        end
    end
    if Peer_Address = NULL then
        if Peer_i ≠ M_1 then
        |   Peer_Address := M_i.Search_EQuery(CPU_query);
        else
        |   Peer_Address := NULL;
        end
    end
    return Peer_Address;
end
```

Algorithm 2. Exact Query Searching Algorithm.

the updating traffic is characterized by small messages sent each T period. Thus, we must fix the value of T by balancing the congestion and the updating of information.

3.2   Range Query Searching Algorithm

The Alg. 3 shows the range query searching algorithm, to which three parameters are passed: the low and high limits of the searched CPU range values and the number of items that the algorithm has to catch. Compared with the Exact Query algorithm, Alg. 3 differs in two points. The first point is that it returns a List of Peer Addresses (LPA), which contains the desired CPU power inside the requested range. The second one is that the algorithm does not finish until it has filled the LPA up with nr_items or it has reached level zero (M_1). Note that the cost of this algorithm is also $\theta(\log_{|Area|}(N_{tree}))$.

4   Rebalancing Mechanism

The churn of peers in a P2P environment can unbalance the tree topology. As a consequence and as we can see in Fig. 2(left), one tree branch can be much longer than another (Case 1 and 2) or one area can remain isolated from the root manager (Case 3). Obviously, the unbalancing of the tree will decrease the performance of the search mechanisms explained above.

In order to solve this problem, a rebalancing mechanism oriented to a tree topology is proposed. Alg. 4 allows areas to be moved from one site to another or lost links to be restored.

function $Peer_i.Search\_RQuery(low\_CPU, high\_CPU, nr\_items)$
Result: $LPA(=$List Peer Addresses$)$
begin
> **foreach** $register \in DB.CPU\_Table$ **do**
> > **if** $register.CPU \geq low\_CPU \wedge register.CPU \leq$
> > $high\_CPU \wedge LPA.size < nr\_items$ **then**
> > > $\mid LPA.add(register.Address);$
> > **end**
> **end**
> **if** $LPA.size < nr\_items \wedge Peer_i \neq M_1$ **then**
> > $\mid LPA+ = M_i.Search\_RQuery(low\_cpu, high\_cpu, nr\_items - LPA.size);$
> **end**
> **return** LPA
end

Algorithm 3. Range Query Searching Algorithm.

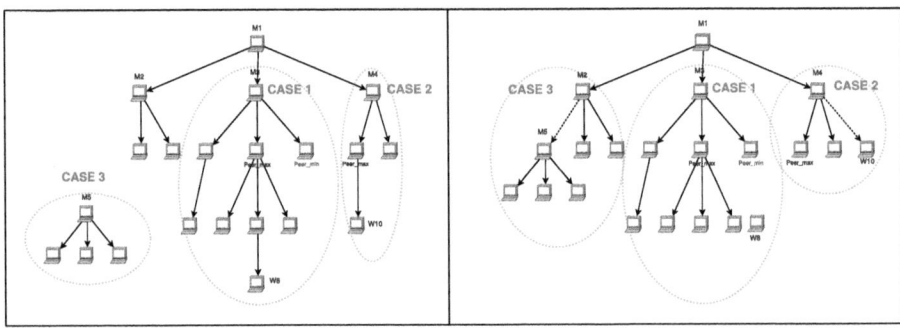

Fig. 2. Tree without rebalancing (left) and with rebalancing (right)

Periodically, each $Peer_i$ of the tree launches the rebalancing algorithm. As we can see in Alg. 4, firstly the algorithm tests if any of the three cases shown in the Fig. 2(left) happens. Whenever it happens, Alg. 4 works as follows in each case:

- **Case 1:** Whenever $Peer_i$ is also a manager M_i and the area A_i is full, M_i checks from the List of its Children (LC) which are their sons with the maximum ($Peer_{max}$) and minimum ($Peer_{min}$) number of levels hanging down from it. If the difference between both values is greater than one level then the tree is considered to be unbalanced and as a consequence the *rebalancing_levels* procedure is called. This procedure moves the branch hanging down from the $Peer_{max}$ son with more levels to the $Peer_{min}$. As we can see in the Case 1 of Fig. 2(left), $M3$ executes the rebalancing mechanism and as a consequence the Peer $W8$ hanging down from the son of $Peer_{max}$ is linked to $Peer_{min}$. The result is shown in Fig. 2(right). Note as the example of Fig. 2 assumes an areas size of 4.

- **Case 2:** This case happens whenever $Peer_{max}$ has levels hanging down from it and the area A_i is not full. This situation means that the tree is unbalanced due to the fact that the capacity of the area is not fulled up. In this case, the half levels below $Peer_{max}$, denoted as *rlevels* in Alg. 4, are linked to $Peer_i$ by means of the same *rebalancing_levels* procedure explained above. As we can see in Case 2 of Fig. 2, $W10$ is linked to $M4$ after applying the rebalancing mechanism.
- **Case 3:** The third and the last part of the Alg. 4 is activated whenever an area A_i is isolated from its above manager. According to this goal, $Peer_i$ checks if there is an above manager (M_i) and if there is not then $Peer_i$ looks for a manager from its List of Top Managers (LTM). If the searching is successful then $Peer_i$ is linked to the new manager. As we can see in Case 3 in Fig. 2, $M5$ is linked to $M2$ after applying the rebalancing mechanism.

Note that all searching operations, whose peers are affected by one of the three rebalancing cases, will be aborted. They will be resumed when the rebalancing procedure was finished.

The cost of this is $\theta(\log_{|Area|}(N_{tree}) - 1)$ because the maximum number of hops is equal to the length of a branch from the root manager M_1 to a leaf node W_i.

5   Experimentation

The performance of our proposals was tested by means of GridSim [8] and Sim-Java [9] simulators. In order to simulate our P2P platform with GridSim, peers were modeled as user entities by means of threads. All entities (peers) were connected by network links, whose bandwidth and latency can be specified at the start time. SimJava features provide the management of events and the mechanism for discovering peers.

All tests were performed with 10,000 peers and a total of 125,000 searches, which follows a Poisson distribution with a mean frequency of 125 *searches/s* by default. According results obtained previously in [4], the updating procedure is continuously executed in periods of 20 seconds. A summary of the experimental results is shown in next section.

5.1   Experimental Results

First of all, we tested the influence of the rebalancing algorithm over the searching algorithms. Likewise, the impact of the number of replicated managers (RM) was also evaluated for both cases, with and without rebalancing. Fig. 3 shows the percentage of unsuccessful searches in relation to the percentage of failed peers for 1 an 3 RMs. In the non-rebalancing case, we can see as the rate of unsuccessful searches scaled consistently according to the number of failed peers and the results ranged up to 40% with only 1 RM and 25% with 3 RMs. When we applied rebalancing, the results were very satisfactory with a rate of unsuccessful searches below 5% for both cases, 1 and 3 RMs. In this case, the number of

procedure $Peer_i.rebalancing()$
Data: $M_i(=\text{Manager of } Peer_i)$, $A_i(=\text{Area of } M_i)$, $LC(=\text{List of Childs})$,
        $LTM(=\text{List of Top Managers})$
begin
    if $Peer_i = M_i$ **then**
        $\{Peer_{max}, Peer_{min}\} := Peer_i.max\_min\_child\_peers(LC);;$
        // Case 1
        if $Peer_{max}.levels\_below - Peer_{min}.levels\_below > 1 \;\; \wedge \;\; A_i.isfull()$
        then
            $rlevels := \lfloor (Peer_{max}.levels\_below - Peer_{min}.levels\_below)/2 \rfloor;$
            $Peer_{max}.rebalancing\_levels(rlevels, Peer_{min});$
        end
        // Case 2
        if $Peer_{max}.levels\_below \geq 1 \;\wedge\;$ **not** $A_i.isfull()$ **then**
            $rlevels := \lfloor Peer_{max}.levels\_below/2 \rfloor;$
            $Peer_{max}.rebalancing\_levels(rlevels, Peer_i);$
        end
    end
    // Case 3
    if $\nexists M_i$ **then**
        for $j := 0$ **to** $LTM.size$ **do**
            $Peer_j := LTM.get(j);$
            if $\exists Peer_j$ **then**
                new $Peer_i \longrightarrow Peer_j;$
                break;
            end
        end
    end
end
procedure $Peer_i.rebalancing\_levels(rlevels, Peer_j)$ **begin**
    if $Peer_i.levels\_below \geq rlevels$ **then**
        $Peer_k := Peer \in LC \mid MAX_{k=1}^{|LC|}(LC.get(k).levels\_below);$
        $Peer_k.rebalancing\_levels(rlevels, Peer_j);$
    else
        new $Peer_i \longrightarrow Peer_j$
    end
end

Algorithm 4. Rebalancing Algorithm

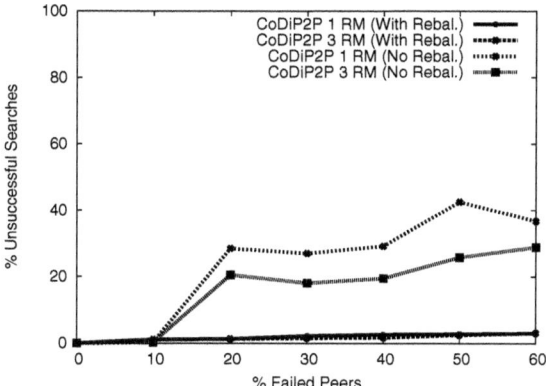

Fig. 3. Comparative of unsuccessful searches in terms of failed peers with and without Rebalancing Algorithm

RMs practically does not affect the behavior of the searches and the influence of the percentage of faulting peers on the system is imperceptible. Therefore, these results prove the well performance of our rebalancing algorithm.

Next, we evaluated the impact of the searching frequency on the exact query case. The performance achieved by our proposals, denoted as CoDiP2P, was compared with Chord. Samples were collected for CoDiP2P with 1 and 3 Replicated Managers (RM) and 1 and 3 successors in the Chord algorithm.

Fig. 4(left) and (right) show the results of the exact query search with a high (12500 $searches/sec$) and low (125 $searches/sec$) searching frequencies, respectively. In general, both plots showed that CoDiP2P obtained better results than Chord, specially when the frequency was high. On the other hand, we saw how the influence of the successors in the Chord case was higher than the use of replicated managers in CoDiP2P. This was because the Chord's successors are active

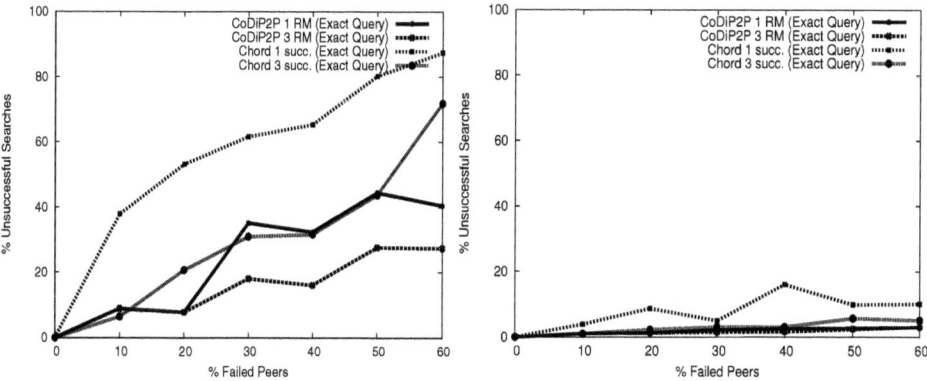

Fig. 4. Exact queries searching Algorithm versus Chord with a $Freq = 12500$ $searches/sec$ (left) and $Freq = 125$ $searches/sec$ (right)

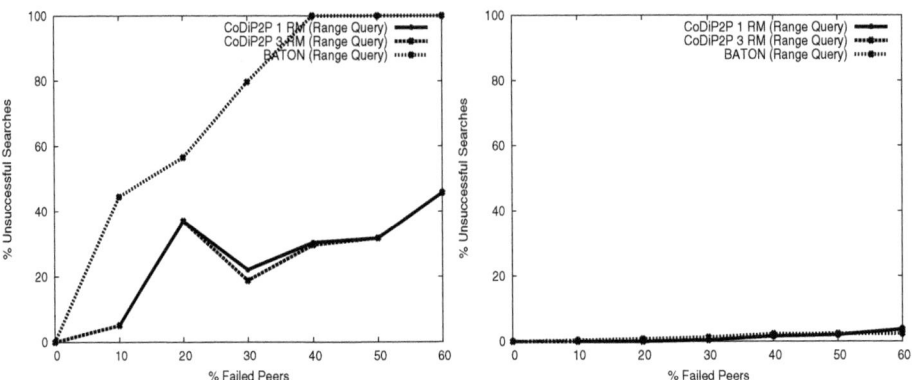

Fig. 5. Range queries searching Algorithm versus Baton with a $Freq = 12500\ searches/sec$ (left) and $Freq = 125\ searches/sec$ (right)

elements in the searching process, whereas in the case of CoDiP2P, the replicated managers do not play such an important role in the searching algorithm. Focusing on the influence of the searching frequency, we saw that the CoDiP2P and Chord results ranged up to 40% and 80% with a high frequency (see Fig. 4(left)), whereas they ranged from 15% to 5% with low frequency respectively (see Fig. 4(right)). This behavior is due to the fact that a low searching frequency gives both systems enough time to recompose system tables and links. However, CoDiP2P continues giving better results than Chord because the system has enough time to rebalance the system completely and thus better performance is obtained. This is due to the maintaining cost of the overall system in CoDiP2P is $\theta(|Area| \cdot \log_{|Area|} N_{tree})$ whereas the maintaining cost of Chord is higher, $\theta(N \cdot \log^2 N)$. In general, we conclude that in any case, the CoDiP2P rebalancing algorithm performs better than the restructuring of the Chord DHT structure.

Our next evaluation was to compare the performance of our range query approach in relation to the BATON algorithm, following the same methodology described above. Fig. 5 shows the percentage of unsuccessful searches in relation to the percentage of failed peers for high (Fig. 5(left)) and low (Fig. 5(right)) searching frequencies. As we can see in both figures, our approach improves the BATON results. This improvement is very significant for the case of high frequency, achieving a maximum difference around 70% (see Fig. 5(left)). This difference is due to the fact that each peer on the BATON system has stored less keys than CoDiP2P and as a consequence BATON needs to do more hops than CoDiP2P for searching a specific set of values. In addition, BATON has a major dependency of the neighbourhood and by this reason the drop of a neighbour has a high repercussion on the searching process.

Next, we compared the response time of CoDiP2P for both cases, exact and range query, in relation to the Chord algorithm. The discovery mechanism needed a minimum interval, called "response time", to update the system completely after a peer fault. According to our aim, we obtained the percentage of unsuccessful searches for an hypothetical massive drop of 50% of the total peers at the same time. Thus, we could compare the robustness of both approaches, CoDiP2P and Chord. From Fig. 6 (left), we can see that CoDiP2P took 255 units of time to recover the parts of the system affected by the faulting peers in both cases, whereas Chord needed more than 800 units of time. This is because Chord updates one entry of its finger table in each updating step and as a consequence, it needs more time to update the total of 160 entries in its finger tables. In contrast, CoDiP2P only needs a number of jumps equal to the number of levels of the tree to update the network completely after a massive drop of peers. In general, we can conclude that CoDiP2P has a response time approximately 3 times faster than Chord.

Finally, Fig. 6 (right) shows the congestion (percentage of messages) achieved throughout the levels of the tree topology for different areas sizes, when the searching mechanism is applied. Level 0 represents the root of the tree. We can see as the congestion increases with the areas size. As it was expected, congestion is critical in the root and decreases when ascending levels. With an areas size of 21 peers, congestion is around 100%, causing a bottleneck in the root node. In order to reduce this congestion, the trees level and the areas size should be limited below a specific threshold. Therefore, in order to maintain the scalability of the system, we should group the peers according to its characteristics in a set of different small trees, which would be connected by a second level topology with good scalability properties. This will be the main goal of future work.

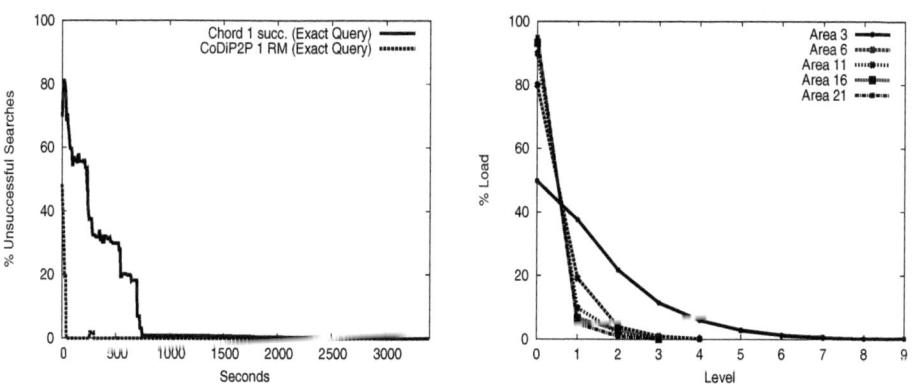

Fig. 6. CoDiP2P and Chord comparative of system time response due to a 50% of Failed Peers (left) and Tree topology congestion (right)

6   Conclusions and Future Work

In this paper, a computing resource discovery mechanism oriented to the CoDiP2P system is presented. Following the CoDiP2P architecture, the lookup mechanism is based on a balanced multi-way tree structure capable of supporting both exact and range queries efficiently. A rebalancing algorithm is also proposed.

The exact query proposal was compared with another exact query algorithm widely used in the literature, Chord. Given that Chord does not implement range queries, we also compared our range query proposal with a binary-tree structure, named BATON. In general, our results show that CoDiP2P performs much better than the other approaches, specially for high frequencies. In this case, CoDiP2P achieves half the rate of unsuccessful searches as the others. Likewise, the results obtained reveal the good performance of our rebalancing algorithm, improving the unsuccessful searches rate by 50%. Robustness was another goal of our work. In doing so, the response time to update the system completely after a massive drop of peers was measured. Our results show that CoDiP2P has a response time approximately 4 times faster than Chord.

The future trend is directed towards extending the tree topology with a second level topology, which will allow to increase the scalability of the platform. Each tree would group a set of peers according to any common characteristic (i.e. locality, computational resources, etc.), whereas the second level would connect the set of trees by means of a Bruijn graph[10], which is characterized by its high scalability and low congestion. Thus, we maintain the effectiveness of the tree topology for searching and overcome its main drawback, the congestion of the root levels for huge systems. Another important trend is testing and monitoring the topology and algorithms under real conditions and network. This will allow to measure the real traffic congestion and bottlenecks caused by the updating algorithm, the communication times taken by the departure algorithm and the effectiveness and overhead messages of the rebalancing algorithm.

References

[1] Foster, I., Iamnitchi, A.: On death, taxes and the convergence of peer-to-peer and grid computing. In: 2nd. Int'l Workshop on P2PSystems (2003)
[2] Gupta, R., Sekhri, V., Somani, A.: CompuP2P: An Architecture for Internet Computing Using Peer-to-Peer Networks. IEEE Transactions on Parallel and Distributed Systems 17(11) (2006)
[3] Kotivalainen, N., Weber, M., Vapa, M., Vuery, J.: Mobile Chedar - A Peer-to-Peer Middleware for Mobile Devices. In: 3rd Int. Conf. on Pervasive Computing and Communications Workshops (PerCom Workshops) (2005)
[4] Castellà, D., Rius, J., Barri, I., Giné, F., Solsona, F.: CoDiP2P: a New P2P Architecture for Distributed Computing. In: Conference on Parallel, Distributed and Network-based Processing (PDP 2009), pp. 323–329 (2009)
[5] Meshkova, E., Riihijrvi, J., Petrova, M., Mhnen, P.: A Survey on Resource Discovery Mechanisms, Peer-to-Peer and Service Discovery Frameworks. Journal of Computer Networks 52, 2097–2128 (2008)
[6] Karger, D., Kaashoek, F., Stoica, I., Morris, R., Balakrishnan, H.: Chord: A Scalable Peer-to-Peer Lookup Service for Internet Applications. In: 2001 ACM SIG-COMM Conference, pp. 149–160 (2001)
[7] Jagadish, H., Ooi, B.C., Vu, Q.H.: BATON: A Balanced Tree Strcuture for Peer-to-Peer Networks. In: The 31st Int'l Conference on Very Large Data Bases (VLDB) Conference (2005)

[8] Buyya, R., Murshed, M.: GridSim: A Toolkit for the Modeling and Simulation of Distributed Resource Management and Scheduling for Grid Computing. Concurrency and Computation: Practice and Experience (CCPE) 14(13-15), 1175–1220 (2002)

[9] SimJava (1998), http://www.dcs.ed.ac.uk/home/hase/simjava/

[10] Loguinov, D., Kumar, A., Rai, V., Ganesh, S.: Graph-theoretic analysis of structured peer-to-peer systems: routing distances and fault resilience. In: Conference on Applications, Technologies, Architectures, and Protocols For Computer Communications (2003)

[11] Caron, E., Desprez, F., Tedeschi, C.: A Dynamic Prefix Tree for the Service Discovery Within Large Scale Grids. In: The Sixth IEEE International Conference on Peer-to-Peer Computing, P2P 2006 (2006)

[12] Harren, M., Hellerstein, J., Huebsch, R., Loo, B., Shenker, S., Stoica, I.: Complex Queries in DHT-based Peer-to-Peer Networks. In: Druschel, P., Kaashoek, M.F., Rowstron, A. (eds.) IPTPS 2002. LNCS, vol. 2429, p. 242. Springer, Heidelberg (2002)

[13] Schmidt, C., Parashar, M.: Squid: Enabling Search in DHT-Based Systems. Journal of Parallel and Distributed Computing 68(7), 962–975 (2008)

[14] Mastroianni, C., Talia, D., Verta, O.: A super-peer model for resource discovery services in large-scale Grids. Future Generation Computer Systems 21, 1235–1248 (2005)

[15] Mastroianni, C., Talia, D., Verta, O.: Designing an information system for Grids: Comparing hierarchical, decentralized P2P and super-peer models. Parallel Computing 34(10), 593–611 (2008)

A Parallel Implementation of the Jacobi-Davidson Eigensolver for Unsymmetric Matrices[*]

Eloy Romero[1], Manuel B. Cruz[2], Jose E. Roman[1], and Paulo B. Vasconcelos[3]

[1] Instituto I3M, Universidad Politécnica de Valencia,
Camino de Vera s/n, 46022 Valencia, Spain
{elroal,jroman}@upvnet.upv.es
[2] Laboratório de Engenharia Matemática
Instituto Superior de Engenharia do Porto,
Rua Dr. Bernardino de Almeida, 431, 4200-072 Porto
mbc@isep.ipp.pt
[3] Centro de Matemática da Universidade do Porto and
Faculdade de Economia da Universidade do Porto,
Rua Dr. Roberto Frias s/n, 4200-464 Porto, Portugal
pjv@fep.up.pt

Abstract. This paper describes a parallel implementation of the Jacobi-Davidson method to compute eigenpairs of large unsymmetric matrices. Taking advantage of the capabilities of the PETSc library —Portable Extensible Toolkit for Scientific Computation—, we build an efficient and robust code adapted either for traditional serial computation or parallel computing environments. Particular emphasis is given to the description of some implementation details of the so-called correction equation, responsible for the subspace expansion, and crucial in the Jacobi-Davidson algorithm. Numerical results are given and the performance of the code is analyzed in terms of serial and parallel efficiency. The developments achieved in the context of this work will be incorporated in future releases of SLEPc —Scalable Library for Eigenvalue Problem Computations—, thus serving the scientific community and guaranteeing dissemination.

Keywords: Message-passing parallel computing, eigenvalue computations, Jacobi-Davidson.

1 Introduction

The Jacobi-Davidson method is a popular technique to compute a few eigenpairs of large sparse matrices, i.e., (λ, x) pairs that satisfy $Ax = \lambda x$, $x \neq 0$. This problem arises in many scientific and engineering areas such as structural dynamics, electrical networks, quantum chemistry and control theory, among others.

[*] This work was partially supported by Fundação para a Ciência e a Tecnologia - FCT, through Centro de Matemática da Universidade do Porto - CMUP, and by the Spanish Ministerio de Ciencia e Innovación under project TIN2009-07519.

J.M.L.M. Palma et al. (Eds.): VECPAR 2010, LNCS 6449, pp. 380–393, 2011.

Its introduction, about 15 years ago, was motivated by the fact that standard iterative eigensolvers often require an expensive factorization of the matrix to compute interior eigenvalues, *e.g.*, shift-and-invert Arnoldi with a direct linear solver. Jacobi-Davidson tries to reduce the cost by solving linear systems only approximately (usually with iterative methods) without compromising the robustness.

General purpose parallel Jacobi-Davidson eigensolvers are currently available in the PRIMME [1], JADAMILU [2] and Anasazi [3] software packages. However, currently PRIMME and JADAMILU can only cope with standard Hermitian eigenproblems, and Anasazi only implements a basic block Davidson method for generalized Hermitian eigenproblems, $Ax = \lambda Bx$, where the user is responsible for implementing the correction equation. There are no freely available parallel implementations for the non-Hermitian case.

There are several publications dealing with parallel Jacobi-Davidson implementations employed for certain applications. For instance, the design of resonant cavities needs to solve real symmetric generalized eigenproblems arising from the Maxwell equations [4]. Other cases result in non-Hermitian problems, as they occur in linear magnetohydrodynamics (MHD): in [5] a complex generalized non-Hermitian problem is solved using the shift-and-invert spectral transformation for searching for the interior eigenvalues closest to some target, and a variant with harmonic Ritz values is presented in [6].

Our aim is to fill the lack of a parallel general purpose non-Hermitian Jacobi-Davidson eigensolver providing a robust and efficient implementation in the context of SLEPc, the Scalable Library for Eigenvalue Problem Computations [7]. Our previous work addresses the Generalized Davidson method for symmetric-definite (and indefinite) generalized problems [8].

In this work we focus on important details of the non-Hermitian version of Jacobi-Davidson: searching for interior eigenvalues using harmonic Ritz values and the real arithmetic management of complex conjugate eigenpairs in problems with real matrices. These improvements will be added to a future fully-fledged Jacobi-Davidson solver in SLEPc. The description of the particular features of the solver implemented in this work are described in Section 2. The analysis of the performance of the code is detailed in Section 3. We conclude with some final remarks.

2   The Jacobi-Davidson Method

The Jacobi-Davidson algorithm combines ideas from Davidson's and Jacobi's methods (see [9]). As all other methods based on projection techniques, it has an extraction phase and another one of subspace expansion (see Algorithm 1).

Regarding the extraction phase, A is projected on a low-dimensional subspace $\mathcal{K}$ of size m with Ritz values and Ritz vectors, respectively, θ_j and $u_j = Vs_j$, $j = 1, ..., m$, being $V = [v_1 v_2 ... v_m]$ the $n \times m$ matrix whose columns form an orthonormal basis of $\mathcal{K}$.

In the expansion phase, a new direction should be considered in such a way that $\mathcal{K}$ is extended providing better information to obtain a new approximation

---

Algorithm 1. Jacobi-Davidson algorithm

1: **procedure** JD($A, u_0, tol, itmax$)
2:      $u \leftarrow u_0/\|u_0\|_2$, $V \leftarrow [u]$, $\theta \leftarrow u^*Au$, $r \leftarrow Au - \theta u$
3:      $m \leftarrow 1$
4:      **while** $\|r\|_2/|\theta| > tol$ & $m < itmax$ **do**
5:           Solve approximately $(I - uu^*)(A - \theta I)(I - uu^*)t = -r$, $t \perp u$
6:           $V \leftarrow$ Orthonormalize$[V, t]$, and $m \leftarrow m + 1$
7:           Compute a desired eigenpair (θ, s) from the projected eigensystem
                 using standard or harmonic techniques
8:           $u \leftarrow Vs$, $r \leftarrow Au - \theta u$
9:      **end while**
10:     **return** θ, u
11: **end procedure**

---

of the selected eigenpair (θ, u). Jacobi [10] proposed to correct u by a vector t, the Jacobi orthogonal component correction (JOCC)

$$A(u + t) = \lambda(u + t), \quad u \perp t. \tag{1}$$

Pre-multiplying (1) by u^*, and considering $\|u\| = 1$, results in

$$\lambda = u^*A(u + t). \tag{2}$$

Projecting (1) onto the orthogonal complement of u, which is done by pre-multiplying by $I - uu^*$, and replacing λ, which is unknown, by the available approximation θ results in the Jacobi-Davidson correction equation,

$$(I - uu^*)(A - \theta I)(I - uu^*)t = -r, \tag{3}$$

being $r = Au - \theta u$ the eigenpair residual. Far from convergence, θ can be set to a target τ.

   Usually, iterative linear solvers are used to solve the equation (3), for instance using Krylov methods. Regarding the required precision for the resolution of (3), a low one is generally accepted. In practice, the performance of the method largely depends on the appropriate tuning of this parameter.

2.1   Computation of Eigenvalues at the Periphery of the Spectrum

For exterior eigenvalues, the Jacobi-Davidson method uses the Rayleigh-Ritz approach to extract the desired approximate eigenpair $(\theta, u = Vs)$, by imposing the Ritz-Galerkin condition on the associated residual,

$$r = Au - \theta u \perp \mathcal{K}, \tag{4}$$

which leads to the low-dimensional projected eigenproblem

$$V^*AVs = \theta s. \tag{5}$$

At each iteration the implementation updates the matrix of the projected system (5), $M = V^*AV$, by computing the mth row and column. Then it computes the Schur decomposition of M ($MQ = QT$, Q orthogonal, T upper-triangular) and sorts it in order to have the desired Ritz value and vector as the first one.

However, the Rayleigh-Ritz method generally gives poor approximate eigenvectors for interior eigenvalues, that is, the Galerkin condition (4) does not imply that the residual norm is small. Moreover, spurious Ritz values can appear and it is difficult to distinguish them from the ones we seek. In that case, the Ritz vector can contain large components in the direction of eigenvectors corresponding to eigenvalues from all over the spectrum, so adding this vector to the search subspace will hinder convergence. The harmonic projection methods are an alternative to solve this problem [11,12].

2.2 Computation of Interior Eigenvalues

For interior eigenpairs, Jacobi-Davidson uses, in general, the harmonic Rayleigh-Ritz extraction, requiring the Petrov-Galerkin condition

$$(A - \tau I)^{-1}u - (\theta - \tau)^{-1}u \perp \mathcal{L}, \tag{6}$$

where $\mathcal{L} \equiv (A - \tau I)^*(A - \tau I)\mathcal{K}$.

This extraction technique exploits the fact that the harmonic Ritz values closest to the target τ are the reciprocals of the largest magnitude Ritz values of $(A - \tau I)^{-1}$, and avoids working with the inverse of a large matrix. The resulting projected generalized eigenvalue problem is then

$$V^*(A - \tau I)^*(A - \tau I)Vs = (\theta - \tau)V^*(A - \tau I)^*Vs. \tag{7}$$

Our implementation of the harmonic extraction is based on the algorithm described in [13, §7.12.3], which maintains W as an orthogonal basis of $(A - \tau I)V$, and updates the matrices $N = W^*W$ and $M = W^*V$ in each iteration. Thus, in this case, step 7 in Algorithm 1 has to solve the eigenproblem associated to the (N, M) matrix pair (see Algorithm 2). The new vectors added to W, as in the case of the V vectors, are orthogonalized with a variant of the Gram-Schmidt process available in SLEPc, which is based on classical Gram-Schmidt with selective reorthogonalization, providing both numerical robustness and good parallel efficiency [14].

2.3 Computing Complex Eigenvalues with Real Arithmetic

Real unsymmetric matrices may have complex eigenvalues and corresponding eigenvectors. It is possible to avoid the complex arithmetic by using real Schur vectors instead of eigenvectors in step 7 of Algorithm 1. The real Schur decomposition consists of an orthogonal matrix and a block upper triangular matrix, which has scalars or two by two blocks on the diagonal. The eigenvalues of two by two blocks correspond to two complex conjugate Ritz values of the original

---

Algorithm 2. Harmonic extraction in Jacobi-Davidson algorithm

---

1: $w \leftarrow (A - \tau I)v_m$, where v_m is the last column of V
2: $W \leftarrow$ Orthonormalize$[W, w]$, with h the orthogonalization coefficients and ρ the norm prior to normalization
3: Update $N \leftarrow \begin{bmatrix} N & h \\ 0 & \rho \end{bmatrix}$,
4: Update M such that $M = W^*V$
5: Compute generalized Schur decomposition $NQ = Z\tilde{N}$, $MQ = Z\tilde{M}$ such that $|\tilde{n}_{1,1}/\tilde{m}_{1,1}|$ is the minimum
6: $s \leftarrow q_1$ and $\theta \leftarrow \overline{\tilde{m}_{1,1}\tilde{n}_{1,1}} + \tau$

---

matrix. The real Schur form requires less storage, since these Schur vectors are always real. Another advantage is that complex conjugate pairs of eigenvalues always appear together.

Our implementation is based on RJDQZ [15], that adapts the extraction process and the correction equation to work with the real Schur form. The changes to the original algorithm are only significant when a complex Ritz pair is selected. In that case, the method works with a real basis $U = [u_1 \; u_2]$ of the invariant subspace associated to the selected complex conjugate pair of Ritz values, $(\theta, \bar{\theta})$, where the corresponding Ritz vectors are $u = u_1 \pm u_2\,i$. The residual is now computed as

$$[r_1 \; r_2] = A[u_1 \; u_2] - [u_1 \; u_2]\begin{bmatrix} \Re(\theta) & \Im(\theta) \\ -\Im(\theta) & \Re(\theta) \end{bmatrix}. \tag{8}$$

Apart from the residual, the correction equation also contains the Ritz value, so it has to be rearranged as

$$P\begin{bmatrix} A - \Re(\theta)I & \Im(\theta)I \\ -\Im(\theta)I & A - \Re(\theta)I \end{bmatrix} P\begin{bmatrix} t_1 \\ t_2 \end{bmatrix} = -\begin{bmatrix} r_1 \\ r_2 \end{bmatrix}. \tag{9}$$

For this correction equation, we propose three different projectors P, that can be selected at runtime in our solver. The first option, P_0, implements the same projector that would be used in complex arithmetic,

$$P_0 = I - \begin{bmatrix} u_1 & u_2 \\ u_2 & -u_1 \end{bmatrix}\begin{bmatrix} u_1^T & u_2^T \\ u_2^T & -u_1^T \end{bmatrix}. \tag{10}$$

P_1 represents the orthogonal projector onto the orthogonal complement of the subspace $\mathcal{U} = \text{span}\{u_1, u_2\}$,

$$P_1 = I - \begin{bmatrix} \hat{u}_1 & \hat{u}_2 & 0 & 0 \\ 0 & 0 & \hat{u}_1 & \hat{u}_2 \end{bmatrix}\begin{bmatrix} \hat{u}_1^T & 0 \\ \hat{u}_2^T & 0 \\ 0 & \hat{u}_1^T \\ 0 & \hat{u}_2^T \end{bmatrix}, \tag{11}$$

where $\{\hat{u}_1, \hat{u}_2\}$ is an orthogonal basis of $\mathcal{U}$. This basis is cheaply obtained from the Schur decomposition of the projected problem. Finally, P_2 implements the following lighter projector

$$P_2 = I - \begin{bmatrix} \hat{u}_1 & 0 \\ 0 & \hat{u}_2 \end{bmatrix} \begin{bmatrix} \hat{u}_1^T & 0 \\ 0 & \hat{u}_2^T \end{bmatrix}. \tag{12}$$

2.4 Preconditioning

In some cases, the convergence of the iterative linear solver for the correction equation (3) can be very slow. Therefore, it is convenient to use a preconditioner. The preconditioner can be chosen as

$$K = (I - uu^*)K_\theta(I - uu^*), \tag{13}$$

where K_θ is some approximation of $A - \theta I$. Since θ changes in each outer iteration, it would be necessary to rebuild K_θ every time. This high overhead can be avoided by using K_τ, where τ is the constant target value, but probably losing effectiveness in the preconditioner. The application of K^{-1} to a given vector is done without building K explicitly, in such a way that only one preconditioner application with K_τ^{-1} is done per inner iteration [9].

2.5 Implementation Details

Our Jacobi-Davidson implementation is intended to be executed on distributed memory parallel machines using PETSc [16]. The problem matrix and the vectors of size n are distributed by blocks of rows and the corresponding operations are parallelized. These include the computation of the matrices of the projected system, the selected Ritz vector and its residual, the solution of the correction equation and the orthogonalization of V and W. The projected problem decomposition and other minor computations are replicated in all nodes.

The algorithm is initialized with a randomly generated vector in parallel, taking care that each processor generates different random sequences. This initial vector must have a nonzero component in the desired eigenvector. To further ensure this feature, and to seek the eigenvalue of largest magnitude, a few iterations of the power method may be applied.

The real Schur decomposition of the projected problem is computed using appropriate LAPACK routines, since they are dense and small matrices. Also, LAPACK is used for sorting the Schur decomposition, keeping the complex conjugate pairs together.

The stopping criterion of the algorithm is currently based on the simple evaluation of the normalized residual $\|Au - \theta u\|_2/|\theta|$. When the convergence is achieved then there exists an exact eigenvalue λ such that $|\lambda - \theta|$ must be sufficiently small, although λ may not be the desired eigenvalue. In the symmetric case, a small relative residual guarantees a small error in the eigenvalue; for the unsymmetric case, however, the norm of the residual may give an indication of the error but it is not guaranteed, especially for highly non-normal matrices.

The correction equations, (3) and (9), are solved using a PETSc Krylov solver (commonly GMRES) without explicitly storing the coefficient matrices of the system, and only defining the matrix-vector product. Also, only applying the left projector is sufficient to guarantee the condition $t \perp u$, provided that a Krylov solver is used with a zero starting vector, as shown in [17].

As mentioned before, the tuning of the stopping criterion for the correction equation is crucial since it greatly influences the convergence behavior of the method. It is widely referred that the required tolerance for the residual on this inner iteration phase does not need to be tight, yet a loose one can provoke a large number of outer iterations or even lead to non-convergence of the Jacobi-Davidson method. A good balance between the required precision for the correction equation and the tolerance of the overall process is thus mandatory. Generally, one can bound the time spent in the correction equation by limiting both the number of iterations and the required precision for the solution. We tested several approaches trying to fulfill the above mentioned ideas as well as to deliver a default option, which can be used by the user as a first approach to solve his eigenproblem. We tested several mechanisms, namely, a fixed small number of GMRES iterations (5, 20 or 40) and a coarse tolerance (10^{-2} or 10^{-3}). We also implemented a dynamic criteria such as a variable tolerance, function of the outer iteration number; for instance, a tolerance of α^{-Its+1} (among others), being Its the outer iteration number, for $\alpha = 1.5, 2.0, 3.0$. The approach proposed in the JDQR code [9] is 0.7^{Its}.

In this work restarting techniques were not considered in order to concentrate on other crucial algorithmic parts and parameters. Details about a restarted variant, together with numerical results on a plasma physics application, can be found in [18]. In general terms, the amount of information remaining after a restart should be neither too small nor too large, in order not to lose too much information and to prevent many time consuming restarts, respectively. A related issue is deflation, required for continuing the computation of other eigenpairs after some of them have converged. Deflation can be implemented by locking converged vectors at restart time [18], so it has not been covered here either.

3   Computational Results

Several experiments have been performed for testing our implementation. The sequential tests use a collection of 136 real unsymmetric matrices of moderate size and density (see Figure 1) chosen from the MatrixMarket[1] database and the University of Florida Sparse Matrix Collection[2]. The sequential tests are primarily intended to evaluate the performance and robustness of our implementation. Also, using such a large collection of matrices allows us to extend previous results comparing standard versus harmonic extraction [6,9] as well as complex versus real arithmetic versions. In order to avoid biased results, the collection is very heterogeneous, and cover the case of large condition number and norm (see Figure 1), clustered eigenvalues and complex conjugate eigenpairs.

---

[1] http://math.nist.gov/MatrixMarket/
[2] http://www.cise.ufl.edu/research/sparse/matrices/

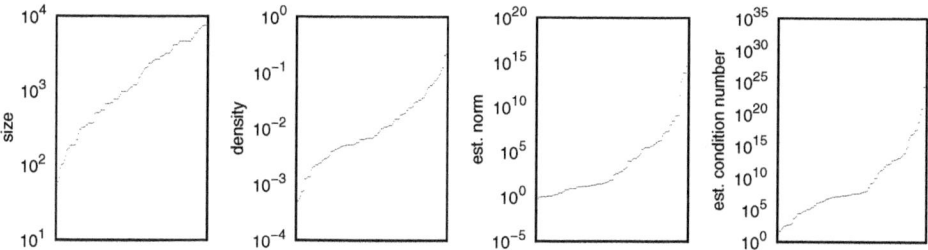

Fig. 1. Size, density, estimated 2-norm and estimated condition number for each matrix used in the sequential tests

Also the Krylov-Schur eigensolver, currently the most powerful method available in SLEPc, is employed and the results will be compared with Jacobi-Davidson.

In the tests to be presented, only one eigenpair is requested, since locking is not available, as mentioned earlier. The convergence criterion is based on the residual norm divided by the absolute value of the approximate eigenvalue, and a tolerance of 10^{-7} was considered. A problem is flagged as unconverged if the desired tolerance was not reached within a maximum of 500 (outer) iterations.

In order to study the parallel performance of our implementation, some large-scale tests were run on two distributed memory machines: (i) Odin, from Universidad Politécnica de Valencia, which is a cluster with 55 dual-processor nodes (however only one process runs per node in the tests), Pentium Xeon processors at 2.8 GHz and 1 GB per node, connected by a high-speed SCI network with 2D torus topology; and (ii) CaesarAugusta, from the Barcelona Supercomputing Center, consisting of 256 JS20 blade computing nodes, each of them with two 64-bit PowerPC 970FX processors running at 2.2 GHz, interconnected with a low latency Myrinet network, where only 90 processors (i.e., 45 nodes running up to 90 processes) are used due to account limitations.

3.1   The Exterior Case

For the case of exterior eigenvalues, the developed method was able to compute, within the specified tolerance, the largest eigenvalue in magnitude and associated eigenvector in nearly all cases, as it is shown in Table 1. Except for the real arithmetic version with P_0 projector, any other configuration achieves (at least) 132 converged problems out of 136, and increasing the number of inner iterations or the power iterations helps convergence in especially difficult problems.

On the other hand, the real arithmetic version with P_0 was clearly the worst one. Since the P_1 and P_2 projectors obtained good results, P_0 will be discarded for subsequent analyses within this subsection.

At this stage we should remark that JDQR[3], a Matlab implementation of a Jacobi-Davidson algorithm as described in [9], solved around 82% of the problems

---

[3] http://www.math.uu.nl/people/sleijpen/JD_software

Table 1. Number of converged problems for different combinations of the variant, the projector, the number of power iterations and the maximum number of iteration for solving the linear equation system

Power its.		0			5	
Max. inner its.	5	20	40	5	20	40
complex arith.	134	135	135	134	136	136
real arith. P_0	134	96	103	118	129	129
real arith. P_1	132	133	132	133	133	134
real arith. P_2	134	133	134	134	133	134

without preconditioning and that Matlab's function *eigs* failed to converge in around 9% of the problems (using default parameters). Although not directly comparable, these results should be read solely to emphasize the degree of difficulty in the numerical solution of the selected testbed.

When considering the time spent by the solvers, Figure 2 shows that the real arithmetic version can be twice as fast as the complex arithmetic counterpart, especially for problems where complex Ritz pairs do not show up often. In contrast, there is no significant difference between the projector P_1 and P_2. However, the plots comparing the number of outer iterations show the penalty of real arithmetic variants. This can be attributed to the fact that the P_1 and P_2 projectors are not mathematically equivalent to the complex case. Still, each iteration is much faster. These conclusions are statistically significant.

Finally, the influence of the maximum number of linear iterations (MLIts) was analyzed. Figure 3 compares the performance of solving the Jacobi-Davidson correction equation with 5, 20 and 40 GMRES iterations at most. It is observed that the value of this parameter does not seem to influence the number of outer iterations. As a consequence, the higher the value of this parameter, the more overall GMRES iterations are required, and the execution time is thus increased significantly. For the computation of eigenvalues at the periphery of the spectrum, taking into account the mechanisms implemented to date in the code and in light of the successful use of a few linear iterations, extensive tests with a dynamic criterion were not done.

One could draw the conclusion that a very small value of MLIts is to be preferred, although this is application dependent, and may not be the case in a restarted implementation.

3.2   The Interior Case

As a test for evaluating the computation of interior eigenvalues, we looked for the smallest magnitude eigenvalue of the matrices in the test battery, that is, we run the harmonic solver with $\tau = 0$. This choice leads to convergence difficulties in some matrices of the test battery, mainly attributable to a large condition number with respect to inversion, see Fig. 1. The Jacobi-Davidson eigensolver is configured to perform 50 iterations of GMRES for solving the correction equation, with or without a preconditioner.

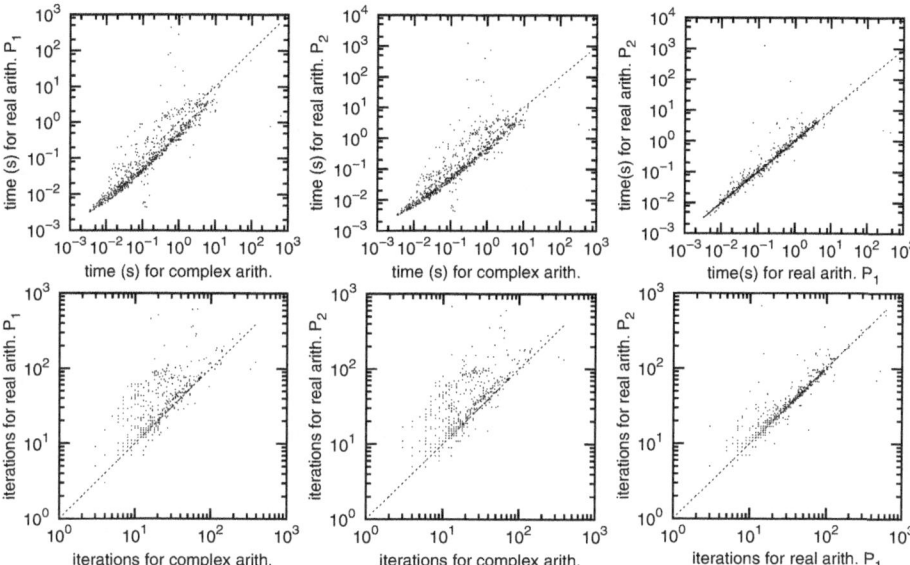

Fig. 2. Plots comparing the execution time (top) and outer iterations (bottom) when solving the problems with the complex arithmetic version and the real arithmetic version with projectors P_1 and P_2

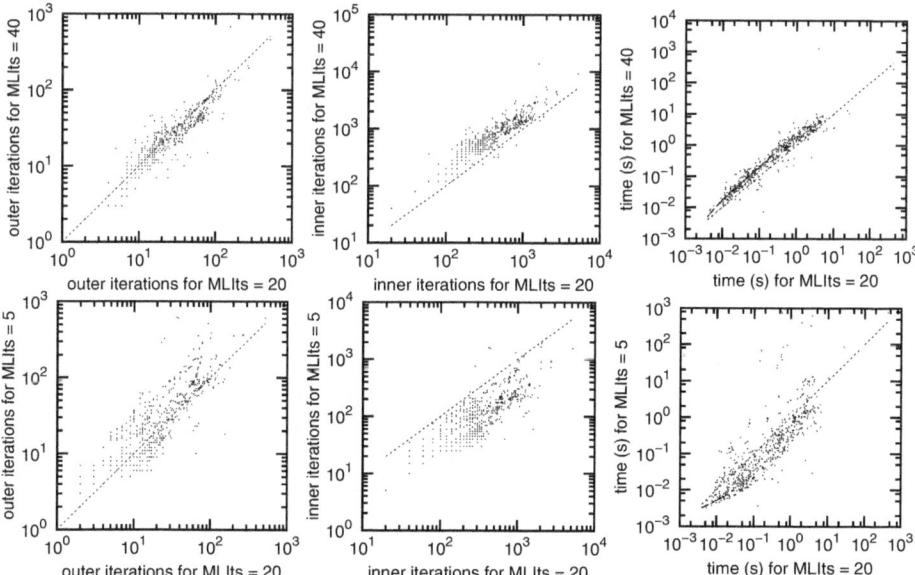

Fig. 3. Plots comparing outer iterations (left), inner iterations (middle), and execution time (right) when solving the problems with 5, 20 and 40 maximum linear iterations (MLIts) of GMRES for solving the Jacobi-Davidson correction equation

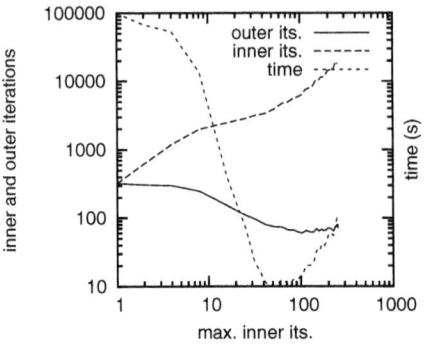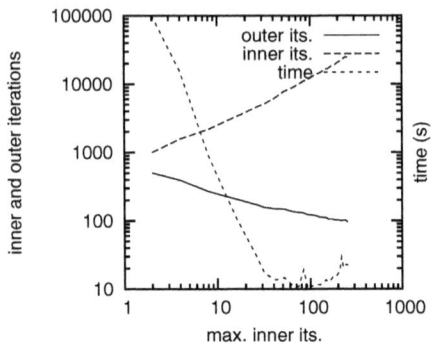

Fig. 4. Plots comparing outer iterations, inner iterations and execution time when obtaining the smallest eigenpairs for the problems *qc324* (left) and *cavity06* (right) with different maximum linear iterations of GMRES for the correction equation

The standard Rayleigh-Ritz extraction version of Jacobi-Davidson is able to reach convergence in 11 problems only, as expected from the theoretical properties previously discussed. The harmonic extraction version (proposed for this case) successfully solved up to 60 problems without using any preconditioning at the correction equation solution, up to 46 problems with a diagonal preconditioner, up to 39 problems with an ILU(0) preconditioner and up to 100 problems using an LU factorization. These results highlight the impact of the preconditioner in the global convergence of the method. However, a more powerful preconditioner is not always better. We have to recall here that the preconditioner is built from matrix $A - \tau I$ (A in this case) and not from $A - \theta I$, as discussed in section 2.4, so its effectiveness may vary widely depending on the problem. Even a complete factorization such as LU should be considered a preconditioner in this case.

Just for reference, JDQZ (the companion to JDQR with harmonic extraction) without preconditioning solved only 15% of the problems.

Regarding the performance, Fig. 4 shows the strong influence of limiting the maximum number of iterations for solving the correction equation in two sample problems. It is observed the rapid decrease of the spent time when allowing more iterations for solving the linear system. Moreover, from a certain point this benefit ends abruptly. A dynamic criteria combining the usual maximum number of iterations and a prescribed tolerance with the requirement of improving the solution quality of the correction equation as the number of outer iterations progresses is, thus, essential. The criteria for leaving the iterative process to improve the solution of the linear systems should be more demanding for the interior case than for the exterior.

3.3   Parallel Performance

The parallel performance is tested with the matrix *tmt_unsym* from the University of Florida Sparse Matrix Collection, arising from a computational electromagnetics application. The dimension of the matrix is 917,825 and it has

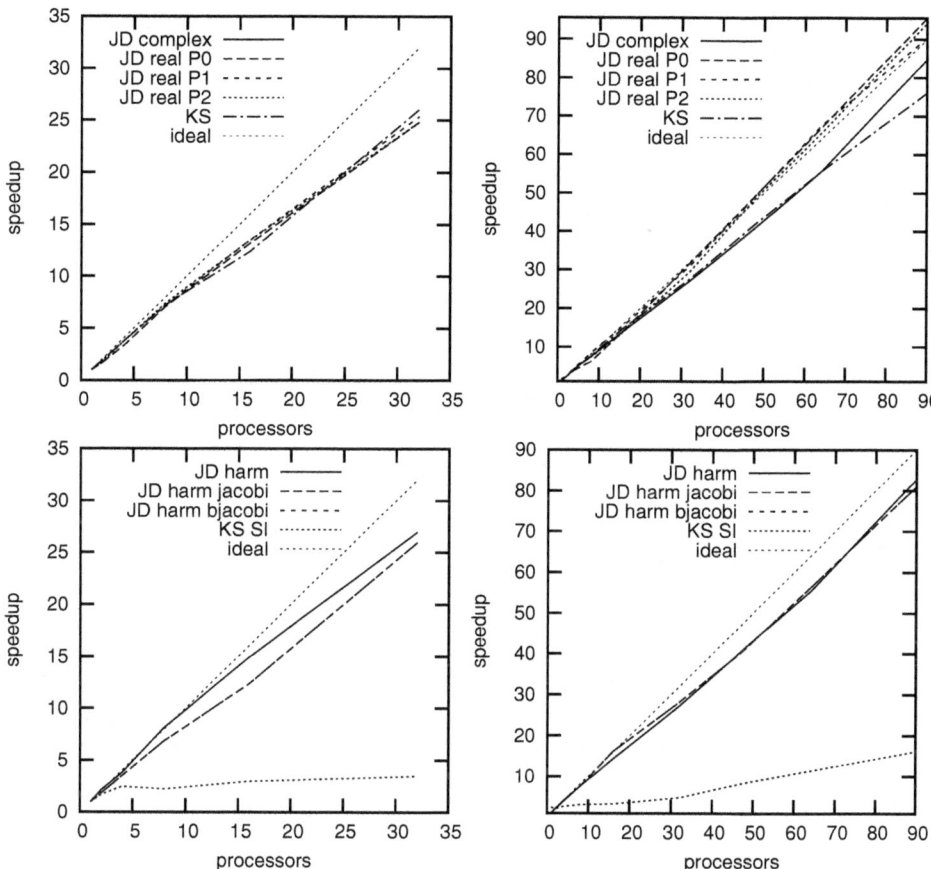

Fig. 5. Speedup on Odin (left column) and CaesarAugusta (right column) when finding the largest magnitude eigenpairs of *tmt_unsym* using the complex and real arithmetic (with projectors P_0, P_1 and P_2) versions of Jacobi-Davidson, and Krylov-Schur (first row); and when finding the eigenpairs closest to 0 of *tmt_unsym* using the harmonic version of Jacobi-Davidson without preconditioner and with Jacobi and block Jacobi preconditioners, and Krylov-Schur using shift-and-invert with MUMPS (second row)

4,584,801 nonzero elements, all of them contained in a narrow band of half-width around 2000.

Generally, the parallel performance of the matrix-vector product and the application of the preconditioner can significantly influence the global performance of an eigensolver. However, the chosen matrix avoids those problems because of the fact that its band structure makes these two operations scalable.

Figure 5 (first row) shows the speedup of the different Jacobi-Davidson versions (real arithmetic using P_0, P_1 and P_2 projectors, and complex arithmetic) and the Krylov-Schur solver in SLEPc, when computing the largest magnitude eigenpair. In the case of CaesarAugusta, speedup is computed relative to the time with two processes (one node). In both machines it can be observed that all tested implementations show good parallel performance, although the complex arithmetic version and Krylov-Schur are slightly worse than the rest in CaesarAugusta.

Regarding the computation of interior eigenvalues, we tried the harmonic Jacobi-Davidson without preconditioner as well as with Jacobi and block Jacobi preconditioners. The latter uses an ILU(0) decomposition in each diagonal block (one per process), although in this particular matrix this preconditioner does not improve over Jacobi. We also compared with SLEPc's Krylov-Schur with shift-and-invert, wherein the associated LU factorization is handled via MUMPS[4]. Figure 5 (second row) shows the obtained speedups. Again all tested implementations reveal speedups not far from the ideal one, except shift-and-invert Krylov-Schur that shows the poor parallel performance of the triangular backward and forward solves, which is particularly bad in this matrix.

4   Conclusions and Future Work

We have presented a parallel implementation of the Jacobi-Davidson method for distributed memory machines for non-Hermitian matrices, using the PETSc and SLEPc libraries. We discuss the design decisions that influence the convergence and the performance of the method, such as the stopping criterion for solving the correction equation, the eigenpair extraction method and the parallelization. The presented results suggest that it is sufficient to perform only a few GMRES iterations when searching for largest eigenvalues, whereas for interior eigenvalues it is necessary to use the harmonic extraction technique combined with much more iterations of the correction equation solver.

Furthermore, a real arithmetic version for real unsymmetric matrices has been implemented also. The results show the sequential and parallel performance improvement of this version with a convergence rate comparable to the complex arithmetic version using our proposed projectors P_1 and P_2.

At this stage of the development, the code is already able to compute exterior as well as interior eigenvalues with a reasonable efficiency despite the lack of a restarting technique for the search subspace. It is able to solve a wide range of difficult test problems, showing good speedups for their parallel performance.

These developments will be added to a fully featured Jacobi-Davidson solver in SLEPc, that will include restarting, locking of converged eigenpairs, and a more elaborate adaptive stopping criterion for the correction equation solver. The possibility of estimating, inexpensively, the residual norm of the eigenvalue problem from the one of the inner linear systems, as presented in [19], is attractive and will be addressed in a future work.

References

1. Stathopoulos, A., McCombs, J.R.: PRIMME: PReconditioned Iterative Multi-Method Eigensolver: Methods and software description. ACM Trans. Math. Software 37(2), 21:1–21:30 (2010)

---

[4] http://graal.ens-lyon.fr/MUMPS/

2. Bollhöfer, M., Notay, Y.: JADAMILU: a software code for computing selected eigenvalues of large sparse symmetric matrices. Comput. Phys. Commun. 177(12), 951–964 (2007)
3. Baker, C.G., Hetmaniuk, U.L., Lehoucq, R.B., Thornquist, H.K.: Anasazi software for the numerical solution of large-scale eigenvalue problems. ACM Trans. Math. Software 36(3), 13:1–13:23 (2009)
4. Arbenz, P., Becka, M., Geus, R., Hetmaniuk, U., Mengotti, T.: On a parallel multi-level preconditioned Maxwell eigensolver. Parallel Comput. 32(2), 157–165 (2006)
5. Nool, M., van der Ploeg, A.: A parallel Jacobi-Davidson-type method for solving large generalized eigenvalue problems in magnetohydrodynamics. SIAM J. Sci. Comput. 22(1), 95–112 (2000)
6. Nool, M., van der Ploeg, A.: Parallel jacobi-davidson for solving generalized eigenvalue problems. In: Hernández, V., Palma, J.M.L.M., Dongarra, J. (eds.) VECPAR 1998. LNCS, vol. 1573, pp. 58–70. Springer, Heidelberg (1999)
7. Hernandez, V., Roman, J.E., Vidal, V.: SLEPc: A scalable and flexible toolkit for the solution of eigenvalue problems. ACM Trans. Math. Soft. 31(3), 351–362 (2005)
8. Romero, E., Roman, J.E.: A parallel implementation of the Davidson method for generalized eigenproblems. In: Chapman, et al. (eds.) Advances in Parallel Computing, vol. 19, pp. 133–140. IOS Press, Amsterdam (2010)
9. Sleijpen, G.L.G., van der Vorst, H.A.: A Jacobi-Davidson iteration method for linear eigenvalue problems. SIAM Review 42(2), 267–293 (2000)
10. Jacobi, C.G.J.: Über ein leichtes verfahren die in der theorie der Säculärstörungen vorkommenden gleichungen numerisch aufzulösen. Crelle's J. 30, 51–94 (1846)
11. Morgan, R.B.: Computing interior eigenvalues of large matrices. Linear Algebra Appl. 154, 289–309 (1991)
12. Paige, C.C., Parlett, B.N., van der Vorst, H.A.: Approximate solutions and eigenvalue bounds from Krylov subspaces. Num. Linear Algebra Appl. 2, 115–134 (1995)
13. Bai, Z., Demmel, J., Dongarra, J., Ruhe, A., van der Vorst, H. (eds.): Templates for the Solution of Algebraic Eigenvalue Problems: A Practical Guide. Society for Industrial and Applied Mathematics, Philadelphia (2000)
14. Hernandez, V., Roman, J.E., Tomas, A.: Parallel Arnoldi eigensolvers with enhanced scalability via global communications rearrangement. Parallel Comput. 33(7-8), 521–540 (2007)
15. van Noorden, T., Rommes, J.: Computing a partial generalized real Schur form using the Jacobi-Davidson method. Num. Linear Algebra Appl. 14, 197–215 (2007)
16. Balay, S., et al.: PETSc Users Manual. Technical Report ANL-95/11 - Revision 3.0.0, Argonne National Laboratory (2008)
17. Sleijpen, G.L.G., van der Vorst, H.A., Meijerink, E.: Efficient expansion of subspaces in the Jacobi-Davidson method for standard and generalized eigenproblems. Electron. Trans. Numer. Anal. 7, 75–89 (1998)
18. Romero, E., Roman, J.E.: A parallel implementation of the jacobi-davidson eigensolver and its application in a plasma turbulence code. In: D'Ambra, P., Guarracino, M., Talia, D. (eds.) Euro-Par 2010. LNCS, vol. 6272, pp. 101–112. Springer, Heidelberg (2010)
19. Hochstenbach, M.E., Notay, Y.: Controlling inner iterations in the Jacobi-Davidson method. SIAM J. Matrix Anal. Appl. 31(2), 460–477 (2009)

The Impact of Data Distribution
in Accuracy and Performance of Parallel
Linear Algebra Subroutines

Björn Rocker[1], Mariana Kolberg[2], and Vincent Heuveline[1]

[1] Karlsruhe Institute of Technology (KIT)
Engineering Mathematics and Computing Lab (EMCL)
Fritz-Erler-Str. 23, 71633 Karlsruhe, Germany
{bjoern.rocker,vincent.heuveline}@kit.edu
[2] Universidade Luterana do Brasil
Av. Farroupilha 8001 Prédio 14, sala 122
Canoas/RS, 92425-900 - Brasil
marianakolberg@gmail.com

Abstract. In parallel computing the data distribution may have a significant impact in the application performance and accuracy. These effects can be observed using the parallel matrix-vector multiplication routine from PBLAS with different grid configurations in data distribution. Matrix-vector multiplication is an especially important operation once it is widely used in numerical simulation (*e.g.*, iterative solvers for linear systems of equations).

This paper presents a mathematical background of error propagation in elementary operations and proposes benchmarks to show how different grid configurations based on the two dimensional cyclic block distribution impacts accuracy and performance using parallel matrix-vector operations. The experimental results validate the theoretical findings.

Keywords: High Performance Computing, Parallel Computing, Rounding Errors, 2-dimensional Blockcyclic Distribution, Verified Computing.

1 Introduction

In many numerical algorithms, problems are reduced to a linear system of equations. Therefore, solving systems like $Ax = b$ with a matrix $A \in \mathbb{R}^{n \times n}$ and a right hand side $b \in \mathbb{R}^n$ is essential in numerical analysis. There are two major ways of solving those systems: by direct solvers, which are mainly based on the Gaussian algorithm, or by iterative solvers which are often based on projections. The second type usually contains one multiplication of a matrix with a vector in each iteration step and the precision of such matrix-vector multiplication has a significant impact on the convergence of the iterative solver [4].

In most modern microprocessors, mathematical operations are performed by using floating point arithmetics. However, the finite floating-point arithmetic

J.M.L.M. Palma et al. (Eds.): VECPAR 2010, LNCS 6449, pp. 394–407, 2011.

can only deliver an approximation of the exact result due to rounding errors. Since the exact result is usually unknown, it is sometimes difficult to measure the quality of these approximations. Besides, as a result of several operations, the accumulation of those errors may have an impact in the accuracy of the results.

There are many papers proposing different solutions to find more accurate results. Some authors concern is to improve the numerical accuracy of the computed results in computers through the use of extra precise iterative refinement [5,6]. Others try to use mixed-precision algorithms [9,14,10,1] to obtain a good accuracy and improve the performance. Another possible way to deal with this unreliability is to use verified computing [11]. Such techniques provide an interval result that surely contains the correct result [12,13]. However, the use of such methods may increase the execution time significantly. This effect is even worse for large linear systems, that may need several days or even more to be solved. Based on these researches, it is possible to notice that there is a tradeoff performance versus accuracy.

Parallel computing is a well-known choice for simulating large problems. Since many numerical problems are solved via a large linear system of equations, a parallel algorithm would be a good approach. In this context, the libraries BLAS [3] and LAPACK [15] seem a good choice, since they have a parallel version (PBLAS and SCALAPACK [2]) that could be used in the case of very large systems. However, it is important to remember that these libraries provide an approximation of the correct result and not a verified result.

It is well known that the data distribution has a major impact in the performance of a parallel application [2]. However, the data distribution can also present an important influence on the accuracy of the numerical results. Sometimes a fixed problem can lead to distinct solutions depending on the data distribution or the number of processes used in its solution. This effect can possibly be explained by the rounding error theory [16].

Based on that, this paper investigates the impact of different grids configuration used in the two-dimensional block cyclic distribution on the accuracy and performance of the parallel matrix-vector multiplication implemented by PBLAS. This particular distribution was chosen since it was proved to be a good choice for parallel matrix distribution on parallel environments with distributed memory [7]. Other interesting data distribution was proposed in [8], however it is also based on block distribution and was consider equivalent to the two-dimensional block cyclic distribution [17].

In this paper, the performance of different grids configuration was measured and compared among them. To evaluate the accuracy of the approximations generated by PBLAS, a comparison with the verified solution provided by C-XSC [12] is done. The experimental results indicate how the grids should be configured to find a compromise between accuracy and performance considering the application needs.

This text is organized as follows. To better understand this problem, section 2 presents two important backgrounds: the theory of rounding errors and the two-dimensional block cyclic distribution scheme. Section 3 introduces the platform, input data and results obtained in the numerical experiments. Finally, section 4 present some final remarks and considerations about future work.

2   Background

This section presents the background used in this paper. Section 2.1 introduces the theoretical background concerning rounding errors, based on a paper of Linz [16]. Section 2.2 describes the two-dimensional block cyclic distribution used by PBLAS.

2.1   Theory of Rounding Errors

Let ϵ be the machine accuracy and $fl(a \circ b)$ the floating point result for an elementary composition of two real numbers a and b. An elementary operation $\circ \in \{+, -, *, /\}$ of a and b can be estimated with $fl(a \circ b) = (a \circ b) + \epsilon(a, b, \circ)$ for the worst case. We assume $A \in \mathbb{R}^{n \times n}, x, y \in \mathbb{R}^n$ and get $y_k = a_k x$ as result for the product $Ax = y$ for every entry $y_k \in y$. Let a_k denote the $k - th$ row of A. For all y_k, the approximation using the floating point arithmetic is $\hat{y}_k$.

Simple approach. The simple strategy for computing each $y_k \in y$ is to add the first entry to the next one and then add the following entries one by one to the previous result. Using floating point arithmetic and the abbreviation $fl(a_{k,i} \cdot x_i) = a_{k,i} \cdot x_i + \epsilon(a_{k,i}, x_i, \cdot) =: \hat{b}_{k,i}$, this strategy can be written as follows:

$$\hat{y}_{k_1} := (\hat{b}_{k,1} + \hat{b}_{k,2}) + \epsilon_1$$

$$\hat{y}_{k_i} := \hat{y}_{k_{i-1}} + \hat{b}_{k,i} + \epsilon_i = y_{k_i} + \sum_{j=1}^{i} \epsilon_j \, , i \in \{2, 3, \ldots n - 1\}$$

Let the representation be the normalized floating-point with binary exponent and q fraction bits and assume the addition to be done by truncating the exact sum to q bits. Let p_i be the exponent of $\hat{y}_{k_i}$ and $\nu = 2^{-q}$. The error for the ith step is then $|\epsilon_i| \leq \nu 2^{p_i}$ and the global error can be written using the estimates $a_{k,i} x_i \leq b, |\hat{y}_{k_i}| \leq ib$ and $2^{p_i} \leq 2ib$ in the following way

$$|y_k - \hat{y}_k| \leq \nu \sum_{i=1}^{n-1} 2^{p_i} \leq 2\nu b \sum_{i=1}^{n} i = \nu b n(n+1).$$

This means the error using this approach grows like $O(n^2)$.

Advanced approach. The second strategy for the summation is the so called "Fan-In" algorithm. The values are added to each other in pairs and the algorithm is then executed recursively. Let us define the notation

$$y_k = \underbrace{\underbrace{a_{k,1}x_1 + a_{k,2}x_2}_{\hat{y}_{n_{1,1}}} + \underbrace{a_{k,3}x_3 + a_{k,4}x_4 +}_{\hat{y}_{n_{2,1}}} \cdots + a_{k,n}x_n}_{}$$

$$\underbrace{\phantom{a_{k,1}x_1 + a_{k,2}x_2 + a_{k,3}x_3 + a_{k,4}x_4}}_{\hat{y}_{n_{1,2}}}$$

$$\underbrace{\phantom{a_{k,1}x_1 + a_{k,2}x_2 + a_{k,3}x_3 + a_{k,4}x_4 + a_{k,n}x_n}}_{\hat{y}_{n_{1,m}}}$$

and

$$\hat{y}_{k_{i,j}} = \hat{y}_{k_{2i-1,j-1}} + \hat{y}_{k_{2i-1,j-1}} + \epsilon_{i,j}$$

where $\epsilon_{i,j}$ is the rounding error when computing $\hat{y}_{k_{i,j}}$. For the global error we have

$$|y_k - \hat{y}_k| = \sum_{\sigma_{1,k}} \epsilon_{i,j} \leq \nu \sum_{\sigma_{1,k}} 2^{p_{\sigma_{1,k}}}$$

where p is the exponent of the result and $\sigma_{1,k}$ is the set of all index pairs needed to get $\hat{y}_{k_{i,j}}$. Assuming that $a_{k,i}x_i \leq b$, we have:

$$\hat{y}_{k_{i,j}} \leq 2^j b \text{ and } 2^{p_{i,j}} \leq 2^{j+1}b.$$

Based on that, the error upper bound can be estimated by

$$|y_n - \hat{y}_k| \leq \nu \sum_{\sigma_{1,k}} 2^{p_{\sigma_{1,k}}} \leq 2\nu b \sum_{j=1}^{k} \sum_{i=1}^{n/2^j} 2^j = 2\nu bkn \leq 2\nu b\, n \log_2 n.$$

For the advanced approach the error grows like $O(n \log_2 n)$. The proof can be extended to cases in which more than two entries are added to each other using the "Fan-In"-algorithm. In that case, the error propagations is bounded by the one presented by the strategies above.

The two approaches differ in a factor of $n/(2 \log_2 n)$. This study suggests that a finer granularity in the summation leads to lower upper boundaries for rounding errors. The proofs presented above show the impact of rounding errors in scalar products, which are commonly part of matrix-vector multiplications.

2.2   Data Distribution in Numerical Algorithms

On distributed memory platforms, the application programmer is responsible for assigning the data to each processor. How this is done has a major impact on the load balance and communication characteristics of the algorithm, and largely determines its performance and scalability [2].

PBLAS routines are implemented supposing the matrices are stored in the distributed memory according to the two-dimensional block cyclic distribution [7]. In this distribution, an M by N matrix is first decomposed into MB by NB blocks starting at its upper left corner. The distribution of a vector is done considering the vector as a column of the matrix. Suppose we have the following 10x10 matrix, a vector of length 10 an MB and NB equal 3. In this case, we would have the following blocks:

$$\left(\begin{array}{ccc|ccc|ccc|c}
A_{0,0} & A_{0,1} & A_{0,2} & A_{0,3} & A_{0,4} & A_{0,5} & A_{0,6} & A_{0,7} & A_{0,8} & A_{0,9} \\
A_{1,0} & A_{1,1} & A_{1,2} & A_{1,3} & A_{1,4} & A_{1,5} & A_{1,6} & A_{1,7} & A_{1,8} & A_{1,9} \\
A_{2,0} & A_{2,1} & A_{2,2} & A_{2,3} & A_{2,4} & A_{2,5} & A_{2,6} & A_{2,7} & A_{2,8} & A_{2,9} \\
\hline
A_{3,0} & A_{3,1} & A_{3,2} & A_{3,3} & A_{3,4} & A_{3,5} & A_{3,6} & A_{3,7} & A_{3,8} & A_{3,9} \\
A_{4,0} & A_{4,1} & A_{4,2} & A_{4,3} & A_{4,4} & A_{4,5} & A_{4,6} & A_{4,7} & A_{4,8} & A_{4,9} \\
A_{5,0} & A_{5,1} & A_{5,2} & A_{5,3} & A_{5,4} & A_{5,5} & A_{5,6} & A_{5,7} & A_{5,8} & A_{5,9} \\
\hline
A_{6,0} & A_{6,1} & A_{6,2} & A_{6,3} & A_{6,4} & A_{6,5} & A_{6,6} & A_{6,7} & A_{6,8} & A_{6,9} \\
A_{7,0} & A_{7,1} & A_{7,2} & A_{7,3} & A_{7,4} & A_{7,5} & A_{7,6} & A_{7,7} & A_{7,8} & A_{7,9} \\
A_{8,0} & A_{8,1} & A_{8,2} & A_{8,3} & A_{8,4} & A_{8,5} & A_{8,6} & A_{8,7} & A_{8,8} & A_{8,9} \\
\hline
A_{9,0} & A_{9,1} & A_{9,2} & A_{9,3} & A_{9,4} & A_{9,5} & A_{9,6} & A_{9,7} & A_{9,8} & A_{9,9}
\end{array}\right)
\left(\begin{array}{c} b_0 \\ b_1 \\ b_2 \\ b_3 \\ b_4 \\ b_5 \\ b_6 \\ b_7 \\ b_8 \\ b_9 \end{array}\right)$$

Suppose we have 4 processors. The process grid would be a 2x2 grid as follows:

$$\left(\begin{array}{c|c} P^0 & P^1 \\ \hline P^2 & P^3 \end{array}\right)$$

These blocks are then uniformly distributed across the process grid. Thus, every processor owns a collection of blocks [2]. The first row of blocks will be distributed among the first row of the processor grid, that means among P_0 and P_1, while the second row will be distributed among P_2 and P_3, and so on. For this example, we would have:

$$\left(\begin{array}{c|c|c|c} P^0 & P^1 & P^0 & P^1 \\ \hline P^2 & P^3 & P^2 & P^3 \\ \hline P^0 & P^1 & P^0 & P^1 \\ \hline P^2 & P^3 & P^2 & P^3 \end{array}\right)
\left(\begin{array}{c} P^0 \\ \hline P^2 \\ \hline P^0 \\ \hline P^2 \end{array}\right)$$

According to this distribution, each processor would have the following data:

$$P^0 : \left(\begin{array}{ccc|ccc}
A_{0,0} & A_{0,1} & A_{0,2} & A_{0,6} & A_{0,7} & A_{0,8} \\
A_{1,0} & A_{1,1} & A_{1,2} & A_{1,6} & A_{1,7} & A_{1,8} \\
A_{2,0} & A_{2,1} & A_{2,2} & A_{2,6} & A_{2,7} & A_{2,8} \\
\hline
A_{6,0} & A_{6,1} & A_{6,2} & A_{6,6} & A_{6,7} & A_{6,8} \\
A_{7,0} & A_{7,1} & A_{7,2} & A_{7,6} & A_{7,7} & A_{7,8} \\
A_{8,0} & A_{8,1} & A_{8,2} & A_{8,6} & A_{8,7} & A_{8,8}
\end{array}\right)
\left(\begin{array}{c} b_0 \\ b_1 \\ b_2 \\ b_6 \\ b_7 \\ b_8 \end{array}\right)
\quad P^1 : \left(\begin{array}{ccc|c}
A_{0,3} & A_{0,4} & A_{0,5} & A_{0,9} \\
A_{1,3} & A_{1,4} & A_{1,5} & A_{1,9} \\
A_{2,3} & A_{2,4} & A_{2,5} & A_{2,9} \\
\hline
A_{6,3} & A_{6,4} & A_{6,5} & A_{6,9} \\
A_{7,3} & A_{7,4} & A_{7,5} & A_{7,9} \\
A_{8,3} & A_{8,4} & A_{8,5} & A_{8,9}
\end{array}\right)$$

$$P^2 : \left(\begin{array}{ccc|ccc}
A_{3,0} & A_{3,1} & A_{3,2} & A_{3,6} & A_{3,7} & A_{3,8} \\
A_{4,0} & A_{4,1} & A_{4,2} & A_{4,6} & A_{4,7} & A_{4,8} \\
A_{5,0} & A_{5,1} & A_{5,2} & A_{5,6} & A_{5,7} & A_{5,8} \\
\hline
A_{9,0} & A_{9,1} & A_{9,2} & A_{9,6} & A_{9,7} & A_{9,8}
\end{array}\right)
\left(\begin{array}{c} b_3 \\ b_4 \\ b_5 \\ b_9 \end{array}\right)
\quad P^3 : \left(\begin{array}{ccc|c}
A_{3,3} & A_{3,4} & A_{3,5} & A_{3,9} \\
A_{4,3} & A_{4,4} & A_{4,5} & A_{4,9} \\
A_{5,3} & A_{5,4} & A_{5,5} & A_{5,9} \\
\hline
A_{9,3} & A_{9,4} & A_{9,5} & A_{9,9}
\end{array}\right)$$

The two dimensional block cyclic distribution is usually not used when the matrix has a sparse data structure. Common storage formats, like CRS (compressed row storage), CCS (compressed column storage) etc., usually lead to distributions where a certain number of rows or columns are given to each process. The CRS presents a row-wise data distribution, that could be seen as a $np \times 1$ grid of processes in the two dimensional block cyclic distribution. In the same way, a $1 \times np$ grid could be interpreted as a column wise distribution, e.g. similar to a strategy taken when the CCS is used.

From the mathematical theory it is clear that the upper bound for the rounding errors occurring in a sparse matrix vector multiplication increases with the number of nonzero elements (nnz) per row. In the proof presented in the last subsection, the variable n should be replaced with nnz, assuming that the matrix data is sparse and the vector data has $O(n)$ entries.

It is important to mention that, for a matrix vector multiplication, no benefit in accuracy can be expected due to parallelization when this is done based on a row-wise data distribution and assuming that the sequential part on each process does not use techniques like "fan-in".

3   Numerical Experiments

This section presents experimental results for different grid compositions. The accuracy and performance are the focus of these tests.

The results of the considered matrix-vector multiplication were computed using first the sequential BLAS-routines to obtain the sequential time. After this sequential test, we used the PBLAS-routines from the MKL package for different grid compositions. All results have been computed three times to avoid effects caused by hard- and software problems.

To evaluate which grid presents the best accuracy among the tested grids we analyzed the accuracy obtained by the parallel implementation through a comparison with a verified result. The library used to obtain the verified result was C-XSC, which stands for "extension for scientific computing", and is a free programming tool for the development of numerical algorithms which provides highly accurate and automatically verified results. C-XSC does computations based on interval arithmetics and direct rounding, providing an enclosure of the exact solution, which is represented by an interval. This means that for a matrix-vector multiplication, C-XSC will deliver a vector of intervals, each entry of the vector containing an interval enclosure of the correct solution. The diameter of the intervals is usually very small, since C-XSC implementation uses techniques to iteratively reduce the interval diameter proofing that the interval result includes the exact result [12].

The average error from the PBLAS result to the C-XSC result is computed as follows. First, for each component of the result vector it is checked if it is in the interior of the interval given from C-XSC. If it is inside the interval, it is considered correct. If it is not inside, the distance to the interval is stored. Then the arithmetic mean over all distances is computed, which we denote as average error.

Next section introduces the platform used for the experiments. Section 3.2 illustrates the input data of four different matrices used in the tests. Finally section 3.3 presents the accuracy and performance results with some considerations.

3.1   Platform

The software platform used for executing the numerical experiments is composed of optimized versions of the library PBLAS (Intel CMKL in version 10.0.2.018 for test case M1 and version 10.1.2.024 for the test cases M2, M3 and M4), C-XSC version 2.2.3 and the standard Message Passing Interface (MPI), more precise the OpenMPI implementation in version 1.2.8. The compiler used in these tests was the Intel compiler in version 10.1.021.

The Institutscluster located at the Steinbuch Computing Centre (SCC) at the Karlsruhe Institute of Technology (KIT) was chosen as hardware platform. It consists of 200 computing nodes each equipped with two Intel quadcore EM64T Xeon 5355 processors running at 2,667 GHZ, 16 GB of main memory and an Infiniband 4x DDR interconnect. 17,57 TFlops is the overall peak performance of the whole system and 15,2 TFlops in the Linpack benchmark.

3.2   Input Data

The results shown in Section 3.3 refer to four different input matrices and vectors. M1, the first test matrix, has dimension 16000 and is filled with pseudo random numbers from the interval $[-0, 5; 0.5]$. Its properties can be seen in table 1.

Table 1. Sparsity plots and properties of the test-matrices M1 and M2

M1 (Random)	M2 (GEMAT11)
problem: Artificial	problem: Power flow.
problem size: $n = 16000$	problem size: $n = 4929$
sparsity: $nnz = 256000000$	sparsity: $nnz = 33185$
cond. number: n.a.	cond. number: 3.74e+08
Frobenius norm: n.a.	Frobenius norm: 8.2e+02

M2 is a square matrix with dimension 4929 used as the initial basis for constrained nonlinear optimization problem represented by GEMAT1 which is the Jacobian matrix for an approximately 2400 bus system in the Western United States. M2, arising in the area of power flow, is a sparse matrix with a medium condition number, as presented in Table 1. It is important to mention that no special data storage is used for sparse matrices. They are always stored as dense matrices.

The third matrix tested was M3. As shows Table 2, it is a dense matrix with dimension 66. This matrix is used in the generalized eigenvalue problem $Kx = \lambda Mx$, where M3 is matrix K and matrix M is BCSSTM02 [4], from BCSSTRUC1[4] set. This matrix arises in dynamic analysis in structural engineering.

Table 2. Sparsity plots and properties of the test-matrices M3 and M4

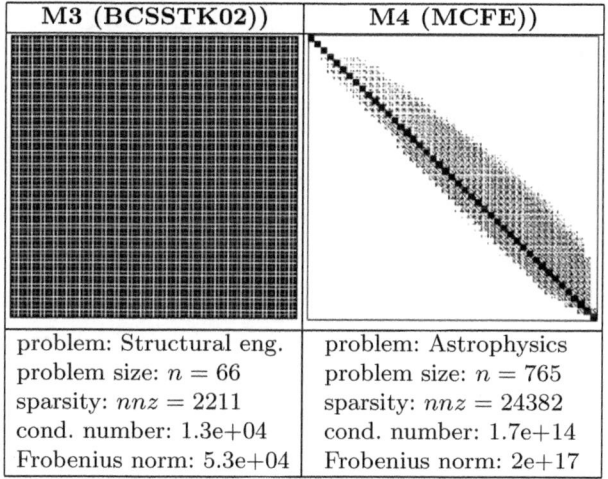

M3 (BCSSTK02))	M4 (MCFE))
problem: Structural eng.	problem: Astrophysics
problem size: $n = 66$	problem size: $n = 765$
sparsity: $nnz = 2211$	sparsity: $nnz = 24382$
cond. number: 1.3e+04	cond. number: 1.7e+14
Frobenius norm: 5.3e+04	Frobenius norm: 2e+17

Table 2 also presents the properties of matrix M4. It is a sparse matrix with dimension 765 which presents a very high condition number. This matrix is used in the real application of nonlinear radiative transfer and statistical equilibrium in astrophysics.

Detailed information about the creation and properties of the test cases M2 to M4 can be found on the Matrix Market website[1].

3.3   Numerical Results

This section discusses the results of a set of experiments using the four different matrices presented in the previous section. The first analysis is based on the accuracy obtained using different grid sizes. After that, the performance results are shown.

---

[1] http://math.nist.gov/MatrixMarket

Accuracy. Figure 1 presents the accuracy obtained using nine processes and different grid configurations for test cases M1 and M2. The comparison between our PBLAS algorithm and the verified results of the C-XSC-algorithm show that there are constellations of processors grids, namely when the grid is $np \times 1$, where the accuracy of the parallel computation is as precise as the sequential one. In all other cases when the grid of processors is not $np \times 1$, the results present a better accuracy, suggesting that the accuracy depends on the grid. Another important observation is that the optimal accuracy for a fixed number of processes can be found by a $1 \times np$ grid. In general we observed that the more columns the processes grid have, the better is the accuracy.

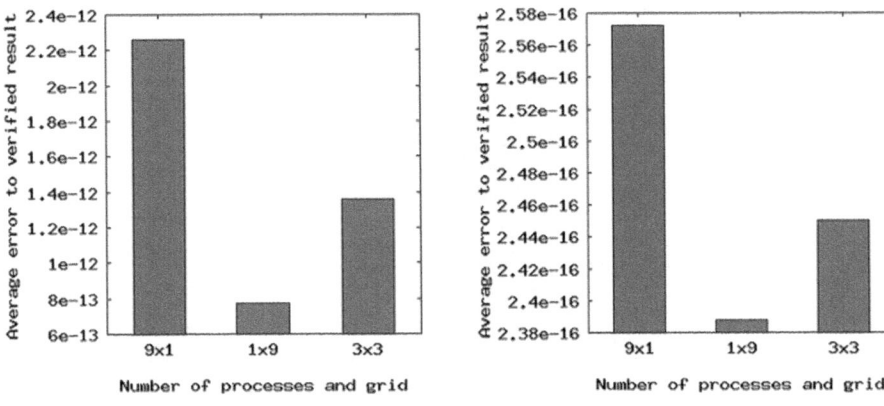

Fig. 1. Average error to the verified result for three different grids of processes and a fixed number of nine processes for the test cases M1 (left plot) and M2 (right plot)

Let us investigate our example in the light of Section 2.1 using, for simplicity, a number of four processes. Let the dimension of the matrix be n, the number of processes $np = 4$, the grid of the processes $nr \times nc$ and the block size $nb := \frac{n}{np}$. Considering the case of an 2×2 grid of processes, the distribution follows the scheme in the example in section 2.2. For a 1×4 and 4×1 grid, let us use the notation P_{Bc}^{a}, where a (from 1 to np) represents the number of the processes containing the data, B denotes if it is a matrix(M) or a vector(v) and c is the number of the data block related to one processor. The data distribution is as presented in tables 3 and 4.

Analyzing table 3 and 4, it is possible to notice that in the computation based on a $np \times 1$ grid, each entry of the result vector is computed in just one process, which means that the summation is done like in the simple approach from Section 2.1.

Table 3. Data distribution for two different grids of four processes

$$\left(\begin{array}{c|c|c|c} P_{M0}^0 & P_{M0}^1 & P_{M0}^2 & P_{M0}^3 \\ \hline P_{M1}^0 & P_{M1}^1 & P_{M1}^2 & P_{M1}^3 \\ \hline P_{M2}^0 & P_{M2}^1 & P_{M2}^2 & P_{M2}^3 \\ \hline P_{M3}^0 & P_{M3}^1 & P_{M3}^2 & P_{M3}^3 \end{array}\right) \left(\begin{array}{c} P_{v0}^0 \\ \hline P_{v1}^0 \\ \hline P_{v2}^0 \\ \hline P_{v3}^0 \end{array}\right) \qquad \left(\begin{array}{c|c|c|c} P_{M0}^0 & P_{M1}^0 & P_{M2}^0 & P_{M3}^0 \\ \hline P_{M0}^1 & P_{M1}^1 & P_{M2}^1 & P_{M3}^1 \\ \hline P_{M0}^2 & P_{M1}^2 & P_{M2}^2 & P_{M3}^2 \\ \hline P_{M0}^3 & P_{M1}^3 & P_{M2}^3 & P_{M3}^3 \end{array}\right) \left(\begin{array}{c} P_{v0}^0 \\ \hline P_{v0}^1 \\ \hline P_{v0}^2 \\ \hline P_{v0}^3 \end{array}\right)$$

(a) Grid (1×4) (b) Grid (4×1)

The structure of the result distribution is shown in table 4. The leading entry is the position of the resulting vector, followed by the explanation of which parts were combined to compute the result.

Table 4. Processes which contain the final result and parts from which it is computed

$$\left(\begin{array}{c} P^0(P_{M0}^0 P_{v0}^0 + P_{M0}^1 P_{v1}^0 + P_{M0}^2 P_{v2}^0 + P_{M0}^3 P_{v3}^0) \\ \hline P^0(P_{M1}^0 P_{v0}^0 + P_{M1}^1 P_{v1}^0 + P_{M1}^2 P_{v2}^0 + P_{M1}^3 P_{v3}^0) \\ \hline P^0(P_{M2}^0 P_{v0}^0 + P_{M2}^1 P_{v1}^0 + P_{M2}^2 P_{v2}^0 + P_{M2}^3 P_{v3}^0) \\ \hline P^0(P_{M3}^0 P_{v0}^0 + P_{M3}^1 P_{v1}^0 + P_{M3}^2 P_{v2}^0 + P_{M3}^3 P_{v3}^0) \end{array}\right) \left(\begin{array}{c} P^0(P_{M0}^0 P_{v0}^0 + P_{M1}^0 P_{v0}^1 + P_{M2}^0 P_{v0}^2 + P_{M3}^0 P_{v0}^3) \\ \hline P^1(P_{M1}^1 P_{v0}^0 + P_{M1}^1 P_{v0}^1 + P_{M1}^1 P_{v0}^2 + P_{M1}^1 P_{v0}^3) \\ \hline P^2(P_{M2}^2 P_{v0}^0 + P_{M2}^2 P_{v0}^1 + P_{M2}^2 P_{v0}^2 + P_{M2}^2 P_{v0}^3) \\ \hline P^3(P_{M0}^3 P_{v0}^0 + P_{M1}^3 P_{v0}^1 + P_{M2}^3 P_{v0}^2 + P_{M3}^3 P_{v0}^3) \end{array}\right)$$

(a) Grid (1×4) (b) Grid (4×1)

The advanced approach presented in section 2.1 can be found on the $1 \times np$ grid where the final result will be placed on process one, but there are intermediate results on every process leading to a higher quality in the computed result. This suggests that the number of columns of the processor grid is responsible for the granularity of the computation - a higher numbers of columns can lead to better accuracy. So it is not astonishing that a symmetric grid produces results with an intermediate precision (bounded by the other grids).

It is also possible to notice that the results for the application based problem M2 show that for all grid sizes the average error is less than $2.58e-16$ which is excellent considering the double precision format.

Figure 2 presents the average error for matrix M3 and M4. Based on the M3 graphic, we can see that even for small problems, the data distribution and the executed computations can have an impact on the result. The average error to the verified result, depending on the grid configuration, differs in about one magnitude.

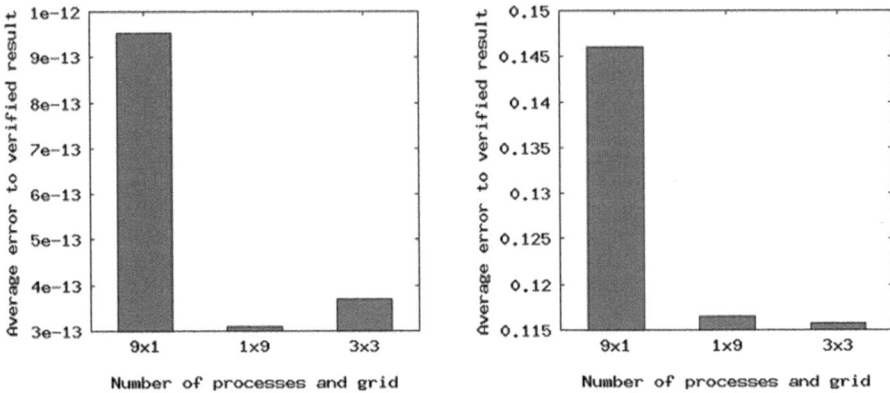

Fig. 2. Average error to the verified result for three different grids of processes and a fixed number of nine processes for the test cases M3 (left plot) and M4 (right plot)

For the test case M4 we observe that the 9×1 grid delivers, analogue to all other experiments, the most inaccurate result. The fact that the 3×3 grid is a little more accurate than the 1×9 grid might be astonishing on the first view but this is possible because the mathematical theory gives only a upper bound for the rounding error propagation.

The results in Figure 3 show for different number of processes and grids of processes the average error to the verified result based on a matrix with dimension 8192 and input data generated like in M1. For all grids $np \times 1$ the accuracy is like in the sequential case and independent of the number of processes. The

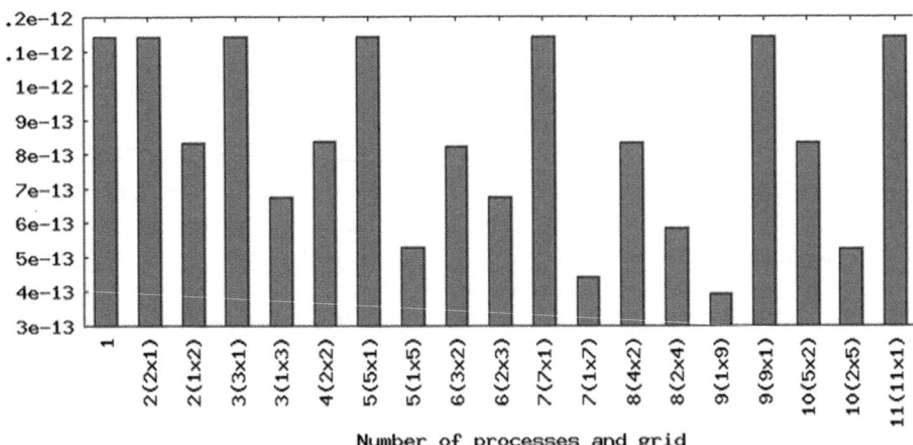

Fig. 3. Average error to the verified result for different grids and different numbers of processes. A matrix similar to M1 but with dimension 8192 was taken for the experiments.

graphic shows that the more processes are used the better is the result. It can also be observed that the larger the number of processes, the better is the accuracy, following a logarithmic behavior, which corresponds to the theoretical findings.

Performance. This section presents the performance analysis considering matrix M1. The performance analysis for matrices 2 to 4 were not discussed since they have small dimensions. In this case it is not worth to parallelize the multiplication, since the program would spend more time communicating among the processors than computing the result. Therefore it would maybe increase the computational time instead of speedup the computation. Since matrix 1 has dimension 16000, it is the natural choice for the performance test.

Figure 4 shows that directly interrelated to the grid is the processing speed. It is possible to notice that the computational time for the same problem size is very different depending on the grid.

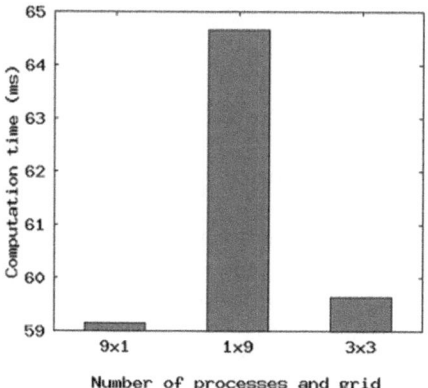

Fig. 4. Commutation time for three different grids of processes and a fixed number of nine processes for the test case M1

This performance variation can be explained by the fact that different grids communicate differently. The amount, length and topology of such communication have a significant impact on the performance [7]. In Table 3 is possible to notice that some of the data that a processes need may be stored in another processes, and therefore the processes need to communicate before the computation to send/receive parts these data. This occurs also after the computation, when parts of the result have to be collected from each processor and accumulated so that the result is found. This is the case of the 1×4 grid in Table 4.

The communication load-balance is optimal if it is equally distributed on all processes. The impact on the performance depends significantly on the underlying hardware (interconnect, memory bandwidth etc.). This means that not a single process is sending or receiving a big bunch of data to all other processes, but that all processes are sending little bunches of data to all other processes.

For the grids shown above, the structure of the communication is:

(a) Grid (1×4)

before computation		after computation	
sender → receiver	size	sender → receiver	size
$P^0 \to P^1$	nb	$P^1 \to P^0$	n
$P^0 \to P^2$	nb	$P^2 \to P^0$	n
$P^0 \to P^3$	nb	$P^3 \to P^0$	n

(b) Grid (4×1)

before computation		after computation	
sender → receiver	size	sender → receiver	size
$P^0 \to P^1$	nb	–	–
$P^0 \to P^2$	nb	–	–
$P^0 \to P^3$	nb	–	–
$P^1 \to P^0$	nb	–	–
$P^1 \to P^2$	nb	–	–
$P^1 \to P^3$	nb	–	–
$P^2 \to P^0$	nb	–	–
$P^2 \to P^1$	nb	–	–
$P^2 \to P^3$	nb	–	–
$P^3 \to P^0$	nb	–	–
$P^3 \to P^1$	nb	–	–
$P^3 \to P^2$	nb	–	–

(c) Grid (2×2)

before computation		after computation	
sender → receiver	size	sender → receiver	size
$P^0 \to P^2$	$2 * nb$	$P^1 \to P^0$	$n/2$
$P^2 \to P^1$	$2 * nb$	$P^3 \to P^2$	$n/2$
$P^2 \to P^3$	$2 * nb$		

4 Final Remarks and Future Work

This paper presents the theory of rounding error propagation for elementary kernels and validates the theoretical findings based on numerical experiments. Beside accuracy the influence on performance of the process-grid using the two-dimensional block cyclic distribution was addressed.

Tests show that the process-grid has a significant impact on both, but in a different way. The experiments suggested that the more columns the grid has, the better is the accuracy. However, this is not true for the performance, in which the effect is the opposite: the more columns the grid has, the worse is the performance. For symmetric grids, the performance achieved was good due to a better balance in the communication process. It presented, however, a little less accuracy then in the best case.

It is important to mention, that the impact of the data distribution on the results of numerical simulations depend strongly on the particular problem as well as on the numerical procedures employed to find the solution.

Ongoing research is the evaluation of different hardware platforms like CPUs from various vendors, GPUs and other accelerators as well as full cluster systems based on different interconnects. A second focus is the impact of the utilized software like compilers, optimization flags and different libraries. In addition, experiments in the context of complete solvers are planned. One goal is to accelerate linear solvers like CG or GMRES by performing a data reordering during the solution process.

Acknowledgement. Mariana Kolberg thanks to FAPERGS for the financial support through the research project 0905026.

References

1. Baboulin, M., Buttari, A., Dongarra, J., Langou, J., Langou, J., Luszcek, P., Kurzak, J., Tomov, S.: Accelerating scientific computations with mixed precision algorithms (2008)

2. Blackford, L.S., Choi, J., Cleary, A., Demmel, J., Dhillon, I., Dongarra, J., Hammarling, S., Henry, G., Petitet, A., Stanley, K., Walker, D., Whaley, R.C.: Scalapack: A portable linear algebra library for distributed memory computers Design Issues And Performance. In: SUPERCOMPUTING 1996. IEEE Computer Society, Los Alamitos (1996)
3. Blackford, L.S., Demmel, J., Dongarra, J., Duff, I., Hammarling, S., Henry, G., Heroux, M., Kaufman, L., Lumsdaine, A., Petitet, A., Pozo, R., Remington, K., Whaley, R.C.: An Updated Set of Basic Linear Algebra Subprograms (BLAS). ACM Transactions on Mathematical Software 28(2), 135–151 (2002)
4. Demmel, J., Dumitriu, I., Holtz, O.: Fast linear algebra is stable. Numer. Math. 108(1), 59–91 (2007)
5. Demmel, J., Hida, Y., Kahan, W., Li, X.S., Mukherjee, S., Jason Riedy, E.: Error bounds from extra-precise iterative refinement. ACM Trans. Math. Softw. 32(2), 325–351 (2006)
6. Demmel, J., Dongarra, J.: LAPACK 2005 prospectus: Reliable and scalable software for linear algebra computations on high end computers. LAPACK Working Note 164, UT-CS-05-546 (February 2005)
7. Dongarra, J., Walker, D.: Lapack working note 58: The design of linear algebra libraries for high performance computers. Technical Report UT-CS-93-188, Knoxville, TN, USA (1993)
8. Edwards, C., Geng, P., Patra, A., Van De Geijn, R.: Parallel matrix distributions: Have we been doing it all wrong (1996)
9. Giraud, L., Haidar, A., Watson, L.T.: Mixed-precision preconditioners in parallel domain decomposition solvers. Technical Report TR/PA/06/84, CERFACS, Toulouse, France, Also appeared as IRIT Technical report ENSEEIHT-IRIT RT/APO/06/08 (2006)
10. Göddeke, D., Strzodka, R., Turek, S.: Performance and accuracy of hardware–oriented native–, emulated– and mixed–precision solvers in FEM simulations (2007)
11. Hammer, R., Ratz, D., Kulisch, U., Hocks, M.: C++ Toolbox for Verified Scientific Computing I: Basic Numerical Problems. Springer, New York (1997)
12. Klatte, R., Kulisch, U., Lawo, C., Rauch, R., Wiethoff, A.: C-XSC- A C++ Class Library for Extended Scientific Computing. Springer, Berlin (1993)
13. Kulisch, U., Miranker, W.: Computer Arithmetic in Theory and Practice. Academic Press, New York (1981)
14. Langou, J., Langou, J., Luszczek, P., Kurzak, J., Buttari, A., Dongarra, J.: Exploiting the performance of 32 bit floating point arithmetic in obtaining 64 bit accuracy (revisiting iterative refinement for linear systems). In: SC 2006: Proceedings of the 2006 ACM/IEEE Conference on Supercomputing, p. 113. ACM, New York (2006)
15. LAPACK. Linear Algebra Package, http://www.cs.colorado.edu/~jessup/lapack/ (visited in February 09, 2009)
16. Linz, P.: Accurate floating-point summation. ACM Commun. 13(6), 361–362 (1970)
17. Sidani, M., Harrod, B.: Parallel matrix distributions: Have we been doing it all right? Technical report, Knoxville, TN, USA (1996)

On a Strategy for Spectral Clustering
with Parallel Computation

Sandrine Mouysset, Joseph Noailles, Daniel Ruiz, and Ronan Guivarch

University of Toulouse, IRIT-ENSEEIHT, France
{sandrine.mouysset,jnoaille,daniel.ruiz,ronan.guivarch}@enseeiht.fr

Abstract. Spectral Clustering is one of the most important method based on space dimension reduction used in Pattern Recognition. This method consists in selecting dominant eigenvectors of a matrix called affinity matrix in order to define a low-dimensional data space in which data points are easy to cluster. By exploiting properties of Spectral Clustering, we propose a method where we apply independently the algorithm on particular subdomains and gather the results to determine a global partition. Additionally, with a criterion for determining the number of clusters, the domain decomposition strategy for parallel spectral clustering is robust and efficient.

1   Introduction

Clustering aims to partition a data set by grouping similar elements into subsets. Two general main issues concern, on the one hand, the choice of a similarity criterion and, on the other hand, the way to separate clusters the one from the other. Spectral methods, and in particular the spectral clustering algorithm introduced by Ng-Jordan-Weiss (NJW) [1], are useful when considering non-convex shaped subsets of points. These methods are widely used in Pattern Recognition and in particular in Bioinformatics and image segmentation. The number of targeted clusters k is usually assumed to be known. From the spectral elements of an affinity normalized matrix, data points are clustered in a low-dimensionnal space made by the first eigenvectors of the normalized affinity matrix. Several approaches about parallel Spectral Clustering [5], [6], [2] were recently suggested, mainly focused on linear algebra techniques to reduce computational costs. However, the authors do not get rid of the construction of the complete affinity matrix and the problem of determining the number of clusters is still open.

In this paper, we propose to cluster on subdomains by breaking up the data set into data subsets with respect to their geometrical coordinates in a straighforward way. With an appropriate Gaussian affinity parameter and a method to determine the number of clusters, each processor applies independently the spectral clustering algorithm on subsets of data points and provide a local partition on these data subsets. Based on these local partitions, a gathering step ensures the connection between subsets of data and determines a global partition. We

J.M.L.M. Palma et al. (Eds.): VECPAR 2010, LNCS 6449, pp. 408–420, 2011.

analyze in particular two different approaches of the type and we experiment on a geometrical particular example and on an image segmentation example. We identify the potential for parallelism of the algorithm as well as numerical behaviour and limitations.

2   Parallel Spectral Clustering: Algorithm and Justification

Spectral clustering uses eigenvectors of a matrix, called Gaussian affinity matrix, in order to define a low-dimensional space in which data points can be clustered (see algorithm 1).

Algorithm 1. Spectral Clustering Algorithm

Input: data set S, number of clusters k

1. Form the affinity matrix $A \in \mathbb{R}^{n \times n}$ defined by:

$$A_{ij} = \begin{cases} \exp\left(-\frac{\|x_i - x_j\|^2}{(\sigma/2)^2}\right) & \text{if } i \neq j, \\ 0 & \text{otherwise,} \end{cases} \qquad (1)$$

2. Construct the normalized matrix: $L = D^{-1/2}AD^{-1/2}$ with $D_{i,i} = \sum_{j=1}^{n} A_{ij}$,
3. Assemble the matrix $X = [X_1 X_2 .. X_k] \in \mathbb{R}^{n \times k}$ by stacking the eigenvectors associated with the k largest eigenvalues of L,
4. Form the matrix Y by normalizing each row in the $n \times k$ matrix X,
5. Treat each row of Y as a point in $\mathbb{R}^k$, and group them in k clusters via the *K-means* method,
6. Assign the original point x_i to cluster j when row i of matrix Y belongs to cluster j.

The Gaussian affinity matrix defined by (1) could be interpreted as a discretization of the Heat kernel [3]. And in particular, it is shown in [8] that this matrix is a discrete representation of the L^2 heat operator onto appropriate connected domains in $\mathbb{R}^p$. Thanks to properties of the heat equation, eigenvectors of this matrix are an asymptotical discrete representation of L^2 eigenfunctions with support included in only one connected component.

Clustering in subdomains resumes in restricting the support of these L^2 particular eigenfunctions. Therefore, we can apply Spectral Clustering on subdomains to identify connected components. The subdomains can be defined in a straightforward way by subdividing original data set according to their geometrical coordinates and a partition can be extracted independently and in parallel from each subset. Then, at the grouping level, spectral clustering algorithm is made on a subset with geometrical coordinates close to the boundaries of the previous subdomains. This partitionning will connect together clusters which belong to different subdomains thanks to the transitive relation: $\forall x_{i_1}, x_{i_2}, x_{i_3} \in S$,

$$\text{if } x_{i_1}, x_{i_2} \in C^1 \text{ and } x_{i_2}, x_{i_3} \in C^2 \text{ then } C^1 \cup C^2 = P \text{ and } x_{i_1}, x_{i_2}, x_{i_3} \in P \quad (2)$$

where S is a data set, C^1 and C^2 two distinct clusters and P a larger cluster which includes both C^1 and C^2.

Two main problems arise from this divide and conquer strategy: the difficulty to choose a Gaussian affinity parameter σ and the number of clusters k which remains unknown and may even vary from one subdomain to the other. We propose two ways to overcome these drawbacks. In the following, let us consider a p-dimensional data set $S = \{x_1, .., x_n\} \subset \mathbb{R}^p$. In the next section, we shall address the proper choice of the parameter σ and in section 2.2, we propose a way to overcome the problem of not knowing the number of clusters a priori.

2.1   Choice of the Affinity Parameter σ

The Gaussian affinity matrix (1) is widely used and depends on a free parameter σ. It is known that this parameter affects the results in spectral clustering and spectral embedding. A global heuristic for this parameter was proposed in [4] in which both the dimension of the problem as well as the density of points in the given p-th dimensional data set are integrated. With an assumption that the p-dimensionnal data set is isotropic enough, the data set S is included in a p-dimensionnal box bounded by D_{max} the largest distance between pairs of points in S: $D_{\max} = \max_{1 \leq i,j \leq n} \|x_i - x_j\|$.

So a reference distance noted σ could be defined: this distance represents the case of an uniform distribution in the sense that all pair of points are separated by the same distance σ in the box of edge size D_{max}:

$$\sigma = \frac{D_{\max}}{n^{\frac{1}{p}}}. \tag{3}$$

From this definition, clusters may exist if there are points that are at a distance no more than a fraction of σ. We could define such parameter for each subdomain. However, with a straighforward decomposition as the one proposed, one can find easily that a local σ in each subdomain will be close to the value of a global σ defined on the whole data set in the same way. This avoids local computations. However, this is not the case for the interface data set where a local σ must be considered. To conclude, we only need to compute two values of σ: one for the interface where the topology of the volume changes drastically, and one common to all the other "cubic" subdomains.

2.2   Number of Clusters k

The problematic of the right choice of k is all the more accurate that this number may vary from one subdomain to the other in such a domain decomposition strategy. We therefore consider in each subdomain a quality measure based on ratios of Frobenius norms, see for instance [4]. For instance, after indexing data points by cluster as followed, for $k = 3$:

$$\hat{L} = \begin{bmatrix} L^{(11)} & L^{(12)} & L^{(13)} \\ L^{(21)} & L^{(22)} & L^{(23)} \\ L^{(31)} & L^{(32)} & L^{(33)} \end{bmatrix},$$

the off-diagonal blocks will represent the affinity between clusters and the diagonal ones the affinity within clusters. The ratios between the Frobenius norm of the off-diagonal blocks and that of the diagonal ones could be evaluated:

$$r_{ij} = \frac{\|L^{(ij)}\|_F}{\|L^{(ii)}\|_F}.$$

By definition, the appropriate number of clusters k corresponds to a situation where points which belong to different clusters have low affinity between each other whereas points in same clusters have higher affinity. Among various values for k, the final number of cluster is defined so that the affinity between clusters is the lowest and the affinity within clusters is the highest as followed:

$$k = arg\min_{k'} \frac{2}{k'(k'-1)} \sum_{\substack{i=1 \\ j=i+1}}^{k'} r_{ij}. \tag{4}$$

This last equation (4) provides an average link of the affinity between clusters. In Fig. 1, the ratio $\eta = \frac{2}{k(k-1)} \sum_{\substack{i=1 \\ j=i+1}}^{k} r_{ij}$, function of the number of cluster k,

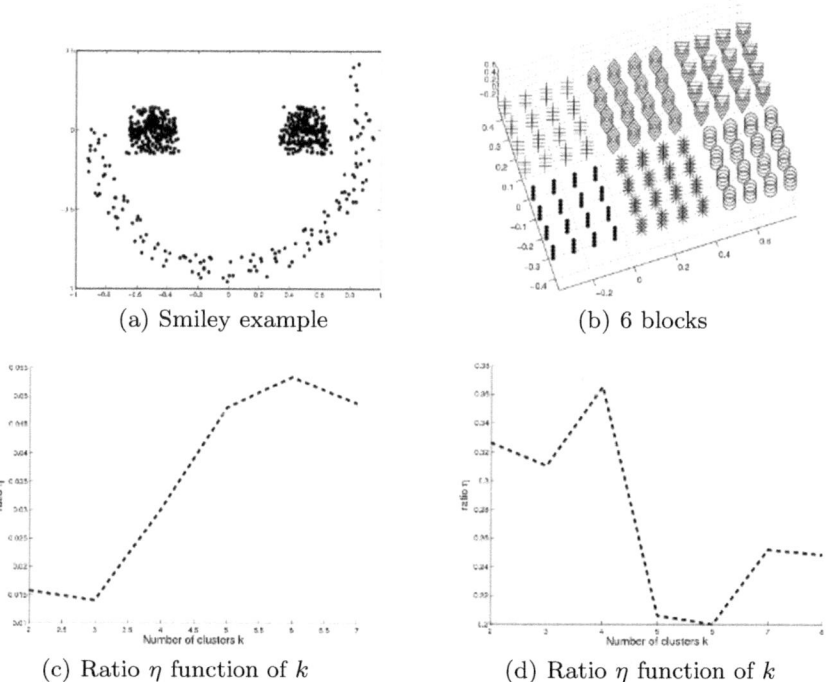

(a) Smiley example   (b) 6 blocks

(c) Ratio η function of k (d) Ratio η function of k

Fig. 1. Examples for determining the number of cluster

is plotted on two examples with various densities among clusters. The minimum of the ratio η is reached for the optimal value of k.

Moreoever, dividing the whole data set in subdomains may lead to situations in which a subdomain contains only one cluster. If the number of clusters k which satisfy (4) is equal to 2 in one subdomain, we then compare the numerator of ratio η to its denominator. Based on a threshold β, if the ratio $\dfrac{\|L_{12}\|_F}{\|L_{11}\|_F}$ is larger than β, we set the value k to 1 instead of 2.

3   Implementation: Algorithm Components

We shall now detail the different steps, described in Fig. 2, of the algorithm with respect to the strategy proposed previously.

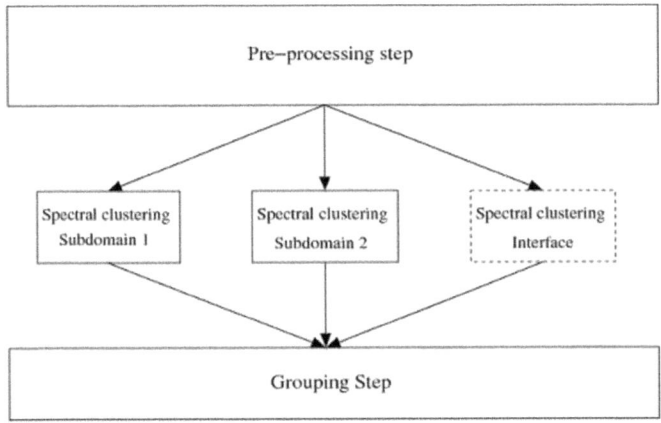

Fig. 2. Principle of parallel Spectral clustering for $q = 2$

3.1   Pre-processing Step: Partition S in q Subdomains

Let us include all data points in a box of edge l_i for the ith-dimension, $i = \{1, .., p\}$ where:

$$l_i = max_{1 < i_1, i_2 \leq n} |x_{i_1}(i) - x_{i_2}(i)|, \forall i \in \{1, .., p\}. \tag{5}$$

According to the maximum length on each dimension, the box is divided in q subboxes where $q = \Pi_{i=1}^{p} q_i$ and q_i denotes the number of subdivisions on the i-th dimension. Then, the affinity parameter σ is computed as indicated in (3). The number of processors is fixed to $nbproc = q + 1$.

3.2   Domain Decomposition: Interface and Subdomains

Interface. It includes all points with a maximum norm distance to the boundaries less than a given γ bandwidth. This interface should help to reconnect together clusters with points in more than one subdomain. Picking up a band-width value $\gamma = 3\sigma$ enables to group together points in the same cluster. As the interface layer does not cover the same volume as the other "cubic" subdomains, the isotropic assumption is not anymore satisfied, and a particular affinity parameter σ^* must be considered. We therefore follow the same idea as in section 2.1 but with an adequate volume measure for the interface:

$$\sigma^* = \frac{Vol(interface)}{n_{interface}^{\frac{1}{p}}}$$

where $Vol(interface)$ represents the real volume of the interface and $n_{interface}$ the number of data points in the interface. The volume of the interface is function of bandwidth γ, the number of cut-size q and $l_1, ..l_p$ the edges of the box in each direction as followed:

$$Vol(interface) = \sum_{i=1}^{p}(q_i - 1)\gamma^{p-1}l_i - \gamma^p \Pi_{i=1}^p (q_i - 1). \tag{6}$$

Subdomains. Each processor from 1 to *nbproc* has a data subset S_i, $i = 1..nbproc$ which coordinates are included in a geometrical subbox. The affinity of all the subdomains have the same global parameter σ defined by (3).

3.3   Spectral Clustering on Subdomains

Some elements of Algorithm 1 are now precised.

Computation of the spectrum of the affinity matrix (1). Classical routines from LAPACK library [7] are used to compute selected eigenvalues and eigenvectors of the normalized affinity matrix A for each subset of data points.

Number of clusters. With an upper bound noted *nblimit*, the number of clusters $k \in \{2, .., nblimit\}$ is chosen to satisfy (4).

Spectral embedding. The centers for k-means in the spectral embedding are initially chosen to be the furthest from each other along a direction.

3.4   Grouping Step

The final partition is formed by grouping partitions from the $nbproc - 1$ independent spectral clustering analyses. The grouping is made with the interface partition and the transitive relation (2). If a point belongs to two different clusters, both clusters are then included in a larger one. As output of the parallel method, a partition of the whole data set S and the final number of clusters k are given.

An example on how our method is applied on a target data set splitted in $q = 4$ subboxes (see Fig. 3). On the left, the clustering result for the interface is plotted. Each color represents a cluster. On the right, the clustering results on the 4 respective subdomains are plotted.

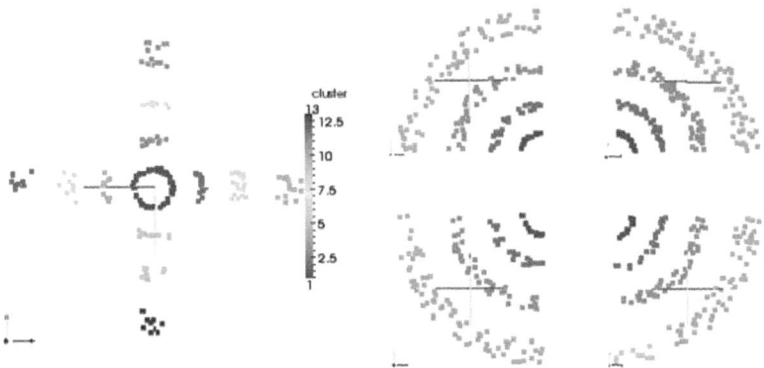

Fig. 3. Target example: interface and subdomains

4   Parallel Experiments

As numerical example, this parallel spectral clustering is tested on a 3D geometrical case which represents 2 non-concentric truncated spherical areas included in a larger one as shown in the left of Fig. 4. On the right of the same figure, one zoom around each included truncated sphere is plotted. It shows the proximity between the small spheres and the big one.

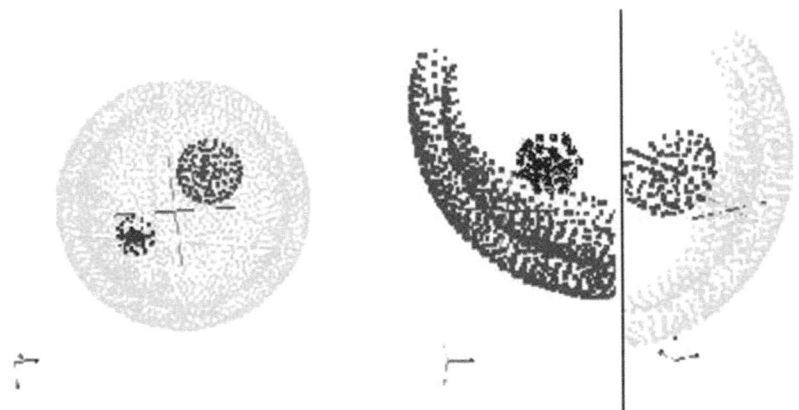

Fig. 4. Geometrical example and zooms: $n = 4361$

The numerical experiments were carried out on the Hyperion supercomputer[1]. Hyperion is the latest supercomputer of the CICT (Centre Interuniversitaire de Calcul de Toulouse). With its 352 bi-Intel "Nehalem" EP quad-core nodes it can

---

[1] http://www.calmip.cict.fr/spip/spip.php?rubrique90

develop a peak of 33TFlops. Each node has 4.5 GB memory dedicated for each of the cores with an overall of 32 GB fully available memory on the node. We vary the number of points of this geometrical example from $n = 4361$ to $n = 15247$ points.

For our tests, the domain is successively divided in $q = \{1, 3, 5, 13\}$ subboxes. The timings for each step of parallel Spectral clustering are measured. We give in Table 1, for each problem size and each distribution the number of data in the interface, the total time and the percentage of this time spent in the spectral clustering computation on the subdomains.

Table 1. 3 truncated spheres with interface

n	Number of processors	Number of data in the interface	Total Time (sec)	% of Total Time for spectral clustering
	1	-	251.12	99.9
4361	5	1596	13.19	97.4
	9	2131	30.6	98.3
	13	2559	54.36	98.8
	1	-	2716.34	99.9
9700	5	3601	156.04	98.4
	9	4868	357.42	99.4
	13	5738	610.89	99.7
	1	-	> 3h	-
15247	3	5532	549.82	99.5
	9	7531	1259.86	99.8
	13	8950	2177.16	99.8

We can retain from these results the following information:

- the main part of our algorithm is the spectral clustering on subdomains;
- the time spent in this part is the time of the processor which gets the more data: there is a synchronization point at the end of this part, before the grouping step;
- with this example, the interface gets the maximum number of data;
- the speed-up is larger than the ratio between the total number of points to the maximum data on one subdomain. For example, with $n = 4361$ points and 5 processors, the ratio is 2.73 and the speed-up is 15.76. This can be explained by the non-linearity of our problem with the computation of eigenvectors from Gaussian the affinity matrix.
- the spectral clustering on subdomains is faster than considering the whole data set. Computation of parameters σ, σ^* and the grouping step doesn't penalize our strategy; the time spent in these parts is negligible (less than 2% of total time).

As remarks, the loop implemented to test several values of k in spectral clustering algorithm until satisfying (4) become less and less costly when the number of processors increase. This is due to eigenvectors computation which is less costly with smaller dense affinity matrices. Also, subdividing the whole data set implicitly reduces the Gaussian affinity to diagonal subblocks (after permutations). However when the data set is subdivided in larger numbers of subdomains, the data set of the interface becomes the most time consuming computational task.

We shall investigate its influence and study the trade-off between subdivisions and interface size.

5   Discussion and Alternative

As shown in the previous examples, using interface which connects all the partitions could present some limitations. In fact, the more the domain is subdivided, the larger is the set in the interface. So to limit this drawback, a threshold, noted τ, should be defined for the number of subdomains in each axis. This threshold τ represents the ratio between the volume covered by the interface and the total volume.

$$\tau = \frac{Vol(interface)}{Vol} \tag{7}$$

where $Vol(interface)$ is defined by (6) and Vol is the total volume function of l_i defined by (5) for $i = \{1, .., p\}$: $Vol = \Pi_{i=1}^{p} l_i$.

To overcome this drawback of considering the interface as a distinct subdomain, the data set of interface could be included in the others subdomains. In fact, the whole data set is subdivided in q subboxes which have a non-empty intersection. This leads to reduce the number of processors ($nbproc = q$) and avoid computing a special parameter σ^* for the interface. The main advantage is that the Spectral clustering method is used on all subdomains with the same topology of volume and does not break the isotropic distribution. However the threshold τ is still preserved in order to reduce the time in grouping step. So the volume of the intersection between subdomains is upbounded by a fraction of the volume of the whole data set. Thus, this strategy with intersection is resumed in Fig. 5 for $q = 2$.

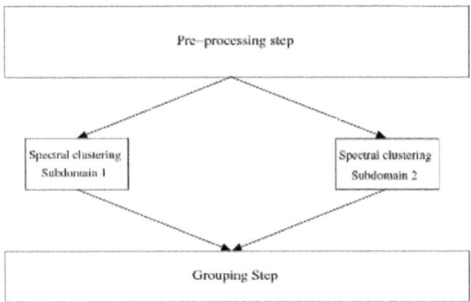

Fig. 5. Principle of alternative parallel Spectral clustering with intersection for $q = 2$

In the same way, Fig. 6 illustrates this alternative on the previous target example divided in $q = 4$ subboxes. On the left, the final clustering results, after the grouping step, is plotted.

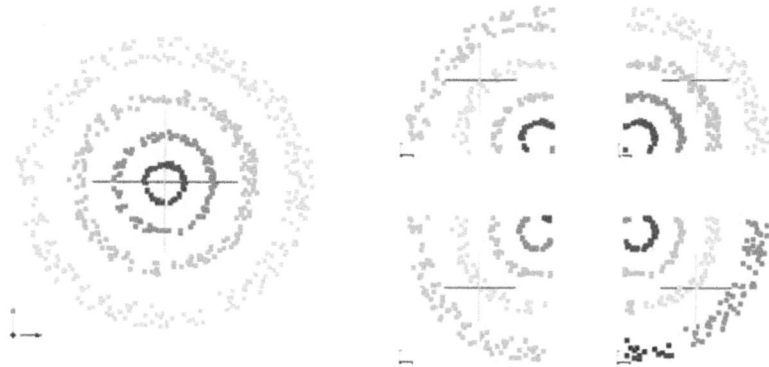

Fig. 6. Target example: subdomains with intersection

5.1   Numerical Experiments: Geometrical Example

The same examples than in section 4 are tested with this new strategy. In the same way, the results are resumed in the Table 2 with the timings for respective steps of parallel spectral clustering with intersection.

Table 2. 3 truncated spheres with intersection between subdomains

n	Number of processors	Maximum of data by processor	Total Time (sec)	% of Total Time for spectral clustering
4361	1	-	232.19	99.8
	4	1662	14.73	97.2
	8	984	3.37	85.2
	12	1004	3.71	82.5
9700	1	-	2716.34	99.9
	4	3712	157.76	99.4
	8	2265	38.89	97
	12	2283	40.43	96.6
15247	1	-	> 3h	-
	4	5760	578.92	99.6
	8	3531	133.79	98.1
	12	3517	131.83	97.9

We can observe that this alternative has the same main behaviours than the one with interface:

– very good speed-up, much larger than the ratio of the total number of data to the maximum number of data on a subdomain;

- the main part of the time is spent in the spectral clustering step;
- the time of the spectral clustering step is the time of the processor with the maximum number of data.

We can express some specific remarks for this strategy:

- the times are better than the interface strategy times with an equivalent number of processor: for example, with $q = 12$ and $n = 15247$, the total time is divided by 16;
- the time is decreasing when the number of subdivisions increases at the condition that the maximum number of data on a processor decreases. We observe, for example, that with $n = 4361$ points and $q = 12$, the processor with the maximum number of points has more points that the equivalent one with $q = 8$. That explains the larger time with $q = 12$ than $q = 8$.

The last remark opens some reflexion about how to divide the domain: a splitting that balances the number of data among the processors will give better results than an automatic splitting of the geometry.

5.2   An Image Segmentation Example

An image segmentation in grayscale is now considered. This kind of example is well designed to the parallel strategy thanks to an uniform distribution with respect to the geometrical coordinates per processor. The affinity matrix is defined as a 3-dimension rectangular box [4] which includes both geometrical coordinates and brightness. The steps between pixels and brightness are about the same magnitude. This means that the image data can be considered as isotropic enough. This approach is tested on an image representing flowers. This image is a 186×230 picture *i.e* $n = 42780$ data points. Due to the large number of data, the parallel spectral clustering is applied on $q = 20$ processors with $nblimit = 20$.

In Fig. 7, the original data set is plotted on the left and the final clustering results on the right. The spectral clustering result has determined 111 clusters.

Original data

Clusters number = 111

Fig. 7. First example of image segmentation tested on Hyperion

Compared to the original data set, the shapes of the different flowers are well-described. Moreover, the details on the lily can be recognized. The total time spent is equal to 398.78 sec for $n = 42780$ which confirms the computational performance with this parallel spectral clustering with intersection. Note that, with less processors, this example fails because of the lack of virtual memory.

Original data

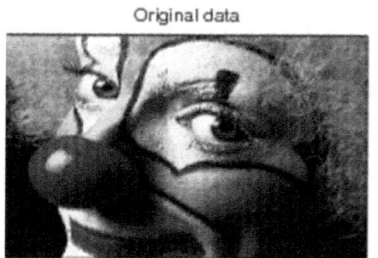

Clusters number = 203

Fig. 8. Second example of image segmentation tested on Hyperion

Let consider a second picture of size 200×320 which presents more contrasts and a larger dataset with $n = 64000$ data points. The parallel method is applied with the same parameters as for the previous example and Fig. 8 represents both original data set and clustering result. After the grouping step, 203 clusters were found in 3110.38 seconds. This total time is larger than the previous example due to the larger dataset and the complexity of the distribution of data points. As result, the clown face is well-recognized as well as its hair. This overlapping strategy is relevant for image segmentation applications but some further investigations should be led specially on color pictures. In fact, considering color picture remains including both 2D geometrical information and 3D colorness information together and an equilibrium between geometrical and color information via some normalization techniques should be studied.

6   Conclusion and Ongoing Works

By exploiting the property of connected components, Spectral clustering could be independently applied on geometrical subdomains without altering the final partition. With an independant way of determining the number of targeted clusters k in each subdomain, the method is completely unsupervised. However, considering an isolated data set for interface presents some limit. It depends on the trade-off between dividing and grouping. The alternative which consists in including this interface in all the subdomains improves the parallel approach.

Futhermore, the strategy could be improved with techniques for distributing uniformily the data per processor and some techniques for sparsifying Gaussian affinity matrix. On sparse data sets, sparse in the sense of the distribution in the enclosing volume, we may also benefit from techniques of graph partitioning, such as Metis techniques. Applied to the graph of nearest neighbours in the data set, we partition in a more equilibrated way the data points in subsets. Some sparsification techniques, such as thresholding the affinity between data points, could also be introduced to speed up the algorithm when the subdomains are still large enough. It will permit reducing the time dedicated to spectral clustering in subdomains.

References

1. Ng, A.Y., Jordan, M.I., Weiss, Y.: On spectral clustering: analysis and an algorithm. In: Proc. Adv. Neural Info. Processing Systems (2002)
2. Chen, W.-Y., Yangqiu, S., Bai, H., Lin, C.-J., Chang, E.Y.: Parallel Spectral Clustering in Distributed Systems. IEEE Transactions on Pattern Analysis and Machine Intelligence (2010)
3. Belkin, M., Niyogi, P.: Laplacian Eigenmaps and Spectral Techniques for Embedding and Clustering. In: Advances in Neural Information Processing Systems (2002)
4. Mouysset, S., Noailles, J., Ruiz, D.: Using a Global Parameter for Gaussian Affinity Matrices in Spectral Clustering. In: Palma, J.M.L.M., Amestoy, P.R., Daydé, M., Mattoso, M., Lopes, J.C. (eds.) VECPAR 2008. LNCS, vol. 5336. Springer, Heidelberg (2008)
5. Song, Y., Chen, W.Y., Bai, H., Lin, C.J., Chang, E.Y.: Parallel spectral clustering. In: Daelemans, W., Goethals, B., Morik, K. (eds.) ECML PKDD 2008, Part II. LNCS (LNAI), vol. 5212, pp. 374–389. Springer, Heidelberg (2008)
6. Fowlkes, C., Belongie, S., Chung, F., Malik, J.: Spectral grouping using the Nystrom method. IEEE Transactions on Pattern Analysis and Machine Intelligence (2004)
7. Anderson, E., Bai, Z., Bischof, C., Blackford, S., Demmel, J., Dongarra, J., Du Croz, J., Greenbaum, A., Hammarling, S., McKenney, A., et al.: LAPACK Users' guide. Society for Industrial Mathematics, Philadelphia (1999)
8. Mouysset, S., Noailles, J., Ruiz, D.: On an interpretation of Spectral Clustering via Heat equation and Finite Elements theory. In: Proceedings of International Conference on Data Mining and Knowledge Engineering (2010)

On Techniques to Improve Robustness and Scalability of a Parallel Hybrid Linear Solver

Ichitaro Yamazaki and Xiaoye S. Li

Lawrence Berkeley National Laboratory, Berkeley, CA 94720, USA

Abstract. A hybrid linear solver based on the Schur complement method has great potential to be a general purpose solver scalable on tens of thousands of processors. For this, it is imperative to exploit two levels of parallelism; namely, solving independent subdomains in parallel and using multiple processors per subdomain. This hierarchical parallelism can lead to a scalable implementation which maintains numerical stability at the same time. In this framework, load imbalance and excessive communication, which can lead to performance bottlenecks, occur at two levels: in an intra-processor group assigned to the same subdomain and among inter-processor groups assigned to different subdomains. We developed several techniques to address these issues, such as taking advantage of the sparsity of right-hand-sides during the triangular solutions with interfaces, load balancing sparse matrix-matrix multiplication to form update matrices, and designing an effective asynchronous point-to-point communication of the update matrices. We present numerical results to demonstrate that with the help of these techniques, our hybrid solver can efficiently solve large-scale highly-indefinite linear systems on thousands of processors.

1 The Schur Complement Method and Parallelization

Modern numerical simulations give rise to large-scale sparse linear systems of equations that are difficult to solve using standard techniques. Matrices that can be directly factorized are limited in size due to large memory requirements. Preconditioned iterative solvers require less memory, but often suffer from slow convergence. To mitigate these problems, several parallel hybrid solvers have been developed based on a non-overlapping domain decomposition idea called the Schur complement method [5,7].

In the Schur complement method, the original linear system is first reordered into a 2×2 block system of the following form:

$$\begin{pmatrix} A_{11} & A_{12} \\ A_{21} & A_{22} \end{pmatrix} \begin{pmatrix} x_1 \\ x_2 \end{pmatrix} = \begin{pmatrix} b_1 \\ b_2 \end{pmatrix},$$ (1)

where A_{11} and A_{22} respectively represent *interior subdomains* and *separators*, and A_{12} and A_{21} are the *interfaces* between A_{11} and A_{22}. By eliminating the unknowns associated with the interior subdomains A_{11}, we obtain

$$\begin{pmatrix} A_{11} & A_{12} \\ 0 & S \end{pmatrix} \begin{pmatrix} x_1 \\ x_2 \end{pmatrix} = \begin{pmatrix} b_1 \\ \widehat{b}_2 \end{pmatrix},$$ (2)

J.M.L.M. Palma et al. (Eds.): VECPAR 2010, LNCS 6449, pp. 421–434, 2011.
© Springer-Verlag Berlin Heidelberg 2011

where S is the Schur complement defined as

$$S = A_{22} - A_{21}A_{11}^{-1}A_{12}, \tag{3}$$

and $\widehat{b}_2 = b_2 - A_{21}A_{11}^{-1}b_1$. Subsequently, the solution of the linear system (1) can be computed by first solving the Schur complement system

$$Sx_2 = \widehat{b}_2, \tag{4}$$

and then solving the interior system

$$A_{11}x_1 = b_1 - A_{12}x_2. \tag{5}$$

For a detailed discussion of the Schur complement method, see [13] and the references therein.

The existing parallel hybrid solvers typically use a direct method to eliminate the unknows associated with the interior subdomains, while a preconditioned iterative method is used to solve the Schur complement system, where most of the fill occurs. These solvers often exhibit great parallel performance since the interior subdomains can be factorized independently from one other, and a direct method is effective for factorizing these relatively small subdomains. Furthermore, for a symmetric positive definite system, the Schur complement has a smaller condition number than the original matrix [13, Section 4.2], and fewer iterations are often needed to solve the Schur complement system. General purpose parallel hybrid solvers based on this Schur complement method have been developed, and their effectiveness has been shown for some applications [5,7]. Unfortunately, for highly-indefinite systems, these solvers can still suffer from slow convergence when solving the Schur complement systems [14].

In this paper, we present some of the challenges encountered in the development of a robust and efficient general purpose hybrid solver targeted for thousands of processors and our approaches to resolving these challenges. Our parallel implementation consists of the following three phases:

1) Extracting and factorizing the interior subdomains. We use a parallel nested disection algorithm implemented in PT-SCOTCH [8] to extract interior subdomains. For an unsymmetric matrix A, PT-SCOTCH is applied to the graph of $|A| + |A|^T$. Then, these interior subdomains are factorized using a direct method.

There are two approaches to assigning processors to factorize subdomains. One approach is to assign a single processor to factorize one or more interior subdomains, which we refer to as a *one-level* parallel approach. An advantage of this approach is that multiple subdomains can be assigned to a processor such that the workload is balanced among the processors. A serious drawback of this approach, however, is that many subdomains must be generated in order to use a large number of processors. This increases the size of the Schur complement, and often leads to slow convergence. An alternative is to assign multiple processors to each interior subdomain, which allows us to increase the processor count without

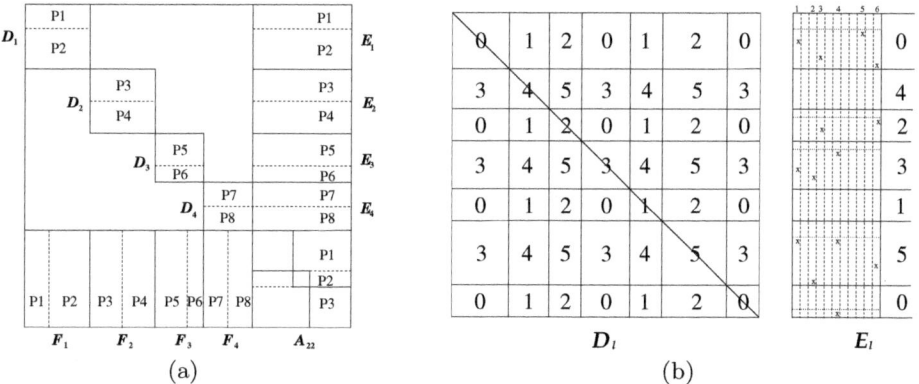

Fig. 1. (a): distribution of the coefficient matrix. Each group g_ℓ contains two processors per subdomain, and the group g_S contains three processors. (b): subdomain D_ℓ stored in a 2D block-cyclic format using a 2×3 process grid, and its corresponding RHSs E_ℓ with a block size of six. An "x" in E_ℓ represents the first nonzero in an individual column of the supernodal block, and a horizontal dotted line represents the first nonzero in the entire block.

increasing the number of subdomains or the size of the Schur complement. This approach is refered to as a *two-level* parallel approach and is the focus of our study in this paper.

When k interior subdomains are extracted, the coefficient matrix of Eq. (1) has the following block-structure:

$$\left(\frac{A_{11} \mid A_{12}}{A_{21} \mid A_{22}} \right) = \begin{pmatrix} D_1 & & & \mid E_1 \\ & D_2 & & \mid E_2 \\ & & \ddots & \mid \vdots \\ & & & D_k \mid E_k \\ \hline F_1 & F_2 & \dots & F_k \mid A_{22} \end{pmatrix}, \qquad (6)$$

where D_ℓ is the ℓ-th subdomain, and E_ℓ and F_ℓ are the interfaces between D_ℓ and A_{22}. In our implementation, each processor is assigned to a processor groups g_ℓ that factorizes the subdomain D_ℓ. Then, the rows of D_ℓ and E_ℓ, and the columns of F_ℓ are distributed among the processors in the processor group g_ℓ. The nonzeros of D_ℓ and F_ℓ are stored in the Compressed Row Storage (CRS) format, while those of E_ℓ are stored in the Compressed Column Storage (CCS) format [3]. Another processor group g_S is created, which consists of a subset of avaliable processors and will be used to solve the Schur complement system. The rows of A_{22} are distributed among the processors in this processor group g_S. The selection of processors to be assigned to g_S will be discussed in Section 2.3. Fig. 1(a) shows an example of a matrix distribution using our two-level approach. Finally, the parallel direct solver SuperLU_DIST [10] is used to factorize each subdomain.

2) Computing an approximate Schur complement. In order to provide the roubustness of solving highly-indefinite systems, our implementation explicitly computes an approximate Schur complement, which is distributed over the processor group g_S. This is the most challenging phase of the parallel solution, especially in a two-level parallel framework. We need to deal with the load imbalance and communication not only within an *intra-processor group* assigned to the same subdomain, but also among the *inter-processor groups* assigned to different subdomains. We have developed a number of techniques to enhance performance of our hybrid solver to compute the approximate Schur complement $\widetilde{S}$. These techniques are the focus of our paper and will be discussed in Section 2.

3) Computing the solution. A preconditioned Krylov method of PETSc [11] is used to solve the Schur complement system (4), where the preconditioner is the exact LU factors of an approximate Schur complement $\widetilde{S}$. SuperLU_DIST is used to compute the preconditioner. At each iteration, the matrix-vector multiplication with S is computed by applying a sequence of the sparse matrix operations (3) on the vector, and hence, S is not stored explicitly. To improve the load balance of the matrix-vector multiplication, the matrices D_ℓ, E_ℓ, and F_ℓ are distributed among the processors in the processor group g_ℓ such that they each own a similar number of nonzeros. Recall that SuperLU_DIST internally uses a 2D block-cyclic format (see Fig. 1(b)). Hence, the performance of SuperLU_DIST to compute the LU factorization of D_ℓ and to solve the corresponding triangular systems is not affected by the initial distribution of the coefficient matrix D_ℓ and the right-hand-sides (RHSs) E_ℓ and F_ℓ. We note that at each iteration, the processors in the processor group g_S are used to apply the preconditioner $\widetilde{S}$, while all the available processors are used for the matrix-vector multiplication with S. The final solution is computed by solving the interior system (5) with the already-computed LU factors.

2   Efficient Computation of an Approximate Schur Complement

In this section, we describe the techniques to enhance the performance of our hybrid solver to compute an approximate Schur complement $\widetilde{S}$. To demonstrate the effectiveness of these techniques, we use the numerical results of a highly-indefinite matrix from the numerical simulation of an accelerator cavity design [1,9]; namely, **tdr455k** of dimension $2,738,556$ with $112,756,352$ nonzeros. For the numerical experiments, we extracted 16 subdomains D_ℓ using PT-SCOTCH. All the experiments were conducted on the Cray XT4 machine at NERSC.

Given the LU factorization $D_\ell = L_\ell U_\ell$ of the interior subdomain D_ℓ,[1] the Schur complement S can be computed as

---

[1] The matrix D_ℓ is scaled and permuted to enhance numerical stability and preseve the sparsity of L_ℓ and U_ℓ. For clarity, the scaling and permutation are not shown in the expression.

$$S = A_{22} - \sum_{\ell=1}^{k} F_\ell D_\ell^{-1} E_\ell \tag{7}$$

$$= A_{22} - \sum_{\ell=1}^{k} (U_\ell^{-T} F_\ell^T)^T (L_\ell^{-1} E_\ell) \tag{8}$$

$$= A_{22} - \sum_{p=1}^{n_p} W^{(p)} G^{(p)}, \tag{9}$$

where n_p is the number of processors used to solve the entire system, and the matrices $G^{(p)}$ and $W^{(p)}$ are given by

$$G^{(p)} = G(j_p : (j_{p+1} - 1), :), \quad W^{(p)} = W(:, j_p : (j_{p+1} - 1)), \tag{10}$$

such that the p-th processor owns the j_p-th through $(j_{p+1} - 1)$-th rows of $G = L_{11}^{-1} A_{12}$ and the corresponding columns of $W = (U_{11}^{-T} A_{21}^T)^T$, where the LU factorization $A_{11} = L_{11} U_{11}$ is given by L_ℓ and U_ℓ, i.e., $L_{11} = \text{diag}(L_1, L_2, \ldots, L_k)$ and $U_{11} = \text{diag}(U_1, U_2, \ldots, U_k)$. Once the matrices $G^{(p)}$ and $W^{(p)}$ are computed, their nonzeros are stored in the CRS format. Then, the p-th processor computes its update matrix $T^{(p)} = W^{(p)} G^{(p)}$. To efficiently manage the required memory, the memory for storing each row of $W^{(p)}$ is freed as soon as the corresponding row of $T^{(p)}$ is computed. The rows of $T^{(p)}$ are sent to the q-th processor which owns the corresponding rows of A_{22}, and the q-th processor computes the corresponding rows of the Schur complement S.

Large amounts of fill may occur in $G^{(p)}$ and $W^{(p)}$. To reduce the memory and computational costs, their approximations $\widetilde{G}^{(p)}$ and $\widetilde{W}^{(p)}$ are computed by discarding nonzeros with magnitudes less than a prescribed threshold, and an approximate update matrix $\widetilde{T}^{(p)}$ is computed by $\widetilde{T}^{(p)} = \widetilde{W}^{(p)} \widetilde{G}^{(p)}$. If the p-th processor belongs to the processor group g_S, to compute its local portion of an approximate Schur complement, it gathers the corresponding rows of $\widetilde{T}^{(q)}$ from all the processors and explicitly computes $\widehat{S}^{(p)} = A_{22}^{(p)} - \sum_q \widetilde{T}^{(q)}(i_p : (i_{p+1}-1), :)$, where the p-th processor owns the i_p-th through $(i_{p+1} - 1)$-th row of A_{22}; i.e., $A_{22}^{(p)} = A_{22}(i_p : (i_{p+1} - 1), :)$. To further reduce the costs, small nonzeros are discarded from $\widehat{S}^{(p)}$ to form its approximation $\widetilde{S}^{(p)}$. Prior to discarding the small nonzeros, we preprocess $\widehat{S}^{(p)}$ to enhance numerical stability by permuting large nonzeros to the diagonal. This preprocessing is performed in a distributed fashion; namely, the p-th processor uses an existing serial code MC64 [4] and computes the permutation of its local matrix that corresponds to the p-th diagonal block of $\widehat{S}$. The off-diagonal blocks are permuted accordingly. This distributed preprocessing technique enhances the numerical stability without forming the global approximate Schur complement $\widehat{S}$ on each processor. See [14] for more details on the preprocessing technique.

We now describe several techniques to enhance the performance of computing the approximate Schur complement $\widetilde{S}^{(p)}$.

2.1 Sparse Triangular Solution with Sparse Right-Hand-Sides

SuperLU_DIST stores the LU factors L_ℓ and U_ℓ in a 2D block-cyclic format based on the supernodal structure of L_ℓ. RHSs are assumed to be dense, and are distributed by block rows conforming to the supernodal partition (see Fig. 1(b)). Since the communication and computation patterns of the triangular solutions do not change between RHSs, a symbolic triangular solution subroutine is invoked once to compute static communication and computation schedules. Then, the triangular systems are solved by a series of scheduled block operations with the supernodal blocks.

Our first performance-enhancing technique is to exploit the sparsity of RHSs E_ℓ and F_ℓ^T when solving the triangular systems to form $G^{(p)}$ and $W^{(p)}$, respectively. For this, we modified the symbolic triangular solution subroutine of SuperLU_DIST so that only non-empty messages are sent and only computations with non-empty blocks are performed. Since the sparsity pattern of each column of E_ℓ or F_ℓ is different, this subroutine is invoked for each triangular solution with each column. This symbolic subroutine sets up the communication and computation schedules with respect to the non-empty supernodal blocks of RHSs, which are typically not dense. Because of the supernodal structure of LU factors, when the fill occurs in the solution vector, it occurs all the way to the boundary of the supernodes. Hence, during numerical solution, we keep track of the first nonzero in each supernodal block of a RHS. Then, the block operations are performed only for the elements below the first nonzero location so that the operations with explicit zeros are eliminated. We have observed that exploiting the sparsity of RHSs leads to an order-of-magnitude speedup in computing $W^{(p)}$ and $G^{(p)}$.

There are typically tens to hundreds of thousands of columns in E_ℓ. Hence, it could be costly to perform the triangular solution one column at a time. Further optimization can be achieved by grouping E_ℓ into blocks of multiple columns and solving one block at a time. There are several advantages with blocking: 1) the symbolic solution only needs to be computed per block, 2) fewer messages need to be sent to compute $W^{(p)}$, and 3) the data locality to access the LU factors may be improved. During numerical solution with the multiple RHSs, we keep track of the first nonzero within each supernodal block of the multiple columns (see Fig. 1(b)). The disadvantage is that we need to pad explicit zeros so that these columns have the same nonzero pattern. The padded zeros occur between the first nonzero position of the multiple columns and that of the individual columns. Hence, blocking introduces a trade-off between the data locality and the number of unwanted padded zeros. Specifically, data locality may be improved by increasing the block size; however, this increases the number of padded zeros. In the special case in which the block size is set to be one, there are no operations with explicit zero operands, but only a small amount of locality is available. For our test matrix **tdr455k**, the advantages outweighted the disadvantages, and the average and maximum speedups of 5.7 and 7.4, respectively, were achieved using our defalut block size of 50 and one preocessor per subdomain.

To reduce the number of padded zeros introduced by blocking, we employ the following technique: We first permute the rows of E_ℓ according to a postorder of the elimination tree of D_ℓ. Then, the columns of E_ℓ are permuted in the descending order of the row indices of their first nonzeros. The columns of F_ℓ^T are similarly permuted. One reason why this ordering reduces the number of padded zero is as follows: When a column has the first nonzero at the location corresponding to the i-th node of the elimination tree, then according to the Gilbert's path theorem [6], this first nonzero will generate the fill in the solution vector at the positions corresponding to the nodes on the path from the i-th node to the root of the elimination tree. After RHSs are sorted based on the postorder of their first nonzero row indices, the paths of the adjacent columns are likely to have their starting nodes close together, and the large parts of the paths overlap in the elimination tree. Hence, the solution vectors in the same column block are likely to have fill at similar locations, reducing the number of padded zeros. Furthermore, during the triangular solution, only the columns of the L-factor corresponding to the nodes on the paths are accessed. Hence, postordering RHSs also improves the data locality to access the L-factor. For our test matrix **tdr455k**, average and maximum speedups of about 1.3 and 1.6 were achieved using this postordering technique and one processor per subdomain. Similar topological orderings have been used for a sparse triangular solution with multiple sparse RHSs [12] and for computing elements of the inverse of a sparse matrix [2]. We have also introduced another ordering technique using a hypergraph model to maximize the similarity of the sparsity patterns among the solution vectors in a column block [15].

2.2   Intra-processor Load Balance

We have developed a technique to improve the intra-processor load balance to compute the sparse matrix-matrix multiplication $\widetilde{T}^{(p)} = \widetilde{W}^{(p)}\widetilde{G}^{(p)}$. This is done by distributing the columns of $\widetilde{W}^{(p)}$ and the rows of $\widetilde{G}^{(p)}$ so that each processor in the same processor group g_ℓ owns a similar number of nonzeros. Specifically, Fig. 2 shows the pseudocode to compute $\widetilde{G}^{(p)}$, where the p-th processor belongs to the processor group g_ℓ, $\widehat{E}_\ell$ contains the non-empty columns of E_ℓ, $\widehat{E}^{(p)}$ is the rows of $\widehat{E}_\ell$ stored by the p-th processor, n_c is the number of columns of $\widehat{E}_\ell$, β is the column block size, and n_b is the number of blocks (i.e., $n_b = \lfloor \frac{n_c}{\beta} \rfloor$). At Step 1.b of the pseudocode, the consecutive rows of $\widehat{E}^{(p)}$ are distributed among all the processors (see Fig. 1(a)), but the solution vector $X^{(p)}$ is distributed into block rows conforming to the supernodal partition and only among the diagonal processors (see Fig. 1(b)). After each triangular solution with a block of RHSs, the diagonal processor compresses each column of $X^{(p)}$ into $\widetilde{X}^{(p)}$, excluding the explicitly padded zeros and discarding small nonzeros (Step 1.c). Then, $\widetilde{X}^{(p)}$ is incrementaly stored in $Y^{(p)}$ using the CCS format (Step 1.d). Once all the solution blocks are computed, $Y^{(p)}$ is redistributed among all the processors

1. Compute the solution vectors for the 1-st through $(n_b\beta)$-th columns
$Y^{(p)} := [\,]$
for $k := 1, \ldots, n_b$ **do**
   a. Extract the next right-hand-side block,
     $B^{(p)} := \widehat{E}^{(p)}(:, ((k-1)\beta + 1) : (k\beta))$
   b. Compute the sparse triangular solution,
     $X^{(p)} \leftarrow L_\ell^{-1} B$
   c. Sparsify the solution vectors,
     $\widetilde{X}^{(p)} \leftarrow X^{(p)}$
   d. Store the solution vectors,
     $Y^{(p)} \leftarrow [Y^{(p)} \widetilde{X}^{(p)}]$
end for
2. Distribute $Y^{(p)}$ from the diagonal processors to all the processors
   $\widetilde{G}^{(p)} \leftarrow Y^{(p)}$
3. Compute the remaining solution vectors
   a. $B^{(p)} := \widehat{E}^{(p)}(:, (n_b\beta + 1) : n_c)$
   b. $X^{(p)} \leftarrow L_\ell^{-1} B$
   c. $\widetilde{G}^{(p)} \leftarrow [\widetilde{G}^{(p)} \widetilde{X}^{(p)}]$

Fig. 2. Pseudocode to compute $G^{(p)}$

in g_ℓ so that each processor owns consecutive rows of the solution vectors and roughly the same number of nonzeros (Step 2). Note that the remaining columns of the solution vectors are computed separately (Step 3). This is because the data structure for the triangular solution inside SuperLU_DIST must be reinitialized when the block size changes, and the redistribution of $Y^{(p)}$ into $\widetilde{G}^{(p)}$ needs to be peformed before the block size changes. These remaining colums of $\widetilde{G}^{(p)}$ and the solution vectors $\widetilde{W}^{(p)}$ of the upper triangular system are distributed into the format that has been set up to load balance the first $(n_b\beta)$ columns of $\widetilde{G}^{(p)}$. In our numerical experiments using four processors for each of the 16 interior subdomains of **tdr455k**, without this load balancing technique, some processors had only a negligible amount of work to compute the matrix-matrix multiplication, and the load imbalance as measured by the computation time was an up to five order of magnitude difference. With the technique described here, the load imbalance became less than a factor of two. As a result, this technique reduced the time to compute the Schur complement by a factor of 2.6 and the total solution time by a factor of 1.7.

2.3   Inter-processor Load Balance

In comparison to the original system, the Schur complement system is typically much smaller in dimension. Hence, only a subset of processors g_S is used to factorize the approximate Schur complement $\widetilde{S}$. Subsequently, for the computation of $\widetilde{S}$, all the processors compute their local update matrices $\widetilde{T}^{(p)} = \widetilde{W}^{(p)} \widetilde{G}^{(p)}$, but only the processors in the processor group g_S compute their local portion $\widehat{S}^{(p)}$ of $\widetilde{S}$ by gathering the corresponding rows of $\widetilde{T}^{(q)}$. In this section, we study two techniques to improve the inter-processor load balance: a strategy to accommo-date this *all-to-subset* communication of $\widetilde{T}^{(p)}$ and one to select processors to solve the Schur complement system.

To accomodate the all-to-subset communication of $\widetilde{T}^{(p)}$, an MPI all-to-all communication subroutine can be used. Even though this simpifies the imple-mentation, there are two shortcomings with this approach. First, there are often large variations in the sizes of the subdomains D_ℓ and in the sparsity of the interfaces E_ℓ and F_ℓ. Even though the intra-processor load balance is improved by the technique described in Section 2.2, this leads to poor load balance among the inter-processor groups to compute $\widetilde{T}^{(p)}$. Since the global all-to-all commu-nication imposes synchronization among all the processors, this load imbalance forces some processors to be idle while waiting for the other processors to com-plete the computation of $\widetilde{T}^{(q)}$. Second, the all-to-all communication requires a large buffer to recieve the corresponding rows of $\widetilde{T}^{(q)}$ from all the processors at the same time.

To mitigate these shortcomings, we designed an asynchronous point-to-point communication protocol to transfer $\widetilde{T}^{(p)}$. Fig. 3 shows the pseudocode of this protocol, where p is the ID of this processor, m_s is the number of processors to which $\widetilde{T}^{(p)}$ needs to be sent, m_r is the number of processors from which nonempty $\widetilde{T}^{(q)}(p) = \widetilde{T}^{(q)}(i_p : (i_{p+1}-1), :)$ will be received, ISend($*, q, t$) and IRecv($*, q, t$) respectively indicate nonblocking send operation to and receive operation from the q-th processor with a tag t, Wait($*, t$) blocks until the nonblocking receive operation with a tag t is processed, and Recv($*, q, t$) is a blocking receive opera-tion from the processor q, where ANY_SOURCE can be used in the place of q to indicate a receive operation from any source, and SENDER_SOURCE indicates the ID of the sender processor. Furthermore, Allocate($U(k)$, size(k)) allocates the buffer $U(k)$ to receive $\widetilde{T}^{(q)}(p)$ in the CRS format, where size(k) is the maxi-mum number of nonzeros that can be stored in the buffer, and Free($U(k)$) frees the buffer $U(k)$. In our implementation, all the matrix operations are performed by taking advantage of their sparsity. For efficient memory management, rows of $\widetilde{T}^{(p)}$ are freed, once they are received, and we alternatly reuse two receiving buffers, $U(0)$ and $U(1)$, whose sizes are stored in size(0) and size(1), respectively.

Our point-to-point communication is desinged to overlap the computation of $\widetilde{T}^{(p)}$ with the communication and summation of $\widetilde{T}^{(q)}(p)$. Hence, there is a greater chance of overlap when the processors with less work to compute $\widetilde{T}^{(p)}$ are assigned to form the approximate Schur complement $\widehat{S}^{(p)}$. To achieve this, we assign processors from relatively small interior subdomains to the Schur comple-ment. This not only increases the potential for the overlap, but also improves the overall load balance of memory requirement. Furthermore, on Line 1 of the pseu-docode, we set $q_{\pi_{m_r}} = p$ so that communication of $\widetilde{T}^{(q)}(p)$ from other processors can be overlapped with the computation and summation of the local $\widetilde{T}^{(p)}(p)$. For our test matrix **tdr455k**, the size of the Schur complement was only about 0.5% of the total dimension, and the summation of the update matrices required only a neglibile amout of time (i.e., less than one second out of 120 seconds spent to compute $\widetilde{S}^{(p)}$ on 16 processors). As a result, this point-to-point communication

```
     /* All the processors perform Lines 1 through 10. */
1.   for q = q_{π_1}, q_{π_2}, ..., q_{π_{m_s}} do
2.       T(q) := W^{(p)}(i_q : (i_{q+1} − 1), :) * G^{(p)}  /* compute the rows of T to be sent to the q-th processor */
3.       if(q == p) then  /* sending to itself */
4.           size(0) = nnz(T̃(q))
5.           U(0) := T̃(q)
6.       else
7.           ISend(nnz(T̃(q)), q, t_0)
8.           if( nnz(T̃(q)) > 0 ) then Isend(T̃(q), q, t_1)
9.       end if
10.  end if


     /* The processor subset responsible for solving the Schur complement perform Lines 11 through 75. */
11.  if( p_0 ∈ g_S ) then
12.      p_0 := p
13.      n_m := 1  /* number of received messages */
14.      while( size(0) == 0 and n_m < m_r ) do  /* find the size of the first nonempty message */
15.          Recv(size(0), ANY_SOURCE, t_0)
16.          n_m := n_m + 1
17.          p_0 := SENDER_SOURCE
18.      end do
19.      if( size(0) > 0 ) then
20.          n_k := 1  /* number of received nonempty messages */
21.          if( p_0 ≠ p ) then
22.              Allocate( U(0), size(0) )
23.              IRecv(U(0), p_0, t_1)
24.          end if
25.      end if
26.      if( n_m < m_r ) then
27.          IRecv(size(1), ANY_SOURCE, t_0)  /* request the size of the next message */
28.      end if
29.      S := A_{22}^{(p)}  /* initialize the Schur complement */
30.      if( n_m < m_r ) then  /* find the size of the second nonempty message */
31.          Wait(size(1), t_0)
32.          n_m := n_m + 1
33.          while( size(1) == 0 and n_m < m_r ) do
34.              Recv(size(1), ANY_SOURCE, t_0)
35.              n_m := n_m + 1
36.          end do
37.      end if
38.      p_1 := SENDER_SOURCE
39.      if( size(1) > 0 ) then
40.          Allocate(U(1), size(1))  /* allocate the buffer for the second message */
41.          IRecv(U(1), p_1, t_1)
42.          if( n_m < m_r ) then   /* request the size of the next message */
43.              IRecv(size(0), ANY_SOURCE, t_0)
44.          end if
45.      end if
46.      if( p_0 ≠ p ) then
47.          Wait(U(0), t_1)
48.      end if
49.      S := S − U(0)  /* sparse update of the local Shur complement */
50.      while( n_m ≤ m_r ) do  /* while there is more nonempty messages */
51.          if( n_m < m_r ) then
52.              Wait(nnz, t_0)
53.              n_m := n_m + 1
54.              while( nnz == 0 and n_m < m_r ) do
55.                  Recv(nnz, ANY_SOURCE, t_0)
56.                  n_m := n_m + 1
57.              end do
58.              p_1 := SENDER_SOURCE
59.              if( nnz > 0 ) then
60.                  if( nnz > size(mod(n_k − 1, 2))) then
61.                      Free(U(mod(n_k − 1, 2)))
62.                      Allocate( U(mod(n_k − 1, 2)), nnz )
63.                      size(mod(n_k − 1, 2)) = nnz
64.                  end if
65.                  IRecv(U(mod(n_k − 1, 2)), p_1, t_1)
66.              end if
67.              if( n_m < m_r ) then
68.                  IRecv(nnz, ANY_SOURCE, t_0)
69.              end if
70.          end if
71.          Wait(U(mod(n_k, 2)), t_1)
72.          S := S − U(mod(n_k, 2))
73.          n_k := n_k + 1
74.      end do
75.  end if
```

Fig. 3. Pseudocode of the point-to-point communication to form S

did not reduce the computation time significantly. However, for other problems with large interfaces, the solution time may be reduced more significantly using this point-to-point communication.

More importantly, though, in comparison to the all-to-all communication, this point-to-point communiation can significantly reduce memory requirements. This is because the point-to-point communication alternately uses only two receiving buffers, instead of a large buffer to hold all the receiving updates in the all-to-all communication. When four processors were used on each of the 16 interior subdomains of **tdr455k**, the point-to-point communication reduced the total number of nonzero elements in the receiving buffers by a factor of about 3.1 on average, and up to 6.3. Notice that the size of the all-to-all communication buffer may increase with the number of processors. As a result when 64 processors were used on each of the 16 interior subdomains of **tdr455k**, the buffer size was reduced by a factor of 55.9 on average, and up to 97.0 using the point-to-point communication.

3   Parallel Performance

We now present parallel performance of our hybrid solver. For the iterative solution of the Schur complement systems, the initial aproximation to the solution was the zero vector, and the computed solution was considered to have converged when the ℓ_2-norm of the initial residual was reduced by at least twelve orders of magnitude. This is the solution accuracy required in the actual simulations.

Fig. 4(a) compares the total solution times required by SuperLU_DIST and our hybrid solver to solve the **tdr455k** linear system. The hybrid solver used a threshold σ_1 to enforce the sparsity of $\widetilde{E}$ and $\widetilde{F}$, σ_2 to enforce the sparsity of $\widetilde{S}$, and unrestarted GMRES for solving the Schur complement system. With the one-level parallel approach of our hybrid solver, the number of interior subdomains was set to be equal to the total number of processors. With the two-level approach, the number of interior subdomains was fixed to be 16, and the

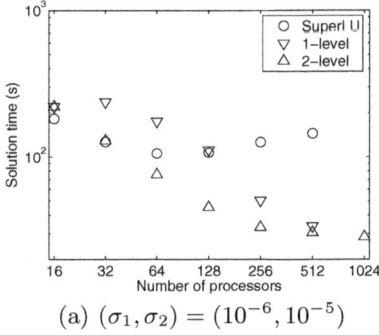
(a) $(\sigma_1, \sigma_2) = (10^{-6}, 10^{-5})$

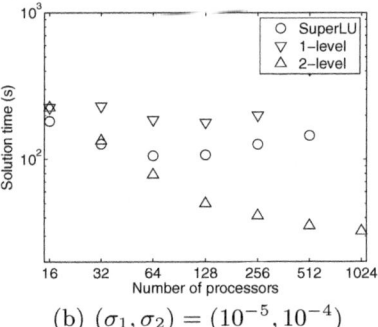
(b) $(\sigma_1, \sigma_2) = (10^{-5}, 10^{-4})$

Fig. 4. Solution times required by SuperLU_DIST and our hybrid solver on **tdr455k**

processors were evenly distributed among the interior subdomains. The figure shows that the solution time with our hybrid solver scaled better than that with SuperLU_DIST.

Fig. 4(a) also shows that with the small dropping thresholds σ_1 and σ_2, the scaling of the one-level and two-level approaches were similar. This is because the number of GMRES iterations was nearly independent of the number of interior subdomains, and GMRES converged within 20 iterations even when more interior subdomains were generated for the one-level approach to use more processors. For comparison, we have tested a state-of-the-art hybrid solver HIPS [5], which implements the one-level parallelization. HIPS computes the preconditioner for solving the Schur complement system based on an ILU factorization of $\widetilde{S}$, where the sparsity of the preconditioner is enforced based on both the numerical values and locations of nonzeros. Specifically, fill is allowed only between separators adjacent to the same subdomain. As a result, the computation of the preconditioner scales to a large number of processors. Unfortunately, this preconditioner was not effective for **tdr455k**; specifically, it required 151 iterations on 16 processors, and it failed to converge within $1,000$ iterations on 32 processors even though the dropping thresholds were set to be zero. Moreover, even when HIPS converged, our solver solved the linear system faster.

Larger dropping thresholds reduce the memory requirement of our hybrid solver. For example, in Fig. 4(b), less memory was needed since the thresholds were increased by an order of magnitude from those in Fig. 4(a). Specifically, in Fig. 4(a), about 15% of the nonozeros were discarded from the matrices E and F, and about 50% of the nonzeros were discarded from the Schur complement $\widetilde{S}$. The respective percentages of the discarded nonzeros for Fig. 4(b) were about 30% and 75%. However, with the large thresholds, the number of GMRES iterations may increase as more interior subdomains are generated. For example, in Fig. 4(b), the number of iterations increased from 32 to 290 when the number of interior subdomains increased from 16 to 256. As a result, the solution time of the one-level approach did not scale. On the other hand, the two-level approach exhibited more robust performance since the processor count can be increased while fixing the size of the Schur complement. These results illustrate the advantage of the two-level approach. For our target of solving larger problems, a more strict sparsity constraint may be needed. Therefore, Fig. 4(b) represents the more practical behavior of our hybrid solver.

Finally, in Table 1, we show the timing results of our hybrid solver to solve another linear system **tdr8cavity** of dimension $17,799,228$ with $727,163,784$ nonzeros. For these experiments, we fixed the number of subdomains to be 64, which resulted in the average subdomain dimension of $277,220$ and the Schur complement dimension of $57,150$. In the table, n_p, n_{g_ℓ}, and n_{g_S} are the numbers of processors used to solve the entire system, to factorize a subdomain, and to solve the Schur complement system, respectively; itrs is the number of BiCGSTAB iterations required for the solution convergence; LU(D_ℓ), Comp($\widetilde{S}$), LU($\widetilde{S}$), Solve, and Total are the times in seconds for factorizing the ℓ-th subdomain, computing $\widetilde{S}$, factorizing $\widetilde{S}$, computing the solution vector, and solving

Table 1. Solution times required by our hybrid solver on **tdr8cavity**

					Time				
n_p	n_{g_ℓ}	n_{g_S}	itrs	LU(D_ℓ)	Comp($\widetilde{S}$)	LU($\widetilde{S}$)	Solve	Total	Speedup
64	1×1	8×8	9	334.0	305.5	4.4	37.7	681.7	
128	1×2	8×8	9	176.0	165.0	4.3	31.8	377.4	1.81
256	2×2	8×8	9	91.3	80.1	4.8	24.2	200.6	1.89
512	2×4	8×8	8	52.5	47.9	4.6	19.3	125.5	1.59
1024	4×4	8×8	8	32.7	28.1	4.7	20.7	86.4	1.47
2048	4×8	8×8	10	22.9	20.9	4.9	22.6	71.6	1.19
4096	8×8	8×8	8	20.3	13.1	4.5	25.2	63.5	1.13

the entire system, respectively; and Speedup is the speedup gained by increasing the processor count by a factor of two. The table shows that the first two phases of the hybrid solver LU(D_ℓ) and Comp($\widetilde{S}$) scaled to thousands of processors. Since n_{g_S} is fixed to be 64, LU($\widetilde{S}$) stayed the same. Even though the last phase Solve did not scale, the total solution time scaled to thousands of processors. We were not able to use SuperLU_DIST to solve this linear system due to the excessive communication required for the triangular solution.

4   Conclusion

We presented several techniques to improve the robustness and scalability of our parallel hybrid solver based on the Schur complement method. Numerical results have shown that our solver can be numerically more robust than another hybrid solver, HIPS, while its solution time scales better than that of the direct solver SuperLU_DIST. We are studying other techniques to further improve the performance of our hybrid solver such as improving initial partition, assigning different numbers of processors to subdomains, distributing the Schur complement based on the separator boundaries, and other parallel preconditioning techniques for the Schur complement system.

Acknowledgements

We gratefully thank Bora Uçar at CNRS, François-Henry Rouet at ENSEEIHT-IRIT, and Esmond Ng at LBNL for helpful discussions. We also thank Lie-Quan Lee at SLAC for providing the test matrix. This research was supported in part by the Director, Office of Science, Office of Advanced Scientific Computing Research, of the U.S. DOE under Contract No. DE-AC02-05CH11231.

References

1. Community Petascale Project for Accelerator Science and Simulation (ComPASS), https://compass.fnal.gov
2. Amestoy, P., Rouet, F.-H., Uçar, B.: On computing arbitrary entries of the inverse of a matrix. In: SIAM Workshop on Combinatorial Scientific Computing (CSC 2009), Monterey (2009)

3. Barrett, R., Berry, M., Chan, T.F., Demmel, J., Donato, J., Dongarra, J., Eijkhout, V., Pozo, R., Romine, C., van der Vorst, H.: Templates for the solution of linear systems: Building blocks for the iterative methods. SIAM, Philadelphia (1994)
4. Duff, I., Koster, J.: The design and use of algorithms for permuting large entries to the diagonal of sparse matrices. SIAM J. Matrix Anal. Appl. 20(4), 889–901 (1999)
5. Gaidamour, J., Henon, P.: HIPS: a parallel hybrid direct/iterative solver based on a schur complement. In: Proc. PMAA (2008)
6. Gilbert, J.: Predicting structure in sparse matrix computations. SIAM J. Matrix Analysis and Applications 15, 62–79 (1994)
7. Giraud, L., Haidar, A., Watson, L.T.: Parallel scalability study of hybrid precon- ditioners in three dimensions. Parallel Computing 34, 363–379 (2008)
8. Laboratoire Bordelais de Recherche en Informatique (LaBRI). SCOTCH - Software package and libraries for graph, mesh and hypergraph partitioning, static mapping, and parallel and sequential sparse matrix block ordering,
   `http://www.labri.fr/perso/pelegrin/scotch/`
9. Lee, L.-Q., Li, Z., Ng, C.-K., Ko, K.: Omega3P: A parallel finite-element eigenmode analysis code for accelerator cavities. Technical Report SLAC-PUB-13529, Stanford Linear Accelerator Center (2009)
10. Li, X., Demmel, J.: SuperLU_DIST: A scalable distributed-memory sparse direct solver for unsymmetric linear systems. ACM Trans. Mathematical Software 29(2), 110–140 (2003)
11. Mathematics and Computer Science Division, Argonne National Laboratory. The portable, extensible, toolkit for scientific computation (PETSc),
    `http://www.mcs.anl.gov/petsc`
12. Slavova, T.: Parallel triangular solution in an out-of-core multifrontal approach for solving large sparse linear systems. PhD thesis, Institut National Polytechnique de Toulouse, Toulouse, France (2009)
13. Smith, B., Bjorstad, P., Gropp, W.: Domain Decomposition. Parallel Multilevel Methods for Elliptic Partial Differential Equations. Cambridge University Press, New York (1996)
14. Yamazaki, I., Li, X., Ng, E.: Preconditioning schur complement systems of highly- indefinite linear systems for a parallel hybrid solver. In: The Proceedings of the International Conference on Preconditioning Techniques for Scientific and Indus- trial Applications, Numer. Math. Theor. Meth. Appl., vol. 3(3), pp. 352–366 (2009)
15. Yamazaki, I., Li, X., Ng, E., Rouet, F.-H., Uçar, B.: Combinatorial issues in a paralle hybrid linear solver. In: SIAM Workshop on Parallel Processing (PP 2010), Seatle (2010)

Solving Dense Interval Linear Systems with Verified Computing on Multicore Architectures

Cleber Roberto Milani[1], Mariana Kolberg[2,*], and Luiz Gustavo Fernandes[1]

[1] GMAP - PPGCC - PUCRS
Av. Ipiranga, 6681 - Prédio 32
CEP 90619-900 - Porto Alegre, Brazil
{cleber.milani,luiz.fernandes}@pucrs.br
[2] Universidade Luterana do Brasil
Av. Farroupilha 8001 Prédio 14, sala 122
Canoas/RS, 92425-900 - Brasil
marianakolberg@gmail.com

Abstract. Automatic result verification is an important tool to reduce the impact of floating-point errors in numerical computation and to guarantee the mathematical rigor of results. One fundamental problem in Verified Computing is to find an enclosure that surely contains the exact result of a linear system. Many works have been developed to optimize Verified Computing algorithms using parallel programming techniques and message passing paradigm on clusters of computers. However, the High Performance Computing scenario changed considerably with the emergence of multicore architectures in the past few years. This paper presents an ongoing research project which has the purpose of developing a self-verified solver for dense interval linear systems optimized for parallel execution on these new architectures. The current version has obtained up to 85% of reduction at execution time and a speedup of 6.70 when solving a $15,000 \times 15,000$ interval linear system on an eight core computer.

1 Introduction and Motivation

In numerical algorithms, the correct implementation of a method does not guarantee that the computed result will be correct. Floating point arithmetic uses finite fractions to represent the real numbers, which are originally defined in Mathematics as infinite fractions. The difference between the true value and the approximation is the *roundoff error*. Floating point operations on computers are considered of *maximum accuracy* if the rounded result differs at most by one unit in the last place from the exact result. Automatic result verification is an important technique to reduce the impact of arithmetic errors in Numerical Computation [1,2]. Verified Computing guarantees the mathematical rigor of the results of a given computation by providing an interval result that surely contains the correct result. This interval result is called an *enclosure* [3].

---

* Financial support: FAPERGS research project number 0905026.

J.M.L.M. Palma et al. (Eds.): VECPAR 2010, LNCS 6449, pp. 435–448, 2011.

Interval Arithmetic provides the mathematical basis for Verified Computing. This arithmetic is based on sets of intervals, rather than sets of real numbers. Typically, the interval evaluation of an arithmetic expression such as a polynomial costs about twice as much as the evaluation of the expression in simple floating point arithmetic. However, using interval evaluation with directed roundings, the algorithm may provide a guarantee of the computed result which could not be achieved even by thousands of floating point evaluations [1,2,3].

Usually the input of a numerical method are point numbers. Engineering and scientific problems, however, are frequently modeled by numerical simulations on computers which are based on real measures that sometimes may be unprecise. To deal with uncertain data, the computer should be able to support interval input data as input data instead of point numbers and to do computations using Interval Arithmetic [3]. Dealing with uncertain data in the context of linear systems means that an interval linear system must be solved. The solution of such a system is not trivial, since an infinite number of matrices contained in the interval should be computed. However, the computation of this solution set is a NP-complete problem [17]. Thus, the only possible way to find a solution is to compute a narrow interval that contains the solution set (interval vectors) and whose overestimation decreases as the widths of the entries in A and b decrease [3,4].

There are many libraries that compute approximate solutions to point linear systems. Widely used for that purpose are the optimized software libraries LAPACK (Linear Algebra PACKage) [5] and SCALAPACK (SCAlable Linear Algebra PACKage) [6]. These libraries present a great performance and can manage to find an approximation of the correct solution, which is needed to compute the error bounds faster than verified libraries [3,18,19]. However, even when using highly optimized libraries to solve part of the verified method, finding the enclosure of an interval linear system still remains a very high computational cost task when dealing with large dense interval systems. Thus, the use of High Performance Computing (HPC) techniques appears as a useful tool to drop down the time needed to solve interval linear systems with Verified Computing [3,4].

2   Parallel and Verified Computing

Many works have been developed for implement self-validated (SV) linear systems solvers using parallel computing on clusters of computers. These works try to combine message passing paradigm programming with linear algebra libraries like ScaLAPACK. Some of these works like [7,8] operate with point input data while others like [3] propose solvers to treat uncertain input data represented by intervals matrices and vectors. A first study in the direction of multicore processors was also presented in [18]. However, this work presents an initial approach specifically for dualcores processors.

Nowadays computers are almost all built with multicore processors which cannot be considered as independently processors. Unlike architectures with multiple independent processors, multicore systems share on-chip resources and therefore cannot be considered as the new SMP [9]). On the Top500 List [10] released in November 2009, 426 systems are using quadcores processors, 59 systems use dualcores and only four systems still use single core processors. Six systems use IBMs advanced Sony PlayStation 3 processor with 9 cores and three systems are using the new 6 cores Shanghai AMD Opteron processor.

Historically, the standard parallelization approach of numerical linear algebra used by the LAPACK and ScaLAPACK libraries relied on parallel implementations of BLAS (Basic Linear Algebra Subprograms) [16]. But, although this approach solves numerous complexity problems, it also enforces a very rigid and inflexible software structure, where, at the level of linear Algebra, the algorithms are expressed in a serial way [9].

Recent research efforts are addressing this critical and highly disruptive situation. In [9], the authors present the Parallel Linear Algebra for Scalable Multicores Architectures (PLASMA) which should succeed LAPACK and ScaLA-PACK. PLASMA relies on tile algorithms, which provide fine granularity parallelism. Standard linear algebra algorithms are represented as Directed Acyclic Graphs (DAG) where nodes represent tasks and edges represent dependencies among them. This programming model enforces asynchronous and out of order scheduling of operations [9,11]. In [12], it is presented the *SuperMatrix*, a runtime system that parallelizes matrix operations for SMP and multicores architectures. The *SuperMatrix* idea is based on a number of insights gained from the FLAME project [13]. Basically, it views matrices hierarchically as blocks that serve as units of data where operations over those blocks are treated as units of computation. Thus, implementation transparently enqueues the required operations (internally tracking dependencies) and then executes the operations using out-of-order execution techniques inspired by superscalar microarchitectures. However, these optimized software libraries do not implement verified computing methods. Additionally, support for uncertain input data and interval linear systems solvers are not provided.

On the other hand, verified computing tools (such as C-XSC [17]) can provide verified results but the execution times for solving the problem are much higher since they are not developed for multicore architectures. Additional performance losses are introduced in the application by the use of special data structures and operations to implement dot scalar products in C-XSC verified methods. [8]. This effect is even worse when dealing with large interval systems [3]. In this context, this paper proposes a self-verified solver for dense interval linear systems optimized for parallel execution on multicore processors.

3   Mathematical Background

Previous researches show the Midpoint-Radius Interval Arithmetic as a good choice for implementations using floating point arithmetic [3,18,19]. The main point in using Midpoint-Radius arithmetic is that this representation allows to employ optimized algorithms and software libraries to implement operations. The use of such libraries have the striking advantages that they are available for almost every computer hardware and that they are individually adapted and tuned for specific hardware and compiler configurations. A Midpoint-Radius interval is defined as follows [14]:

$$\langle a, \alpha \rangle := \{x \in \Re / |x - a| \leq \alpha\} \quad for \quad a \in \Re, 0 \leq \alpha \in \Re \qquad (1)$$

Interval operations always satisfy the fundamental property of isotonicity. That is, if X is contained in another interval X', and Y is contained in Y', then the combination of X and Y is contained in the interval computed by combining the bigger intervals X' and Y' [2]. Sometimes the standard definition of Midpoint-Radius arithmetic causes overestimation. However, it was proved by Rump [14] that the overestimation of Midpoint-Radius Arithmetic is uniformly bounded by 1.5 for the basic arithmetic operations as well as for vector and matrix operations over $\Re$. In the case of an interval presenting a not too large radius, the factor is quantified to be near 1.

For interval vectors and interval matrices the relations $=$, $\overset{\circ}{\subset}$, and $\subseteq$ are defined component wise. The inner inclusion relation is defined by $[x] \overset{\circ}{\subset} [y] \Leftrightarrow [x]_i \overset{\circ}{\subset} [y]_i$, $i = 1, \ldots, n$ for $[x], [y] \in I\Re^n$. On the other hand, the proper subset relation is defined by $[x] \subset [y] \Leftrightarrow ([x] \subseteq [y] \, and \, [x] \neq [y])$. The *midpoint* and the *diameter* of an interval vector or matrix are also defined component wise. For example, $m([x]) = (m([x]_i))$, and $d([A]) = \left(d\left([a]_{ij} \right) \right)$, for $[x] \in I\Re^n, [A] \in I\Re^{n \times n}$.

Many algorithms for numerical verification are based on the application of well known fixed point theorems to interval sets. As an example, the Brouwer's Fixed Point Theorem is used to guarantee the convergence. Let $X = [x] \in I\Re^n$ be a machine interval vector. As a box in n-dimensional space, $[x]$ satisfies the conditions of Brouwer's Fixed Point Theorem. Supposing its possible to find a box with $f([x]) \subseteq [x]$, then $[x]$ is proved to be an enclosure of at least one fixed point x^* of f. The assertion remains valid replacing f by its floating point interval evaluation $f_\diamond$ because $f_\diamond([x]) \subseteq [x]$ implies $f([x]) \subseteq [x]$ since $f_\diamond([x])$ is a superset of $f([x])$ [2,4].

In order to achieve validated enclosures, the algorithm must enclose all sources of error that can be generated during the computation. The basic approach of many SV-methods consists of the computation of an approximate solution, local linearization and estimation of linearization and numerical errors by means of suitable theorems the assumptions of which are verified on the computer [21]. A simple mechanism for implementing these idea follows the principle of *iterative refinement*. However, it is important to mention that an interval algorithm differs significantly from the corresponding point algorithm.

The method we chose is based on the *Residual Iteration Scheme* of the *Newton-like Method*. The main reason for this choice is that besides being a well-established SV-method, it also allows the use of optimized libraries because it can be implemented with Midpoint-Radius arithmetic. The description of the method, fully given by [2], is summarized on the following. Let $Ax = b$ be a real system of equations, finding a solution of the system $Ax = b$ is equivalent to finding a zero of $f(x) = Ax - b$. Hence, Newton's method gives the following fixed point iteration scheme, where $x^{(0)}$ is some arbitrary starting value [2]:

$$x^{(k+1)} = x^{(k)} - A^{-1}\left(Ax^{(k)} - b\right), \quad k = 0, 1, \ldots \tag{2}$$

In general, the inverse of A is not exactly known. Thus, instead of A^{-1}, an approximate inverse $R \approx A^{-1}$ of A is used. Replacing the real iterates $x^{(k)}$ by interval vectors $[x]^{(k)} \in I\Re^n$, if there exists an index k with $[x]^{(k+1)} \subset [x]^{(k)}$, then, by Brouwer's Fixed point Theorem, the equation has at least one fixed point $x \in [x]^{(k)}$. Supposing R is regular, then this fixed point is also a solution of $Ax = b$. However, considering the diameter of $[x]^{(k+1)}$ the following is obtained: $d\left([x]^{(k+1)}\right) = d\left([x]^{(k)}\right) + d\left(R\left(A\,[x]^{(k+1)} - b\right)\right) \geq d\left([x]^{(k)}\right)$. Thus, in general, the subset relation will not be satisfied. For this reason, the right-hand side is modified to the following equation, where I denotes the n x n identity matrix:

$$x^{(k+1)} = Rb + (I - RA)\,x^{(k)}, \quad k = 0, 1, \ldots, \tag{3}$$

It was proved that if there exists and index k with $[x]^{(k+1)} \overset{\circ}{\subset} [x]^{(k)}$, then the matrices R and A are regular, and there is a unique solution x of the system $Ax = b$ with $x \in [x]^{(k+1)}$. This result remains valid for any matrix R. However, it is an empirical fact that the better R approximates the inverse of A, the faster the inclusion relation will be satisfied. Additionally, it is a well-known numerical principle that an approximate solution $\tilde{x}$ of $Ax = b$ may be improved by solving the system $Ay = d$, where $d = b - A\tilde{x}$ is the residual of $A\tilde{x}$. Since $y = A^{-1}(b - A\tilde{x}) = x - \tilde{x}$, the exact solution of $Ax = b$ is given by $x = \tilde{x} + y$. Therefore, the Residual Iteration Scheme is presented on Equation 4.

$$y^{(k+1)} = \underbrace{R\,(b - A\tilde{x})}_{=:z} + \underbrace{(I - RA)}_{=:C}\,y^{(k)}, \quad k = 0, 1, \ldots \tag{4}$$

The residual equation $Ay = d$ has a unique solution $y \in [y]^{(k+1)} \overset{\circ}{\subset} [y]^{(k)}$ for the corresponding interval iteration scheme. Moreover, since $y = x - \tilde{x} \in [y]^{(k+1)}$, a verified solution of the unique solution of $Ax = b$ is given by $\tilde{x} + [y]^{(k+1)}$. These results remain valid if replace the exact expressions for z and C in (4) by interval extensions. However, to avoid overestimation effects, it is highly recommended to evaluate $b - A\tilde{x}$ and $I - RA$ without any intermediate rounding [2].

4   Proposed Approach

The Residual Iteration Scheme adaptation to solve interval linear systems using Verified Computing led to Algorithm 1, proposed on [17]. Its result is a high accuracy interval vector that surely contains the correct result (enclosure). Verification process is composed by steps 5 to 15. These steps use the Midpoint-Radius arithmetic with direct roundings [2,3].

Algorithm 1. Enclosure of a square interval linear system

1: $R \approx mid\,([A])^{-1}$ {Compute an approximate inverse using LU-Decomposition algorithm}
2: $\tilde{x} \approx R.mid\,([b])$ {Compute the approximation of the solution}
3: $[z] \supseteq R\,([b] - [A]\,\tilde{x})$ {Compute enclosure for the residuum}
4: $[C] \supseteq (I - R\,[A])$ {Compute enclosure for the iteration matrix}
5: $[w] := [z]\,, k := 0$ {Initialize machine interval vector}
6: **while not** $([w] \subseteq int\,[y]\,\mathbf{or}\,k > 10)$ **do**
7:      $[y] := [w]$
8:      $[w] := [z] + [C]\,[y]$
9:      $k + +$
10: **end while**
11: **if** $[w] \subseteq int\,[y]$ **then**
12:     $\sum\,([A]\,, [b]) \subseteq \tilde{x} + [w]$ {The solution set $(\sum)$ is contained in the solution found by the method}
13: **else**
14:     No Verification
15: **end if**

4.1   Initial Implementation

An initial version of Algorithm 1 using Midpoint-Radius arithmetic was implemented and used to obtain the computational costs of each step. This implementation was developed in C++ using the Intel MKL 10.2.1.017 [15] library for optimized LAPACK and BLAS routines for Intel processors. In order to achieve better performance, the approximate inverse R and approximate solution x are calculated using only traditional floating point operations and only the midpoint matrix. Later, to compute the residuum, interval arithmetic is applied using original interval matrix $[A]$ and interval vector $[b]$ to ensure the accuracy of the result [3].

Step 1 (approximate inverse calculation using LU-Decomposition) uses the following LAPACK routines: *dgetrf*, *dlange*, *dgecon* and *dgetri*. Step 2 (approximation of the solution) is implemented by BLAS *dgemv* routine. Steps 3 and 4 compute respectively the enclosure for the residuum and enclosure for the iteration matrix. Since $[A]$ and $[b]$ as well as $[C]$ and $[z]$ are interval matrices and vectors, the enclosure computation must employ interval algorithms as defined on [14]. Let $A = \langle \tilde{a}, \alpha \rangle \in I^{+}F$ and $B = \left\langle \tilde{b}, \beta \right\rangle \in I^{+}F$ be two Midpoint-Radius intervals, the operations of addition and subtraction $C := A \circ B \in I^{+}F$, with $\circ \in \{+, -\}$ and $C = \langle \tilde{c}, \gamma \rangle$ are implemented in IEEE 754 Standard for Binary Floating point Arithmetic [22] by Algorithm 2. Similarly, the multiplication is defined by Algorithm 3. The symbols $\square$, ∇ and Δ indicate respectively the directed roundings for nearest, downward and upward.

As previously mentioned, the major advantage of Midpoint-Radius Arithmetic is to allow calculation with pure floating point operations without making any changes in the rounding mode on interim operations. Therefore, although $[C]$ and $[z]$ are intervals, they are calculated with *dgemv* and *dgemm* BLAS routines with

---

Algorithm 2. IEEE 754 Midpoint-Radius Interval Addition and Subtraction.

1: $\tilde{c} = \square \left(\tilde{a} \circ \tilde{b} \right)$

2: $\tilde{\gamma} = \Delta \left(\epsilon' \, |\tilde{c}| + \tilde{\alpha} + \tilde{\beta} \right)$

Algorithm 3. IEEE 754 Midpoint-Radius Interval Multiplication.

1: $\tilde{c} = \square \left(\tilde{a}.\tilde{b} \right)$

2: $\tilde{\gamma} = \Delta \left(\eta + \epsilon' \, |\tilde{c}| + (|\tilde{a}| + \tilde{\alpha}) \, \tilde{\beta} + \tilde{\alpha}\tilde{\beta} \right)$

---

directed roundings. The rounding mode is manipulated by *fesetround* C++ function from *fenv.h* header which supports four rounding modes: *FE_UPWARD*, *FE_DOWNWARD*, *FE_TONEAREST*, and *FE_TOWARDZERO*.

An error will be generated in the midpoint evaluation. This error should be compensated using the relative error unit. According to [14], denote the relative rounding error unit by ϵ, set $\epsilon' = \frac{1}{2}\epsilon$, and denote the smallest representable (unnormalized) positive floating point number by η. In IEEE 754 double precision $epsilon = 2^{-52}$ and $\eta = 2^{-1074}$. Therefore, the evaluation of C midpoint $(\tilde{c})$ and radius $(\tilde{\gamma})$ is given by Algorithm 4.

Algorithm 4. IEEE 754 Matrix-matrix Midpoint-Radius Interval Multiplication.

1: $\tilde{c}_1 = \nabla \left(R.mid\left(A \right) \right)$

2: $\tilde{c}_2 = \Delta \left(R.mid\left(A \right) \right)$

3: $\tilde{c} = \Delta \left(\tilde{c}_1 + 0.5 \left(\tilde{c}_2 - \tilde{c}_1 \right) \right)$

4: $\tilde{\gamma} = \Delta \left(\tilde{c} - \tilde{c}_1 \right) + |R| \, .rad\left(A \right)$

At last, steps from 5 to 15 implement the iteration to obtain the enclosure. Again, Midpoint-Radius Arithmetic and direct roundings are employed. Step 8 ($[C]$ and $[y]$ multiplication) uses BLAS *dgemv* with directed roundings. The while loop verifies if the new result is contained in the interior of the previous result. If it is true, the while loop is finished, and the enclosure was found. If not, it tries for 10 iterations to find the enclosure. It is an empirical fact that the inner inclusion is satisfied nearly after a few steps or never [2].

4.2   Initial Approach Evaluation

Two kinds of evaluations were considered around the initial implementation: accuracy and performance. Aiming at verifying the accuracy, we used an ill-conditioned matrix generated by the well known Boothroyd/Dekker formula (Equation 5) with dimension 10, which has a condition number $1.09.10^{+15}$. The radius for both matrix A and vector b were defined as $0.1.10^{-10}$.

$$A_{ij} = \binom{n+i-1}{i-1} \times \binom{n-1}{n-j} \times \frac{n}{i+j-1}, b_i = i, \forall (i,j) = 1, 2, \ldots, N \qquad (5)$$

The results found by the solver are presented in Table 1. It is important to highlight that despite the implemented solver uses Midpoint-Radius Arithmetic to do the computation, the results in Table 1 were converted to Infimum-Supremum notation to facilitate the visualization. *Exact Result* column indicates the known exact point result, *Infimum* and *Supremum* columns contain the interval bounds of the enclosure.

Table 1. Results found by implemented solver for a 10x10 Boothroyd/Dekker formula interval linear system

	Exact Result	Infimum	Supremum
0	0.0	-0.0000119	0.0000108
1	1.0	0.9998992	1.0001113
2	-2.0	-2.0005827	-1.9994736
3	3.0	2.9979758	3.0022444
4	-4.0	-4.0070747	-3.9936270
5	5.0	4.9826141	5.0193190
6	-6.0	-6.0472924	-5.9574730
7	7.0	6.9045776	7.1061843
8	-8.0	-8.2221804	-7.8004473
9	9.0	8.6064275	9.4384110

As expected, the exact result is contained in the interior of the solution set our solver computed. The interval diameter varies between 2.27×10^{-5} and 8.319835×10^{-1}. This was expected, since the Boothroyd/Dekker formula creates a very ill-conditioned system. Experiments of well-conditioned systems randomly generated with values between 0 and 1 were also performed. In these cases, the diameter was between 0 and 1×10^{-7}. The average condition number of these systems was around 6.12×10^{1}.

Performance experiments were carried out over a Intel Core 2 Duo T6400 2.00 GHz processor with 2MB L2 and 3GB of DDR2 667MHz RAM operating in dual channel. The operating system is Linux Ubuntu 9.04 (32 bits version, kernel 2.6.28-13-generic). The compiler used was gcc v. 4.3.3 with the MKL library v.10.2.1.017. The input for these experiments were linear systems randomly generated with values between 0 and 1 for A and b and a radius of 0.1×10^{-10} on both cases. The execution times of each step of the algorithm was computed. For simplicity reasons, steps from 6 to 15 were joined into one step. Table 2 presents the average execution times for each step for solving a system with dimension $n = 5,000$.

Table 2 shows that the computation of the approximate inverse R and the computation of the interval matrix C (steps 1 and 4 respectively) are the two most computational intensive operations in this algorithm. Step 1 takes more

Table 2. Average exec. times (sec) for a randomly generated system with $n = 5,000$

Task Description	Execution times
Computation of approximate inverse R (Step 1)	144.652161
Computation of approximate solution x (Step 2)	0.535467
Computation of enclosure for the residuum z of x (Step 3)	1.949728
Compute enclosure for the iteration matrix C (Step 4)	109.452704
Machine interval vector initialization (Step 5)	0.000117
Iterative refinement and inner inclusion verification (Steps 6 to 15)	4.635784
Total execution time including E/S operations	262.710470

then 55% (144.65 seconds) of the total time while Step 4 takes 42% (109.45 seconds). These two steps correspond to 97% of total processing time, and therefore, they must be carefully parallelized aiming at a better performance.

4.3 Optimized Parallel Approach

As presented in the previous subsection, steps 1 and 4 are the most time consuming operations in the algorithm. Thus, the proposed parallelization focused on these two steps as follows.

Optimization of the Approximate Inverse Calculation: Since the Newton Like Iteration requires only an approximation of R (inverse matrix of A) and once our approach employs Midpoint-Radius Interval Arithmetic, R can be computed using highly optimized software libraries. In [3], the *pdgetri* routine of ScaLAPACK was employed for R calculation. Our initial approach was implemented using analogous LAPACK routine *dgetri*. However, although MKL implementation of LAPACK is highly optimized for Intel processors, LAPACK algebra algorithms are not efficient on multicore. Hence, as expected LAPACK routines had no performance gain when increasing the number of cores.

Therefore, our strategy for Step 1 is to explore fine granularity parallelism as well as asynchronous and out of order scheduling of operations by employing the PLASMA library. However, the most actual version of PLASMA does not provide yet a matrix inversion routine. In fact, when dealing with multicore processors there are no libraries available that can be directly employed for optimized matrix inversion. Thus, the idea is exploit PLASMA *dgesv* routine.

The *dgesv* routine was developed to compute the solution of a system of linear equations. However, it is possible to operate the right hand side b of *dgesv* as a matrix and it is a well-known mathematical property that multiplying a matrix by its inverse results in the identity matrix. Considering that, we employed PLASMA *dgesv* routine passing to A and b parameters, respectively, matrix A and its identity matrix. Thus, the result computed by *dgesv* is the approximate inverse R.

It is important to mention that while packages like LAPACK and ScaLAPACK exploit parallelism within multithreaded BLAS, PLASMA uses BLAS only for high performance implementations of single core operations (often referred to as kernels). PLASMA exploits parallelism at the algorithmic level above the level of BLAS. For that reason, PLASMA must be linked with a sequential BLAS library or a multithreaded BLAS library with multithreading disabled. PLASMA must not be used in conjunction with a multithreaded BLAS, as this is likely to create more threads than actual cores, which annihilates PLASMA's performance [23]. Since our approach takes advantage of multithreaded BLAS in operations executed by other steps (like matrices multiplication) we used multithreaded MKL. To avoid affecting PLASMA performance, the function *mkl_set_num_threads* is used to dynamically control the number of threads.

Optimization of the Iteration Matrix Computation: Concerning Step 4, the computation of the iteration matrix $[C]$, the adopted strategy is to use half of the available processors to compute the interval upper bound and the other half to compute the interval lower bound. A similar strategy was successfully employed in [18] where threads were used to compute the interval bounds on a dual core processor. In that case, however, synchronization is simpler and it was not necessary to deal with load balancing.

The main idea is to utilize different threads to execute the operations in each rounding mode. This strategy avoids the frequent switching of rounding mode which is a time expensive operation. Additionally, since the cache is shared between cores, computing distinct bounds over the same data in parallel optimizes data locality. Threads are created and managed using the standard POSIX threads library [20]. Shared memory and POSIX semaphores primitives are applied for inter-thread synchronization.

Initially, a routine verifies the number of available cores and distributes the bound threads among them. Cores identified by odd numbers are assigned to upper bound computation and the even numbers to lower bound. If the number of total cores available is odd, then upper bound will be computed with one more thread than lower bound. The *cpu_set_t* variables of *sched.h* header are used to create the core pools. Threads are then statically attributed to cores by calling *sched_setaffinity* function. It is important to highlight, that defining the processor affinity instructs the operating system kernel scheduler to do not change the processor used by one particular thread.

After threads are assigned to processors they start setting their rounding modes and get blocked by semaphores until the main flow releases them all at once. On the sequence, each thread calls the *dgemm* BLAS routine for the matrix-matrix multiplication. The main flow blocks itself with a semaphore until the computation of upper and lower bounds ends. Once the computation of $[C]$ is completed, threads send signals to unblock main flow semaphore, which then follows to next step.

4.4 Optimized Approach Evaluation

In order to verify the benefits of employed optimizations, two kind of experiments were performed. The first concerns the correctness of the result. The second experiment was done to evaluate the speedup improvement brought by the proposed method. The evaluations were executed in a 2 processors quad-core Intel Xeon E5520 2.27 GHz with 128 KB L1, 1MB L2, 8MB L3 shared and 16 GB of DDR3 1066 MHz RAM. The operating system is Linux Ubuntu 9.04 (kernel 2.6.28-11-server). The compiler used was gcc v. 4.3.3 along with the libraries MKL 10.2.2.025 and PLASMA 2.1.0.

Once modifications were done in the algorithm, we conducted some experiments with the same well-conditioned and ill-conditioned matrices solved by our initial approach to confirm that there were no accuracy loss on the result. The tests generated by the Boothroyd/Dekker formula presented almost the same accuracy on both versions (initial and optimized). As required by the algorithm, both interval results contain the exact result. For well-conditioned matrices, both implementations give exactly the same results.

We carried out performance experiments for matrices dimensions from 1,000 to 15,000. Table 3 presents the execution times in seconds for each algorithm step when solving a random 15,000 × 15,000 interval linear system varying the number of cores. Column *Imp.* refers to the approach where *In.* is the initial implementation and *Op.* is the optimized version. *Cores* column indicates the number of cores employed in that execution and columns *Step 1..15* refer to the algorithms steps in the same way as in Table 2. As we had a small standard deviation, we just run the solver 10 times for each situation. The highest and lowest execution times were removed and the final times were obtained by calculating the arithmetic mean of remaining times.

Table 3. Execution times in seconds to solve a 15,000 × 15,000 interval linear system

Imp.	Cores	Step 1	Step 2	Step 3	Step 4	Step 5	Step 6–15	Total
In.	1	1,905.4969	8.3916	23.6316	2,204.0620	0.0002	73.0728	4,488.7467
Op.	1	1,147.8716	5.6718	19.1374	2,218.5130	0.0002	70.4101	3,461.6043
Op.	2	575.8929	5.7024	19.3800	1,169.3131	0.0002	64.8068	1,835.0956
Op.	3	387.9773	5.6246	18.1846	1,058.4973	0.0002	68.9738	1,539.2580
Op.	4	292.6839	5.6923	19.4538	646.0218	0.0002	32.4533	996.3056
Op.	5	249.2505	5.5699	18.1954	626.2732	0.0002	34.9536	934.2430
Op.	6	209.5168	5.7400	19.3016	493.2460	0.0002	33.6801	761.4850
Op.	7	182.3047	5.5987	17.8858	474.7029	0.0002	34.1734	714.6659
Op.	8	160.8892	5.6867	18.9293	451.5206	0.0002	32.9934	670.0196

Figure 1 shows the speedups obtained from the execution times presented in Table 3. Line *Sp T.T.Seq.* is the speedup of total execution time comparing optimized implementation running in n cores to the initial approach in 1 core (i.e., $\frac{T\_Op.(n)}{T\_In.(1)}$). *Sp T.T.Par.* concerns to optimized total time in n cores compared to optimized algorithm executing in 1 core (i.e., $\frac{T\_Op.(n)}{T\_Op.(1)}$). *Sp Inv. Seq.* and

Sp Inv. Par. illustrate speedups obtained in an analogous manner considering only the Step 1 execution time. *Sp S4. Par.* presents the speedup for Step 4 of algorithm.

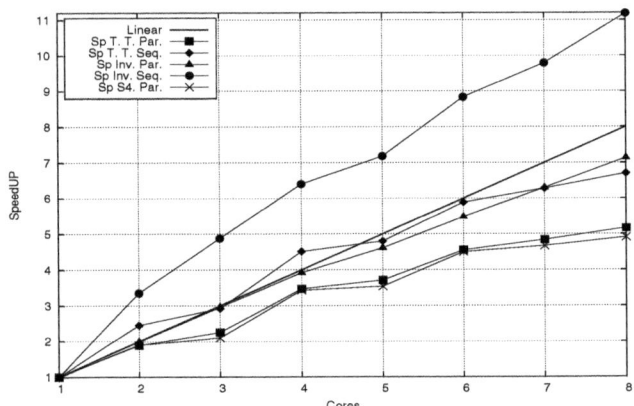

Fig. 1. *Speedups* obtained solving an interval linear system of size 15,000 × 15,000

In Table 3 and Figure 1 it is possible to see a significant reduction in the execution time. *Sp T.T.Seq.* initially presented super linear speed up and slowly decreased until 6.70 for 8 cores, which is a expected result due to scalability issues like as the influence of sequential portions of code. *Sp T.T.Par.* also presented high speedups and a similar behavior. The main reason for this difference is the Step 1 of the algorithm. The optimized implementation running in one core spent only 60% of the time spent by the initial approach. This happens because PLASMA optimizations not boil down only to the parallelism but also to new algorithmic approaches for data management and tasks scheduling which are more suitable for multicore architectures.

Sp Inv. Seq. and *Sp Inv. Par.* can be explained by these same reasons. It is important to note that Step 1 computed with LAPACK *dgetri* routine on 8 cores spent 1,864.1687 seconds, which means a speedup of 1.02 and confirms that this is not suitable for multicore.

Sp S4. Par. presented good speedups too. We suppose that this is due to cache effects. In the sequential version, all matrix elements must be loaded in the cache to compute [C] with rounding-up, and after that, again, to compute it with rounding-down. If the entire matrix does not fit in the cache, there will be many cache misses for each rounding mode. Multithreaded version allows a more effective utilization of the available cache memory because more threads use the same data at the same time resulting in a better speedup as expected.

At last, verification steps (6–15) although not explicit parallelized showed performance gains too. The reason is that the use of *dgemm* routine benefits from multithreaded MKL.

5   Considerations and Future Work

This paper presented the current version of a self-verified solver for dense interval linear systems optimized for parallel execution on multicore architectures. The implementation delivered enclosures of the correct solutions for interval input data with considerable accuracy. The computational costs of each of its intermediate steps were computed and the main time consuming steps among them were optimized aiming at obtaining performance gain on multicore processors. The proposed solution led to a scalable implementation which has achieved up to 85% of reduction at execution time when solving a 15,000 × 15,000 interval linear system over an eight core computer.

Its important to mention that the presented solver was written for dense systems. However, sparse systems are also supported although they will be treated as a dense system. No special method or data storage is used concerning the sparsity of these systems. Many performance related issues still remain under investigation. There is a clear tradeoff between the overhead incurred by thread synchronization and the performance gain, which affects the solver scalability. Therefore, future directions includes the investigation on how to optimize the parallelized steps, the identification of other parts of the algorithm to parallelize and the exploitation of new architectures as the hybrid computers that mix GPUs and multicore processing.

The ability of finding verified results for dense linear systems of equations increases the result accuracy. The possibility to perform this computation in multicore architectures reduces the computational time that verified computing need through the benefits of high performance computing. Therefore, the use of verified and high performance computing together appears as a suitable way to increase the reliability and performance of many applications, specially when those applications deal with uncertain data.

References

1. Hayes, B.: A Lucid Interval. American Scientist 91(6), 484–488 (2003)
2. Hammer, R., Ratz, D., Kulisch, U., Hocks, M.: C++ Toolbox for Verified Scientific Computing I: Basic Numerical Problems. Springer, New York (1997)
3. Kolberg, M., Dorn, M., Bohlender, G., Fernandes, L.G.: Parallel Verified Linear System Solver for Uncertain Input Data. In: Proceedings of 20th SBAC-PAD - International Symposium on Computer Architecture and High Performance Computing, pp. 89–96 (2008)
4. Kearfott, R.: Interval Computations: Introduction, Uses, and Resources. Euromath Bulletin 2(1), 95–112 (1996)
5. Demmel, J.: LAPACK: A Portable Linear Algebra Library for Supercomputers. In: Proceedings of IEEE Control Systems Society Workshop on Computer-Aided Control System Design, pp. 1–7 (1989)
6. Choi, J., Demmel, J., Dhillon, I., Dongarra, J., Ostrouchov, S., Petitet, A., Stanley, K., Walker, D., Whaley, R.: ScaLAPACK: a Portable Linear Algebra Library for Distributed Memory Computers – Design Issues and Performance. Computer Physics Communications 97(1-2), 1–15 (1996)

7. Kolberg, M., Baldo, L., Velho, P., Fernandes, L.G., Claudio, D.: Optimizing a Parallel Self-verified Method for Solving Linear Systems. In: Kågström, B., Elmroth, E., Dongarra, J., Waśniewski, J. (eds.) PARA 2006. LNCS, vol. 4699, pp. 949–955. Springer, Heidelberg (2007)
8. Kolberg, M., Fernandes, L.G., Claudio, D.: Dense Linear System: A Parallel Self-verified Solver. International Journal of Parallel Programming 36(4), 412–425 (2008)
9. Buttari, A., Dongarra, J., Kurzak, J., Langou, J., Luszczek, P., Tomov, S.: The Impact of Multicore on Math Software. In: Kågström, B., Elmroth, E., Dongarra, J., Waśniewski, J. (eds.) PARA 2006. LNCS, vol. 4699, pp. 1–10. Springer, Heidelberg (2007)
10. TOP 500 Supercomputing Home Page, http://www.top500.org/ (accessed on December 11, 2009)
11. Agullo, E., Hadri, B., Ltaief, H., Dongarra, J.: Comparative Study of One-Sided Factorizations with Multiple Software Packages on Multi-Core Hardware. LAPACK Working Note 217, ICL, UTK (2009)
12. Chan, E., Van Zee, F., Bientinesi, P., Quintana-Orti, E., Quintana-Orti, G., van de Geijn, R.: SuperMatrix: a Multithreaded Runtime Scheduling System for Algorithms-by-blocks. In: Proceedings of the 13th ACM SIGPLAN Symposium on Principles and Practice of Parallel Programming, pp. 123–132 (2008)
13. Gunnels, J., Gustavson, F., Henry, G., van de Geijn, R.: FLAME: Formal Linear Algebra Methods Environment. ACM Transactions on Mathematical Software 27(4), 422–455 (2001)
14. Rump, S.M.: Fast and Parallel Interval Arithmetic. BIT Numerical Mathematics 39(3), 534–554 (1999)
15. Intel Math Kernel Library Home Page, http://software.intel.com/en-us/intel-mkl/ (accessed on December 11, 2009)
16. Lawson, C., Hanson, R., Kincaid, D., Krogh, F.: Basic Linear Algebra Subprograms for Fortran Usage. ACM Transactions on Mathematical Software 5(3), 308–323 (1979)
17. Klatte, R., Kulisch, U., Lawo, C., Rauch, R., Wiethoff, A.: C-XSC - A C++ Class Library for Extended Scientific Computing. Springer, Heidelberg (1993)
18. Kolberg, M., Cordeiro, D., Bohlender, G., Fernandes, L.G., Goldman, A.: A Multithreaded Verified Method for Solving Linear Systems in Dual-Core Processors. In: 9th PARA - International Workshop on State-of-the-Art in Scientific and Parallel Computing, Trondheim - Noruega. PARA 2008 - Revised Selected Papers. LNCS. Springer, Heidelberg (2008) (to appear)
19. Kolberg, M., Bohlender, G., Claudio, D.M.: Improving the Performance of a Verified Linear System Solver Using Optimized Libraries and Parallel Computing. In: Palma, J.M.L.M., Amestoy, P.R., Daydé, M., Mattoso, M., Lopes, J.C. (eds.) VECPAR 2008. LNCS, vol. 5336, pp. 13–26. Springer, Heidelberg (2008)
20. Butenhof, D.R.: Programming with POSIX Threads. Addison-Wesley Longman Publishing Co., Inc., Boston (1997)
21. Rump, S.M.: Self-validating Methods. Linear Algebra and Its Applications 324(1-3), 3–13 (2001)
22. ANSI/IEEE. A Standard for Binary Floating-point Arithmetic, Std.754-1985. American National Standards Institute / Institute of Electrical and Eletronics Engineers. USA (1985)
23. PLASMA README, http://icl.cs.utk.edu/projectsfiles/plasma/html (accessed on April 8, 2010)

TRACEMIN-Fiedler: A Parallel Algorithm for Computing the Fiedler Vector

Murat Manguoglu[1], Eric Cox[2], Faisal Saied[2], and Ahmed Sameh[2]

[1] Istanbul Technical University,
Informatics Institute,
Computational Science and Engineering Program
34469-Maslak, Istanbul, Turkey
[2] Purdue University,
Department of Computer Science,
305 N. University Street,
West Lafayette, IN 47907

Abstract. The eigenvector corresponding to the second smallest eigenvalue of the Laplacian of a graph, known as the Fiedler vector, has a number of applications in areas that include matrix reordering, graph partitioning, protein analysis, data mining, machine learning, and web search. The computation of the Fiedler vector has been regarded as an expensive process as it involves solving a large eigenvalue problem. We present a novel and efficient parallel algorithm for computing the Fiedler vector of large graphs based on the Trace Minimization algorithm. We compare the parallel performance of our method with a multilevel scheme, designed specifically for computing the Fiedler vector, which is implemented in routine MC73_FIEDLER of the Harwell Subroutine Library (HSL).

1 Introduction

The second smallest eigenvalue and the corresponding eigenvector of the Laplacian of a graph have been used in a number of application areas including matrix reordering [10,9,8,1], graph partitioning [12,13], machine learning [11], protein analysis and data mining [5,16], and web search [4]. The second smallest eigenvalue of the Laplacian of a graph is sometimes called *the algebraic connectivity of the graph*, and the corresponding eigenvector is known as the *Fiedler vector*, due to the pioneering work of Fiedler [3].

For a given $n \times n$ sparse symmetric matrix A, or an undirected weighted graph with positive weights, one can form the weighted-Laplacian matrix, L_w, as follows:

$$L_w(i,j) = \begin{cases} \sum_{\hat{j}} |A(i,\hat{j})| & \text{if } i = j, \\ -|A(i,j)| & \text{if } i \neq j. \end{cases} \tag{1}$$

One can obtain the unweighted Laplacian by simply replacing each nonzero element of the matrix A by 1. In this paper, we focus on the more general weighted case; the method we present is also applicable to the unweighted Laplacian. Since the Fiedler vector can be computed independently for disconnected graphs, we assume that the graph is

J.M.L.M. Palma et al. (Eds.): VECPAR 2010, LNCS 6449, pp. 449–455, 2011.

connected. The eigenvalues of L_w are $0 = \lambda_1 < \lambda_2 \leq \lambda_3 \leq ... \leq \lambda_n$. The eigenvector x_2 corresponding to smallest nontrivial eigenvalue λ_2 is the sought Fiedler vector.

A state of the art multilevel solver [7] called MC73_FIEDLER for computing the Fiedler vector is implemented in the Harwell Subroutine Library(HSL) [6]. It uses a series of levels of coarser graphs where the eigenvalue problem corresponding to the coarsest level is solved via the Lanczos method for estimating the Fiedler vector. The results are then prolongated to the finer graphs and Rayleigh Quotient Iterations (RQI) with shift and invert are used for refining the eigenvector. Linear systems encountered in RQI are solved via the SYMMLQ algorithm.

We describe a novel parallel solver: TRACEMIN-Fiedler based on the Trace Minimization algorithm (TRACEMIN) [15,14] in Section 2 and present results in Section 3 comparing it to MC73_FIEDLER.

2 The TRACEMIN-Fiedler Algorithm

We consider solving the standard symmetric eigenvalue problem

$$\mathbf{L}x = \lambda x \tag{2}$$

where L denotes the weighted Laplacian, using the TRACEMIN scheme for obtaining the Fiedler vector. The basic TRACEMIN algorithm [15,14] can be summarized as follows. Let $\mathbf{X}_k$ be an approximation of the eigenvectors corresponding to the p smallest eigenvalues such that $\mathbf{X}_k^T \mathbf{L}\mathbf{X}_k = \Sigma_k$ and $\mathbf{X}_k^T \mathbf{X}_k = \mathbf{I}$, where $\Sigma_k = diag(\rho_1^{(k)}, \rho_2^{(k)}, ..., \rho_p^{(k)})$. The updated approximation is obtained by solving the minimization problem

$$\min \operatorname{tr}(\mathbf{X}_k - \Delta_k)^T \mathbf{L}(\mathbf{X}_k - \Delta_k), \quad \text{subject to } \Delta_k^T \mathbf{X}_k = 0. \tag{3}$$

This in turn leads to the need for solving a saddle point problem, in each iteration of the TRACEMIN algorithm, of the form

$$\begin{bmatrix} \mathbf{L} & \mathbf{X}_k \\ \mathbf{X}_k^T & 0 \end{bmatrix} \begin{bmatrix} \Delta_k \\ \mathbf{N}_k \end{bmatrix} = \begin{bmatrix} \mathbf{L}\mathbf{X}_k \\ 0 \end{bmatrix}. \tag{4}$$

Once Δ_k is obtained $(\mathbf{X}_k - \Delta_k)$ is then used to obtain $\mathbf{X}_{k+1}$ which forms the section $\mathbf{X}_{k+1}^T \mathbf{L}\mathbf{X}_{k+1} = \Sigma_{k+1}, \mathbf{X}_{k+1}^T \mathbf{X}_{k+1} = \mathbf{I}$. The TRACEMIN-Fiedler algorithm, which based on the basic TRACEMIN algorithm, is given in Figure 1.

In step 4 the columns of the matrix $\mathbf{X}_k$ are orthonormal because columns of $\mathbf{V}_k$ and $\mathbf{Y}_k$ are orthonormal. The most time consuming part of the algorithm is solving the saddle-point problem in each outer TRACEMIN iteration. This involves, in turn, solving large sparse symmetric positive semi-definite systems of the form $\mathbf{L}\mathbf{W}_k = \mathbf{X}_k$ using the Conjugate Gradient algorithm with a diagonal preconditioner. Although it is possible to use other preconditioners, we chose diagonal preconditioner for: (i) its scalability and (ii) effectiveness. Our main enhancement of the basic TRACEMIN scheme are contained in step 8, solving systems involving the Laplacian, and step 7 concerning the deflation process. In the TRACEMIN-Fiedler algorithm, not only is the coefficient matrix L guaranteed to be symmetric positive semi-definite, but that its diagonal

---

Algorithm 1:

Data: $\mathbf{L}$ is the $n \times n$ Laplacian matrix defined in Eqn.(1) , ε_{out} is the stopping criterion for the $||.||_\infty$ of the eigenvalue problem residual

Result: x_2 is the eigenvector corresponding to the second smallest eigenvalue of $\mathbf{L}$

$p \longleftarrow 2; \quad q \longleftarrow 3p$;

$n_{conv} \longleftarrow 0; \quad \mathbf{X}_{conv} \longleftarrow [\quad]$;

$\hat{\mathbf{L}} \longleftarrow \mathbf{L} + ||\mathbf{L}||_\infty 10^{-12} \times \mathbf{I}$;

$\mathbf{D} \longleftarrow$ the diagonal of $\mathbf{L}$;

$\hat{\mathbf{D}} \longleftarrow$ the diagonal of $\hat{\mathbf{L}}$;

$\mathbf{X}_1 \longleftarrow rand(n,q)$;

for $k = 1, 2, \ldots max\_it$ **do**

    1. Orthonormalize $\mathbf{X}_k$ into $\mathbf{V}_k$;

    2. Compute the interaction matrix $\mathbf{H}_k \longleftarrow \mathbf{V}_k^T \mathbf{L} \mathbf{V}_k$;

    3. Compute the eigendecomposition $\mathbf{H}_k \mathbf{Y}_k = \mathbf{Y}_k \Sigma_k$ of $\mathbf{H}_k$. The eigenvalues Σ_k are arranged in ascending order and the eigenvectors are chosen to be orthogonal;

    4. Compute the corresponding Ritz vectors $\mathbf{X}_k \longleftarrow \mathbf{V}_k \mathbf{Y}_k$;

    Note that $\mathbf{X}_k$ is a section, i.e. $\mathbf{X}_k^T \mathbf{L} \mathbf{X}_k = \Sigma_k, \mathbf{X}_k^T \mathbf{X}_k = \mathbf{I}$;

    5. Compute the relative residual $||\mathbf{L}\mathbf{X}_k - \mathbf{X}_k \Sigma_k||_\infty / ||\mathbf{L}||_\infty$;

    6. Test for convergence: If the relative residual of an approximate eigenvector is less than ε_{out}, move that vector from $\mathbf{X}_k$ to $\mathbf{X}_{conv}$ and replace n_{conv} by $n_{conv} + 1$ increment. If $n_{conv} \geq p$, stop;

    7. Deflate: If $n_{conv} > 1, \mathbf{X}_k \longleftarrow \mathbf{X}_k - \mathbf{X}_{conv}(\mathbf{X}_{conv}^T \mathbf{X}_k)$;

    8. **if** $k = 1$ **then**

        | Solve the linear system $\hat{\mathbf{L}} \mathbf{W}_k = \mathbf{X}_k$ approximately via the PCG scheme using the diagonal preconditioner $\hat{\mathbf{D}}$;

    else

        | Solve the linear system $\mathbf{L} \mathbf{W}_k = \mathbf{X}_k$ approximately via the PCG scheme using the diagonal preconditioner $\mathbf{D}$;

    9. Form the Schur complement $\mathbf{S}_k \longleftarrow \mathbf{X}_k^T \mathbf{W}_k$;

    10. Solve the linear system $\mathbf{S}_k \mathbf{N}_k = \mathbf{X}_k^T \mathbf{X}_k$ for $\mathbf{N}_k$ directly;

    11. Update $\mathbf{X}_{k+1} \longleftarrow \mathbf{X}_k - \Delta_k = \mathbf{W}_k \mathbf{N}_k$;

Fig. 1. TRACEMIN-Fiedler algorithm

(the preconditioner) is guaranteed to have positive elements. On the other hand, in MC73_FIEDLER there is no guarantee that the linear systems, arising in the RQI with shift and invert, are symmetric positive semi-definite with positive diagonal elements. Hence, MC73_FIEDLER uses SYMMLQ without any preconditioning to solve linear systems in the Rayleigh Quotient Iterations.

We should note here that the matrix $\mathbf{L}$ is symmetric positive semi-definite with one zero eigenvalue. As soon as the first eigenvalue has converged, however, the right hand side $\mathbf{X}_k$ is orthogonal to the null space of $\mathbf{L}$ due to the deflation step 7. Since the smallest (i.e. 0) eigenvalue converges after the first iteration of the algorithm we add a small diagonal perturbation for the first iteration of the algorithm only in order to ensure PCG will not fail.

The order of the linear system in step 10 is $q \times q$ where $q = 6$, therefore we solve these small systems directly. We note that our algorithm can easily compute additional eigenvectors of the Laplacian matrix by setting p to be the number of desired of smallest eigenpairs.

3 Parallel Implementation of TRACEMIN-Fiedler

The parallel TRACEMIN-Fiedler algorithm consists of the same basic steps as the serial algorithm 1. The matrix and vectors are partitioned and distributed in block rows across the processors. Our parallel implementation is based on the MPI communication library.

One critical part of the parallelization is the matrix vector product. Due to the block nature of the TRACEMIN algorithm, the matrix L is applied to a set of vectors at a time, which leads to greater efficiency. The amount of communication needed in the matrix vector product is problem dependent. The scalability of this operation and therefore of the overall parallel TRACEMIN-Fiedler algorithm varies depending on the number of non-zeros in L and their location. The parallel matrix-vector multiplication operation is performed in Step 2, for the computation of $\mathbf{H}_k$, in Step 5, for computing the residuals, and once in each iteration of the PCG solve in Step 8.

The other type of communication needed in the parallel TRACEMIN-Fiedler algorithm is the AllReduce operation. This is required in the computation of dot products and norms. In particular, the AllReduce communication is performed in Step 1, for the orthonormalization step, in Step 2, for the computation of $\mathbf{H}_k$, on Step 5, in the computation of the residual norms, in Step 7, for the deflation operation and in Step 9, in the computation of the Schur complement matrix. The AllReduce communication operation is performed three times in each iteration of the PCG solve in Step 8. In our implementation, most AllReduce operations are applied to a set of vectors, which is more efficient than doing more reductions one at a time.

4 Numerical Results

We implement the TRACEMIN-Fiedler algorithm in Figure 1 in parallel using MPI. We compare the parallel performance of MC73_FIEDLER with TRACEMIN-Fiedler using a cluster with Infiniband interconnection where each node consists of two quad-core Intel Xeon CPUs (X5560) running at 2.80GHz (8 cores per node). For both solvers we set the stopping tolerance for the $\infty - norm$ of the eigenvalue problem residual to 10^{-5}. In TRACEMIN-Fiedler we set the inner stopping criterion as $\varepsilon_{in} = 10^{-1} * \varepsilon_{out}$, and the maximum number of the preconditioned CG to 30. For MC73_FIEDLER, we use all the default parameters.

The set of test matrices are obtained from the University of Florida (UF) Sparse Matrix Collection [2]. A search for matrices in this collection which are square, real, and which are of order $2,000,000 < N < 5,000,000$ returns four matrices listed in Table 1. If a matrix, A, is nonsymmetric we use $(|A| + |A^T|)/2$, instead. Furthermore, if the adjacency graph of A has any disconnected single vertices we removed them since those vertices are independent and have trivial solutions. We apply both MC73_FIEDLER and TRACEMIN-Fiedler to the weighted Laplacian generated from the adjacency graph

Table 1. Matrix size (N), number of nonzeros (nnz), and type of test matrices

Matrix Group/Name	N	nnz	symmetric
1. Rajat/rajat31	$4,690,002$	$20,316,253$	no
2. Schenk/nlpkkt	$3,542,400$	$95,117,792$	yes
3. Freescale/Freescale1	$3,428,755$	$17,052,626$	no
4. Zaoui/kktPower	$2,063,494$	$12,771,361$	yes

of the preprocessed matrix where the weights are the absolute values of matrix entries. After obtaining the Fiedler vector x_2 returned by each algorithm, we compute the corresponding eigenvalue λ_2,

$$\lambda_2 = \frac{x_2^T L x_2}{x_2^T x_2}. \tag{5}$$

We report the relative residuals $\|Lx_2 - \lambda_2 x_2\|_\infty / \|L\|_\infty$ in Table 2.

Table 2. Relative residuals $\|Lx - \lambda x\|_\infty / \|L\|_\infty$ for TRACEMIN-Fiedler and MC73_FIEDLER where $\varepsilon_{out} = 10^{-5}$

Matrix/Cores	TRACEMIN-Fiedler				MC73_FIEDLER
	1	8	16	32	1
rajat31	1.1×10^{-12}	1.1×10^{-12}	1.1×10^{-12}	1.1×10^{-12}	3.03×10^{-9}
nlpkkt	9.1×10^{-6}	9.1×10^{-6}	9.1×10^{-6}	9.1×10^{-6}	6.49×10^{-7}
Freescale1	7.5×10^{-12}	7.5×10^{-12}	7.5×10^{-12}	7.5×10^{-12}	1.03×10^{-7}
kktPower	3.1×10^{-24}	3.1×10^{-24}	3.1×10^{-24}	3.1×10^{-24}	4.07×10^{-8}

Table 3. Total time in seconds (rounded to the first decimal place) for TRACEMIN-Fiedler and MC73_FIEDLER

Matrix/Cores	TRACEMIN-Fiedler				MC73_FIEDLER
	1	8	16	32	1
rajat31	5.6	1.4	0.7	0.4	81.5
nlpkkt	100.5	24.9	15.3	10.8	83.9
Freescale1	61.5	23.5	16.0	12.5	52.8
kktPower	4.8	1.0	0.7	0.5	341.6

The total time required by TRACEMIN-Fiedler using 1, 8, 16, and 32 mpi processes (on 1,1,2,4 nodes, respectively) are presented in Table 3. We note that there is one core used per mpi processes. We emphasize that the parallel scalability results for TRACEMIN-Fiedler is preliminary and that there is more room for improvement. Since MC73_FIEDLER is purely sequential we have used it on a single core. The speed improvements realized by TRACEMIN-Fiedler on 1, 8, 16, and 32 cores over MC73_FIEDLER on a single core are depicted in Figure 2, with the actual solve times and the speed improvement values are given in Tables 3 and 4. Note that on 32 cores, our scheme realizes speed improvements over MC73_FIEDLER that range between 4 and 641 for our four test matrices. A component of the algorithm that has significant

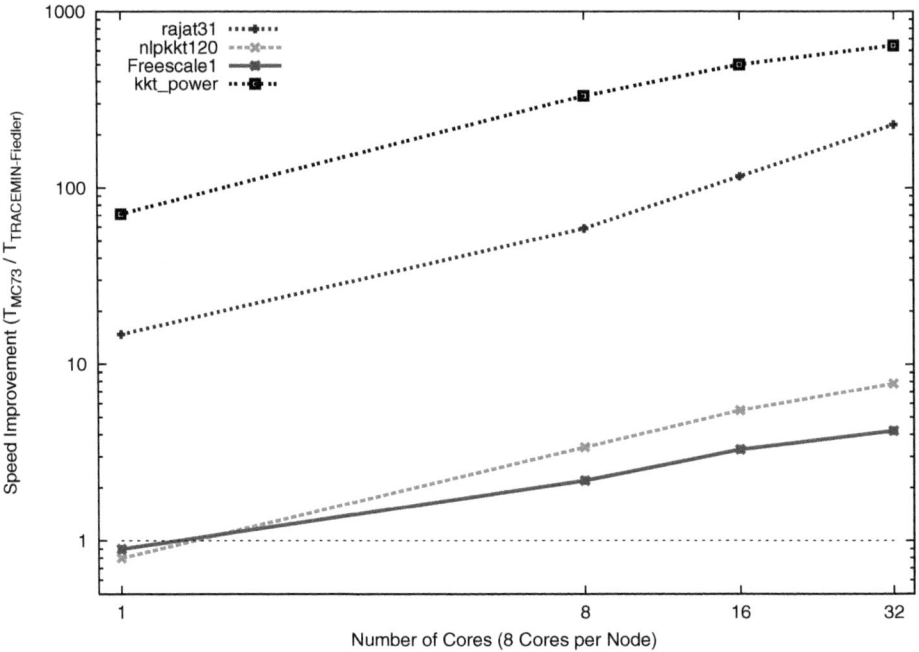

Fig. 2. Speed improvement of TRACEMIN-Fiedler compared to uniprocessor HSL_MC73 for four test problems

Table 4. Speed improvement over MC73_FIEDLER ($T_{MC73\_FIEDLER}/T$)

Matrix/Cores	TRACEMIN-Fiedler				MC73_FIEDLER
	1	8	16	32	1
rajat31	14.5	59.2	116.5	227.5	1.0
nlpkkt	0.8	3.4	5.5	7.8	1.0
Freescale1	0.9	2.2	3.3	4.2	1.0
kktPower	71.2	332.3	501.0	641.4	1.0

influence on the scalability is the sparse matrix-vector multiplication routine. Therefore, due to the differences in the sparsity patterns, test examples exhibit varying degrees of scalability.

5   Conclusions

We have presented a new algorithm for computing the Fiedler vector on parallel computing platforms, and have shown its effectiveness compared to the well-known scheme given by routine MC73_FIEDLER of the Harwell Subroutine Library for computing the Fiedler vector of four large sparse matrices.

References

1. Barnard, S.T., Pothen, A., Simon, H.: A spectral algorithm for envelope reduction of sparse matrices. Numerical Linear Algebra with Applications 2(4), 317–334 (1995)
2. Davis, T.A.: University of Florida sparse matrix collection. NA Digest (1997)
3. Fiedler, M.: Algebraic connectivity of graphs. Czechoslovak Mathematical Journal 23(2), 298–305 (1973)
4. He, X., Zha, H., Ding, C.H.Q., Simon, H.D.: Web document clustering using hyperlink structures. Computational Statistics & Data Analysis 41(1), 19–45 (2002)
5. Higham, D.J., Kalna, G., Kibble, M.: Spectral clustering and its use in bioinformatics. Journal of Computational and Applied Mathematics, Special issue dedicated to Professor Shinnosuke Oharu on the occasion of his 65th birthday 204(1), 25–37 (2007)
6. HSL. A collection of Fortran codes for large-scale scientific computation (2004), `http://www.cse.scitech.ac.uk/nag/hsl/`
7. Hu, Y.F., Scott, J.A.: HSL_MC73: a fast multilevel Fiedler and profile reduction code. Technical Report RAL-TR-2003-036 (2003)
8. Manguoglu, M.: A parallel hybrid sparse linear system solver. In: Computational Electromagnetics International Workshop, 2009. CEM 2009, pp. 38–43 (July 2009)
9. Manguoglu, M., Koyutürk, M., Sameh, A.H., Grama, A.: Weighted matrix ordering and parallel banded preconditioners for iterative linear system solvers. SIAM Journal on Scientific Computing 32(3), 1201–1216 (2010)
10. Manguoglu, M., Sameh, A.H., Schenk, O.: PSPIKE: A Parallel Hybrid Sparse Linear System Solver. In: Sips, H., Epema, D., Lin, H.-X. (eds.) Euro-Par 2009. LNCS, vol. 5704, pp. 797–808. Springer, Heidelberg (2009)
11. Ng, A.Y., Jordan, M.I., Weiss, Y.: On spectral clustering: Analysis and an algorithm. In: Advances in Neural Information Processing Systems 14, pp. 849–856. MIT Press, Cambridge (2001)
12. Pothen, A., Simon, H.D., Liou, K.-P.: Partitioning sparse matrices with eigenvectors of graphs. SIAM J. Matrix Anal. Appl. 11(3), 430–452 (1990)
13. Qiu, H., Hancock, E.R.: Graph matching and clustering using spectral partitions. Pattern Recognition 39(1), 22–34 (2006)
14. Sameh, A., Tong, Z.: The trace minimization method for the symmetric generalized eigenvalue problem. J. Comput. Appl. Math. 123(1-2), 155–175 (2000)
15. Sameh, A.H., Wisniewski, J.A.: A trace minimization algorithm for the generalized eigenvalue problem. SIAM Journal on Numerical Analysis 19(6), 1243–1259 (1982)
16. Shepherd, S.J., Beggs, C.B., Jones, S.: Amino acid partitioning using a fiedler vector model. Journal European Biophysics Journal 37(1), 105–109 (2007)

Applying Parallel Design Techniques to Template Matching with GPUs

Robert Finis Anderson, J. Steven Kirtzic*, and Ovidiu Daescu**

University of Texas at Dallas
Richardson, TX, USA
{rfa061000,jsk061000,daescu}@utdallas.edu

Abstract. Designing algorithms for data parallelism can create significant gains in performance on SIMD architectures. The performance of General Purpose GPUs can also benefit from careful analysis of memory usage and data flow due to their large throughput and system memory bottlenecks. In this paper we present an algorithm for template matching that is designed from the beginning for the GPU architecture and achieves greater than an order of magnitude speedup over traditional algorithms designed for the CPU and reimplemented on the GPU. This shows that it is not only desirable to adapt existing algorithms to run on GPUs, but also that future algorithms should be designed with the GPU architecture in mind.

1 Introduction

The advent of massively multiprocessor GPUs has opened a floodgate of opportunities for parallel processing applications, ranging from cutting-edge gaming graphics to the efficient implementation of classic algorithms [1]. In this paper please note that we will often refer to the machine containing the GPU as the "host" and the GPU itself as the "device".

Figure 1 depicts the structure of the NVIDIA GeForce 8800 series as an example of a typical GPGPU (General Purpose GPU) device. The GeForce 8800 contains 16 multiprocessors, each containing 8 semi-independent cores for a total of 128 processing units. Each of the 128 processors can run as many as 96 threads concurrently, for a maximum of 12,288 threads executing in parallel. The computing model is SIMD (single instruction multiple data), and the memory model is a NUMA (non-uniform memory access) with a semi-shared address space. This stands in contrast to a modern desktop or server PC's CPU, which is typically either SISD (single instruction single data) or MIMD (multi-instruction multiple data), in the case of a multi-processor or multi-core machine. Additionally, from the perspective of the programmer, all memory is explicitly shared (in multi-threading environments) or explicitly separate (in multi-processing environments) on a desktop machine.

---

* Kirtzic's research was supported by NSF award 0742477.
** Daescu's research was partially supported by NSF awards CCF-0635013 and CNS-1035460.

J.M.L.M. Palma et al. (Eds.): VECPAR 2010, LNCS 6449, pp. 456–468, 2011.

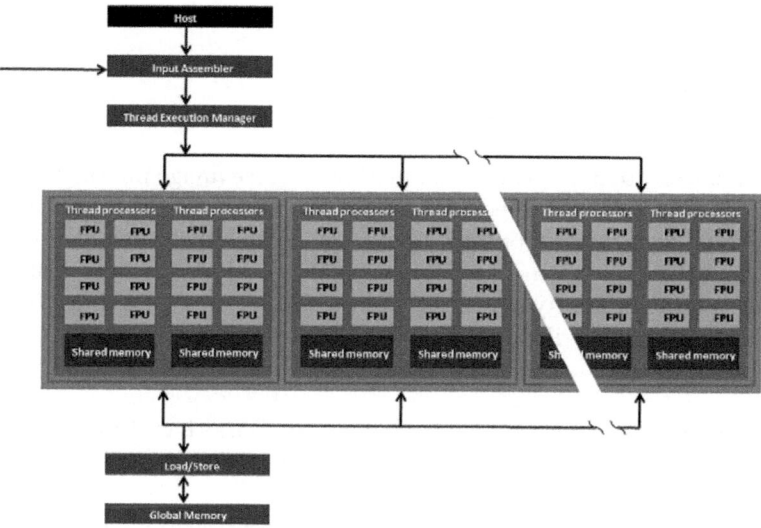

Fig. 1. NVIDIA GeForce 8800 Architecture

These differences in processor architectures lead to different programming models, with different optimal algorithm designs. For an example of an algorithm design under similar architectural constraints, see [2]. Likewise, for a good introduction to the differences in algorithm analysis for various architectures, which must take into account not only running time, but also the amount of idle processing power and the amount of extra work done in a parallel setting over the best serial algorithms, see [3].

In addition to these considerations, the GPGPU has one more unique constraint: the connection bandwidth between the CPU and the GPU is quite limited compared to the bandwidth of the GPU's internal memory [4,5]. In fact, given that the GPU cannot directly access the host's main memory, hard drives, or peripherals, and modern hosts can contain multiple interconnected GPU units, dealing with the GPU can be thought of as distributed computation on a small local network with the host acting as a control node.

In this paper we present a GPU-based algorithm design for image template matching, which is a building block for many high-level Computer Vision applications, such as face and object detection [6,7], texture synthesis [8], image compression [9,10], and video compression [11,10]. Algorithms of this type are often infeasibly slow in raw form [12], and there has been much research into methods for accelerating template matching for various applications.

To date, there have been several attempts at adapting sequential algorithms to the data-parallel GPU architecture [13,14,15,16] rather than designing with data-parallelism in mind. In contrast, we designed an algorithm for GPGPU execution from the ground up, while analyzing the unique steps taken in the design process.

1.1   Template Matching Background

Some template matching acceleration methods ignore image information deemed irrelevant or unnecessary to reduce run time, or make use of statistical analysis to produce a likely answer, but are unable to guarantee finding the best match according to the chosen error measure e.g. [17,18,19]. A second set of methods which has emerged recently makes use of bounds on the error measure to achieve acceleration without sacrificing accuracy, although the choice in error measures is somewhat more limited [12,20,21]. Our proposed algorithm falls into this second set.

Throughout this paper we make use of the l_1 norm-based distance measure (i.e. the sum of absolute differences) between the template and the image sub-window. We denote the l_1 norm of a vector x by $|x|$.

Let vector $x \in \Re^n$ represent the template we are matching. This vector is formed by concatenating the rows of the template image together into one long sequence. Let I represent the image we are searching, which is larger in all dimensions than the template image. We consider each template-sized subwindow y_i in I a potential match. The subwindows often overlap, and each of them contains n pixels. Each of these subwindows is converted into a vector using the same process as for x. For convenience we define $Y = \{y_1, y_2, \ldots y_m\}$ to be the set of all potential match vectors. In practice, m (the number of potential matches) is slightly less than the number of pixels in I.

The error for the ith candidate (or sub-window) is: $E_i = |x - y_i|$. Given x and I, a template matching algorithm attempts to find the y_i which minimizes E_i. In accelerating template matching, we place bounds on the value of E_i, which we denote as $l_i \le E_i \le u_i$. We define those bounds using the triangle and Cauchy-Schwarz inequalities: $|y_i| - |x| \le |y_i - x| \le |y_i| + |x|$ Note that if we define an orthogonal set of masking vectors m_j, described in Fig. 2, we can define a tightening series of bounds on E_i by taking the major diagonal of the outer product of m_j with x and y_i to get x^j and y_i^j, where j is the index of the masking vector m. This is analogous to the "image strips" of [12]. Using these values we define a recursive relation on the series of bounds on E_i in Fig. 3.

$$
\begin{array}{r|l}
m_0 & 0\ 0\ 0 \cdots 0\ 0\ 0 \cdots 0\ 0\ 0 \cdots 0\ 0\ 0 \cdots \\
m_1 & 1\ 1\ 1 \cdots 0\ 0\ 0 \cdots 0\ 0\ 0 \cdots 0\ 0\ 0 \cdots \\
m_2 & 0\ 0\ 0 \cdots 1\ 1\ 1 \cdots 0\ 0\ 0 \cdots 0\ 0\ 0 \cdots \\
m_3 & 0\ 0\ 0 \cdots 0\ 0\ 0 \cdots 1\ 1\ 1 \cdots 0\ 0\ 0 \cdots \\
\vdots &
\end{array}
$$

Fig. 2. The set of masking vectors m_j. The length of the sections of 1s and 0s is typically some constant fraction of image width * image height.

$$\text{diff}_0 = |x^0 - y_i^0|$$
$$s_x^0 = x^0$$
$$s_y^0 = y_i^0$$
$$u_i^0 = \text{diff}_0 + |x - s_x^0| + |y_i - s_y^0|$$
$$l_i^0 = \text{diff}_0 + |x - s_x^0| + |y_i - s_y^0|$$
$$\text{diff}_j = \text{diff}_{j-1} + |x^j - y_i^j|$$
$$s_x^j = s_x^{j-1} + x^j$$
$$s_y^j = s_y^{j-1} + y_i^j$$
$$u_i^j = \text{diff}_j + |x - s_x^j| + |y_i - s_y^j|$$
$$l_i^j = \text{diff}_j + |x - s_x^j| - |y_i - s_y^j|$$

Fig. 3. Definition of the progressive bounds on E_i

2   Case Study: Full Search and On-Card Memory

We first consider the case of the Full Search Method of template matching, otherwise known as a brute force method. We have selected as that feature set the pixel values of x and y_i. For our purposes, we define E_i as the distance between the total pixel values of x and y_i. The traditional Full Search Method calculates E_i for all $y_i \in Y$, and returns $y_{opt} = \arg\min_{y_i} E_i$. The algorithm is straightforward: as each E_i is calcuated, the algorithm compares it to a global minimum, updating as necessary. The first step in adapting an existing serial algorithm for implementation on a GPU is to analyze the algorithm and determine which parts (if any) can benefit from parallelization. Our GPU adaptation is very similar to the original, with the exception that after computing E_i at all locations simulatenously, the algorithm uses the "reduce" subroutine [2,16], commonly used in data parallel environments, to find a minimum or maximum. Given that m is the number of subwindows, and the template x contains n pixels, this approach runs in $O(mn)$ time, which comes to $\approx 4 * 10^{10}$ operations. GPU implementation of similar methods has been explored in [16]. The straightforward GPU implementation should run in $O(\frac{mn}{p} + \log m)$ time, where p is the number of processors, assuming that $1 \ll n$. This bound comes from mn work being done on p processors, and the reduction step which takes $\log m$ time. We present the actual results in Table 1. Compilation of the CPU code was performed by MS Visual C++ 2008 with all optimzations turned on, while the GPU code was compiled by NVIDIA's nvcc and optimized by open64 [22]. Given the considerable differences in architecture, one can see that the ratio of overall runtimes of the CPU to naive GPU implementation (which we define as "speedup", S) is only $23290/3042 = 7.66$. Given the number of processing units p is 128, this is clearly not a cost optimal solution, as it yields an efficiency of .060 (we define "efficiency" as $E = \frac{S}{p}$). The majority of this is due to communication overhead, as main memory on the GPU is uncached. Experimentation confirms that the instruction throughput is only .034.

Most GPGPU architectures include a limited, local, user-controlled cache. This local cache (which is called "shared memory") is typically too small to hold an entire image (in our case it is 16KB in size). Therefore the image must be loaded a portion at a time, and the threads sharing a given piece of memory synchronized. The groups of threads which can access a given piece of shared memory are organized into "blocks". Threads within a block can typically use shared memory to communicate and synchronize with one another, but are unable to do so (directly) with threads outside of that block. Therefore, the input data should be broken up according to thread blocks when possible. In the case of template matching this is relatively easy, given that inputs y_i though y_{i+n} are the only information required to compute E_i through E_{i+n}. However, the values y_i and y_{i+1} overlap considerably, leading to a certain amount of replication. The results of this approach appear in Table 1 as "GPU Shared". While it represents a vast improvement, the instruction throughput (processor utilization) is still only around .5 due to synchronization, bank conflict, and redundant loading issues.

A fast, cached, read-only memory called "texture memory" is also available on most GPUs, which in practice operates at nearly the speed of the shared memory. This memory is effectively a cached version of the the GPU's main memory, which becomes read-only to prevent cache inconsistency. Using this memory eliminates the expensive synchronization step and its associated processor idle time. Using the texture memory to hold the template and the image, we see a speedup of $S = 216.89$. [1] Furthermore, experimentation yielded the instruction throughput of this approach to be .966, and given that this method has only a factor of log m excess computation over the serial algorithm, this means that theoretical efficiency is near 1. This also gives us our theoretical run time of $O(\frac{mn}{p} + \log m)$. This texture method is fast when compared to Full Search Methods on the CPU, but performs a great deal of excess computation when compared to the best serial methods (i.e. accelerated methods), giving it a low efficiency $E = \frac{S}{p} = \frac{T_s}{pT_p}$, where T_s is serial execution time, and T_p is parallel execution time. In other words, it is not strictly necessary to compute E_i at all locations. Our algorithm attempts to address this fact, while maintaining efficient parallel execution.

Table 1. Run time in ms for Full Search Method template matching on a 512x512 image and a 64x64 template. Times are in ms.

	Run Time	Copy Time
CPU	23290	N/A
GPU	3042	217.7
GPU Shared	200.68	217.7
GPU Text.	107.38	2.361

---

[1] Noting again that $p = 128$, this would appear to be super-linear, especially considering that the clock speed of the GPU is considerably slower than that of the CPU.

The third column in Table 1 represents the amount of time required to copy the image data from the host to the GPU under these various approaches. As can be seen, the copy time of this step cannot be ignored. We further explore this issue in Table 2, where we compare the memory allocation and copy times for varying sizes of data. We conclude from this that it may be beneficial to perform some tasks serially on the host if they can reduce the amount of data that must be transferred to the GPU.

Table 2. Average results over 1000 trials of basic CUDA memory operations. "malloc" and "malloc 2D" refer to allocating an array and a byte aligned 2 dimensional array on the GPU, respectively. "copy" and "copy 2D" refer to copying data from the CPU's global memory to the GPU's global memory into the respective data structures. The first column refers to the amount of data used for that experiment, in bytes. All times are in ms.

size	malloc	copy	malloc 2D	copy 2D
$4 * 10^3$	0.067567	0.005253	0.116700	0.014929
$4 * 10^5$	0.118616	0.291486	0.122187	0.296680
$4 * 10^6$	0.141160	2.576290	0.180513	2.713126
$4 * 10^7$	0.241793	23.344471	0.629537	24.801236

3   GPU Acceleration Method

In designing the algorithm in Figure 4, we wanted to off-load as much of the computation that could be conducted in parallel onto the GPU as possible, while still minimizing the amount of memory transfer that had to be done. In addition, we wanted to minimize the total work done by the algorithm, to reduce the level of excess computation as compared to the best serial algorithms. Lastly, but with equal importance, we needed to use data parallel design methodologies in the algorithm.

The unique points of our algorithm when compared to the Full Search Method are a) the combination of the upper bounds of [21] with the very fast bounding methods of [12], and more importantly b) the division of steps between the CPU and GPU such that the CPU deals with the largest amount of memory, and the largest number of subwindows, while also doing as little real computation as possible, leaving the GPU to do extensive computation on only a minimal number of subwindows. The second point has the combined effect of minimizing memory transfer and excess computation.

Essentially, the algorithm begins by performing an initial scan of the data on the CPU, performing approximately 5 operations per subwindow to find initial upper and lower bounds on the match value of each location in the image using the base case of Fig. 3, as explained in Sectin 1.1. The masking vectors (or image-strips) were chosen in particular because they reduce the amount of excess computation over other bounding methods used in template matching, i.e. [21,20,17]. Every time the bounds of y_i are updated, the computed values

PARALLELTEMPLATEMATCH(x, Y)

```
1   InitBounds(Y, x)
2   E_guess, y_best ← FindBestInitMatch(Y, x)
3   Y ← Prune(Y, E_guess)
        ▷ From here onwards, the code is executed on the GPU by many
        ▷ threads in parallel.
4   while |Y| > 1
5       do
                ▷ Tighten the bounds on the remaining members of Y.
6           i ← ThreadID
7           UpdateBounds(y_i, x)
8           if l_i < E_guess
9               then
10                  if u_i < E_guess
11                      then l_i, u_i ← ComputeE(y_i, x)
12                          E_guess ← u_i
13                          y_best ← y_i
14                  else  break
15  if E_i < E_guess
16      then E_guess ← E_i, y_best ← y_i
17  return y_best, E_guess
```

Fig. 4. The main method of our GPU based template matching algorithm

can be reused directly for computing E_i. Reduction of excess computation is especially important in GPU programming, as it is replicated over each processor.

The next step is a single run of the "Prune" method (see Figure 5) on the CPU before beginning the run of the algorithm on the GPU. The Prune step reduces execution time because it drastically reduces the number of locations that the GPU must consider (and therefore the amount of data transfer from host to GPU), often by 99% or more. Yet this step does only a very small fraction of the overall work of the algorithm (on the order of a single comparison operation per y_i). Experimentation has shown, however, that as image noise levels increase, fewer candidates are pruned, resulting in more calculations to be done, which requires the remaining calculations to be done on the GPU versus the CPU.

Some of these initial steps could benefit from parallel execution, except that in our experiments the cost of transferring the full image meta-data from host to GPU memory more than cancels the benefits. These steps could, however, be implemented to run on a multicore CPU and one should expect to see a significant increase in speed. The pruning method is examined in more depth in Figure 5. All steps after this point take place on the GPU.

We chose to transfer to GPU at this point because the workload increases dramatically here, as the algorithm begins comparing pixel values directly to tighten the bounds on the individual y_i. The pixel values of the y_i are held in texture memory as opposed to shared memory, as are those of the template,

FindBestInitMatch(Y, x)

1   l_{min}, y_{min}
2   **for** $y_i \in Y$
3   **do**
4   **if** $l_i < l_{min}$
5   **then** $l_{min} \leftarrow l_i$
6                   $y_{min} \leftarrow y_i$
7   $l_{min}, u_{min} \leftarrow$ ComputeE(y_{min}, x)
8   **return** l_{min}, y_{min}

Prune(Y, E_{guess})

1   **for** $y_i \in Y$
2   **do**
3   **if** $l_i > E_{guess}$
4   **then** $Y \leftarrow \{Y - y_i\}$
5   **return** Y

Fig. 5. The relevant subroutines called by our main method

since they are not modified during the run of the algorithm. This allows for a great increase in access and copy speeds. Furthermore, very little data is actually shared between concurrent threads at run time. This, combined with the very limited size of the shared memory, led to our decision to only use it to store pointers to the candidates. The upper and lower bounds of the candidates are held in global memory initially, but since we have chosen a one-to-one candidate-to-thread mapping, each thread copies the bounds to local memory (registers) and performs their calculations there, avoiding costly global memory access. With the CUDA architecture, threads are organized into blocks that can be of one, two or three dimensions in geometry. These blocks are then organized into grids that can likewise be one, two, or three dimesions. Our grids of thread blocks are two-dimensional grids consisting of three-dimensional thread blocks. Experimentally we did not notice any significant difference in performance due to differences in grid and thread block geometries. The sizes of our grids and blocks were determined based upon the size of the input data. Although branching is typically avoided in SIMD programming, we stop those threads whose candidates are no longer possible matches (that is, $l_i > E_{guess}$). These threads wait at a synchronization barrier, allowing the multiprocessor to allocate more time to the threads that still contain potential matches. Each thread then compares its current distance value against a global minimum to allow for a degree of synchronization between multiprocessors.

The combination of these steps to reduce the memory footprint, memory copy time, and execution workload on the GPU result in our algorithm's accelerated performance. This design is scalable and not hardware specific, and can be ported to any CUDA GPU with similar results.

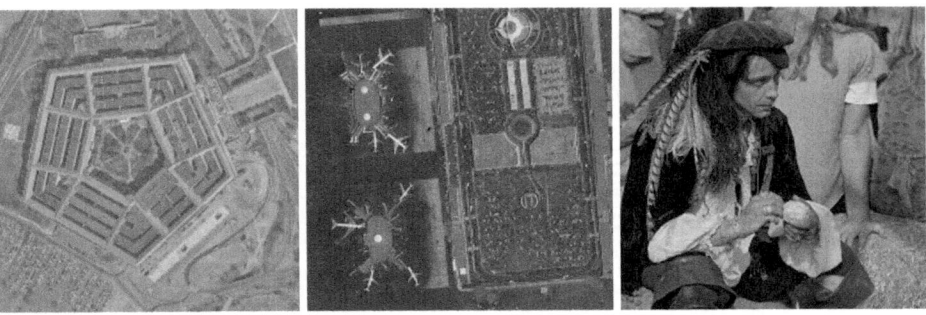

Fig. 6. The standard images "pentagon" (512x512), "airport" (512x512), and "man" (1024x1024)

Fig. 7. The digital camera images "second" (2306x1535) and "rob ref" (3072x2304)

4   Results

Our experimental design consisted of averaging the results of running our algorithm over a number of trials with a variety of images of different sizes and resolutions. We first tested with a few standard images ("pentagon" at 512x512, "airport" at 512x512, and "man" at 1024x1024), and then considered a few images captured on a modern digital camera ("second" at 2306x1535 and "rob ref" at 3072x2304. The standard images can be seen in Figure 6, and the digital camera images can be seen in Figure 7.

We extracted a template from each and tested with noise levels ranging from noiseless ($\sigma = 0$) to very noisy ($\sigma = 70$). We then ran the Full Search Method (using textures as described above) for the same number of trials on the same GPU using the same input.

Our experimentation yielded the following performance results: When comparing the performance of our algorithm to the Full Search Method on small images (512 x 512) at zero to low noise levels, our algorithm has better performance than the Full Search Method. However, as the amount of noise increases to extreme levels, our algorithm begins to slow down, while the Full Search Method remains unchanged. This is due to the fact that at high noise levels, the Prune step executed on the CPU eliminates fewer candidates and effectively becomes excess computation or overhead instead of contributing efficiently to returning a result. The results for these experiments run with the pentagon and airport images are shown in Figure 8, where we report on the speedup factor compared to increasing noise levels.

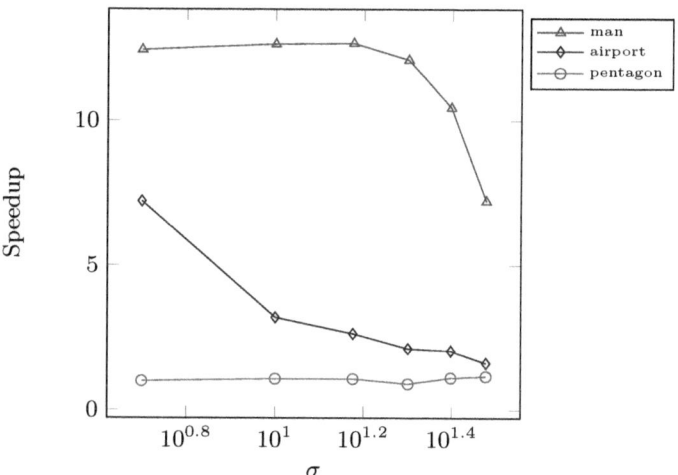

Fig. 8. Ratio of speedup versus noise level σ of our algorithm for different images of different sizes

When comparing our algorithm's performance to that of the Full Search Method on medium to large images one can see the tremendous performance increase of our algorithm. With an image size of 1024x1024 and a template size of 256x256, our algorithm experiences a 12 times performance increase over the Full Search Method. Note that the size of the template chosen corresponds to the man's face in the image, which is a commonly used template matching method. The comparison of the performance increase for the 1024x1024 man image to the performance increase of the smaller 512x512 images can also be seen in Figure 8. Furthermore, with a noiseless image size of 2306x1535 and a template size of 304x280, our algorithm performed 7 times faster, and nearly 39 times faster with a noiseless 3072x2304 image and a template size of 584x782 (again with the templates corresponding to faces in their respective images). The results for the runnning times on these large images are summarized in Table 3.

4.1 Analysis

The worst case run time of the algorithm is actually no better than the naive algorithm described in Section 2. In practice, however, the expected run time of the

Table 3. The images "second" and "rob ref" were taken with a modern digital camera, and are of size 2306x1535 and 3072x2304 respectively. These larger images allow for comparatively large improvements in run time. The run times for the Parallel and Full Search algorithm implementations are expressed in milliseconds.

Image	Noise	Parallel	Full Search	Improvement
second	0	3979.879	27930.175	7.018
rob ref	0	6123.839	237215.515	38.736

algorithm is significantly lower than this. This is not uncommon in accelerated template matching techniques, GPU or host based [12,20,21].

Along similar lines, the fact that our algorithm does not make use of a great deal of the GPU during the final stages of its runtime to avoid excess comptuation means that the instruction throughput is actually quite low (around .05). This has positive and negative consequences. The obvious negative is that much of the GPU is idle, and current GPUs do not allow multiple host threads to use the GPU simultaneously. The positive consequence is that it means the algorithm is very efficient, and since next generation GPU architectures do allow multiple host threads to use the GPU simultaneously [23,24], our algorithm will leave more of the GPU open to other threads. This would be advantageous in computer vision settings where template matching is used as a low level algorithm since it would "leave room" on the GPU for higher level processes.

5   Conclusions and Future Work

We have shown here that while adapting existing algorithms to run on GPUs can provide considerable increases in performance, an algorithm that is designed specifically to run on a GPU can have a nearly 39 times performance increase over algorithms that are simply adapted to run on GPUs. We have shown that in addition to considerations of data parallel algorithm design and analysis, one must also carefully consider the unique memory structure and transfer costs of GPUs to fully harness their power. That power is increasing, with CPU and GPU manufacturers preparing to release next generation GPU architectures, which will include features such as full C++ support, error correcting memory, double precision support, and a chip-wide high-speed communication [23,24].

The work done here could very well be extended to multimedia database searching, as our algorithm's ability to eliminate many candidates before calling the GPU would allow searching a very large database without overwhelming the GPU's limited memory. Additionally, using a clever memory copy algorithm, one could adapt this algorithm to search extremely large images, such as those generated by astronomical surveys, by loading only image regions representing likely matches onto the GPU.

References

1. Owens, J., Houston, M., Luebke, D., Green, S., Stone, J., Phillips, J.: GPU computing. Proceedings of the IEEE 96(5), 879–899 (2008)
2. Hillis, W.D., Steele, J.G.L.: Data parallel algorithms. Commun. ACM 29(12), 1170–1183 (1986)
3. Kumar, V., Grama, A., Gupta, A., Karypis, G.: Introduction to Parallel Computing, 2nd edn. Addison-Wesley Longman Publishing Co., Inc., Amsterdam (2002)
4. NVIDIA Corp.: NVIDIA CUDA programming guide v2.3.1 (August 2009)
5. AMD Inc.: ATI stream computing user guide rev1.4.0a (April 2009)
6. Jin, Z., Lou, Z., Yang, J., Sun, Q.: Face detection using template matching and skin-color information. Neurocomput. 70(4-6), 794–800 (2007)
7. Brunelli, R., Poggio, T.: Face recognition: features versus templates. IEEE Transactions on Pattern Analysis and Machine Intelligence 15(10), 1042–1052 (1993)
8. Efros, A.A., Freeman, W.T.: Image quilting for texture synthesis and transfer. In: Proceedings of the 28th Annual Conference on Computer Graphics and Interactive Techniques, pp. 341–346. ACM, New York (2001)
9. Luczak, T., Szpankowski, W.: A suboptimal lossy data compression based on approximate pattern matching. IEEE Transactions on Information Theory 43(5), 1439–1451 (1997)
10. Rodrigues, N., da Silva, E., de Carvalho, M., de Faria, S., da Silva, V.: On dictionary adaptation for recurrent pattern image coding. IEEE Transactions on Image Processing 17(9), 1640–1653 (2008)
11. Li, R., Zeng, B., Liou, M.: A new three-step search algorithm for block motion estimation. IEEE Transactions on Circuits and Systems for Video Technology 4(4), 438–442 (1994)
12. Tombari, F., Mattoccia, S., Stefano, L.D.: Full-Search-Equivalent pattern matching with incremental dissimilarity approximations. IEEE Transactions on Pattern Analysis and Machine Intelligence 31(1), 129–141 (2009)
13. Abate, A., Nappi, M., Ricciardi, S., Sabatino, G.: GPU accelerated 3D face registration / recognition. In: Lee, S.-W., Li, S.Z. (eds.) ICB 2007. LNCS, vol. 4642, pp. 938–947. Springer, Heidelberg (2007)
14. Huang, J., Ponce, S.P., Park, S.I., Cao, Y., Quek, F.: GPU-accelerated computation for robust motion tracking using the CUDA framework. In: 5th International Conference on Visual Information Engineering, VIE 2008, pp. 437–442 (2008)
15. Stefano, L.D., Mattoccia, S., Tombari, F.: Speeding-up NCC-based template matching using parallel multimedia instructions. In: Proceedings of Seventh International Workshop on Computer Architecture for Machine Perception, CAMP 2005, pp. 193–197 (2005)
16. Massachusetts Institute of Technology: IAP09 CUDA@MIT 6.963 (2009)
17. Goshtasby, A., Gage, S.H., Bartholic, J.F.: A Two-Stage cross correlation approach to template matching. IEEE Transactions on Pattern Analysis and Machine Intelligence PAMI-6(3), 374–378 (1984)
18. Rosenfeld, A., Vanderburg, G.: Coarse-Fine template matching. IEEE Transactions on Systems, Man and Cybernetics 7(2), 104–107 (1977)
19. Pele, O., Werman, M.: Robust Real-Time pattern matching using bayesian sequential hypothesis testing. IEEE Transactions on Pattern Analysis and Machine Intelligence 30(8), 1427–1443 (2008)

20. Hel-Or, Y.: Real-time pattern matching using projection kernels. IEEE Transactions on Pattern Analysis and Machine Intelligence 27(9), 1430–1445 (2005)
21. Schweitzer, H., Anderson, R.F., Deng, R.A.: A near optimal Acceptance-Rejection algorithm for exact Cross-Correlation search. In: Proceedings of the IEEE International Conference on Computer Vision, Kyoto, Japan (2009) Poster Session
22. Murphy, M.: NVIDIA's experience with open64. In: Open64 Workshop at Intl. Symposium on Code Generation and Optimization (CGO), Boston, Massachusetts, United States (April 2008)
23. NVIDIA Corp.: NVIDIAs next generation CUDA compute architecture: Fermi (September 2009)
24. Seiler, L., Carmean, D., Sprangle, E., Forsyth, T., Abrash, M., Dubey, P., Junkins, S., Lake, A., Sugerman, J., Cavin, R., Espasa, R., Grochowski, E., Juan, T., Hanrahan, P.: Larrabee: a many-core x86 architecture for visual computing. ACM Trans. Graph. 27(3), 1–15 (2008)

Author Index

GPSR Compliance

The European Union's (EU) General Product Safety Regulation (GPSR) is a set of rules that requires consumer products to be safe and our obligations to ensure this.

If you have any concerns about our products, you can contact us on ProductSafety@springernature.com

In case Publisher is established outside the EU, the EU authorized representative is:

Springer Nature Customer Service Center GmbH
Europaplatz 3
69115 Heidelberg, Germany

Batch number: 09473985

Printed by Printforce, the Netherlands